EVOLUTION OF THE EARTH

Treatise on Geophysics

EVOLUTION OF THE EARTH

Editor-in-Chief

Professor Gerald Schubert

Department of Earth and Space Sciences and Institute of Geophysics and Planetary Physics,
University of California Los Angeles, Los Angeles, CA, USA

Volume Editor

Dr. David Stevenson

California Institute of Technology, Pasadena, CA, USA

ELSEVIER

AMSTERDAM • BOSTON • HEIDELBERG • LONDON • NEW YORK • OXFORD
PARIS • SAN DIEGO • SAN FRANCISCO • SINGAPORE • SYDNEY • TOKYO

Elsevier B.V.
Radarweg 29, 1043 NX Amsterdam, the Netherlands

First edition 2009

Notice
No responsibility is assumed by the publisher for any injury and/or damage to persons
or property as a matter of products liability, negligence or otherwise, or from any use
or operation of any methods, products, instructions or ideas contained in the material
herein. Because of rapid advances in the medical sciences, in particular, independent
verification of diagnoses and drug dosages should be made

British Library Cataloguing in Publication Data
A catalogue record for this book is available from the British Library

Library of Congress Control Number: 2009929987

ISBN: 978-0-444-53464-4

For information on all Elsevier publications
visit our website at elsevierdirect.com

Printed and bound in Spain

09 10 11 12 10 9 8 7 6 5 4 3 2 1

Contents

Preface

Geophysics is the physics of the Earth, the science that studies the Earth by measuring the physical consequences of its presence and activity. It is a science of extraordinary breadth, requiring 10 volumes of this treatise for its description. Only a treatise can present a science with the breadth of geophysics if, in addition to completeness of the subject matter, it is intended to discuss the material in great depth. Thus, while there are many books on geophysics dealing with its many subdivisions, a single book cannot give more than an introductory flavor of each topic. At the other extreme, a single book can cover one aspect of geophysics in great detail, as is done in each of the volumes of this treatise, but the treatise has the unique advantage of having been designed as an integrated series, an important feature of an interdisciplinary science such as geophysics. From the outset, the treatise was planned to cover each area of geophysics from the basics to the cutting edge so that the beginning student could learn the subject and the advanced researcher could have an up-to-date and thorough exposition of the state of the field. The planning of the contents of each volume was carried out with the active participation of the editors of all the volumes to insure that each subject area of the treatise benefited from the multitude of connections to other areas.

Geophysics includes the study of the Earth's fluid envelope and its near-space environment. However, in this treatise, the subject has been narrowed to the solid Earth. The *Treatise on Geophysics* discusses the atmosphere, ocean, and plasmasphere of the Earth only in connection with how these parts of the Earth affect the solid planet. While the realm of geophysics has here been narrowed to the solid Earth, it is broadened to include other planets of our solar system and the planets of other stars. Accordingly, the treatise includes a volume on the planets, although that volume deals mostly with the terrestrial planets of our own solar system. The gas and ice giant planets of the outer solar system and similar extra-solar planets are discussed in only one chapter of the treatise. Even the *Treatise on Geophysics* must be circumscribed to some extent. One could envision a future treatise on Planetary and Space Physics or a treatise on Atmospheric and Oceanic Physics.

Geophysics is fundamentally an interdisciplinary endeavor, built on the foundations of physics, mathematics, geology, astronomy, and other disciplines. Its roots therefore go far back in history, but the science has blossomed only in the last century with the explosive increase in our ability to measure the properties of the Earth and the processes going on inside the Earth and on and above its surface. The technological advances of the last century in laboratory and field instrumentation, computing, and satellite-based remote sensing are largely responsible for the explosive growth of geophysics. In addition to the enhanced ability to make crucial measurements and collect and analyze enormous amounts of data, progress in geophysics was facilitated by the acceptance of the paradigm of plate tectonics and mantle convection in the 1960s. This new view of how the Earth works enabled an understanding of earthquakes, volcanoes, mountain building, indeed all of geology, at a fundamental level. The exploration of the planets and moons of our solar system, beginning with the Apollo missions to the Moon, has invigorated geophysics and further extended its purview beyond the Earth. Today geophysics is a vital and thriving enterprise involving many thousands of scientists throughout the world. The interdisciplinarity and global nature of geophysics identifies it as one of the great unifying endeavors of humanity.

The keys to the success of an enterprise such as the *Treatise on Geophysics* are the editors of the individual volumes and the authors who have contributed chapters. The editors are leaders in their fields of expertise, as distinguished a group of geophysicists as could be assembled on the planet. They know well the topics that had to be covered to achieve the breadth and depth required by the treatise, and they know who were the best of their colleagues to write on each subject. The list of chapter authors is an impressive one, consisting of geophysicists who have made major contributions to their fields of study. The quality and coverage achieved by this group of editors and authors has insured that the treatise will be the definitive major reference work and textbook in geophysics.

Each volume of the treatise begins with an 'Overview' chapter by the volume editor. The Overviews provide the editors' perspectives of their fields, views of the past, present, and future. They also summarize the contents of their volumes and discuss important topics not addressed elsewhere in the chapters. The Overview chapters are excellent introductions to their volumes and should not be missed in the rush to read a particular chapter. The title and editors of the 10 volumes of the treatise are:

Volume 1: Seismology and Structure of the Earth

> Barbara Romanowicz
> University of California, Berkeley, CA, USA
> Adam Dziewonski
> Harvard University, Cambridge, MA, USA

Volume 2: Mineral Physics

> G. David Price
> University College London, UK

Volume 3: Geodesy

> Thomas Herring
> Massachusetts Institute of Technology, Cambridge, MA, USA

Volume 4: Earthquake Seismology

> Hiroo Kanamori
> California Institute of Technology, Pasadena, CA, USA

Volume 5: Geomagnetism

> Masaru Kono
> Okayama University, Misasa, Japan

Volume 6: Crust and Lithosphere Dynamics

> Anthony B. Watts
> University of Oxford, Oxford, UK

Volume 7: Mantle Dynamics

> David Bercovici
> Yale University, New Haven, CT, USA

Volume 8: Core Dynamics

> Peter Olson
> Johns Hopkins University, Baltimore, MD, USA

Volume 9: Evolution of the Earth

> David Stevenson
> California Institute of Technology, Pasadena, CA, USA

Volume 10: Planets and Moons

> Tilman Spohn
> Deutsches Zentrum für Luft-und Raumfahrt, GER

In addition, an eleventh volume of the treatise provides a comprehensive index.

The *Treatise on Geophysics* has the advantage of a role model to emulate, the highly successful *Treatise on Geochemistry*. Indeed, the name *Treatise on Geophysics* was decided on by the editors in analogy with the geochemistry compendium. The *Concise Oxford English Dictionary* defines treatise as "a written work dealing formally and systematically with a subject." Treatise aptly describes both the geochemistry and geophysics collections.

The *Treatise on Geophysics* was initially promoted by Casper van Dijk (Publisher at Elsevier) who persuaded the Editor-in-Chief to take on the project. Initial meetings between the two defined the scope of the treatise and led to invitations to the editors of the individual volumes to participate. Once the editors were on board, the details of the volume contents were decided and the invitations to individual chapter authors were issued. There followed a period of hard work by the editors and authors to bring the treatise to completion. Thanks are due to a number of members of the Elsevier team, Brian Ronan (Developmental Editor), Tirza Van Daalen (Books Publisher), Zoe Kruze (Senior Development Editor), Gareth Steed (Production Project Manager), and Kate Newell (Editorial Assistant).

G. Schubert
Editor-in-Chief

Contributors

Z. Ben-Avraham
Tel Aviv University, Ramat Aviv, Israel

G. F. Davies
Australian National University, Canberra, ACT, Australia

A. N. Halliday
University of Oxford, Oxford, OX, UK

H. J. Melosh
University of Arizona, Tucson, AZ, USA

F. Nimmo
University of California, Santa Cruz, CA, USA

W. R. Peltier
University of Toronto, Toronto, ON, Canada

G. J. Retallack
University of Oregon, Eugene, OR, USA

D. C. Rubie
Universität Bayreuth, Bayreuth, Germany

N. H. Sleep
Stanford University, Stanford, CA, USA

V. Solomatov
Washington University, St. Louis, MO, USA

M. Stein
Geological Survey of Israel, Jerusalem, Israel

D. J. Stevenson
California Institute of Technology, Pasadena, CA, USA

Q. Williams
University of California, Santa Cruz, CA, USA

B. J. Wood
Macquarie University, Sydney, NSW, Australia

EDITORIAL ADVISORY BOARD

1 Earth Formation and Evolution

D. J. Stevenson, California Institute of Technology, Pasadena, CA, USA

1.1 Introduction

1.1.1 How Should We Think of Earth and Earth Evolution?

Evolutionary science is for the most part based on observation and indirect inference. It is not experimental science, even though experiments can certainly play a role in our understanding of processes. We can never hope to have the resources to build our own planet and observe how it evolves; we cannot even hope (at least in the foreseeable future) to observe an ensemble of Earth-like planets elsewhere in the universe and at diverse stages of their evolution (though there is certainly much discussion about detection of such planets; e.g., Seager (2003)). There are two central ideas that govern our thinking about Earth and its history. One is 'provenance': the nature and origin of the material that went into making Earth. This is our cosmic heritage, one that we presumably share with neighboring terrestrial planets, and (to some uncertain extent) we share with the meteorites and the abundances of elements in the Sun. The other is 'process': Earth is an engine and its current structure is a consequence of those ongoing processes, expressed in the form it takes now. The most obvious and important of these processes is plate tectonics and the inextricably entwined process of mantle convection. However, this central evolutionary process cannot be separated from the nature of the atmosphere and ocean, the geochemical evolution of various parts of Earth expressed in the rock record, and life.

Figure 1 shows conceptually the ideas of Earth evolution, expressed as a curve in some multidimensional space that is here simplified by focusing on two variables ('this' and 'that'), the identities of which are not important. They could be physical variables such as temperature, or chemical variables (composition of a particular reservoir) or isotopic tracers. The figure intends to convey the idea that we have an initial condition, an evolutionary path, and a present state. The initial condition is dictated not only by provenance but also by the physics of the formation process. By analogy, we would say that the apples from an apple tree owe much of their nature not only to the genetics of apples (the process of their formation) but also, to some extent, the soil and climate in which the tree grew. We are informed of this initial condition by astronomy, which tells us about how planets form in other solar systems, by geochemistry (a memory within Earth of the materials and conditions of Earth formation), and by physical modeling: simulations and analysis of what may have occurred. Notably, we do not get information on the initial condition from geology since there are no rocks or landforms that date back to the earliest history of Earth. Geology, aided by geochemistry and geobiology, plays a central role informing us about Earth history. Though some geophysicists study evolution, nearly all geophysical techniques are directed toward understanding a snapshot of present Earth, or a very short period prior to present Earth, and it is only through modeling (e.g., of geological data) that the physical aspects of evolution are illuminated.

In **Figure 2**, another important idea is conveyed: for many purposes, we should think of time logarithmically. This is in striking contrast to the way many geoscientists think of time, because they focus (naturally enough) on where the record is best. As a result, far more geological investigations are carried out for

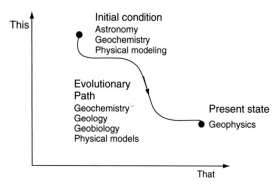

Figure 1 Conceptual view of Earth evolution, identifying the three crucial elements (the initial condition, the evolution path, and the present state) and the sciences that contribute to their understanding. The axes are unimportant, since the diagram is merely a 2-D slice of a multidimensional phase space. They might represent temperature or composition, for example.

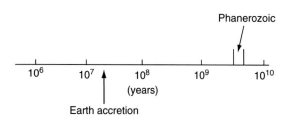

Figure 2 The logarithmic representation of geologic time. The energy budget and rapidity of processes at early times motivates this perspective. A similar view is often taken of cosmology.

the Phanerozoic (10% of Earth history) than the entire period before this. More importantly, the processes that govern early history are very energetic and fast. As a consequence, more could have happened in the first millions to hundreds of millions of years than throughout all of subsequent geologic time.

Table 1 develops this idea further by identifying some of the important timescales of relevance to Earth history and prehistory (here taken to mean the important events that took place even before Earth formed). From this emerges the subdivision of geologic time into the accretion phase (the aggregation of bodies to make Earth), lasting a hundred million years at most, an early evolution in which the high-energy consequences of the accretion (the stored heat) and possibly later impacts still play a role, perhaps lasting as long as half a billion years, and the rest of geologic time in which the energetics of Earth is strongly affected by the long-lived radioactive heat sources.

In this overview chapter, an attempt is made to identify the main themes of Earth history, viewed geophysically, and to provide a context for appreciating the more detailed following chapters. At the end, some of the outstanding issues are revisited, reminding us that this is very much a living science in which there are many things not known or understood.

1.1.2   History and Themes

Hopkins (1839) in his "Preliminary observations on the refrigeration of the globe" illustrates well the prevailing

Table 1   Some important timescales

Process	Timescale	Comments
Formation of Earth	10^7–10^8 years	Infrequent large impacts; background flux of small impacts
Cooling of Earth after a giant impact	1000 years (deepest part of magma ocean) 100–1000 years (condensation of a silicate vapor atmosphere) $\sim 10^6$ years (condensation of steam atmosphere, assuming no additional energy input)	A wide range of timescales, some of which are very fast but nonetheless important. Formation of a water ocean can be fast
Elimination of heat in excess of the thermal state that allows convective heat transport in equilibrium with radioactive heat production	$\sim 10^8$ years	Earth loses most of the thermal memory of a possible very hot beginning
Decline of impact flux	$(1–7) \times 10^8$ years	The late heavy bombardment at 3.8 Ga may have been a spike rather than part of a tail in the impact flux
Current timescale to cool mantle by 500 K	5×10^9–10^{10} years	Very slow because of the high mantle viscosity

view of that time when Earth started hot and was cooling over time. This hot beginning now seems natural to us as a consequence of the gravitational energy of Earth formation, and it has been a consistently popular view even when the justifications for its advocacy were imperfectly developed. Famously, Lord Kelvin (**Figure 3**) took the hot initial Earth and applied conduction theory to the outermost region to estimate the age of Earth at 100 My or less. Burchfield (1975) in his book, *Lord Kelvin and the Age of the Earth*, documents Kelvin's various estimates and the conflict with the Victorian geologists of the time who believed that Kelvin's estimates could not be sufficient to explain the features we see. Kelvin's confidence was bolstered by the similar estimate he obtained for the age of the Sun. Indeed, astrophysicists refer to the 'Kelvin time' as a characteristic cooling timescale for a body, defined as the heat content divided by luminosity. We now know that Kelvin was wrong about the Sun because he was unaware of the additional (and dominant) energy source provided by fusion of hydrogen to helium. Ironically, Kelvin could have obtained a correct order of magnitude estimate for Earth's age had he evaluated

Figure 3 Lord Kelvin played a major role in ideas about Earth evolution and the age of Earth. Although he got the wrong answer for Earth's age, he could have obtained roughly the right age had he used for Earth the method that he used for the Sun.

Earth's Kelvin time. For a plausible estimate of mean internal temperature of $\sim$2000 K (a number that would have seemed perfectly reasonable to Kelvin), a heat capacity of $700\,\mathrm{J\,kg^{-1}}$, a mass of $6 \times 10^{24}\,\mathrm{kg}$ and an energy output of $4 \times 10^{13}\,\mathrm{W}$, he could have obtained

$$\tau_{\text{Kelvin}} \approx \frac{(6 \times 10^{24}) \times (700) \times (2000)}{4 \times 10^{13}} \approx 2 \times 10^{17}\,\mathrm{s}$$
$$\approx 7\,\mathrm{Ga} \tag{1}$$

It should not have been unreasonable for him to suppose that this was physically sensible since at that time the fluidity of Earth's interior was still in doubt and the concept of efficient convective transport already existed. The subsequent discovery of fission and long-lived radioactive heat sources was 'not' the reason he got the wrong answer. Indeed, we now think that Earth could have a significant part of its heat outflow and dynamics even if those radioactive heat sources did not exist. Davies discusses this at far greater length in his chapter.

Plate tectonics and mantle convection is a central theme as the primary controlling principle of most of Earth evolution. Mobility of Earth was proposed by Wegener (1912) and the connection to deep-seated motions was also suggested long ago, for example, Bull (1921). The acceptance of these ideas was delayed, especially in the geophysical community, by the perceived absence of compelling evidence together with doubts about physical process: Could rocks flow in the way that was needed? So much has been written on this that any attempt to summarize briefly here would be superfluous. However, one aspect deserves comment – our current view of Earth evolution is not merely the physical picture of how Earth loses heat and thereby drives the plates; it is the profound interconnection of this to all the other aspects of Earth: (1) the nature of the atmosphere, (2) the existence and persistence of the hydrosphere, (3) the maintenance of the magnetic field, (4) the evolution of the continents, and (5) the evolution of life. Most importantly, these are coupled systems, not one where the mantle is dictating all else. For example, life influences the sediments on Earth, which in turn influence what is cycled back into the mantle, which in turn influences volcanism and pate motions. These interconnections are illustrated in **Figure 4**.

There are some other themes that are important and yet sometimes escape critical attention. The first is the central role of 'common processes' rather than 'special processes'. In the early, perhaps more speculative, days

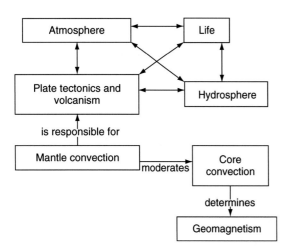

Figure 4   The interconnected aspects of Earth evolution.

of scientific theory-building, it was acceptable to invoke a special process for planet formation, including Earth, one that is rare for stars in general. Beginning in the 1950s, the Soviet school of cosmogony developed the view that planet formation is a natural process, building on much older Laplacian ideas of a nebula around the Sun, but placing that in the context of a universal concept of a disk around a forming star. Safronov (1972) played a central role in this development. Contemporaneously, the astronomical community developed a star formation scenario that naturally developed an attendant disk designed to handle the angular momentum budget of the originating interstellar cloud of gas and dust. Support for this picture grew from the 1980s onward because of astronomical observations. We now have abundant evidence for planets around Sun-like stars (of order 200 examples), though it is not yet clear how many of these systems possess terrestrial planets. This absence of evidence is not surprising given the insensitivity of the Doppler technique used to find most planets. Still, the prevailing view now is that there is nothing very special about how Earth came to be. Similarly, we are reluctant to suggest something very special about how Earth evolved, even though there is clearly a major difference between the nature of Earth and the nature of Venus, the most Earth-like of our neighbors in size, though not the most Earth-like in habitability.

Another important theme comes from 'meteoritics'. It is widely accepted, yet not entirely obvious, that meteorites inform us about terrestrial planets and about the building blocks for Earth. Certainly we have a remarkable amount of information about meteorites: (1) their composition, (2) the timing of their formation, and (3) the conditions that they

encountered (both physical and chemical). Chapter 2 provides much information on this. We must ask, however, whether the validity of this approach is as self-evident as it is sometimes portrayed or whether it is more an example of using what we have, as in the classic story of the drunk looking for his lost car keys under the lamppost because it is the only place where he can see. The relationship of materials in the Earth-forming zone to those that formed asteroids (meteorite parent bodies) remains poorly understood.

Also central to current ideas is the role of large impacts rather than dust or merely small bodies in the accumulation of Earth. Unlike the other themes listed here, this one is more theoretical, though it is consistent with the 'clearing' of dusty disks around stars, taken to infer the formation of planetesimals (though not necessarily requiring the formation of mostly large bodies). This theme originated in work in the 1970s (Hartmann and Davis, 1975; Cameron and Ward, 1976) motivated in large part by ideas of lunar formation, but also consistent with the Safronov model of planet formation as developed further by Wetherill (1976) in particular. Our current ideas of planet formation (e.g., Chambers, 2004) retain the feature, first noted by Wetherill, that the material that goes into making Earth comes from a wide range of heliocentric distances, but that there is nonetheless some expected variation in final composition simply because of the small number of large bodies that participate. This important idea is illustrated in **Figure 5**. It is not known whether this is in fact true; the test lies in a better understanding of the compositions (including isotopic make-up) of the other terrestrial planets.

1.2 Physical and Chemical Constraints

1.2.1 Important Ideas

We cannot figure out origin and early evolution except to the extent that Earth has a 'memory'. The most obvious memory of a planet is its total mass: we do not know of any way of significantly modifying this after the planet forms. In respect of major elements, the planet is a closed system. The application of cosmochemistry and meteoritics to planets is heavily dependent on this idea of closed systems, especially when we speak of 'provenance' (the reservoir of material that was available for forming a planet). It is less obvious but true that a planet will not change its orbital radius significantly. Some other physical attributes such as spin or obliquity or orbital eccentricity can vary

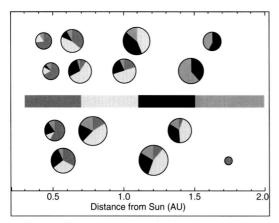

0.5   1.0   1.5   2.0
Distance from Sun (AU)

Figure 5 The results of four outcomes for the final stages of accretion of the terrestrial planets. The shaded bar graph represents an initial distribution of some tracer (e.g., oxygen isotopes or some compositional variable). The circles and their relative sizes give the final outcome in number spacing and relative masses of the resulting planets. The pie diagrams within each circle represent the relative amounts of material arising from each part of the initial distribution. This shows how substantial mixing is typical, that is, the material that ends up in Earth is not typically the material that was originally at 1 AU from the Sun. Reproduced from Chambers JE (2004) Planetary accretion in the inner solar system. *Earth and Planetary Science Letters* 223: 241–252, with permission from Elsevier.

significantly (they are not conserved quantities). Aspects of the history of Earth rotation are in Chapter 10.

The retention of a memory is not guaranteed, especially for physical attributes, even when conservation principles apply. For example, the properties of an ice cube are independent of whether the water molecules were once in the vapor phase or once in a liquid form, at least if the material is in strict thermodynamic equilibrium. The water molecules have no memory. More subtle analysis would however discern whether the isotopic mix of the oxygen on the ice cube were similar or the same as some other ice cube. One could in this way decide whether the ice cube came from Mars or Earth.

Thermal memories can be elusive. For example, consider a body that cools according to this equation

$$\frac{dT}{dt} = -kT^{a+1} \qquad [2]$$

where T is temperature, t is time, and k is a constant. The exponent a is assumed to be greater than unity, and in many realistic cases (e.g., mantle convection) it may be substantially larger than unity. The solution

$$T(t) = [akt + T(0)^{-a}]^{-1/a} \approx (akt)^{-1/a} \qquad [3]$$

retains poor memory of its initial condition if $a > 1$ and substantial cooling has taken place, that is, $T(t)$ is substantially smaller than $T(0)$. This is a common situation in planets. The loss of memory can be even more striking if there is a long-lived heat source (radioactivity). Examples of this are displayed in the thermal histories discussed in Chapter 8. By contrast, a planet that differentiates (e.g., into core and mantle) may retain some memory of initial thermal state should the deeper reservoir (e.g., the core) be considerably hotter than the near-surface reservoir. This would appear to be the case for Earth's core.

Planets are not in thermodynamical equilibrium; that is, one part of the planet is often prevented (by finite diffusivity) from equilibrating with other parts, especially if at least one of the reservoirs is solid. This is a central tenet of many aspects of geochemistry. For this reason, chemical reservoirs are a very important source of memory.

1.2.2 Some Useful Estimates

It is useful to have an appreciation of various energy budgets, timescales, and dimensionless numbers. Among terrestrial planets, solar energy is large compared with the energy normally available from the interior of a planet. For example, the incident sunlight on Earth exceeds the current heat flow from Earth's interior by a factor of 5000. This means that the thermal state of a planet surface is determined by the Sun. However, nonsolar energy (e.g., accretion) can be comparable or more important for short periods of time during the planet formation phase.

The total time of planet accumulation can be long compared to free fall times because the starting material is so widely dispersed. This initial dilution is due in turn to the requirements of angular momentum conservation. For example, suppose we allowed the mass of Earth to be initially dispersed in a volume that is of order one-tenth of a cubic astronomical unit, that is, $0.1 \, AU^3$ (where 1 AU is the Earth–Sun separation). Then if an embryo Earth were plowing through this material with a relative velocity of 1 km s^{-1} (a significant fraction of orbital velocity), it would grow at a rate given by

$$4\pi \rho_o R^2 \frac{dR}{dt} = \pi R^2 v\rho \qquad [4]$$

where ρ_o is the density of solid matter, $\rho \sim 10 M_\oplus / (1 \, AU)^3$ is 12 orders of magnitude smaller, $v \sim 1$ km s^{-1}, R is the radius, and a geometrical cross section is assumed. This predicts dR/dt of a few centimeters per year and a planet formation time of order 100 Ma. Notice that this is independent of whether the accreting material is in the form of dust or larger bodies.

The formation of a planet takes dispersed matter and aggregates it into a mass M of radius R. As a consequence, the gravitational energy changes from a small value to about $-GM^2/R$. By the first law of thermodynamics, this energy release must go somewhere. A small part at most is consumed breaking up material but most of this energy is converted into heat. If all that energy were stored internally then the temperature rise is

$$\Delta T \sim \frac{GM}{RC_p} \sim 40\,000 \left(\frac{R}{R_\oplus}\right)^2 \qquad [5]$$

where C_p is the specific heat of terrestrial materials and $R_\oplus$ is Earth radius. For the larger bodies, this is so large that some of the rock and iron is converted into vapor; indeed, $GM/RL_v \sim 1$ at Earth mass, where L_v is the latent heat of vaporization.

If a planet is initially undifferentiated and then separates into core and mantle, then the resulting energy released as heat (from the resulting larger but more negative total gravitational energy) is evidently some fraction of this amount. For a core that is one-third of the mass but twice the density of rock and settles an average of one-half the planetary radius, the resulting temperature increase is reduced from eqn [5] by a factor of roughly 10, implying a large effect (thousands of degrees potentially) for Earth. As with accretional heating, this energy may not be uniformly distributed internally, but unlike that in eqn [5] it is not readily radiated to space.

Suppose we were to radiate away all of the energy of accretion in time t at a temperature of T_r (ignoring the effect of the Sun or local nebula environment) as infrared (IR) black body radiation. Accordingly,

$$4\pi R^2 \sigma T_r^4 t \sim \frac{GM^2}{R}$$
$$\Rightarrow \quad T_r \sim 350 \left(\frac{10^7 \text{yr}}{t}\right)^{1/4} \left(\frac{R}{R_\oplus}\right)^{3/4} \qquad [6]$$

This equation requires careful interpretation. If it were indeed true that Earth formed from small particles (e.g., dust) over 10 or 100 My, then it would not get hot – the energy of accretion is similar to delivered sunlight (at current solar luminosity) over that period. However, the delivery of mass is highly nonuniform and much of the mass is delivered in giant impacts. This same equation would say that if you had to radiate away one-tenth of Earth's energy of formation

(the energy arising from impact with a Mars-sized impacting body) at $T \sim 2000$ K appropriate to a silicate vapor atmosphere, then it would take ~ 1000 years. This is far longer than the delivery time of the energy in a giant impact ($\sim$a day or less) implying that very high temperatures are unavoidable.

The early state of the planet depends on the partitioning of energy input between surficial (available for prompt radiative loss) and deep seated (available for storage and only eliminated if there is efficient heat transfer from the interior). A crude but useful way to estimate this relies on the introduction of a parameter f, the fraction of input energy that is delivered to the deep interior. In the simple case of little energy transport from interior to surface, eqn [5] might then be replaced by

$$\Delta T \approx \frac{fGM(t)}{R(t)} \qquad [7]$$

where ΔT is the temperature difference between the near-surface interior and surface at the time when the planet has mass $M(t)$ and radius $R(t)$. It is common practice to think of $f \sim 0.1$ or even less, but the physical basis for this choice is unclear since the heat loss will be high if eqn [7] predicts a high temperature. In other words, f is a function, not a number. However, the prevailing view of the initial state of Earth is that it is set by a giant impact, presumably the impact that made our Moon. For this view, eqn [7] is not useful and one must instead appeal to the outcome of impact simulations, for example, Canup (2004). These show that temperature increases of many thousands of degrees are likely, though the temperature distribution is very heterogeneous. This is discussed further in Chapter 3 and also figures prominently in the initial condition for the discussion of the resulting magma ocean in Chapter 4.

An enormous range of heat fluxes F from Earth's interior is possible, both in the accretion epoch and subsequently: in fully molten medium, convective transport can be enormous if the medium is even only slightly superadiabatic. For a superadiabatic temperature difference δT, convective length scale L, and coefficient of thermal expansion α, mixing length theory (*see* Chapter 4) predicts $F \sim \rho C_p \delta T (g\alpha \delta T L)^{1/2}$, assuming viscosity is small enough to be unimportant. For layer of thickness L, the time to cool by ΔT is accordingly

$$\tau_{cool} \sim 10 \left(\frac{1000}{\Delta T}\right)^{1/2} \left(\frac{\Delta T}{\delta T}\right)^{3/2} \left(\frac{L}{g}\right)^{1/2} \qquad [8]$$

Since $(L/g)^{1/2}$ is a short timescale (e.g., hours), we see immediately that the cooling time is short, even for very small temperature fluctuations driving the convection (e.g., of order one degree).

By contrast, cooling times in a system that is mostly solid (and therefore very viscous) can be enormously longer. The corresponding formula for heat flux in this case is $F \sim 0.1 \rho C_p \delta T (g \alpha \delta T \kappa^2 / v)^{1/3}$, where κ is the thermal diffusivity and v is the kinematic viscosity (*see* Chapter 8). For the usually realistic choice of $\kappa \sim 0.01\ cm^2\ s^{-1}$, this predicts a cooling time,

$$\tau_{cool} \sim 10^{10} yr \left(\frac{L}{1000\ km}\right) \left(\frac{\Delta T}{1000}\right) \left(\frac{100}{\delta T}\right)^{4/3}$$
$$\times \left(\frac{v}{10^{20}\ cm^2 s^{-1}}\right)^{1/3} \qquad [9]$$

where the result has been scaled to a choice of viscosity that is roughly like that expected for silicate material near its melting point. The precise value is not important because the real significance of this result lies not in the specific numbers predicted but in the profound difference between cooing of a liquid and cooling of a solid. A planet that melts can lose heat prodigiously, but when Earth is mostly solid, the cooling rate is far slower. One consequence of this is that the time that elapses between the very hot Earth and the formation of a water ocean can be small. Evidence of an early ocean is discussed in Chapter 5 and also figures in the discussion by Stein and Ben-Avraham on the origin and evolution of continents. It may also be relevant to the initiation of plate tectonics (*see* Chapter 6).

We also note that

$$\frac{GM^2}{R \int_0^\infty Q_{radio}(t) dt} \sim 10 \left(\frac{R}{R_\oplus}\right)^2 \qquad [10]$$

where Q_{radio} is the chondritic radiogenic heat production, excluding short-lived sources (^{26}Al, ^{60}Fe). This emphasizes the importance of gravity as setting the stage for subsequent evolution on Earth.

Work done breaking up materials is small because the strength of materials is small compared to GM^2/R^4, the typical energy per unit volume associated with gravity. On the other hand, $GM^2/R^4 K \sim (R/R_\oplus)^2$ where K is the bulk modulus. This means that for Earth mass planets, gravitational self-compression leads to a higher density (smaller radius) than a body comprising the same constituents but zero internal pressure. This effect is enough to affect significantly the mineral assemblage within larger

terrestrial planets, thereby affecting melting behavior, differentiation, core properties, and possible mantle layering. It also means that Earth's interior can be heated by adiabatic compression alone.

The biggest effect of adiabatic compression is in a possible massive atmosphere, since gas is highly compressible. As a consequence, the radiative temperature of a planet can be much less than the surface temperature of the planet even when the atmospheric mass is a small fraction of the total mass. This assumes that it is opaque (e.g., as in the greenhouse effect, but the argument provided here is not limited to that effect). For example, an adiabatic atmosphere that radiates at a pressure P_e at temperature T_e and has a basal pressure $P_s \gg P_e$ will have a surface temperature T_s given by

$$T_s \sim T_e \left[10^{12} \left(\frac{R}{R_\oplus}\right)^2 \left(\frac{1\ bar}{P_e}\right) \left(\frac{M_{atm}}{M}\right)\right]^\gamma \qquad [11]$$

where $T \propto P^\gamma$ is the adiabatic relationship and M_{atm} is the atmospheric mass. The factor of 10^{12} demonstrates the remarkable blanketing effect that is possible even for an atmospheric mass that is less than the planet mass by a factor of a million.

Planetary embryos form early enough that they are imbedded in the solar nebula. Gravitational attraction increases the nebula gas density near the embryo surface. For an isothermal atmosphere of negligible mass,

$$\frac{\rho(r=R)}{\rho(r \to \infty)} = \exp\left[\frac{GM}{Rc^2}\right] \qquad [12]$$

where c is the speed of sound for the nebula (primarily hydrogen). Since the nebula is very low density, this is only of interest for surface conditions (e.g., ingassing of nebula at the surface of a magma ocean) if GM/Rc^2 is larger than ~ 5 or 10. For a Mars mass and $T \sim 300\ K$, $GM/Rc^2 \sim 12$. For such an embryo, the nebula might be $\rho \sim 10^{-9}\ g\ cm^{-3}$ and the surface density could then be $\rho \sim 10^{-4}\ g\ cm^{-3}$, potentially optically thick, and with a surface pressure (~ 1 bar with a large uncertainty) that allows modest ingassing. Thus, embryos larger than Mars, including the growing Earth, could have had a massive atmosphere of near-solar composition. However, the planetary evidence suggests that this effect is small. For example, the neon-to-argon ratio for Earth is much lesser than the solar nebula ratio even though the ingassing ability (predicted by Henry's law) is similar. This is discussed further in Chapter 5. Presumably, the nebula was eliminated early in the period of growth of large embryos.

1.3 Commentary on Formation Models

Astronomical observations of newly forming stars indicate the presence of disks of gas and dust. The material in the disk is typically a few percent of a solar mass, more than sufficient to explain the observed planetary mass of our system or other systems discovered thus far. If the disk has solar composition, then the amount of condensable material (as rock or iron) internal to a few astronomical units is sufficient to explain the masses of the terrestrial planets. However, the disks extend to tens of astronomical units or more, and this is enforced by the angular momentum budget of the originating interstellar cloud. This angular momentum is responsible for the possibility of planet formation but also guarantees that the terrestrial zone is a small part of a much bigger picture. This means that we probably cannot understand formation of terrestrial planets without some understanding of the formation of the gas giants, especially Jupiter. The disks have a radial variation in temperature, wholly or partly because of the energy release of the central body (the forming star). As a consequence, the region within a few astronomical units of the star is hundreds of degrees kelvin or more, sufficient to avoid the condensation of water ice. The absence of large amounts of water in the terrestrial zone in our solar system is interpreted as evidence for terrestrial planet formation internal to the 'snow line' (the place outward of which water ice can condense). Typically, this temperature is of order 160 K (low compared to 270 K because of the very low vapor pressures characteristic of such a nebula). Chapter 5 discusses the source of water on Earth, which is presumably external to the terrestrial planetary zone.

The central ideas in current models of terrestrial planet accretion are three: (1) planetesimal formation, (2) runaway and oligarchic growth, and (3) late stage accumulation. Planetesimal formation refers to the accumulation of bodies of order kilometers in size. Runaway and oligarchic growth are two dynamical stages of a process that we can (for our purpose) lump into the rapid-accumulation planetary embryos of up to order Mars in mass but in closely spaced low eccentricity and inclination orbits. The last and slowest stage proceeds for hundreds of Moon- and Mars-sized objects to the terrestrial planets that we see. According to astronomical observations, this last stage is probably occurring after the solar nebula has been removed (by accretion onto the Sun or expulsion to the interstellar medium). Chapter 2 connects this physical picture to the cosmochemical evidence.

Planetesimal formation remains mysterious even though there is no doubt that it occurred, since its consequences are expressed among the meteorites arising from differentiated asteroids that must have formed in the first million years after the formation of the solar nebula. They may have formed by the poorly understood process of sticking of dust particles during very slow velocity collisions in the dusty gas. They may alternatively have formed by a gravitationally mediated process as the dust settled toward the mid-plane of the disk. Gravitational instabilities are an attractive mechanism, especially given the uncertainties of the physics of sticking, but the relevant fluid dynamics is still debated (Weidenschilling, 2006). This remains one of the major puzzles of planet formation.

The next stage, from planetesimals to Moon and Mars mass embryos, is perhaps better understood theoretically but less well founded in direct observation. This stage, lasting 10^5–10^6 years, is rapid because of gravitational focusing. Bodies encounter each other at velocities considerably less than the escape velocity from the larger body, and, as a consequence, the cross section for collision is far larger than the geometric cross section, perhaps by as much as a factor of a thousand. The planetesimal swarm is 'cold' (i.e., the random velocities of the bodies are very small compared to Keplerian orbital velocities.) This process of embryo growth is terminated by isolation: when the orbits are nearly circular, the bodies reach a stage where there are no crossing orbits (i.e., no overlap of their gravitational spheres of influence). From the point of view of understanding the bodies that we see, this stage is of great interest in three ways. (1) These embryos form quickly and therefore may be hot, both because of possible short-lived radioactive elements and also because of accretional heating (cf. the scaling discussed in Section 1.2). They could therefore form cores and primordial crusts. In this sense, the primordial crusts and current cores of the terrestrial planets have the potential to predate the formation of the planets. (2) The largest of these embryos could conceivably be a surviving planet, most plausibly Mars. In this picture, Mars is special as an isolated outlier in the formation process. (3) Embryos form while the solar nebula is still present and could therefore have a component of solar composition gas. Equation [12] suggests that this, however, is small.

The late-stage aggregation is conceptually like that envisaged long ago in the work of Safronov (1972), whereby the scattering of bodies causes growth of eccentricity and inclination allowing the orbits to cross. It has received a lot of attention in recent times because of the development of N-body codes that can handle the outcome of a population of bodies scattering from one another gravitationally and occasionally colliding (Chambers, 2004). The main shortcoming of these calculations is the failure to take full account of what actually happens when two bodies have a very close encounter (a quite common occurrence) or collide. Tidal disruption is possible in a close encounter (i.e., the outcome is not necessarily the two intact bodies that existed before encounter) and collisions do not always lead to a clean merging of the two bodies. Close encounters can also create additional bodies through tidal disruption.

It is easy to see by order of magnitude (a slightly more sophisticated version of eqn [4]) that this late-stage process can take as long as 100 Ma. However, it is stochastic and it involves large bodies. As a consequence, it is not possible (and may perhaps never be possible) to say exactly what sequence of major events took place in the formation of a particular planet.

1.4 Commentary on Early Evolution Models

The high-energy events of late-stage accumulation are expected to play a central role in setting the stage for Earth structure and evolution. The building blocks are a set of planetary embryos that almost certainly form iron-rich cores because of the combined effects of gravitational energy release in a relatively short time and the possible effects of ^{26}Al heating. This early but important differentiation event is characterized by relatively modest temperatures ($\sim$2000 K or less) and pressures (20 GPa or less, perhaps a lot less) appropriate to bodies in this size range. It is likely and perhaps significant for some constituents (e.g., noble gases) that this phase occurs in the presence of the solar nebula. In highly idealized numerical models (Chambers, 2004), it is usually assumed that nearly all of Earth's mass accumulates in the subsequent merging of these massive embryos. In reality, it is not known how much material arrives in small bodies (planetesimals), whose

origin could be either the bodies that were not swept up during runaway, or debris created during frequent close encounters in the final orbit-crossing phase. Giant impacts will certainly create extensive melting and at least a transient magma ocean, especially if one assumes that the interiors of the embryos prior to impact are at or even above the solidus of mantle minerals. This is a reasonable assumption because of the limited time available for elimination of earlier heating, together with the higher radioactive heating of that epoch. It is also possible that there is a persistent magma ocean, even in the long intervals of time ($\sim$10^7 years) between the giant impacts, sheltered by a steam atmosphere greenhouse that is sustained by delivery of a 'rain' of small bodies that heat its base. This kind of magma ocean is less certain than the transient high-energy ocean that is present immediately after a giant impact. These magma ocean scenarios are analyzed in detail in Chapter 4 and also figure prominently in the discussion of core formation in Chapter 3.

Of course, short-lived high-energy events could have a bigger role than sustained lower-energy events in determining the composition of current core and mantle because they can create extensive melting in which core–mantle separation is especially efficient. There are two central questions that arise in this picture:

1. To what extent is Earth's core formation accomplished by merging the cores of embryos rather than by separation of iron alloy from the mantle within the Earth?
2. To what extent is the separation of core from mantle and possible internal differentiation of mantle accomplished in a magma ocean environment?

These issues figure prominently in Chapter 3.

In simulations designed to understand the origin of Earth's Moon, smoothed particle hydrodynamics (SPH) is most commonly used to describe the outcome of an impact involving a Mars-sized projectile and a mostly formed Earth (Canup, 2004). This represents perhaps the last (and perhaps most important) giant impact. Core and mantle are tracked and one can follow the extent to which these constituent parts of projectile and target 'mix'. The outcome is essentially stochastic; that is, it varies considerably as a function of details, such as impact parameter and mass ratio of projectile to target, parameters that we can never hope to establish deterministically. However, in typical simulations, only around 10% of the projectile (and thus at most a few percent of an

Earth mass) is placed in orbit and available for making the Moon. Very little of Earth is placed into orbit. Nearly all of the rest of the projectile either merges immediately with Earth or crashes back into Earth within several hours of the initial impact. The iron core of the projectile may be stretched into filaments or broken into blobs. These simulations are low resolution, so even the finest scale features of SPH are hundreds of kilometers in size. It seems likely that much of the iron is emulsified to small scales and equilibrates with the mantle, but this is not certain.

The transitional phase between a mostly molten mantle and an almost entirely solid mantle is still poorly understood. In respect of thermal history, it could be argued that the details of this phase are unimportant (cf. Chapter 8). In respect of the geochemical evolution, it is probably very important for setting up the conditions for formation or preservation of the earliest crust (*see* Chapter 2).

The initially hot core will be allowed to cool by the overlying mantle and will therefore convect, allowing for the possibility of magnetic field generation. Nimmo deals with core evolution in Chapter 9, particularly the puzzle of whether there is enough energy in core cooling and inner growth to explain the energy budget demanded by the persistence of a dynamo throughout nearly all of geologic time. The answer to this puzzle remains uncertain.

What about the origin of plate tectonics? On the one hand, this could be viewed as a geologic question, provided there was agreement about the distinctive signatures of plate tectonics in the rock record. On the other hand, this could be viewed as a fluid dynamical question (albeit one that is unavoidably coupled to the crustal evolution and rheological puzzle of a fragmenting lithosphere). Sleep discusses both aspects of this unresolved question in Chapter 6.

Finally we come to life. The origin of life on Earth remains one of the central scientific problems of our time, even though many in the scientific community have adopted the view that this origin was 'easy', by which they mean that the physical conditions, amounts of material, energy budget, and abundance of liquid water and surfaces allows for conditions that are suitable. Less controversial but perhaps as important, it is abundantly clear that biological evolution has affected the physical aspects of Earth evolution profoundly, just as the physical conditions have mediated that biological evolution. In many ways this geobiological question is the most exciting frontier of Earth science, along with the more immediate

issues of climate change (Earth evolution on a much smaller timescale).

1.5   Outstanding Questions

What are the main issues that arise in our understanding of Earth origin and evolution? We end by providing below a list of 10 questions, linked to the subsequent chapters in this volume. In some cases, these emerge naturally from the perspectives of the various authors while in other cases they show the interconnections that might be less evident from a single chapter.

1. *What is the legacy of Earth formation in the planet as we see it?* What range of accretion scenarios is consistent with Earth? What is the role of chance? What is the role of distance from the Sun? How does geochronology compare with physical estimates of formation times? (*see* Chapter 2)

2. *How is core formation in Earth expressed in the composition of Earth now and the thermal structure and evolution?* Can we figure out the core formation conditions from the abundances of siderophiles in the mantle? Or will we find that the pattern is a mix of low-temperature, low-pressure and high-temperature, high-pressure events? (*see* Chapter 3)

3. *What was the duration of the early magma ocean phase of Earth and what memory (if any) is there of this phase in the early mantle differentiation or crust formation?* Is there an opportunity for melt–solid separation in the early mantle and can this be reconciled with current models of mantle structure and composition? What of the role of an early dense atmosphere? How long did it take for a magma layer within Earth (e.g, in the transition zone) to completely freeze? (*see* Chapter 4)

4. *Where did Earth's water come from? How much do we have and how much of it has cycled between mantle and hydrosphere over geologic time?* Did water play an essential role in the onset of plate tectonics? (*see* Chapter 5)

5. *Why does Earth have plate tectonics?* Mantle convection is easily understood; plate tectonics still resists understanding because of our imperfect knowledge of how the lithosphere fails. The perspective of geologic time helps us solve this puzzle. (*see* Chapter 6)

6. *Why continents? When continents?* Are continents a sideshow in the overall dynamics of Earth evolution (as passengers on plates) or do they play a central role in how the system evolves? (*see* Chapter 7)

7. *How did Earth's mantle temperature evolve through time? Can we reconcile simple convective models with the radiogenic heat supply, the geologic evidence, and our understanding of plate tectonics?* The mismatch between known heat generation from long-lived radioactivity and the actual heat flow out of Earth persists and is incompletely resolved in simple cooling models, suggesting the evolution is more complex. (*see* Chapter 8)

8. *Can we reconcile Earth thermal evolution with the persistence of the geomagnetic field throughout geologic time? When did the inner core form?* The mantle governs the rate at which the core cools and models acceptable on the mantle side suggest a possible problem for core heat flows, including the possibility that the inner core was absent for a major part of early Earth history. But if so, can we supply enough energy to run Earth's magnetic field? (*see* Chapter 9)

9. *What do variations in Earth rotation tell us about the nature of Earth and its evolution?* Have we made sufficient use of rotation history, including true polar wander, in our attempts to reconstruct how Earth evolved? (*see* Chapter 10)

10. *How are the geophysical and biological evolutions of Earth interrelated?* What does the history of life tell us about plate tectonics and the history of water on Earth? To what extent is the co-evolution of life and Earth's physical attributes so strongly coupled that we connect major events in biological evolution to major events in geological history? (Chapter 11)

References

Bull AJ (1921) A hypothesis of mountain building. *Geological Magazine* 58: 364–367.

Burchfield JD (1975) *Lord Kelvin and the Age of the Earth*, 260pp. New York: Science History Publications.

Cameron AGW and Ward WR (1976) The origin of the moon (Abstract). In: Hartmann WK, Phillips RJ, and Taylor GJ (eds.) *Lunar Science VII*, pp. 120–122. Houston: Lunar Science Institute.

Canup RM (2004) Simulations of a late lunar-forming impact. *Icarus* 168: 433–456.

Chambers JE (2004) Planetary accretion in the inner solar system. *Earth and Planetary Science Letters* 223: 241–252.

Hartmann WK and Davis DR (1975) Satellite-sized planetesimals and lunar origin. *Icarus* 24: 504–515.

Hopkins W (1839) Preliminary observations on the refrigeneration of the globe. *Philosophical Transactions of the Royal Society of London* 129: 381–385.

Safronov VS (1972) *Evolution of the Protoplanetary Cloud and Formation of the Earth and the Planets*, 206pp. Jerusalem: Israel Program for Scientific Translations.

Seager S (2003) The search for Earth-like extrasolar planets. *Earth and Planetary Science Letters (Frontiers)* 208: 113–124.

Wegener A (1912) Die Entstehung der Kontinente. *Geologische Rundschau* 3: 276–292.

Weidenschilling SJ (2006) Models of particle layers in the midplane of the solar nebula. *Icarus* 181: 572–586.

Wetherill GW (1976) The role of large bodies in the formation of the Earth. *Proceedings of the Lunar Science Conference* VII: 3245–3257.

2   The Composition and Major Reservoirs of the Earth Around the Time of the Moon-Forming Giant Impact

A. N. Halliday, University of Oxford, Oxford, OX, UK

B. J. Wood, Macquarie University, Sydney, NSW, Australia

2.1   Introduction

A treatise on geophysics needs some summary explanation of the current thinking on how the Earth achieved its current composition and formed its primary reservoirs. This subject has taken enormous strides in the past decade, thanks to significant improvements in mass spectrometry in particular. However, additional discoveries are being fuelled by exciting developments in other areas and are leading to joint research at the interfaces between disciplines traditionally viewed as scientifically distinct. The most significant of these are as follows:

- Accretion dynamics – theoretical calculations of the behavior of the gas and dust in the solar nebula, the construction of planetesimals, planetary embryos and planets, and the fluid dynamics of primordial mixing and differentiation.
- Cosmochemistry – the isotopic and chemical history of the solar system deduced from meteorites and lunar samples.

- Mineral physics and experimental petrology – the simulation experimentally and theoretically of the phases present in the Earth's interior, as well as their physical properties and behavior.
- Observations of planets, stars, disks, and exosolar planetary systems.

In this chapter we summarize briefly these lines of evidence and present the current thinking on the formation and primordial differentiation of the Earth–Moon system. We focus on constraints from cosmochemistry and experimental petrology in particular because other chapters provide complementary information. In particular, the introductory chapter by Stevenson (Chapter 1) covers aspects of the accretion dynamics as do the chapters by Rubie, Melosh, and Nimmo (Chapters 3 and 9). There are no observations of terrestrial planets around other stars at the present time although this is expected to change in the coming years and represents a major thrust in experimental astrophysics. There are however observations of disks and larger exosolar planets, as well as information from other parts of our own solar system. These will be referred to where appropriate. Indeed the links between this area of Earth sciences and astrophysics have probably never been stronger given the current search for exosolar Earth-like planets.

2.2 Key Features of the Earth and Moon

Any theory for how the Earth formed and acquired its present composition and distribution of component reservoirs after the Moon-forming Giant Impact has to explain several things:

1. rocky terrestrial planets grew from the same disk as the gas giants Jupiter and Saturn although plausibly at different stages in its development;
2. terrestrial planets and differentiated asteroidal objects are depleted in moderately volatile elements relative to refractory elements;
3. the Moon is larger relative to the size of its host planet than any other moon in the solar system;
4. the Moon carries most of the angular momentum of the Earth–Moon system;
5. the oxygen isotopic composition of the Earth and Moon are identical despite their different chemical compositions;
6. Earth has a larger core proportionally speaking than does the Moon, Mars, or Asteroid Vesta;
7. Earth has a more oxidized mantle than the Moon, Mars, or Asteroid Vesta;
8. highly siderophile (metal-loving) elements in Earth's mantle are inconsistent with low-pressure core formation;
9. Earth has significant water;
10. Earth, unlike the Moon, Mars, or Asteroid Vesta, lacks a geological record for the period prior to 4.0 Ga; and
11. time-integrated parent/daughter ratios determined from isotopic data provide evidence of changes in composition during the accretion history of the protoplanetary material that built the Earth and Moon.

Most of the above are still a matter of some uncertainty and study. The first is particularly problematic. To tackle these subjects this chapter first provides a brief and elementary summary of how the Sun and solar system formed and what we know about terrestrial planet bulk compositions. We then give a synopsis of the range of dynamic and isotopic constraints on the formation of the Earth and Moon, following this with a summary of the huge field of experimental and theoretical petrology that provides insights into how the Earth must have changed during its accretion history.

2.3 The Birth of the Solar System

The Hubble Space Telescope has provided us with fascinating images of numerous young solar mass stars forming in giant molecular clouds such as the Eagle and Orion Nebulae. It is thought likely that our Sun originated in a similar kind of environment. New evidence confirms a long-held suspicion that the Sun did not form spontaneously but that cloud collapse was triggered by a shock wave originating from a nearby star. Such shock waves might be produced, for example, from supernovae – the explosive finale to the short life of giant stars. This idea, developed by the late Al Cameron at Harvard (e.g., Cameron and Truran, 1977) among others, appeared to be consistent with the discovery of the daughter products of a range of short-lived nuclides in meteorites (**Table 1**). The half-lives of these nuclides (100 ky–100 My) are such that they must have been synthesized in stars shortly before the start of our solar system (**Figure 1**). The most important paper outlining the likely explanation for the synthesis of the various nuclides in stars is still the work of Burbidge, Burbidge *et al.* (1957). A variety of more recent papers explore the growing body of evidence from short-lived nuclides regarding the later inputs to the solar nebula (Arnould *et al.*, 1997; Busso

Table 1   Short-lived nuclides that have been demonstrated to have existed in the early solar system as a result of anomalies in the abundance of their daughter isotopes

Nuclide	Half-life (My)	Daughter
^{36}Cl	0.3	^{36}Ar, ^{36}S
^{41}Ca	0.1	^{41}K
^{26}Al	0.73	^{26}Mg
^{10}Be	1.5	^{10}B
^{60}Fe	1.5	^{60}Ni
^{53}Mn	3.7	^{53}Cr
^{107}Pd	6.5	^{107}Ag
^{182}Hf	8.9	^{182}W
^{205}Pb	15	^{205}Tl
^{129}I	16	^{129}Xe
^{92}Nb	36	^{92}Zr
^{244}Pu	80	^{136}Xe
^{146}Sm	103	^{142}Nd

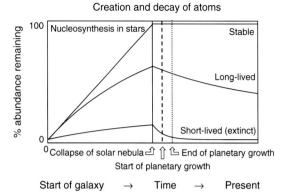

Figure 1 Most solar system nuclides heavier than hydrogen and helium were produced in stars over the history of our galaxy. This schematic figure shows the difference between nuclides that are stable, those that have very long half-lives (such as ^{238}U used for determining the ages of geological events and the solar system itself) and those that have short half-lives of <108 years, assuming all were produced at a constant rate through the history of the galaxy. The short-lived nuclides decay very fast and provide crucial insights into the timescales of events, including planet formation, immediately following their incorporation into the solar nebula.

et al., 2003; Meyer and Clayton, 2000; Meyer and Zinner, 2006; Truran and Cameron, 1978; Wasserburg *et al.*, 1994, 1996, 1998). The early Sun may have been associated with relatively energetic irradiation (Feigelson *et al.*, 2002a, 2002b). It has been proposed (Shu *et al.*, 1997) that the conveyor belt of material being swept in toward the Sun may have led to some light short-lived nuclides being generated locally and heterogeneously (Lee *et al.*, 1998). These might then have been scattered across the disk in early formed condensates (Shu *et al.*, 1997). Although local production in the early Sun is feasible for a range of nuclides (Gounelle *et al.*, 2001; Leya *et al.*, 2003), for most the abundances would be too low to be a match for those found in the meteorites. Furthemore, some of these nuclides could only have been made in a large star that predated the solar system. They were then introduced to the gas and dust that ultimately became our Sun and planets. It is possible that this explosion actually triggered collapse of the cloud material onto a core that became the solar nebula and nascent Sun.

Problems have arisen with this theory. First, a supernova is so powerful that rather than causing collapse it may in fact shred a molecular cloud unless located at sufficient distance that much of the energy and material are already dispersed (Stone and Norman, 1992; Foster and Boss, 1996, 1997; Vanhala

and Cameron, 1998). The time taken to reach the location of the potential new star then becomes sufficiently long (hundreds of thousands of years) that it would be difficult, though not impossible, to detect the former presence of a nuclide such as ^{41}Ca with a half-life of just 100 ky, because most of it would have decayed.

The second problem is that the relative abundances of many of the short-lived nuclides are not well explained in terms of the production ratios expected in a supernova (Wasserburg *et al.*, 1998). Therefore, attention has turned to other stellar sources, the most likely of which is an asymptotic giant branch (AGB) star. AGB stars are relatively common (<8 solar mass) stars that have been shown spectroscopically to be enriched in some heavy elements such as lead (Pb) (Ryan *et al.*, 2001) supporting the idea that they represent the most likely 'factory' for *s*-process nuclides. This was confirmed by the discovery of technetium (Tc) in the spectrum of an AGB star (Merrill, 1952). With no stable nuclei Tc can only be present if it is being synthesized. AGB stars shed material into space but are much-less energetic than supernovae so could be located close to the site of Sun formation. The relative proportions of many of the former short-lived nuclides made better sense with an AGB star model. *r*-process nuclides such as ^{129}I, ^{182}Hf, and ^{244}Pu have to have been produced in a larger star but they possess relatively long half-lives (**Table 1**). The stars which produced them may therefore have been unrelated to the formation of the Sun. The principal problem with the AGB model is ^{60}Fe which has a half-life of just 1.5 My. Although it is not an *r*-process nuclide ^{60}Fe is difficult to make in large quantities in an AGB star. The AGB star model appeared to be consistent with the low initial solar system abundance of ^{60}Fe (Wasserburg *et al.*, 1998) inferred from the first discoveries of this nuclide by Shukolyukov and Lugmair (1993). The initial abundance of ^{60}Fe is now known to have been higher than previously suspected, however (Mostefaoui *et al.*, 2004). With this new discovery a supernova, or at least a very large star, is back on the science agenda as a viable trigger.

The short-lived radionuclides discussed above provide additional information relevant to the early solar system. First, they provide a method for determining timescales from isotopic measurements of the daughter isotopes. Second, they provide a source of heat that would have been available for melting the earliest planetesimals. Both of these are very relevant to understanding how the Earth formed.

2.4 Meteorites

The Sun contains 99.9% of the mass of the solar system and spectroscopic measurements of its chemistry provide an indication of the composition of the circumsolar disk from which the planets grew. However, the compositions of the planets are highly variable and they have been left as a heterogeneous collection of parts of a small fractionated residue after much of the disk had dissipated or been consumed by the Sun. A better set of archives for studying the early solar system and its average composition is found in meteorites – solid extraterrestrial materials that have survived passage through the Earth's atmosphere. These have the additional advantage that they can be studied in the lab at much higher precision than can the composition of the Sun. A detailed description of meteorites is beyond the scope of this chapter but there are two excellent reviews of the main undifferentiated and differentiated meteorite classes that can be found elsewhere (Brearley and Jones, 1998; Mittlefehldt *et al.*, 1998). In order to explain how we know certain things about the Earth's composition and origins, it is necessary to provide a brief explanation of this important archive.

The exact origins of meteorites are varied and the subject of some debate. The range and number of examples of each type has expanded significantly with the acquisition of Antarctic and Saharan meteorites. We refer to the body from which a meteorite came as its 'parent body'. Meteorites are usually divided into three main types:

Chondrites. These are complicated mixtures of dust, glass, and metal thought to represent accumulations of solar system debris that was floating in the circumstellar disk when the Sun formed. The average composition of chondrites is generally similar to that expected from unprocessed circumstellar solids, little affected by melting and loss of volatiles or metal. The most primitive are carbon rich and are called 'carbonaceous chondrites'. 'Ordinary chondrites' are noncarbonaceous chondrites containing metal. 'Enstatite chondrites' are a less-common group of highly reduced chondrites. In some cases a well-observed meteor trajectory has been linked to the meteorites found on the ground and from these it has been established that at least some ordinary chondrites come from the asteroid belt between Mars and Jupiter. Most chondrites contain 'refractory inclusions' (calcium–aluminum inclusions ('CAIs')) enriched in elements, such as Ca and Al, expected to condense at very high temperatures from a hot nebular gas. These may have been ejected from the inner portions of the disk and are the oldest objects identified as having formed in the solar system. CAIs from the Efremovka Chondrite have been dated by $^{235/238}U-^{207/206}Pb$ at $4.567\,2 \pm 0.000\,6\,Ga$ (Amelin *et al.*, 2002), defining the canonical start to the solar system (**Figure 2**). Most chondrites also contain 'chondrules',

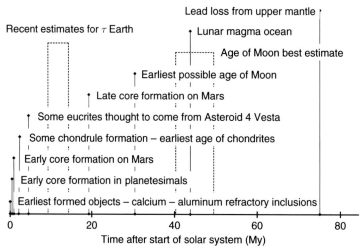

Figure 2   The current best estimates for the timescales over which very early inner solar system objects and the terrestrial planets formed. The approximated mean life of accretion is the time taken to achieve 63% growth at exponentially decreasing rates of growth. The dashed lines indicate the mean lives for accretion deduced for the Earth based on W isotopes. Based on a figure that first appeared in Halliday AN and Kleine T (2006) Meteorites and the timing, mechanisms and conditions of terrestrial planet accretion and early differentiation. In: Lauretta D, Leshin L, MacSween H (eds.) *Meteorites and the Early Solar System II*, pp. 775–801. Tucson: University of Arizona Press.

which are spheroidal objects with textures thought to reflect rapid heating, melting, and quenching of pre-existing silicate material. Many of these formed as much as 2 My after the start of the solar system (Russell *et al.*, 1996; Bizzarro *et al.*, 2004, 2005; Kita *et al.*, 2006).

Achondrites. These are silicate-rich mafic and rare ultramafic igneous rocks not too dissimilar from those formed on Earth but with different chemistry and isotopic compositions. They mainly represent basaltic rocks and cumulates from other planets and asteroids. It is possible to group these and distinguish which planet or asteroid they came from. This is done using their oxygen isotopic compositions. Oxygen is extremely heterogeneous isotopically in the solar system and meteorites with similar chemistry and petrography seem to have specific compositions such that it is thought they can be linked to parent bodies from different regions (**Figure 3**). This is one way in which 'Martian' meteorites are grouped although the more specific link to Mars is established through the dominance of young meteorite ages requiring a geologically active body and the similarity in volatile compositions to the Martian atmosphere, as sampled by Viking. The 'eucrites, howardites, and diogenites' are all thought to come from Asteroid 4 Vesta. Other important groups are the 'angrites' – which are extremely depleted in volatile elements (**Figure 4**) and the 'ureilites' which appear to have been produced by melting of a carbonaceous chondrite-like object that has not undergone core formation. Achondrites and lunar samples are immensely important for providing evidence for how differentiated planets other than the Earth formed. An important feature of all of these samples of differentiated planetary bodies is that they indicate more extreme volatile depletion than is found in chondrites. The same is true of the Earth and Moon.

Irons. These are exactly what one would expect – metallic iron (Fe) with large amounts of nickel (Ni) and some blebs of iron sulfide. Most represent the remains of the metallic cores of early formed small asteroidal bodies. Based on their cooling rates determined from Fe–Ni exsolution features, some ('nonmagmatic') types are thought to have formed at shallow depths from impact melting of asteroidal surfaces (Wasson, 1985). Magmatic group irons typically are more volatile depleted and appear to be derived from the cores of objects of up to hundreds of kilometers in size. They provide

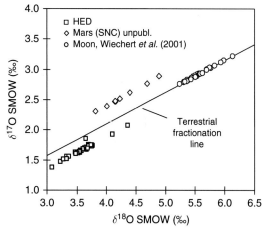

Figure 3 The oxygen isotopic composition of the components in chondrites, in particular CAIs, is highly heterogeneous for reasons that are unclear. The net result of this variability is that different planets possess distinct oxygen isotopic compositions that define an individual mass fractionation line as shown here for eucrites, howardites, and diogenites, which come from Vesta and Martian (SNC) meteorites, thought to come from Mars. The Moon is thought to have formed from the debris produced in a giant impact between the proto-Earth when 90% formed and an impacting Mars-sized planet sometimes named 'Theia'. The fact that the data for lunar samples are collinear with the terrestrial fractionation line could mean that the Moon formed from the Earth, or the planet from which it was created was formed at the same heliocentric distance, or it could mean that the silicate reservoirs of the two planets homogenized during the impact process, for example, by mixing in a vapor cloud from which lunar material condensed. From Halliday AN (2003) The origin and earliest history of the Earth. In: Holland HD and Turekian KK (eds.) *Meteorites, Comets and Planets*, vol. 1, pp. 509–557. *Treatise on Geochemistry*, Oxford: Elsevier-Pergamon.

evidence on the very earliest stages of core formation in the solar system.

The exposure age of a meteorite to cosmic rays can be determined from the nature of the cosmogenic nuclides it contains. Iron meteorites can have very long exposure ages – up to hundreds of millions of years, attesting to a long history of survival in space. Stony meteorites have much-shorter exposure ages. They do not survive well, which explains why we have no examples of silicate mantle material that must have been complementary to the magmatic irons. All of the stony meteorites appear to have been knocked off their parent bodies from shallow depths in the relatively recent past. We do however have some meteorites that contain metal mixed in with silicate that may have formed at some kind of core–mantle boundary in a small object. These are

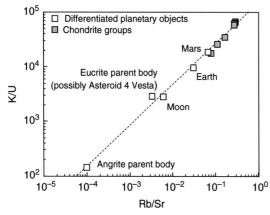

Figure 4   Comparison between the K/U and Rb/Sr ratios of the Earth and other differentiated objects compared with chondrites. The alkali elements K and Rb are both relatively volatile compared with U and Sr which are refractory. Therefore, these trace-element ratios provide an indication of the degree of volatile element depletion in inner solar system differentiated planets relative to chondrites, which are relatively primitive. It can be seen that the differentiated objects are more depleted in moderately volatile elements than are chondrites. Based on a figure that first appeared in Halliday AN and Porcelli D (2001) In search of lost planets – The paleocosmochemistry of the inner solar system. *Earth and Planetary Science Letters* 192: 545–559.

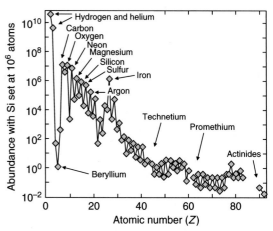

Figure 5   The abundance of elements in our Sun and solar system is estimated from the spectroscopic determination of the composition of the Sun and the laboratory analysis of primitive meteorites called carbonaceous chondrites – thought to represent unprocessed dust and other solid debris from the circumsolar disk. To compare the abundance of different elements it is customary to scale the elements relative to one million atoms of silicon. The pattern provides powerful clues to how the various elements were created. See text for details. Based on a figure in W.S. Broecker, 1985, 'How to build a habitable planet? ELDIGIO press', New York, 291 pp, with kind permission.

called 'pallasites'. 'Mesosiderites' are also metal–silicate mixtures but are thought to be the products of impact melting.

2.5   Meteorites and the Composition of the Earth and Its Primary Reservoirs

The Earth is depleted in those components that are expected to have partially resided in the gas phase of the solar nebula at low temperatures, such as H, N, O, and all of the noble gases. Although this is obvious, the magnitude of the depletion is in part gauged from comparing the Sun's composition with those of meteorites. The most primitive kind of chondrite is called a CI carbonaceous chondrite. If one compares compositional data for the Sun with those for CI carbonaceous chondrites, the agreement is very good for all but the most volatile elements. Thus, these primitive meteorites provide examples of the materials which went to produce the volatile-depleted terrestrial planets.

The estimated abundances of the elements in the solar system, dominated by the Sun (**Figure 5**), range through 13 orders of magnitude and are most easily compared by plotting on a log scale such that the number of atoms of Si is 10^6. Hydrogen and helium dominate over all other elements, reflecting production in the Big Bang and in first-generation stars. Other features reflect the processes of nucleosynthesis in stars, as first explained by Burbidge *et al.* (1957). The essential features are as follows.

First, the abundances of the elements decrease with increasing atomic number. This is because most of the elements are made from lighter elements. There is an anomalous peak in abundance at Fe, which is about 1000 times more abundant than its neighbors, reflecting the maximum binding energy per nucleon at ^{56}Fe. There is a big dip in the abundance of Li, Be, and B, because these elements are not stable in stellar interiors. There is also a saw-toothed variability that is superimposed on the overall curve, reflecting the relatively high stability of even-numbered nuclides compared with odd-numbered nuclides. Finally, all of the elements in the periodic table are present in the solar system except those with no long-lived or stable nuclides. Technetium (Tc), promethium (Pm), and the trans-uranic elements fall into this category.

The four most abundant elements in the terrestrial planets are O, Mg, Si, and Fe. Broadly speaking, the density of the Earth and the size of its core provide a first-order indication that the ratio of Fe to Mg is at least similar to the values found in chondrites. In fact it is slightly higher, as discussed below, but to a first approximation the Earth has chondritic proportions of elements that are refractory or only slightly volatile. Since carbonaceous chondrites appear to represent undifferentiated protoplanetary material, it seems likely that the bulk Earth contains the same relative proportions of all refractory elements as carbonaceous chondrites. This is the basis of the 'chondritic reference model'. In detail there are significant differences relative to chondrites, particularly with respect to the volatile elements. Furthermore, core formation has redistributed the Earth's elements internally. During and after growth of the Earth, 'siderophile' (literally 'metal-loving') elements were partitioned into the metallic core, leaving 'lithophile' (literally 'rock-loving') elements behind in the silicate mantle.

The composition of the bulk silicate Earth (BSE), otherwise known as the primitive mantle (PM), can be deduced in more detail by determining the compositions of a wide range of basalts and/or peridotite xenoliths and using these to identify specific features that are systematic. Determining BSE in this way requires that one makes an assumption of a homogeneous mantle. This assumption is still hotly debated by some and for more detail. However, two key observations lend confidence that the chemistry of the mantle is broadly speaking homogeneous, in striking contrast to Mars, for example (Halliday and Lee, 1999). First, no isotopic heterogeneity older than about 2 Gy has survived in the oceanic mantle pointing to widespread efficient mixing processes over geological time. Second, the ratios of the refractory lithophile elements Nb/U are relatively uniform but nonchondritic and fractionated by crust production (Hofmann et al., 1986). There is no reason why this ongoing fractionating process should have left the mantle so homogeneous unless the mantle is also relatively well mixed.

Assuming these features are representative of the total mantle one can combine estimates of mantle composition with those of the average composition of the continental crust to define the composition of the PM (=BSE). For example, mantle peridotites are fractionated residues of partial melting but ratios of moderately or highly compatible elements such as Mg/Fe or concentrations such as of Ni or Mg

can be compared for a wide range of samples and a representative estimate made for the composition of the BSE.

A different approach is adopted for the highly incompatible elements, which are lost from peridotite during partial melting. Basaltic magmas are the products of partial melting of broad regions of the peridotitic mantle. They fractionate compatible elements one from another but have lesser effects on highly incompatible elements such as the large ion lithophile elements. The latter are essentially completely extracted into the basaltic melt, yielding constant ratios one to another unless the degree of melting is very small. Element ratios that are not significantly fractionated by partial melting, as evidenced by little or no variation in ratio with change in concentration, are particularly useful for determining the BSE depletion in moderately volatile elements. The volatile-element depletion can be determined from their ratio relative to a refractory element of similar incompatibility if both elements are lithophile and hence not partitioned into the core. Relevant examples include Rb/Ba (Hofmann and White, 1983), K/U (Jochum et al., 1983), and In/Y (Yi et al., 1995, 2000). A similar approach can be adopted for siderophile elements. Refractory siderophile/refractory lithophile element ratios for pairs of elements of similar incompatibility provide a clear indication of depletion caused by core formation, for example, Mo/Pr and W/Ba (Newsom et al., 1986). Finally, one can in the same way deduce the primitive mantle concentrations of some elements that are depleted because of both core formation and volatile loss, for example, ^{204}Pb/Ce (Hofmann et al., 1986), Sn/Sm (Jochum et al., 1993; Yi et al, 2000), Cd/Dy (Yi et al., 2000). As with peridotite data the budgets for the continental crust need to be added back in with those determined for the composition of the (depleted) mantle, to derive a value for the primitive mantle (which is the same as the BSE).

The conclusion reached from determining relative proportions of less-incompatible lithophile elements in peridotites such as heavy rare earth elements (REEs), Ca, Sc, and Ti (e.g., Lee et al., 1996), is that, although the bulk Earth does not have an exactly CI composition (Drake and Righter, 2002), refractory lithophile elements are present in approximately the same relative proportions in the BSE as in the carbonaceous chondrites (Allègre et al., 1995b; McDonough, 2003; McDonough and Sun, 1995).

The moderately volatile elements such as potassium (K) are those predicted to have condensed 50% of their mass from a gas of solar composition at temperatures of 700–1100 K (Lodders, 2003). They are depleted in most chondrite groups when compared to solar or CI compositions (**Figures 4** and **6**). There has been much discussion in the literature of the cause of this, usually related to imperfect mixing of CAI, chondrule, and matrix fractions (e.g., Zanda *et al.*, 2006). Moderately volatile elements are strongly depleted in the Earth and other differentiated objects, more so even than in chondrites (**Figures 4** and **6**). By plotting volatile siderophile/refractory lithophile element ratios (e.g., S/Sr) in the BSE and chondrites versus ratios of lithophile moderately volatile to refractory lithophile elements (e.g., Rb/Sr), one can estimate the likely partitioning of volatile elements into the core (Halliday and Porcelli, 2001) (**Figure 6**).

The very refractory lithophile elements are enriched relative to chondrites by a factor of ~3 (McDonough and Sun, 1995), more than can be explained by separation of the core. A plot of concentrations of elements in the BSE divided by those in CI chondrites, normalized to $[Mg]_{Earth}/[Mg]_{CI} = 1.0$ in order to correct for the high volatile contents of CI chondrites is shown in **Figure 7**. These elemental ratios are plotted as a function of the temperature by which 50% of the element of interest would have condensed during cooling of a gas of solar system composition (Lodders, 2003). Note that condensation may not be the mechanism of volatile loss.

If one compares the BSE composition determined as described above, with that of the Sun or CI chondrites (**Figures 4** and **6–8**), several features stand out:

1. There is an overall trend of decreasing chondrite-normalized abundance of lithophile elements as a function of decreasing half-mass condensation temperature (**Figure 7**). Losses of volatiles were important.

2. Some elements, for example, the halogens, are depleted (**Figure 7**) for reasons that are less clear. Even the alkalis do not display a clear relationship between depletion and volatility. Therefore, volatile losses may have involved additional controls beyond nebular condensation temperature.

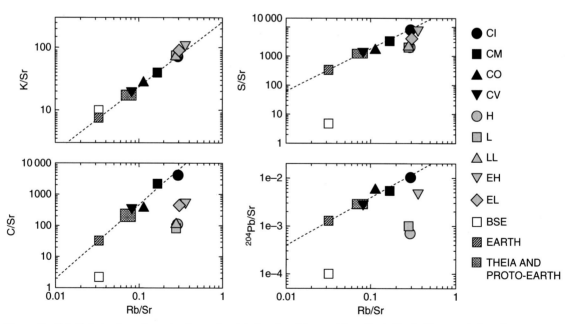

Figure 6   Volatile/refractory element ratio – ratio plots for chondrites and the silicate Earth. The correlations for carbonaceous chondrites can be used to define the composition of the Earth, the Rb/Sr ratio of which is well known because the Sr isotopic composition of the bulk silicate Earth represents the time-integrated Rb/Sr. The values for Theia are time-integrated compositions, assuming time-integrated Rb/Sr deduced from the Sr isotopic composition of the Moon (**Figure 8**) can be used to calculate other chemical compositions from the correlations in carbonaceous chondrites (Halliday AN and Porcelli D (2001) In search of lost planets – The paleocosmochemistry of the inner solar system. *Earth and Planetary Science Letters* 192: 545–559). Other data from Newsom (1995).

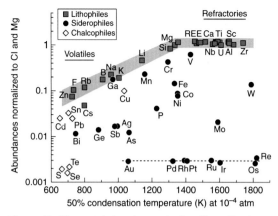

Figure 7 Elemental abundances in the silicate Earth ratioed to those in CI carbonaceous chondrites and normalized to $[Mg]_{Earth} / [Mg]_{CI} = 1.0$ are plotted on the ordinate. The abundances are plotted against the temperature by which 50% of the element of interest would have condensed from a gas of solar composition at a total pressure of 10^4 bar (Lodders, 2003). The silicate Earth is depleted in volatile elements (which condense at low temperatures) due to a combination of incomplete accretion and late volatile loss. Depletions of the silicate Earth in siderophile and some chalcophile elements are due to sequestration in the core. McDonough WF and Sun S-S (1995) The composition of the Earth. *Chemical Geology* 120: 223–253.

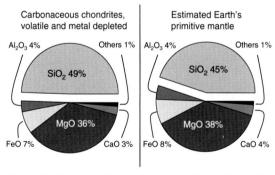

Figure 8 Pie chart showing the composition of the bulk silicate Earth compared with CI chondrites after correction for differences in volatiles and subtraction of a 'core' component. Note the depletion in Si which may be caused by partitioning into the core or a nonchondritic Earth.

3. The siderophile elements are depleted, even if they are refractory, presumably because they are partitioned into the Earth's core. The most depleted, hence presumably most siderophile are the platinum group elements (PGEs), rhenium (Re), sulfur (S), selenium (Se), and tellurium (Te) (**Figure 7**). They are present in roughly chondritic relative proportions, a striking feature that relates to the history of accretion and core formation, discussed below.

4. The Si/Mg ratio of the BSE is lower than is found in chondrites (**Figure 8**). A variety of reasons have been proposed to account for this, including partitioning of Si into the core (Gessman *et al.*, 2001; Wade and Wood, 2005; Takafuji *et al.*, 2005), losses of Si to space (Ringwood, 1989b), and the idea that Earth did not actually accrete from material that was the same as chondrites (Hewins and Herzberg, 1996; Drake and Righter, 2002).

5. The Mg/Fe ratio of the total Earth also appears to be slightly lower than is found in chondrites (Palme, 2000). The reason for this is also unclear. It could simply be that the Earth formed from non-chondritic material but it is also possible that this reflects some level of erosion of the outer part of the Earth during accretion, as has been proposed for Mercury, with its high density hence proportionally bigger core (Benz *et al.*, 1987).

The reasons for these patterns in composition of the Earth remain the focus of considerable debate. They provide evidence that the growth of the Earth was far from simple and may have changed with time, a view reinforced by isotopic data. However, a part of the composition probably relates to Earth's location within the circumstellar disk from which the raw materials were obtained. This is therefore discussed first before moving on to other aspects related to the development of the Earth itself.

2.6 The Circumstellar Disk and the Composition of the Earth

About 200 years ago Kant and Laplace considered that the most likely explanation for the formation of the solar system was planetary accretion from a swirling disk resulting from the angular momentum of the cloud of gas and dust that collapsed to form the Sun. A number of the above-compositional features of the Earth are thought to relate to early processes associated with such a circumstellar disk.

The general explanation for the depletion of the Earth in highly volatile elements such as H and He is that most of the growth of terrestrial planets, unlike the gas giants, postdated the loss of nebular gases from the disk. However, this is far from certain. For example, a recent model for the origin of Earth's water is that it was formed from nebular hydrogen trapped by the Earth as it grew (Ikoma and Genda, 2006). It is not possible to trap significant gas in this way until the Earth reaches >30% of its current mass, which given the short timescales for nebula dispersal is hard to reconcile with protracted accretion (see below). Ikoma and Genda point out that the nebula can be much less dense than the nominal value, that is, captured gas is only logarithmically sensitive to nebula density, although the laws governing nebula dispersal and declining gas density are not well established.

In support of the Ikoma and Genda model there is clear isotopic evidence that some solar noble gases were trapped in the Earth (Porcelli *et al.*, 2001) and Mars (Marty and Marti, 2002), probably as a vestige of the nebula still present at the time of Earth's accretion (Hayashi *et al.*, 1979). How much is left today and whether some is stored in the core is unclear (Porcelli and Halliday, 2001). There is Xe isotopic evidence that the vast majority (>99%) of Earth's noble gases were lost subsequently (Porcelli and Pepin, 2000) (**Figure 9**). The dynamics and timescales for accretion are dependent on the presence or absence of nebular gas as discussed below.

As already pointed out the moderately volatile elements are also depleted in the Earth (**Figures 4, 6,** and **7**) (Gast, 1960; Wasserburg *et al.*, 1964; Cassen, 1996). The traditional explanation for this is that the inner 'terrestrial' planets accreted where it was hotter, within the so-called 'ice line' (Cassen, 1996; Humayun and Cassen, 2000). It was long assumed

for several reasons that the solar nebula in the terrestrial planet-forming region started as a very hot, well-mixed gas from which all of the solid and liquid Earth materials condensed. The geochemistry literature contains many references to this hot nebula, as well as to major T-Tauri heating events that may have further depleted the inner solar system in moderately volatile elements (e.g., Lugmair and Galer, 1992). Some nebula models predict early temperatures that were sufficiently high to prevent condensation of moderately volatile elements (Humayun and Cassen, 2000) which somehow were lost subsequently.

Understanding how the volatile depletion of the inner solar system occurred can be greatly aided by comparisons with modern circumstellar disks. These were first detected using ground-based interferometry. Circumstellar protoplanetary disks (termed 'proplyds') now are plainly visible around young stars in the Orion Nebula thanks to the Hubble Space Telescope (McCaughrean and O'Dell, 1996). With such refined observations combined with theoretical considerations, it is possible to trace the likely early stages of development of our own solar system (Ciesla and Charnley, 2006).

A solar T-Tauri stage during development of the circumstellar disk could have had a profound effect on volatile depletions. T-Tauri stars are pre-main sequence stars. They are a few times 10^5 to a few times 10^6 years in age and have many of the characteristics of our Sun but are much brighter. Many have disks. It has long been argued that the T-Tauri effect causes an early phase of heating of the inner portions of the disk. The T-Tauri stage may last a few million years and is often linked to the Earth's depletion in moderately volatile elements. Because it heats the disk surface however, it is unsure whether it would have a great effect on the composition of the gas and dust in the accretionary midplane of the disk where planetesimal accretion is dominant. Heating of inner solar system material in the midplane of the disk will, however, occur from compressional effects. The thermal response can be calculated for material in the disk being swept into an increasingly dense region during migration toward the Sun during the early stages of disk development. Boss (1990) included compressional heating and grain opacity in his modeling and showed that temperatures in excess of 1500 K could be expected in the terrestrial planet-forming region. The main heating takes place at the midplane, because that is where most of the mass is concentrated. The surface of the disk is much cooler.

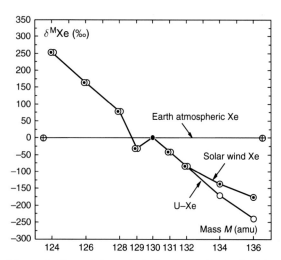

Figure 9 Xenon is isotopically fractionated in the Earth's atmosphere relative to the estimated composition of the material from which it formed. Porcelli D and Pepin RO (2000) Rare gas constraints on early earth history. In: Canup RM and Righter K (eds.) *Origin of the Earth and Moon.* pp. 435–458. Tucson: The University of Arizona Press.

This process would certainly be very early but how it would have evolved with time and tie into the timescales for disk heating deduced from cosmochemistry is unclear.

Chondrules and CAIs provide us with the most important information about the processes of heating and melting in the disk that occurred within the first few million years of the solar system. The exact nature of these events has been hard to establish but chondrules are most likely produced by flash melting of the dusty disk (Kita *et al.*, 2006). What role they played in the formation of planetesimals has been unclear but some models predict that they might have accumulated preferentially in certain stagnant regions of the disk, facilitating planet formation (Cuzzi *et al.*, 2001). Hewins and Herzberg (1996) have proposed that this is reflected in the Earth's nonchondritic Si/Mg (**Figure 8**).

2.7   Dynamics of Planet Formation

In broad terms the rates of accretion of planets from disks are affected by the amounts of mass in the disks themselves. If there is considerable nebular gas present at the time of accretion the rates are faster (e.g., Hayashi *et al.*, 1985). If gas is present then it is important to know what Jupiter is doing at the time of terrestrial planet accretion. At present this is underconstrained and model dependent. The absence of nebular gas is also calculated to lead to eccentric orbits (Agnor and Ward, 2002). Recently however, it has been shown that accretion in the presence of swarms of planetesimals also decreases eccentricity (O'Brien *et al.*, 2006).

The most widely accepted model of terrestrial planet formation is the planetesimal theory (Chambers, 2004). In the simplest terms accretion of terrestrial planets is envisaged as taking place in four stages:

1. settling of circumstellar dust to the midplane of the disk;
2. growth of planetesimals up to ~1 km in size;
3. runaway growth of planetary embryos up to ~10^3 km in size; and
4. stochastic growth of larger objects through late-stage collisions.

Stage 1 takes place over timescales of thousands of years if there is no turbulence but much longer otherwise. It provides a relatively dense plane of material from which the planets grow. The second stage is poorly understood but is necessary in order to build objects that are of sufficient mass for gravity to play a major role. Planetesimals would need to be about a kilometer in size in order for the gravitationally driven stage 3 to start. Although we do not know how stage 2 happens, somehow it must be possible. Fluffy aggregates of dust have been made in the lab but these are typically less than a centimeter in size (Blum, 2000). Larger objects are more problematic. One obvious suggestion is that some kind of glue was involved. Volatiles would not have condensed in the inner solar system, however. Not only were the pressures too low, but the temperatures were probably high because of heating as material was swept into the Sun (Boss, 1990). Organic material or molten droplets such as chondrules (Wiechert and Halliday, 2007) may have played an important role in cementing material together. An alternative to glues that has long been considered (Ward, 2000) is that within a disk of dust and gas there is local separation and clumping of material as it is being swept around. This leads to gravitational instabilities whereby an entire section of the disk has relatively high gravity and accumulates into a zone of concentrated mass (Goldreich and Ward, 1973; Weidenschilling, 2006). This is similar to some models envisaged by some for Jupiter formation (Boss, 1997).

Once stage 2 has occurred, runaway growth builds the 1-km-sized objects into 1000-km-sized objects. This mechanism exploits the facts that: (1) larger objects exert stronger gravitational forces, (2) the velocity dispersion is small so the cross-section for collision can be much larger than the physical cross-section for the largest bodies in the swarm, (3) collisions resulting in growth are favored if the material is not on an inclined orbit, and (4) larger objects tend not to take inclined orbits. The net result is that the bigger the object the larger it becomes until all of the material available within a given feeding zone or heliocentric distance is incorporated into planetary embryos. This is thought to take place within a few hundred thousand years (Kortenkamp and Wetherill, 2000; Kortenkamp *et al.*, 2000; Lissauer, 1987, 1993). The ultimate size depends on the amount of material available. Using models for the density of the solar nebula it is possible that a body of the size of Mars (~0.1 M_E) could originate in this fashion at its current heliocentric distance. However, in the vicinity of the Earth the maximum size of object would be Moon-sized (~0.01 M_E) or possibly even smaller.

Building objects that are as large as the Earth is thought to require a more protracted history of collisions between the 1000-km-sized planetary embryos. This is a stochastic process that is hard to predict in a precise manner. The Russian thoretician Safronov (1954) proposed that, in the absence of a nebula the growth of the terrestrial planets would be dominated by such a history of planetary collisions. The late George Wetherill (1980) took this model and ran Monte Carlo simulations of terrestrial planetary growth. He showed that indeed some runs generated planets of the right size and distribution to be matches for Mercury, Venus, Earth, and Mars. He monitored the timescales involved in these 'successful' runs and found that most of the mass was accreted in the first 10 My but that significant accretion continued for much longer. Wetherill also tracked the provenance of material that built the terrestrial planets and showed that, in contrast to runaway growth, the feeding zone concept was flawed. The planetesimals and planetary embryos that built the Earth came from distances that extended over more than 2 AU. More recent calculations of solar system formation have yielded similar results (Canup and Agnor, 2000).

Planetary collisions of the type discussed above would have been catastrophic. The energy released would have been sufficient to raise the temperature of the Earth by thousands of degrees. The most widely held theory for the formation of the Moon is that there was such a catastrophic collision between a Mars-sized planet and the proto-Earth when it was approximately 90% of its current mass. The putative impactor planet sometimes named 'Theia' (the mother of 'Selene' who was the goddess of the Moon) struck the proto-Earth with a glancing blow generating the angular momentum of the Earth–Moon system.

These dynamics models can be tested with geochemistry and petrology, which provide six principal kinds of information relevant to the earliest history of the Earth:

1. Isotopic heterogeneity in the solar system can be used to trace the sources of the components that built the Earth and Moon.
2. Timescales for accretion, volatile loss, and core formation can be determined from short- and long-lived decay systems.
3. Conditions of core formation can be determined from metal-silicate partitioning behavior of indicative 'siderophile' elements.
4. Time-integrated parent/daughter ratios can be determined from isotopic compositions of the

daughter elements yielding insights into the (paleo)chemistry of precursor materials.
5. Mass-dependent stable isotopic fractionations provide constraints on the prevalent processes and conditions.
6. Geochemistry and petrology provide our principal constraints on the state of the Earth at the end of accretion.

This represents a huge area of science and a continually evolving landscape as new and improved techniques are brought to bear on meteorites and lunar samples. The key aspects as they relate to the early Earth are as follows.

2.8   The Age of the Earth

The age of the Earth represents a problem that in many respects is solved. Clair Patterson in 1956 published the results of Pb isotopic analyses of meteorites that lay on an isochron passing through the composition of the silicate Earth and indicating an age of 4.5–4.6 Ga (Patterson, 1956). Following this work many refinements to the age of the Earth were suggested based on better estimate of the average Pb isotopic compositon of the BSE (references galore) but the essential result remains unchanged. These further studies highlight the fact that there are a variety of inherent complications to take into account in determining an isotopic age or rate. These are very relevant to the issue of how well we know the exact timescales for Earth's formation. (see reviews by Halliday (2003, 2006)).

In most isotopic dating one has to make allowance for the initial daughter isotope that was present in the mineral, rock, or reservoir when it formed. This can be corrected for adequately or ignored in certain minerals with very high parent-element/daughter-element ratios. For example, the most accurate ages of ancient rocks in the continental crust are obtained by U–Pb dating of the common zirconium silicate mineral zircon. This mineral has high U/Pb ratio when it forms and is resistant to the effects of natural resetting via diffusion. Lead does not fit readily into the lattice when zircon grows so any initial Pb is minor and corrections can be made with sufficient confidence that the ages are extremely accurate. The oldest portion of a terrestrial zircon grain thus far found yields an age of 4.40 Ga (Wilde et al., 2001).

There are no rocks surviving from the first 500 My of Earth's history and no zircons from the first 100 My. The age of the Earth and solar system is determined using a different approach. Most minerals and rocks carry a significant proportion of initial Pb relative to that formed by radioactive decay. It is necessary to monitor the initial abundance of the daughter isotope precisely by measuring the parent and daughter isotopes relative to a nonradiogenic isotope of the same daughter element. This is commonly done using isochrons which allow one to determine an age without knowing the initial composition *a priori*. There are two isotopes of U which decay to two isotopes of Pb. By comparing the Pb isotopic ratios in reservoirs that all formed at the same time it is even unnecessary to know the parent/daughter ratios, since the ratio $^{238}U/^{235}U$ is constant in nature (137.88). It was using these techniques that Clair Patterson determined that the age of iron and silicate meteorites, and the Earth, was between 4.5 and 4.6 Gy. This defines the time at which the major variations in U/Pb between U-depleted iron meteorites and (variably) Pb-depleted silicate reservoirs, such as the Earth's primitive mantle, were established by volatile loss and core formation.

Although Pb–Pb dating was at one time used for dating a wide variety of terrestrial samples, this method is all but obsolete except for studying extra-terrestrial samples. Most chondrites contain CAIs, which, as previously explained, are enriched in elements expected to condense at very high temperatures from a hot nebular gas, such as U. These are the oldest objects yet identified that formed in the Solar System. CAIs from the Efremovka chondrite have been dated by $^{235/238}U–^{207/206}Pb$ at 4.5672 ± 0.0006 Ga (Amelin *et al.*, 2002) and this is the current best estimate of the age of the solar system (**Figure 2**) which defines a more precise slope to the meteorite isochron (called the 'Geochron') first established by Patterson (**Figure 10**).

Since Patterson published his work various researchers have tried to estimate the average Pb isotopic composition of the bulk silicate portion of the Earth. This provides information on the timing of U/Pb fractionation caused by terrestrial core formation (**Figure 11**). The results of these various estimates are shown in **Figure 10**. It can be seen that none plot on the Geochron and this provide evidences that accretion of the Earth, or core formation, or both, was either late or protracted. These

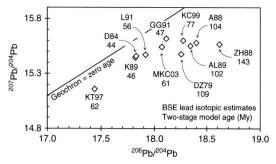

Figure 10 Estimates of the lead isotopic composition of the bulk silicate Earth (BSE) plotted relative to the Geochron defined as the slope corresponding to the start of the solar system. All estimates plot to the right of this line, which, if correct, indicate protracted accretion and/or core formation. The times indicated in million years are the two-stage model ages of core formation assuming the same values for bulk Earth parameters given in Halliday (2004) and Wood and Halliday (2005). Data from Doe and Zartman (1979), Davies (1984), Zartman and Haines (1988), Allègre *et al.* (1988), Allègre and Lewin (1989), Kwon *et al.* (1989), Liew *et al.* (1991), Galer and Goldstein (1991), Kramers and Tolstikhin (1997), Kamber and Collerson (1999), and Murphy *et al.* (2003). These Pb isotope estimates are all significantly longer than the $^{182}Hf–^{182}W$ estimate of 30 My (Kleine *et al.*, 2002; Yin *et al.*, 2002).

results need to be interpreted with caution for two reasons. Firstly, the average Pb isotopic composition of the bulk silicate portion of the Earth is very hard to estimate because the long-lived decay of U has resulted in considerable isotopic heterogeneity. All of the estimates could in principle be wrong. Secondly, converting these results into timescales is strongly model dependent involving assumptions about how the Earth formed. The results of one simple model are shown in **Figure 10**. In this model it is assumed that no U/Pb fractionation took place until a single point in time. So for example, this could describe an Earth that formed its entire core in a single event (**Figure 12**). The ages defined by the Pb isotopic compositions correspond to tens of millions of years after the start of the solar system (**Figure 10**). Although such data were at one time used to 'date' core formation, it is now recognized that the processes of accretion and core formation are more complex than can be described by such an approach. The data do, however, provide evidence of a protracted history of accretion and core formation. To sort out that history it is necessary to use short-lived nuclides, dynamic simulations of planet formation, and petrological constraints on likely core-formation scenarios.

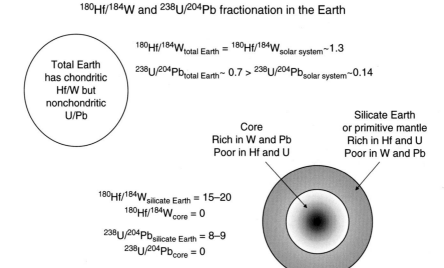

$^{180}Hf/^{184}W$ and $^{238}U/^{204}Pb$ fractionation in the Earth

Total Earth has chondritic Hf/W but nonchondritic U/Pb

$^{180}Hf/^{184}W_{total\ Earth} = {}^{180}Hf/^{184}W_{solar\ system} \sim 1.3$

$^{238}U/^{204}Pb_{total\ Earth} \sim 0.7 > {}^{238}U/^{204}Pb_{solar\ system} \sim 0.14$

Core
Rich in W and Pb
Poor in Hf and U

Silicate Earth
or primitive mantle
Rich in Hf and U
Poor in W and Pb

$^{180}Hf/^{184}W_{silicate\ Earth} = 15\text{--}20$
$^{180}Hf/^{184}W_{core} = 0$

$^{238}U/^{204}Pb_{silicate\ Earth} = 8\text{--}9$
$^{238}U/^{204}Pb_{core} = 0$

Figure 11 Hafnium–tungsten chronometry provides insights into the rates and mechanisms of formation of the solar system whereas U–Pb chronometry provides us with an absolute age of the solar system. In both cases the radioactive parent/radiogenic daughter element ratio is fractionated by core formation, an early planetary process. It is this fractionation that is being dated. The Hf/W ratio of the total Earth is chondritic (average solar system) because Hf and W are both refractory elements. The U/Pb ratio of the Earth is enhanced relative to average solar system because approximately >80% of the Pb was lost by volatilization or incomplete condensation mainly at an early stage of the development of the circumstellar disk. The fractionation within the Earth for Hf/W and U/Pb is similar. In both cases the parent (Hf or U) prefers to reside in the silicate portion of the Earth. In both cases the daughter (W or Pb) prefers to reside in the core.

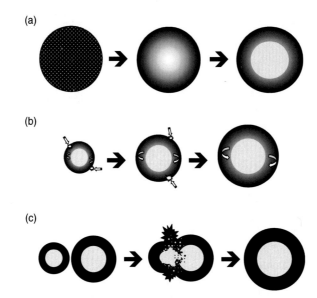

(a)

(b)

(c)

Figure 12 Schematic representation of core formation scenarios and how they translate into isotopic models and timescales. (a) 'Instantaneous core formation'. In this model an age is calculated as the point in time that a fully formed planet segregated into core and silicate reservoirs. (b) 'Continuous core formation'. The planet accretes from material with a chondritic Hf/W ratio and W isotopic composition. Newly accreted material is continuously mixed with W in the outer silicate portion prior to the segregation of new iron and siderophile elements (including W) into the core. Most models assume an exponentially decreasing rate of overall growth for the planet and its core. (c) 'Core–core mixing'. Terrestrial planets appear to have formed from material that was already differentiated into silicate and metal. How much equilibration between accreted metal and the planet's silicate reservoir occurs (an explicit requirement in calculating an age with model (b)) is not well understood at present. The net result if calculated with continuous core formation and equilibration as in model (b) is that the rate of growth of the planet appears faster than it really was. From Halliday AN (2006) The origin of the Earth – What's new? *Elements* 2: 205–210.

2.9 Short-Lived Nuclides and Early Processes

Since Patterson's study it has become clear that one can resolve relatively small time differences between early solar system objects using more precise and different techniques. These provide us with fascinating information on accretion rates and permit the determination of which dynamic models of solar system formation are most probably correct. Progress in this field has been huge in the past 10 years. It is now becoming clear that different isotopic systems applied to the Earth most probably 'date' different processes. Volatile loss, metal segregation, sulfide segregation, mantle melting, atmosphere formation, and continental growth all affect particular elements in distinctive ways. As such, the isotope geochemistry of an expanding range of elements is facilitating the first comprehensive view of the geochemical evolution of the early Earth.

The short-lived nuclides listed in **Table 1** provide a set of powerful tools for unraveling a precise chronology of the early solar system. The advantage of these isotopic systems is that the changes in daughter isotope can only take place over a restricted early time window and there is no correction for the effects of decay over the past 4.5 Gy. They were present in the early solar system because of production in other stars, many of them massive, shortly before (within a few half-lives of) the formation of the Sun (**Figure 1**). Being extinct, the initial abundance can only be inferred. This is achieved by using differences in the atomic abundance of the daughter isotope in early objects (meteorites) of known absolute age.

To utilize any of the nuclides of **Table 1** for deducing early solar system timescales it is necessary to measure precisely the present-day isotopic composition of the daughter element. Therefore, the development of these chronometers is strongly linked to breakthroughs in mass spectrometry. In the simplest system the present-day isotopic composition of the daughter element is a function of the initial abundance of the parent and daughter nuclide at the start of the solar system, the parent/daughter element ratio, and how much time elapsed. In effect, the isotopic composition of the daughter represents a time-integrated parent/daughter elemental ratio. If an object forms without a change in parent/daughter ratio, its time of formation will be undetectable with isotopic chronometry. It is the large

change in parent/daughter ratio that accompanies certain processes associated with planetary accretion that renders some such isotopic systems useful. In fact it is this change that is being dated. If there is no change that accompanies accretion there is no way to distinguish when an object formed relative to the start of the solar system.

The ^{182}Hf–^{182}W chronometer (**Table 1**) has been particularly useful in determining the timescales over which the planets formed. The principles are as follows. Excess ^{182}W comes from the radioactive decay of ^{182}Hf. Every atom of ^{182}Hf decays to a daughter atom of ^{182}W therefore:

$$\left(^{182}W\right)_{today} = \left(^{182}W\right)_{original} + \left(^{182}Hf\right)_{original} \quad [1]$$

Because it is easier to measure these effects using isotopic ratios rather than absolute numbers of atoms we divide by another isotope of W:

$$\left(\frac{^{182}W}{^{184}W}\right)_{today} = \left(\frac{^{182}W}{^{184}W}\right)_{original} + \left(\frac{^{182}Hf}{^{184}W}\right)_{original} \quad [2]$$

However, the ^{182}Hf is no longer extant and so cannot be measured. For this reason we convert eqn [2] to a form that includes a monitor of the amount of ^{182}Hf that would have been present as determined from the amount of Hf today. Hafnium has several stable nuclides and ^{180}Hf is usually deployed. Hence eqn [2] becomes

$$\left(\frac{^{182}W}{^{184}W}\right)_{today} = \left(\frac{^{182}W}{^{184}W}\right)_{original}$$
$$+ \left\{\left(\frac{^{182}Hf}{^{180}Hf}\right)_{original} \times \left(\frac{^{180}Hf}{^{184}W}\right)_{today}\right\} \quad [3]$$

which represents the equation for a straight line. A plot of ^{182}W/^{184}W against ^{180}Hf/^{184}W for a suite of co-genetic samples or minerals will define a straight line the slope of which gives the ^{182}Hf/^{180}Hf at the time the object formed. This can be related in time to the start of the solar system with Soddy and Rutherford's equation for radioactive decay:

$$\left(\frac{^{182}Hf}{^{180}W}\right)_{original} = \left(\frac{^{182}Hf}{^{180}Hf}\right)_{BSSI} \times e^{-\lambda t} \quad [4]$$

in which BSSI is the bulk solar system initial ratio, λ is the decay constant (or probability of decay in unit time) and t is the time that elapsed since the start of the solar system. Using this method and the $(^{182}Hf/^{180}Hf)_{BSSI}$ of $\sim 1 \times 10^{-4}$ (**Table 1**) it has been possible to determine the age of a wide range of early solar system objects. Note that some early papers utilized a $(^{182}Hf/^{180}Hf)_{BSSI}$ of $\sim 1 \times 10^{-5}$ (Jacobsen and Harper, 1996). Later papers used a value of $\sim 2 \times 10^{-4}$ (e.g., Lee and Halliday, 1997). Most recently it has been made clear that the value must be close to $\sim 1 \times 10^{-4}$ (Kleine *et al.*, 2002; Schonberg *et al.*, 2002; Yin *et al.*, 2002).

There have been several reviews of ^{182}Hf–^{182}W chronometry, none of which is complete because the field is changing fast. Two large recent reviews with somewhat different emphases are provided by Jacobsen (2005) and Halliday and Kleine (2006). The principle of the technique is that Hf/W ratios are strongly fractionated by core formation because W normally is moderately siderophile whereas Hf is lithophile (**Figure 11**). Therefore, in the simplest of models, the W isotopic composition of a silicate reservoir such as the Earth's primitive mantle is a function of the timing of core formation. This could be determined from eqns [3] and [4] which provides the time (usually referred to as a two-stage model age) that the Hf/W ratio was fractionated by core formation.

For the ^{182}Hf–^{182}W system this time is given as

$$t_{CHUR} = \frac{1}{\lambda} \ln \left[\left(\frac{^{182}Hf}{^{180}Hf} \right)_{BSSI} \times \left(\frac{\left(\frac{^{182}W}{^{184}W}\right)_{SAMPLE} - \left(\frac{^{182}W}{^{184}W}\right)_{CHONDRITES}}{\left(\frac{^{180}Hf}{^{184}W}\right)_{SAMPLE} - \left(\frac{^{180}Hf}{^{184}W}\right)_{CHONDRITES}} \right) \right] \quad [5]$$

Where t_{CHUR} is the time of separation from a CHondritic Uniform Reservoir. If, however, a planet grows over tens of millions of years and the core grows as the planet gets larger, as is nowadays assumed to be the case for the Earth, the W isotopic composition of the primitive mantle is also a function of the rate of growth or the longevity of accretion and core formation. In this case the equations need to be more complex (Jacobsen, 2005) or numerical models need to be established that calculate the W isotopic effects as a result of each stage of accretion or growth (Halliday, 2004).

As already stated, the earliest precisely dated objects that are believed to have formed within our solar system are the CAIs found in many chondrites (**Figure 2**). It would appear that at least some chondrules formed a few million years after the start of the solar system. Therefore, isotopic dating provides evidence that the disk from which the planets grew

was still a dusty environment with rapid melting events, say 2 My after the start of the solar system (**Figure 2**). This means that chondrites must also have formed after this time.

Differentiated asteroids and their planetary cores and silicate reservoirs, as represented by some iron and basaltic achondrite meteorites formed 'before' chondrites. A variety of chronometers have been used to demonstrate that iron meteorite parent bodies, the eucrite parent body and the angrite parent body formed appear to have formed in the first few million years of the solar system (**Figure 2**) (Kleine *et al.*, 2005a; Schersten *et al.*, 2006; Markowski *et al.*, 2006). With extensive replication and better mass spectrometers very high precision can now be achieved on the W isotopic compositions of iron meteorites. The latest data for iron meteorites provide evidence that accretion and core formation were very short-lived (Kleine *et al.*, 2005; Markowski *et al.*, 2006; Schersten *et al.*, 2006). By correcting for cosmogenic effects it can be demonstrated that some magmatic iron meteorites, thought to represent planetesimal cores, formed within 500 000 years of the start of the solar system (Markowski *et al.*, 2006). Therefore, they appear to represent examples of early planetary embryos, as predicted from dynamic theory.

In a similar manner to differentiated asteroids, Mars also seems to have accreted quickly. The most recent compilations of data for Martian meteorites (Lee and Halliday, 1997, Kleine *et al*, 2004; Foley *et al.*, 2005; Halliday and Kleine, 2006) provide a clear separation of W isotopic compositions between nakhlites and shergottites and confirm the view that accretion and core formation on Mars were fast. Some recent models (Halliday and Kleine, 2006) place the timescale for formation of Mars at less than one million years (**Figure 2**). If this is correct, and the ages 'are' based on assumptions about the model Hf/W of the Martian mantle, Mars probably formed by a mechanism such as runaway growth, rather than by protracted collision-dominated growth. In other words Mars may represent a unique example of a large primitive planetary embryo with a totally different accretion history from that of the Earth.

The conclusion from these studies of short-lived isotope systems in meteorites is that chondrites, though widely considered the most primitive objects in the solar system and the building blocks for the planets, actually did not form for at least 2 My, longer than the timescales for growth of planetary embryos determined theoretically. Where and how they formed is currently unclear. The chronometry is consistent with the chemistry in indicating that chondrites do not represent the actual building blocks of the Earth. In fact it has long been thought that Earth's growth was dominated by accretion of differentiated objects (Taylor and Norman, 1990). Chondrites almost certainly represent a good approximation of average disk material but models that use chondrite variability to extrapolate to the bulk composition of the Earth, or attempt to build the Earth from chondritic components need to be treated with some awareness of this issue.

2.10 Rates of Earth Accretion and Differentiation

Although Patterson determined the age of the Earth there is in fact no such thing as a precise time when all of the constituents were amalgamated. It is more reasonable to instead assume a style of growth and use the isotopic data to define a rate of change of planetary growth from the integrated history of the chemical fractionation of the parent/daughter ratio. The latter is provided by the isotopic composition of the daughter.

The most powerful chronometers for estimating the growth rate of the Earth are ^{182}Hf–^{182}W and ^{235,238}U–^{207,206}Pb (Allegre et al., 1995a; Galer and Goldstein, 1996; Halliday, 2003). In both cases the parent/daughter ratio is strongly fractionated during core formation (**Figure 11**). As such, the W and Pb isotopic compositions in the silicate portion of the planet reflect how fast it grew and internally fractionated by metal segregation. Although it was once thought that accretion was rapid (Hanks and Anderson, 1969) and formation of the core was protracted (Solomon, 1979), there exists strong evidence that the process of separating core material from mantle material and delivering it to the core after it arrives by impact is a faster one (Sasaki and Nakazawa, 1986) than the process of accretion itself; this is discussed in Chapter 3. As such the core grows proportionally, limited in size by the growth of the planet. On this basis Pb and W isotopic data yield constraints on the rates of growth of the planet, provided certain assumptions are valid.

In practice the ^{182}Hf–^{182}W system provides a more powerful constraint on how soon accretion and core formation started whereas ^{235,238}U–^{207,206}Pb provides a better indication of how long it can have persisted. The ^{182}Hf–^{182}W system is better constrained in terms of the values of key parameters in the Earth. To

determine accretion rates one needs to know the parent/daughter ratio in the total planet, as well as in the reservoir, the formation of which is being dated. This is sometimes complex because we do not know independently how much resides in the inaccessible core. Some elements like Hf and W are refractory and were not affected by heating in the inner portions of the disk. Therefore, the Hf/W ratio of the Earth is well known from independent measurements of chondrites (Newsom, 1996), even if the Earth is not exactly chondritic. However, Pb is moderately volatile and therefore variably depleted in the Earth, Moon, Vesta, Mars, and almost certainly in Venus and Mercury. Therefore, two distinct processes have produced depletions of Pb in the silicate Earth – core formation and volatile loss. Two different methods are used to estimate the amount of Pb in the total Earth. The first is a comparison between the amount of Pb depletion in chondrites and that of other volatile elements that are not partitioned into planetary cores (Allègre et al., 1995a) (**Figure 6**). The second is based on estimates of the amount of volatile depletion expected given the average condensation temperature (Galer and Goldstein, 1996) (**Figure 7**). These two approaches are in broad agreement for Pb.

In addition to parent/daughter ratio, isotopic methods require a well-defined value for the isotopic composition of the silicate Earth relative to the rest of the solar system. The W isotopic composition of the silicate Earth is indeed extremely well defined because after 4.5 billions of convective mixing the Earth's silicate reservoir has homogenized any early isotopic variability. The W isotopic age of the Earth could in principle be determined from the composition of a tungsten carbide drill bit! However, with W the problem has been to get the correct value for the average solar system. Earliest Hf–W papers assumed a certain value (Jacobsen and Harper, 1996). Early measurements of chondrites were variable (Lee and Halliday, 1996, 2000). Although most of the early measurements for iron meteorites, achondrites, lunar samples, Martian meteorites, and enstatite chondrites have reproduced well with more modern high-precision methodologies, it is now recognized that the early W isotopic measurements for carbonaceous chondrites in particular, reported in Lee and Halliday (1996), are incorrect and the timescales have had to be reassessed in the light of this (Kleine et al., 2002; Schonberg et al. 2002; Yin et al., 2002; Halliday, 2004). The problem for U–Pb is almost the opposite. The present-day value for the silicate

Earth is very hard to constrain. The reason is that [235]U and [238]U are still alive in the Earth and capable of producing isotopic variations in [207]Pb/[204]Pb and [206]Pb/[204]Pb. Therefore, the Earth is very heterogeneous in its Pb isotopic composition (Galer and Goldstein, 1996). This is shown in **Figure 10** in which 11 estimates of the average Pb isotopic composition of the silicate Earth are presented. Each of these would imply a different timescale for Earth accretion, as discussed below and shown in **Figures 10** and **13**.

The simplest kind of age calculation derives a time of fractionation or two-stage model age by assuming that the fractionating process was instantaneous (**Figure 12(a)**). This is the equivalent of assuming that the core formed instantaneously from a fully formed Earth. The [182]Hf–[182]W model age so determined is ~30 My (Kleine *et al.*, 2002; Schonberg *et al.*, 2002; Yin *et al.*, 2002), whereas the [235,238]U–[207,206]Pb ages are more protracted (**Figure 10**). The possible reasons for this are discussed in more detail below. More complex models calculate an accretion rate (Jacobsen and Harper, 1996; Halliday, 2000, 2004) by making assumptions about the style of growth. Generally speaking, this is

an exponentially decreasing rate of growth that emulates what has arisen from dynamic simulations (**Figure 14**). This is consistent with the kinds of curves produced by the late George Wetherill who modeled the growth of the terrestrial planets using Monte Carlo simulations. To quantify the accretion timescale with an exponentially decreasing rate of growth requires the use of a time constant for accretion as follows:

$$F = 1 - e^{-(1/\tau) \times t} \qquad [6]$$

where F is the mass fraction of the Earth that has accumulated, τ is the mean life for accretion (or inverse of the time constant) and t is time (in the same units). The mean life corresponds to the time taken to achieve 63% growth. The accretionary mean life for [182]Hf–[182]W is ~11 My (Yin *et al.*, 2002) but is again much longer for [235,238]U–[207,206]Pb (Halliday, 2003) (**Figure 13**). Some of the most recent models (Halliday, 2004) also try to simulate the effect of a more stochastic style of accretion with growth from sporadic large (Moon- to Mars-sized) impactors (**Figure 14**). The staggered growth model shown in **Figure 14** assumes that the Earth grew from 90% to 99% of its current mass as a result of a

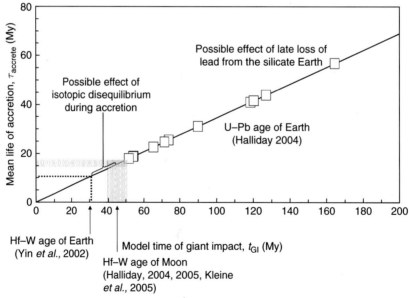

Figure 13   Calculated values for the Earth's mean life of accretion (τ) and time of the giant impact (T_{GI}) given in My as deduced from the different estimates of the Pb isotopic composition of the BSE using the type of accretion model shown in **Figure 14** (see Halliday, 2004). Note that, whilst all calculations assume continuous core formation and total equilibration between accreted material and the BSE, the accretion is punctuated, as predicted from the planetesimal theory of planetary accretion. This generates more protracted calculated timescales than those that assume smooth accretion (Halliday 2004). The [238]U/[204]Pb values assumed for the Earth = 0.7 (Allègre *et al.*, 1995a). Even the Hf–W timescales using the exact same style of model are significantly shorter.

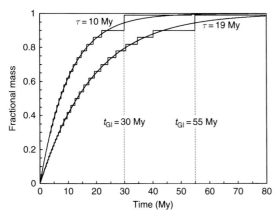

Figure 14 The mean life of accretion of the Earth (τ) is the inverse of the time constant for exponentially decreasing oligarchic growth from stochastic collisions between planetary embryos and planets. The growth curves corresponding to several such mean lives are shown including the one that most closely matches the calculation made by the late George Wetherill based on Monte Carlo simulations. The mean life determined from tungsten isotopes is in excellent agreement with Wetherill's predictions. From Halliday AN (2004) Mixing, volatile loss and compositional change during impact-driven accretion of the Earth. *Nature* 427: 505–509.

collision with an impacting Mars-sized planet called 'Theia' that produced the Moon. The changes in the rate of accretion with time are calculated from the timing of this event. Therefore, the model time of the Giant Impact is directly related to the mean life of accretion in these models (**Figures 13 and 14**). Regardless of the details of the accretion processes this last major impact would have had a dominant effect on the present-day W isotopic composition of the BSE. Therefore, predictions about the last major accretion event to affect the Earth which are based on the W isotopic composition of the BSE, can be tested by determining the age of the Moon.

2.11 The Age of the Moon

From the W isotopic compositions of lunar samples it has been possible to determine that the Moon must have formed tens of million of years after the start of the solar system. The first attempts at this (Lee *et al.*, 1997) revealed high ^{182}W in many lunar samples. Subsequently it was shown that in some of these a portion of the effect was due to cosmogenic production of ^{182}Ta which decays to ^{182}W. This can be corrected in some cases by using exposure ages or

internal mineral systematics (Lee *et al.*, 2002). A novel and in many respects more straightforward and reliable approach has been proposed by Kleine *et al.* (2005) who use carefully separated metals with low Ta/W. All of these approaches yield similar time-scales of between about 30 and 55 My after the start of the solar system (**Figure 2**). This provides strong support for the Giant Impact Theory since such a late origin for an object of the size of the Moon is not readily explicable unless it is derived from a previously formed planet. Defining the age of the Moon in this way provides an excellent example of the effectiveness of ^{182}Hf–^{182}W chronometry. Despite considerable efforts using long-lived chronometers the age of the Moon was for decades only known to be some time in the first 150 My of the solar system (Wasserburg *et al.*, 1977). The power of short-lived chronometers like ^{182}Hf–^{182}W is that there exists a very narrow time window within which any radiogenic effects are possible.

Strictly speaking, the process being dated with ^{182}Hf–^{182}W chronometry of metals is the differentiation of the lunar mantle as a high Hf/W reservoir. This probably took place in the lunar magma ocean. An upper limit for the age of the Moon is however provided by two additional constraints. First, if the Moon formed from an undifferentiated chondritic object or equilibrated isotopically with an entire planet, the age is about 30 My (Kleine *et al.*, 2002; Yin *et al.*, 2002). This is the earliest age it could reasonably have. However, this is hard to reconcile with a Moon that was largely formed from the silicate reservoir of a differentiated planet as in Giant Impact simulations (Canup and Asphaug, 2001). A second method of calculating an age is to use the lunar initial W isotopic composition inferred from the lower limit of the clustering of the W isotopic compositions. On this basis the age would be no earlier than 40 My (Halliday, 2003) (**Figure 2**). This would be in better agreement with the $^{235/238}$U–$^{207/206}$Pb age of Pb loss from the material that formed the Moon corresponding to 4.5 Ga (Hanan and Tilton, 1987).

2.12 First Principles of Chemical Constraints on Core Formation

Having used the ^{182}Hf–^{182}W system to constrain the timing of core formation the physical conditions of the core-separation process can be estimated provided chemical compositions of the BSE, the core, and the bulk Earth are known. As discussed earlier the BSE exhibits a depletion in volatile elements with respect to the chondritic reference model. It also shows depletions in refractory siderophile elements such as W, Mo, Re, and Os. In these latter cases the depletions cannot be due to volatility and must be to the result of partial extraction into the core. The extent of extraction can be estimated by comparing silicate Earth concentrations (normalized to CI chondrites) to those of refractory lithophile elements such as the REEs. This enables us to calculate, for each element, a core–mantle partition coefficient D_i defined as follows:

$$D_i = [i]_{core} / [i]_{silicate\ Earth} \qquad [7]$$

where $[i]$ is the concentration of element i. Lithophile elements have D values close to zero. **Table 2** shows that siderophile elements exhibit a very wide range of core–mantle partitioning behavior reflecting their different chemical properties and the conditions under which core segregation took place.

Core–mantle partitioning is best defined for those refractory elements that are weakly or moderately siderophile and which, like Ni and Co are compatible in solid mantle silicates. The concentrations of these elements vary little in mantle samples, which means that D_i values based on the chondrite reference model have small uncertainties. Highly siderophile refractory elements such as the Pt group (**Figure 7**) are slightly less well constrained. The abundances of these elements in mantle samples are very low and rather variable. It is generally accepted,

however, that their ratios one to another in the silicate Earth are, as shown in **Figure 7**, approximately chondritic.

There are a number of volatile siderophile elements such as S, Se, Te, P, and possibly Si whose core concentrations are difficult to estimate because their bulk Earth contents are not well constrained. This means, for example that depletions in Mn, Si, and Cr (**Figure 7**) could be due to volatility or to sequestration in the core or both. The lack of constraint means that the uncertainties in D_i are large.

Segregation of the reduced core from the oxidized mantle took place at high temperatures and, since transfer from mantle to core involves reduction, element partitioning between the two must have depended on oxygen fugacity. From the current concentration of FeO in the mantle of about 8 wt.% and of Fe in the core of 85% (Allègre *et al.* 1995b; McDonough 2003; McDonough and Sun, 1995), it is fairly straightforward to estimate that core segregation took place at approximately 2 log fO_2 units below Fe–FeO (IW) equilibrium (e.g., Li and Agee, 1996).

Table 2 shows a comparison of the calculated core(metal)–mantle(silicate) partition coefficients and those obtained experimentally at low pressures (0–2 GPa), high temperatures ($\sim$1800 K), and oxygen fugacity corresponding to 8 wt.% FeO in the mantle (Newsom, 1990; O'Neill and Palme, 1998). As can be seen, the observed core–mantle partition coefficients are, for many of the siderophile elements, much

Table 2   Core-Mantle partition coefficients

Element	McDonough and Sun (1995) McDonough (2003)	Allègre et al. (1995)	Likely range	Low pressure experimental D_i (Newsom, 1990; O'Neill and Plame, 1998)
D_{Fe}	13.66	13.65	13.65	13.65
D_{Ni}	26.5	24.4	23–27	4900
D_{Co}	23.8	24.7	23–27	680
D_V	1.83		1.5–2.2	0.02
D_W	16		15–22	3
D_{Pd}	800		600–1000	7×10^5
D_{Ir}	800		600–1000	10^{11}
D_{Pt}	800		600–1000	4×10^6
D_{Nb}			0.2–0.8	
D_{Cr}	3.4	2.9	0.5–3.5*	0.2
D_{Mn}	0.29	5	0.2–2.0*	0.006
D_{Si}	0.29	0.34	0.1–0.35*	10^{-5}
D_S	76		50–100*	50
D_{Ga}			0–1.5*	15
D_P	22	45	20–50*	30

*large uncertainty due to volatility.

(orders of magnitude) smaller than those determined experimentally. This manifests itself in the concentrations of many elements of concern (e.g., Ni, Co, Pd) being much greater in the mantle than would be predicted from the experimental data, the so-called 'excess siderophile problem'. Of equal importance are the weakly siderophile elements such as V and Cr for which the mantle concentrations are 'lower' than predicted from the experimental data. Clearly, the mantle concentrations of these two groups of elements cannot be explained by equilibrium with metallic core at the fixed pressure, temperature, and oxygen-fugacity conditions of **Table 2**. Early core segregation in small planetesimals and planetary embryos, including proto-Earth, should, however, have resulted in siderophile-element abundances generally consistent with the D values given in **Table 2**. The Earth, apparently, has inherited little of the geochemistry of these earlier events.

2.13   Explanations for the 'Excess Siderophile Problem'

Given that the 'excess siderophile problem' was not inherited, but arose from the manner of core formation on Earth, several possible explanations have been proposed:

1. *Inefficient core formation* (Arculus and Delano, 1981; Jones and Drake, 1986). In this model, some fraction of the core was left behind in the mantle, together with its siderophile-element budget. Later reoxidation and remixing of the metal led to the observed elevated mantle concentrations of siderophile elements. Principal objections to this model are that it can only explain the mantle abundances of some elements and that an enormous amount of water, about 15 times the current mass of the hydrosphere would be required to reoxidise the 'retained' metal (O'Neill, 1991).

2. *Heterogeneous accretion* (Chou, 1978; O'Neill, 1991; Wänke, 1981) . In this model it is suggested that the conditions of core segregation changed from reducing at the beginning of accretion to more oxidized toward the end. During the earliest, reduced, phase, highly and moderately siderophile elements (Ni, Co, and Pt group for example) would have been completely extracted to the core, together with some fraction of the weak siderophiles such as Si, V, and Cr. As conditions became more oxidizing, core extraction of siderophile elements would have ceased

progressively, beginning with the weakest siderophiles and ending with more strongly metallic elements. The final phase, after core formation had ceased, was the addition of a 'late veneer' of about 0.5% of chondritic material. The latter raised the concentrations of highly siderophile elements in the mantle and, since core formation had ceased, ensured their chondritic proportions one to another (**Figure 7**). This model solves the 'excess siderophile problem', but, being a disequilibrium model with multiple steps, is very difficult to test.

3. *High-pressure core formation.* Murthy (1991) proposed that the partition coefficients for siderophile elements between metal and silicate should approach unity at high temperatures. This led to a variety of important new studies and recent measurements of metal-silicate partitioning at high pressures and temperatures (Gessmann and Rubie, 1998; Li and Agee, 1996; Li and Agee, 2001; Righter and Drake, 2000; Righter *et al.*, 1997; Thibault and Walter, 1995). They have shown that the partition coefficients of some siderophile elements change such that their mantle abundances, inexplicable by low-pressure equilibrium, may be explained by metal–silicate equilibrium at very high pressures and temperatures (e.g., Li and Agee, 1996; Righter and Drake, 1997). These observations led to the 'deep magma ocean' hypothesis to explain the 'excess siderophile problem'. As the Earth grew, it is argued (e.g., Righter and Drake, 1999), droplets of metallic liquid descended through a 700–1200-km-deep (28–40 GPa) magma ocean, ponding at its base in equilibrium with the magma ocean. The liquid metal subsequently descended in large diapirs to the growing core without further reaction with the surrounding silicate (**Figure 15**). High-pressure core formation can explain, by metal-silicate equilibrium, the partitioning of many elements between core and mantle. It does still require, however, the 'late veneer' of chondritic material to explain the concentrations and proportions of the highly siderophile elements in the silicate Earth.

2.14   The 'Deep Magma Ocean' Model of Core Formation

As the Earth grew the gravitational energy deposited by accreting planetesimals would have increased and at about 10% of its current size (Stevenson, 1981) was sufficient for significant melting to occur. After this point the Earth would have periodically had an extensively molten outer layer (a magma ocean) of variable

High-pressure core segregation model

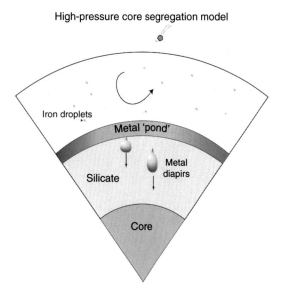

Figure 15 Core formation in the 'deep magma ocean' model. Impacting planetesimals disaggregate and their metallic cores break into small droplets due to Rayleigh–Taylor instabilities. These droplets descend slowly, re-equilibrating with the silicate until they reach a region of high viscosity (solid), where they pond in a layer. The growing dense metal layer eventually becomes unstable and breaks into large blobs (diapirs) which descend to the core without further interaction with the silicate. The liquidus temperature of the silicate mantle should correspond to pressure and temperature conditions at a depth above the lower solid layer and plausibly within the metal layer.

thickness. In some cases the impact energy would have been sufficient to melt both the impactor and the proto-Earth (Cameron and Benz, 1991; Canup and Asphaug, 2001). From W isotope evidence it is clear that planetary cores segregated from their silicate mantles very early. As a result most incoming planetesimals and planets are likely to have been already differentiated into silicate and metal like Vesta, Mars, and the Moon-forming planet Theia (Taylor and Norman, 1990; Yoshino *et al.*, 2003). The fates of pre-existing cores would have been dependent on the sizes of the Earth and impactor. Continuous-core-formation isotopic models normally assume that all of the metal and silicate in the impactor mixes and equilibrates with the silicate portion of the Earth before further growth of the Earth's core (Kleine *et al.*, 2002; Schoenberg *et al*, 2002; Yin *et al*, 2002). However, the degree to which the impactor's core fragments depends on the angle of impact and relative size of the bodies. (*see* Chapter 3). Large fragments of metal could, under some circumstances, have mixed directly with the Earth's core (Karato and Murthy, 1997).

However, the size of those droplets depends on the fluid dynamics of the process.

Rubie *et al.* calculate the size of droplets of liquid metal raining out of a magma ocean and conclude that these should have been about 1 cm diameter. Droplets of this small size would re-equilibrate rapidly with the silicate melt as they fell (Rubie *et al.*, 2003). Liquid metal and silicate would therefore have continued to re-equilibrate until the former either reached the core–mantle boundary (if the mantle were completely molten) or collected at a level above a solid, high-viscosity lower layer (**Figure 15**). In the latter case the metal layer would cease re-equilibrating after it had reached about 5 km in thickness (Rubie *et al.*, 2003) and because of the high viscosity of the lower layer would segregate as large diapirs to the core (Karato and Murthy, 1997). The droplet model can apply in the case of small bodies accreting to the growing Earth, but may not apply after a giant impact such as the Moon-forming impact (Chapter 3). In the latter case isotopic and chemical equilibration of metal from the impactor with the silicate portion of the Earth depends on the extent to which the core of the impactor is fragmented.

The fact that the silicate Earth has a W isotopic composition that is so similar to average solar system (chondrites) means that in general terms metal-silicate equilibration must have been the norm. However, the small difference that is now known to exist between the W isotopic composition of chondrites and the BSE could reflect an event like the Giant Impact that involved particularly rapid amalgamation of large fragments of impactor metal with the Earth's core with little chance for equilibration with the silicate portion of the Earth (Halliday, 2004).

If we know the age of the Moon we know the timing of the last accretion stage in the history of the Earth (**Figures 2** and **14**). From this we can determine the extent to which the W isotopic composition of the silicate Earth reflects planetary-scale disequilibrium of added material during accretion. Preliminary attempts at calculating this (Halliday, 2004; Kleine *et al.*, 2004b; Halliday and Kleine, 2006) show that on average the degree of equilibration between incoming metal and the Earth's silicate must have been high (70–90%) (**Figure 16**). Equilibration would have been even higher if the Moon formed before 45 ± 5 My after the start of the solar system. Therefore, full equilibration with incoming material is the generally accepted norm.

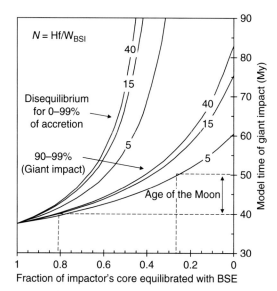

Figure 16 The effect on calculated Hf–W timescales for the Earth's formation of incomplete mixing and equilibration of impacting core material is illustrated. This plot shows the true time of the Giant Impact that generates $\varepsilon^{182}W_{BSE}$ of zero as a function of various levels of incomplete mixing of the impacting core material with the BSE. The lower curves are for disequilibrium during the Giant Impact alone. The upper curves correspond to disequilibrium during the entire accretion process up to and including the Giant Impact. The different curves for different Hf/W$_{BSI}$ (the Hf/W in the silicate portion of the impactor) are also shown. All curves are calculated with $\varepsilon^{182}W_{BSSI} = -3.5$. Assuming an age for the Moon in the range 40–50 My after the start of the solar system it would seem likely that there was significant isotopic disequilibrium during the Giant Impact (Halliday, 2004).

However, the data also allow for the possibility that a very large event, such as the Giant Impact, resulted in sizeable disequilibrium. If equilibration was the norm then between 40% and 60% of the incoming W in metal did not equilibrate with the silicate Earth during the Giant Impact. This might be more easily reconciled with dynamic simulations. However, it should be emphasized that greater resolution of the simulations and a more comprehensive understanding of the physics of metal-silicate mixing under such extreme conditions is needed before one can say which isotopic model is more consistent with the likely physical processes.

In the event that the energetics of impact and core segregation lead to a completely molten mantle, numerical simulations (Abe, 1997; Chapter 4) indicate that such a deep magma ocean would be short-lived and that the lower mantle would crystallize in a few thousand years. A shallower partially molten

layer would crystallize much more slowly, however, and could remain as a mixture of crystals and melt for 100 My (Abe, 1997; Solomatov, 2000, this volume). The energetics of impact, core segregation, and mantle crystallization lead to a dynamic view of the growing Earth in which the outer molten part deepened and shallowed many times after episodic impact. The pressures and temperatures recorded by core–mantle partitioning are therefore values averaged over numerous cycles of metal accumulation and segregation such as that depicted in **Figure 15**.

If one considers the magma ocean model of **Figure 15** and takes account of the likelihood that the magma ocean shallowed and deepened many times during accretion and core segregation, several principles emerge. Firstly, accumulation of metal at the base of a completely or partially molten silicate layer, as shown in **Figure 15** implies temperatures close to or slightly below the liquidus of the mantle peridotite. Secondly, average pressure should have increased as the planet grew. Thirdly, since the silicate liquidus has a positive pressure–temperature slope, average temperature of core segregation should also have increased as the planet grew. Having accepted these principles, we can use the dependences of D_i on pressure and temperature (Chabot and Agee, 2003; Li and Agee, 2001; Righter et al., 1997; Wade and Wood, 2005) to develop constraints on the core segregation process.

2.15 Core Segregation during Growth of the Earth

If we consider the case of instantaneous core formation (**Figure 12(a)**), then current partitioning data for the elements of **Table 2** yield average pressures and temperatures of ~40 GPa and ~3750 K, respectively (Chabot et al., 2005; Wade and Wood, 2005). Given that the mantle has superchondritic Mg/Si these conditions imply that a substantial amount of Si, more than required to give the bulk Earth a chondritic Mg/Si ratio, resides in the core (Wade and Wood, 2005). The reason is that Si partitioning into the metal is strongly temperature dependent and the calculated temperature is very high, in fact 650° above the silicate liquidus. This seems unlikely because the base of the magma ocean should be at or below the mantle liquidus (**Figure 15**).

Continuous accretion and core-formation models (**Figure 12(b)**) do not change the result,

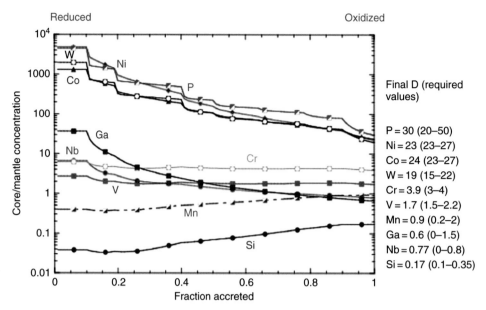

Figure 17   Figure showing one possible path of accretion and core segregation for the Earth. Earth was accreted and the core segregated in 1% intervals. At each step the silicate Earth was homogenized and the core segregated at the base of the magma ocean of **Figure 15**. At 100% accretion (right hand side) the core/mantle partitioning of each element must match the 'required value' of **Table 2**. In order to achieve this and maintain temperatures on the silicate liquidus, the Earth must become more oxidized during accretion. In the example shown, the oxygen fugacity increases in small steps from about 3.5 log units below the Fe–FeO (IW) buffer to 2 log units below IW during accretion and the magma-ocean depth corresponds to 30–40% of the depth to the core–mantle boundary in the growing planet.

temperatures remaining above the silicate liquidus. Apart from invoking wholesale disequilibrium, the simplest solution (Wade and Wood, 2005), as illustrated in **Figure 17**, is to allow oxygen fugacity (as represented by the FeO content of the mantle) to increase as accretion and core segregation progress.

Figure 17 illustrates a continuous process of accretion and core segregation in 1% intervals. New material added to the Earth was assumed to be mixed with the pre-existing mantle, but isolated from the protocore. The core was segregated in 1% steps, each aliquot being in equilibrium with the entire mantle at the time of segregation at pressures constrained by Ni and Co partitioning and temperatures on the peridotite liquidus (Wade and Wood, 2005). The latter were found to be consistent with magma oceans extending to 30–40% of the depth of the growing mantle. A large number of oxygen-fugacity paths yield the partitioning of V, W, Nb, Ni, Co, Mn, Cr, Ga, P, and Si which are consistent with the estimates of **Figure 17**. All require, however, that most of the Earth formed under conditions more reducing than those implied by the current FeO content of the mantle and that the Earth became more oxidized toward the end of accretion. The

combined effects of increasing temperature and pressure during accretion led to a progressive increase in core Si content (**Figure 17**) such that the final core contains about 4% Si. This brings the Mg/Si ratio of the bulk Earth much closer to chondritic than implied in **Figure 8**. As required by all accretion models, 0.5% of chondritic material must be added after cessation of core formation in order to generate the chondritic ratios of the highly siderophile elements such as Pd, Ir, and Pt (**Figure 17**). The current 'best explanation' is therefore an amalgam of 'deep magma ocean' and 'heterogeneous accretion' hypotheses, that is, core segregation at the base of a deepening magma ocean under progressively more oxidizing conditions (Wade and Wood, 2005).

The idea that the Earth became more oxidized during accretion has been proposed on a number of occasions (e.g., O'Neill, 1991), but this leads to two other important questions. Firstly, the Earth has a substantially higher core/mantle mass ratio than Mars and is hence, as a planet, more reduced than this smaller body. What caused the Earth to be more reduced than smaller bodies? Was it due to part of the silicate Earth being removed during accretion as proposed for Mercury (Benz *et al.*, 1997)? If the

Earth later became more oxidized, as evidenced by the relatively oxidized nature of terrestrial as opposed to lunar or Martian basalts,(Karner, et al., 2006) and the constraints of core–mantle partitioning discussed above, then what was the mechanism of oxidation and when did it operate?

2.16   Oxidation State of the Earth during and after Accretion

As planetesimals accreted during the first few million years of solar system history, they would have been surrounded by a hydrogen-rich nebula gas of solar composition. This gas would have had little effect on the chemical compositions of small bodies because the pressure was very low (10^{-4} bar). As protoplanets grew, however, their masses might have become high enough for them to capture a nebula gas atmosphere (Hayashi et al., 1979, 1985; Sasaki and Nakazawa, 1990; Ikoma and Genda, 2006) such that at a mass >20% of the current mass of the Earth hydrogen pressures >1 bar are plausible.

There is a problem with these models because by the time a body that is sufficiently massive to retain a large hydrogen atmosphere has formed, the nebula may have already dissipated. Ikoma and Genda (2006) point out that the atmosphere thereby accreted is only logarithmically sensitive to the nebula density, which implies that a massive hydrogen atmosphere can be accreted even when the nebula has largely departed. However, the nebula may not decline exponentially with time; the favored process suggests it is highly superexponential. That is, the nebula declines to essentially zero pressure in finite time, perhaps a mere 5 My. Accretion rates based on W isotopes are agnostic in the earliest stage of Earth's growth but are ususally modeled with longer timescales (**Figure 14**). However, noble-gas data do provide evidence of solar components within the Earrth which could have been acquired directly from the solar nebula (Porcelli et al., 2001).

So long as the nebula persisted, therefore, the episodic magma ocean at the Earth's surface would have reacted with hydrogen gas, raising the Fe/FeO ratio of the bulk Earth and increasing the mass fraction of its metallic core. After about 10^7 years the nebula dissipated and the oxidation state of the mantle would have been free to increase, particularly if late-arriving planetesimals had higher FeO/Fe ratio than the early ones (O'Neill, 1991). This mechanism, impossible to quantify, would have generated a gradual increase in FeO/Fe ratio of the Earth, but would not have produced any of the Fe^{3+} present in peridotite and mid-ocean ridge basalt (MORB) (Wood, et al., 1990). The latter can be explained if the Earth self-oxidized through crystallization of magnesium silicate perovskite (Frost et al., 2004; Wade and Wood, 2005).

At depths deeper than 660 km in the present-day Earth, the stable phases are $(Mg, Fe)SiO_3$, magnesium silicate perovskite (79% by volume (Wood, 2000)), $(Mg, Fe)O$ magnesiowüstite (16%) and $CaSiO_3$ perovskite (5%). An important property of magnesium silicate perovskite is that it accommodates the 5% Al_2O_3 which it dissolves in peridotite compositions by a coupled substitution with Fe^{3+} (McCammon, 1997; Wood and Rubie, 1996) as follows:

$$Mg^{2+} + Si^{4+} \leftrightarrow Fe^{3+}Al^{3+} \qquad [8]$$

It has recently been discovered that this substitution mechanism is so stable that it forces ferrous iron to disproportionate to ferric iron plus iron metal (Frost et al., 2004).

$$3Fe^{2+} \leftrightarrow 2Fe^{3+} + Fe^0 \qquad [9]$$

Or, in terms of oxide components of the lower mantle:

$$3FeO + Al_2O_3 = 2FeAlO_3 + Fe^0 \qquad [10]$$

This means that, as perovskite began to crystallize in the extensively molten Earth, it dissolved ferric iron as $FeAlO_3$ component and produced Fe metal. Since the perovskite is stable throughout virtually the entire lower mantle, this process took place during most of accretion and core formation on Earth. The implications are that metal sinking through the lower mantle to the core would inevitably have dissolved some of this internally produced metallic iron, resulting in a perovskitic layer which was relatively oxidized. Given the gravitational instability of any metal layer (**Figure 15**), accretional energies, and heat loss, it is inevitable that the depth of the magma ocean fluctuated continuously, thereby generating fronts of dissolution and precipitation at the lower boundary. Perovskite dissolution and re-precipitation therefore released ferric iron to the magma ocean during dissolution and produced more during precipitation. This raised the oxygen fugacity of the mantle. Any later droplets of metal falling through the magma ocean would have reacted with ferric iron, producing more Fe^{2+} which would then have dissolved in the silicate melt. Hence the content of oxidized iron of the mantle (magma ocean) increased naturally as a consequence of perovskite crystallization. In the very final stages of Earth accretion this mechanism may have caused sufficient oxidation to halt metal segregation, setting the stage for the 'late veneer' of chondritic or similar material to add the highly siderophile elements exclusively to the mantle. Note that the oxidation-state change required by partitioning data is a simple consequence of the size of the Earth. Any planet in which magnesium silicate perovskite is an important crystallizing phase should undergo the same process. It could explain why terrestrial basalts are more oxidized than those of Mars (Herd et al., 2002), a planet in which perovskite can only crystallize at depths close to the core–mantle boundary. Because of its small size, Mars cannot have undergone the same period of extensive perovskite crystallization and self-oxidation during accretion as did the Earth.

2.17   Isotopic Evidence for Volatile Losses from the Earth during Accretion

Isotope geochemistry provides two kinds of information on the changes that may have affected a planet during growth. The first is mass-dependent stable isotopic fractionation, which can leave a marked imprint of processes that affected the Earth. The second is isotopic compositons that leave a record of time-integrated parent/daughter elemental ratios that are distinct from those found in the Earth today.

The most dramatic example of mass-dependent fractionations is provided by Xe isotopes. Xenon isotopes provide evidence that the noble gases were more abundant in the early Earth and that a major fraction has been lost, probably by shock-induced blow-off of the atmosphere (**Figure 9**). This provided much of the evidence for loss of a large protoatmosphere by hydrodynamic escape (Walker, 1986; Hunten et al., 1987). However, it is also clear that a late-stage major parent/daughter fractionation clearly affected the noble gases. The Xe constraints are based on the decay of formerly live ^{129}I ($T_{1/2} = 16$ My) and ^{244}Pu (fission $T_{1/2} = 82$ My) (**Table 1**). Put simply, the small difference in composition between the isotopic composition of xenon in the Earth relative to its initial starting composition is consistent with low parent/daughter ratios. The amount of xenon in the Earth today is thought to be at least two orders of magnitude lower than is necessary to satisfy these Xe isotope data (Porcelli et al.,

2001). That is, the calculated ratio of I to Xe is huge; at least two orders of magnitude higher. Similarly one can deduce the former abundance of Pu because it is refractory and lithophile. From meteorite studies we know its early solar system abundance fairly well and can determine how much heavy xenon (e.g., ^{136}Xe) its decay should have generated within the Earth during the first few hundred million years of Earth history. This xenon is missing (Porcelli and Pepin, 2000). Therefore, there must have been a major fractionation in parent/daughter ratio at a relatively late stage.

Because Xe is a gas that has a strong preference for residing in the atmosphere and because similar effects are found in both the I–Xe and Pu–Xe systems, it is clear that there was a significant amount of xenon loss from the Earth, as opposed to selective I or Pu gain. The best estimate for the age of this loss was 50–80 My (Ozima and Podosek, 1999; Porcelli and Pepin, 2000) and it was proposed that this may have been associated with the Moon-forming Giant Impact. However, this is looking increasingly unlikely. Longer timescales are required to satisfy the ^{136}Xe constraint and provide evidence that Earth's Xe was lost over the first few hundred million years. Pepin and Porcelli (2006) have recently proposed two epsiodes of atmospheric blow-off related to giant impacts, one early – and related to Moon formation – and the other significantly later. It is also possible that other processes operated in the Earth's atmosphere most likely involving ionization of Xe (Zahnle et al., 2006).

In addition to losses of elements like the noble gases which are clearly atmophile, the question has arisen as to what extent the depletions in moderately volatile elements also reflect accretion phenomena. Evidence for loss of moderately volatile elements during accretion was hard to reconcile with the uniform K isotopic compositions recorded by inner solar system objects (Humayun and Clayton, 1995). However, the precision on these K isotopic measurements was not at the kind of level that would resolve small evaporation effects. Furthermore, if K were atmophile because it was degassed from the mantle into a hot atmosphere nebular atmosphere which was blown off with Xe (O'Neill, 1991) the losses would not necessarily record a major isotopic fractionation (Halliday, 2004). Pritchard and Stevenson (2000) show that in any case it is possible to have large volatile loss without large isotope effects. They present an explicit calculation showing how the K-41 isotopic effect can be small even when the volatilization is large. The main point is that it is evaporation into a dense atmosphere not into vacuum.

Supporting evidence for accretionary losses of volatile elements is found in oxygen isotopic compositions (**Figure 3**). Oxygen isotopes provide a monitor of the degree of mixing of dust and gas in the disk (Clayton and Maycda, 1975; Wiechert *et al.*, 2001, 2004). The oxygen isotopic composition varies among different classes of meteorites and is therefore assumed to be highly heterogeneous in general among inner solar system objects (Clayton *et al.*, 1973; Clayton 1986, 1993). This is not a normal mass-dependent fractionation so much as a mass-independent effect, the origin of which is still unclear. Regardless of the reason behind the heterogeneity it is thought that the oxygen isotopic composition provides a monitor of the origins of the material in a planetary object. The provenance of material in an object the size of the Earth is expected to have been quite broad (Wetherill, 1994). As such, the oxygen isotopes merely provide some average.

The Earth is distinct in oxygen isotopes from all classes of chondrites except enstatite chondrites, which may have formed in the more reducing inner regions of the solar system. The provenance should be much broader than this. Similarly, the provenance appears to be distinct from that of the material forming the Asteroid 4 Vesta, or Mars, as judged from studies of meteorites thought to be derived from these objects (Clayton, 1986, 1993). The Moon however, has an oxygen isotopic composition that is identical to that of the Earth (Clayton and Mayeda, 1975), recently demonstrated to persist to extremely high precision (Wiechert *et al.*, 2001) (**Figure 3**). This has long provided a powerful line of evidence that the Moon formed from the Earth (Wanke and Dreibus, 1986; Ringwood, 1989) or that the Earth and Moon formed from the same material (Wiechert *et al.*, 2001), even though these two objects have demonstrably different chemical composition.

In simulations of the Giant Impact (Cameron and Benz, 1991) roughly 80% of the Moon forms from the debris derived from Theia (Canup and Asphaug, 2001). If this is correct, the oxygen isotopic evidence provides evidence that Theia and the Earth had similar average provenance. Yet the Moon is far more depleted in volatile elements than the Earth (Taylor and Norman, 1990) (**Figure 4**). A likely explanation for this is that there were major reductions in volatile constituents during the Moon-

forming Giant Impact (O'Neill, 1991b) (**Figure 18**). Support for this is found in the initial $^{87}Sr/^{86}Sr$ of early lunar rocks from the Highlands (Halliday and Porcelli, 2001). These yield an initial $^{87}Sr/^{86}Sr$ for the Moon that is clearly resolvable from the initial composition of the solar system (**Figure 19**), consistent with radiogenic growth in an environment with relatively high Rb/Sr. This time-integrated Rb/Sr can be determined to be ∼0.07, similar to the Rb/Sr of Mars but an order of magnitude less depleted in moderately volatile Rb than the Moon today (Rb/Sr ∼ 0.006) (**Figure 19**). Carlson and Lugmair (1988) noted a similar value for the time-integrated Rb/Sr for the Moon but ascribed it to that of the Earth. However, this is more than double the value independently known for the Earth. Therefore, it appears as though Mars-like objects may have once existed in the vicinity of the Earth.

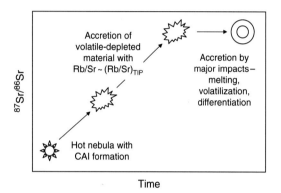

Figure 18 Schematic illustration of the time integrated effects of Rb loss on Sr isotopic compositions in early planetary objects.

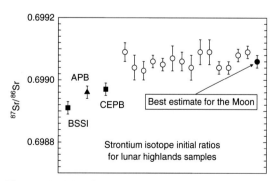

Figure 19 Initial Sr isotope composition of early lunar highland rocks relative to other early solar system objects. From compilation in Halliday and Porcelli (2001). APB, angrite parent body; CEPB, cumulate eucrite parent body; BSSI, bulk solar system initial.

Tungsten isotopes paint a strikingly similar picture (Halliday, 2004). The Hf/W ratio of silicate reservoirs in the terrestrial planets is variable and is thought to depend on the level of oxidation of the mantle because this controls how 'metal loving' W is during core formation. If the mantle is relatively reducing the vast majority of the W is segregated into the core and vice versa. Just as the time-integrated Rb/Sr can be deduced from Sr isotopes, the time-integrated Hf/W can be deduced from W isotopes. Tungsten isotope data for the Moon provide evidence that the precursor objects that provided the raw materials for the Moon had a time-averaged Rb/Sr and Hf/W that was within error identical to the actual Rb/Sr and Hf/W of Mars (**Figure 20**).

Therefore, the lunar data provide evidence of planetesimals and planets in the vicinity of the Earth that were more like Mars in terms of chemical composition. The fact that the oxygen isotopic composition of the Earth and Moon are identical provides evidence that such protoplanetary compositions may well have been a normal feature of inner solar system or terrestrial protoplanetary compositions. This however, assumes that oxygen isotopes provide a comparative monitor of the mix of raw ingredients

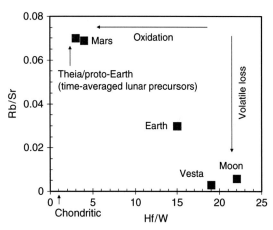

Figure 20 Hf/W appears to be negatively correlated with Rb/Sr in the primitive mantles of inner solar system planetesimals and planets. A possible explanation for this is that the loss of moderately volatile elements was linked to loss of other volatiles during planetary collisions such that the mantles changed from more oxidizing to more reducing. The Moon is an extreme example of this. The fact that the mixture of material from the proto-Earth and Theia that is calculated to have formed the Moon is so like Mars provides evidence that such volatile-rich objects may have been common in the inner solar system during the early stages of planetary accretion. See Halliday and Porcelli (2001) and Halliday (2004) for further details.

in the inner solar system. The isotopic evidence for impact-driven chemical changes in the terrestrial planets during growth is however well supported by evidence that the planet Mercury lost a major fraction of its silicate by impact erosion (Benz *et al.*, 1987).

The mismatch between U–Pb and Hf–W chronometry may also be related to late volatile losses. The rates of accretion of the Earth based on the model of **Figure 14** and the various estimates of the BSE Pb isotopic composition (**Figure 10**) are shown in **Figure 13**. Also shown are the results for W. It is clear that the kinds of timescales involved are on the order of 10^7–10^8 years, providing strong vindication of the Safronov–Wetherill style of protracted accretion via major impacts (Halliday, 2003). It is also immediately apparent however that a discrepancy exists between the W results and every estimate based on Pb isotope data (Halliday, 2004). Either all of these Pb isotope estimates are incorrect or there is some basic difference in the way these clocks operate. There are a number of parameters that play into these models that have a significant level of uncertainty associated.

The small difference between the ^{182}Hf–^{182}W age of the Moon determined from lunar samples and that predicted from Earth's accretion rate does not explain the much-larger difference with $^{235/238}$U–$^{207/206}$Pb. When the two chronometers are compared using the same style of model a variable but unidirectional offset is found between the timing that is based on W and that based on any of the 11 estimates of the Pb isotopic composition of the silicate Earth (**Figure 10**). For example, the two-stage ^{182}Hf–^{182}W model age of the Earth is 30 My whereas the same model applied to $^{235/238}$U–$^{207/206}$Pb yields ages ranging between >40 and >100 My (**Figure 14**). It is of course not unlikely that all the Pb isotope estimates are wrong. Given the difficulties in defining meaningful average Pb isotopic ratios for the silicate Earth this is certainly a possibility. In fact some more recent estimates of Pb isotopic composition of the BSE have been made based on the results of W isotopes so there is circularity involved. Assuming, however, that some of the Pb isotope estimates are close to correct one can consider what kind of processes might result in a difference between W and Pb ages of the core.

One explanation (Wood and Halliday, 2005) is that the transfer of W and Pb to the core changed, but not together, during the accretion history of the Earth. Tungsten is moderately siderophile but not chalcophile, whereas Pb is thought to tend toward the opposite. Following the cooling and oxidation of the Earth after the Giant Impact sulfide is likely to have formed. Removal of this sulfide to the core may have been responsible for a late-stage increase in U/Pb that defines the Pb isotopic compositions observed. This being the case, the Pb isotopic composition of the BSE provides information on the timescales over which the Earth cooled and core segregated following the Giant Impact. The timescales inferred depend on which Pb isotopic estimate is employed. However, they are all of the order of tens of millions of years after the Giant Impact.

There is another possible explanation for the apparent discrepancy between accretion rates determined from ^{182}Hf–^{182}W and $^{235/238}$U–$^{207/206}$Pb. Lead is moderately volatile and as such the younger ages could reflect loss of Pb from the Earth during accretion. Loss of such a heavy element from the Earth is dynamically difficult. However, if the surface of the Earth was very hot with a magma ocean Pb might have been atmophile rather than lithophile. The atmosphere may then have been blown off at a late stage (Walker, 1986). The depletion of the terrestrial planets in moderately volatile elements has long been considered to be a result of incomplete condensation because it was believed that the inner solar system formed from a hot solar nebular gas. Perhaps some of this depletion is, in fact a late-stage phenomenon. Support for this model is found in results for the new ^{205}Pb–^{205}Tl chronometer (Nielsen *et al.*, 2006) which are hard to explain without major losses of Tl from the Earth at a late stage.

With the advent of multiple collector inductively coupled plasma mass spectrometry (MC-ICPMS) (Halliday *et al.*, 1998) it has become possible to measure small mass-dependent isotopic differences between early solar system objects that may have been induced by the processes of accretion and evaporation of silicate and metal. Of particular interest are the less volatile but not highly refractory elements that would not have formed part of the Earth's protoatmosphere. Such elements include Li, Mg, Si, Cr, Fe, and Ni (Lodders, 2003). Precise stable isotopic compositions for mantle-derived samples from the Earth, Moon, Mars, Vesta, and iron meteorite parent bodies have now been published for Li (Magna *et al.*, 2006; Seitz *et al.*, 2006), Mg (Wiechert and Halliday, 2007) and Fe (Poitrasson *et al.*, 2004; Weyer *et al.*, 2005; Williams *et al.*, 2004, 2006). The Fe isotopic data for 'normal' basalts from the Earth and Moon are very slightly heavy (roughly 40 ppm per amu) relative to data for Mars and Vesta (Poitrasson

et al., 2004, Weyer *et al.*, 2005). This would be consistent with the loss of small amount of evaporated Fe during the Giant Impact. However, Williams *et al.* (2004) have shown that isotopic fractionations may also result from oxidation effects. Furthermore, it is conceivable that the silicate Earth, being larger than Mars or Vesta, may have generated a heavy Fe isotopic composition during partitioning ferric iron into perovskite in equilibrium with metal that was transferred to the core, as discussed above (Williams *et al.*, 2006).

However, the similarity between the Earth and Moon in terms of Fe isotopic composition would then be hard to explain unless there was some form of isotopic equilibration between the two bodies during the Giant Impact. Such a model has recently been proposed by Pahlevan and Stevenson (2005) to explain the O isotopic similarities between the Earth and Moon. If correct it has far-reaching implications for interpreting lunar geochemistry. In particular many of the above-mentioned features inferred for Theia would at some level reflect a portion of the proto-Earth instead. Evidence that the Moon does not carry a heavy isotopic composition as a result of boiling during the Giant Impact is found in the Li and Mg isotopic compositions (Magna *et al.*, 2006; Seitz *et al.*, 2006; Wiechert and Halliday, 2007) which provide no hint of a difference between any of the inner solar system differentiated planetary objects. Interestingly however, they do provide evidence that the terrestrial planets, including the Earth, accreted from material that was not exactly chondritic. With new high-precision isotopic data for meteorites evidence for a subtle but important difference between chondrites and the Earth has been growing (Boyet and Carlson, 2005). However, the interpretation has spawned new questions about whether the Earth is simply nonchondritic (Drake and Righter, 2002) or whether there are hidden reservoirs inside the Earth that balance its composition.

2.18 Hidden Reservoirs, Impact Erosion, and the Composition of the Earth

In all of the foregoing discussion it is assumed that the composition of the BSE as deduced from basalts is representative of the entire silicate Earth. In fact it may be different from chondrites or our best estimates of its true composition if there are 'hidden reservoirs'. It has been argued that this is indeed the case, on the basis of oxygen isotopes. The oxygen isotopic heterogeneity of the solar system is not only of importance for defining meteorite parent bodies. It provides a fingerprint for the material that formed the Earth. The similarity between the Earth and Moon in terms of oxygen isotopes provides powerful evidence for a similar average provenance. It is perhaps significant that the only chondrites with an oxygen isotopic composition like that of the Earth are the enstatite chondrites. This has led to a variety of Earth models based on a total composition that is highly reduced and similar to that of the enstatite chondrites (e.g., Javoy, 1999). Because the mantle and crust of the Earth sampled by magmas and xenoliths are inconsistent with this composition, the 'enstatite chondrite Earth' has to have hidden reservoirs (customarily located in the lower mantle and hence not sampled) that are compositionally distinct from anything we can readily measure.

Hidden-reservoir models are of little use scientifically because they are untestable. Further evidence of a link between the Earth and enstatite chondrites has, however, come from nitrogen (Marty et al., 2003) and chromium (Shukolyukov and Lugmair, 2004) isotopic measurements, adding support to the 'hidden reservoir' concept. Recently a hidden reservoir of different composition has been proposed by Tolstikhin and Hofmann (2005) to explain noble gas isotope systematics. More compelling is the new evidence that the BSE as sampled is slightly different from chondrites in terms of ^{142}Nd atomic abundance, possibly reflecting a different ^{146}Sm/^{142}Nd in the early solar system (Boyet and Carlson, 2005). One explanation given is that the Earth contains a hidden primordial reservoir with complementary lower ^{146}Sm/^{142}Nd balancing the higher ^{146}Sm/^{142}Nd of the silicate Earth as sampled by basalts and continents. The apparent model age of this reservoir is 30 My after the start of the solar system, that is, close in age to the Moon-forming Giant Impact.

Melting as a result of the giant impact, followed by settling of majoritic garnet or silicate perovskites in the magma ocean has been proposed as a mechanism for forming the primitive upper mantle with high, nonchondritic Mg/Si ratio (e.g., Ringwood, 1979; Ohtani et al., 1986; Herzberg et al., 1988) and a complementary 'hidden' lower-mantle reservoir. If substantial fractional crystallization of the magma ocean occurred, however, subsequent remixing must have removed most of the material segregated to the lower mantle because of the observation that the ratios of the refractory lithophile elements in

the upper mantle are chondritic. The question is, however, could a small amount of undetected fractionation have shifted the Nd/Sm ratio of the silicate Earth sufficiently to generate the observed ^{142}Nd anomaly? Crystal-melt partition coefficients for majorite and for $CaSiO_3$ and $(Mg,Fe)SiO_3$ perovskites have been measured by Corgne and Wood (2004), Liebske et al. (2005), and Walter et al. (2004) and applied to calculate the maximum volumes of 'hidden' majorite or perovskite which could be present in the lower mantle, given the compositional constraints derived from the upper mantle. In the case of majorite, the maximum allowable hidden reservoir is 14% of the mantle volume (Corgne and Wood, 2004) while the maximum extent of combined calcium and magnesium silicate perovskite fractionation is about 10% (Liebske et al., 2005; Walter et al., 2004). In neither case is the 'hidden' amount sufficient to explain the nonchondritic Mg/Si ratio of the observable BSE (**Figure 21**).

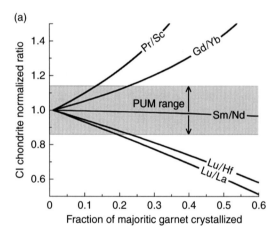

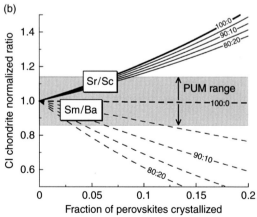

Figure 21 (a, b) The effects of perovskite fractionation on Sm/Nd and Lu/Hf ratios.

Interestingly, the hidden reservoir would in both cases have a higher Sm/Nd ratio than the BSE, which is the wrong direction to explain the ^{142}Nd anomaly of the BSE. This leads, however to the possibility that dense high-pressure melts (complementary to solid majorite and perovskite) could remain isolated at great depths and generate hidden reservoirs (e.g., Suzuki and Ohtani (2003) and references therein). Recent molecular-dynamics simulations in the MgSiO$_3$ system (Stixrude and Kirko, 2005) suggest that the density contrast between melt and crystals decreases with increasing pressure, such that a density crossover near the base of the mantle could occur. Hofmann *et al.* (2005) have proposed that downward migration of such dense melts formed in the deep mantle in equilibrium with Ca-perovskite could have created a deep 'missing' reservoir unradiogenic in Pb (with a U/Pb ratio lower than that of Primitive Upper Mantle (PUM)) as required to balance the Pb composition of the bulk MORB and ocean-island basalt (OIB)-source reservoirs. It would have high Nd/Sm ratio and could hence be the missing subchondritic ^{142}Nd/^{144}Nd reservoir (Boyet and Carlson, 2005).

There are, however, problems preserving any distinct mantle reservoir from this early in Earth history. After the Moon formed there would have been tidal interactions with the Earth that would have caused strong heating of the lower mantle in particular (Zahnle *et al*, 2006). Mantle overturn and melting should have been especially vigorous eliminating any early heterogeneity. Therefore, explanation for the Nd isotopic composition based on hidden reservoirs is at present very tentative.

Assuming that the BSE as sampled is representative of the silicate Earth one can, in principle, use its composition to infer the composition of the complementary core by mass balance, as discussed above. There is, however, indirect evidence that the core may not have grown in equilibrium with material that was compositionally equivalent to the BSE. If the oxygen isotopic composition of the Earth and Moon reflects a similar provenance of material, or if it reflects some isotopic equilibration between the Earth and the material that formed the Moon, the initial Sr and W isotopic composition of the Moon should define the Earth's time-integrated Rb/Sr and Hf/W between the start of the solar system and the time the Moon formed. These are respectively higher and lower than those of the present BSE and very similar to the composition of the present day bulk silicate Mars (Halliday and Porcelli, 2001; Halliday,

2004). This means that core compositions may have been established in equilibrium with reservoirs that no longer exist or have been modified during the process of accretion.

As outlined in the previous section, modification during accretion is particularly important for volatile and moderately volatile elements that may have been blown off relatively easily. However, it may also apply to other elements that were concentrated in the silicate outer portions of the proto-Earth. It is well known that the Earth has a nonchondritic Si/Mg ratio, one possible explanation for which is partitioning of Si (**Figure 7**) into the core. However, the Earth also has a Fe/Mg ratio of at least 2.11 which is 10% higher than the value of 1.90 for CI chondrites (Palme, 2000). A possible explanation for this is impact erosion of the outer portions of the Earth (Halliday *et al*, 2001). This being the case the Earth's potassium, uranium, and thorium budgets would have been reduced because these elements are all highly incompatible and should have been concentrated in the Earth's outer zones. Similarly impact erosion is likely to generate a Sm/Nd that is higher than chondritic through loss of low-Sm/Nd crust. The observed ^{142}Nd effect discussed above is explicable with just a 6% increase in Sm/Nd in the BSE relative to chondrites.

2.19 Concluding Overview

1. The solar system formed because the angular momentum of the collapsing portion of a molecular cloud was transferred to a swirling disk of gas and debris left over from the proto-Sun. The disk experienced strong heating from the T-Tauri stage of irradiation and from frictional heating as material was being rapidly swept in toward the hot new Sun. This led to strong depletions in moderately volatile elements close to the Sun.

2. The Earth was built from differentiated planetary embryos that formed from this volatile depleted dust in the inner solar system. Such embryos and their metallic cores formed extremely early. The latest ^{82}Hf–^{182}W chronometry provides evidence of timescales of <1 My for iron meteorite parent bodies. Mars may also have formed this quickly. These bodies almost certainly melted and differentiated as a result of the decay of short-lived nuclides in particular 26Al and 60Fe.

3. Chondrites are the most abundant form of meteorites that strikes the Earth. They have relative

concentrations of nonvolatile elements that are broadly similar to those found in the Sun. They represent a convenient reference for the bulk composition of the Earth. However, though primitive they are not early. They must have formed at least 1–3 My after the start of the solar system because the chondrules they contain provide evidence of an epoch of formation in the dusty disk or solar nebula at this time.

4. The Earth accreted via a stochastic series of planetary collisions over tens of millions of years. The final major accretion stage was the Moon-forming Giant Impact between 30 and 50 My after the start of the solar system, when the last roughly 10% of the Earth was added.

5. The chemical composition of the solar system is defined using spectroscopic measurements of the Sun combined with laboratory measurements of chondritic meteorites. All of the stable elements and isotopes are present within our solar system and their relative abundance provides powerful evidence in support of the theories of nuclear stability in stars and of stellar nucleosyntheis.

6. The density of the Earth combined with estimates of its core size leads to the conclusion that it must be approximately chondritic in its relative concentrations of refractory elements. Similarly, the chemical composition of the BSE, which is well constrained from measurements of mantle peridotites and basalts, exhibits ratios of refractory lithophile elements which are the same as in chondritic meteorites, the Moon, and the silicate portion of Mars.

7. The BSE is depleted in volatile elements relative to chondrites, and it has Mg/Si slightly greater and Mg/refractory lithophile elements slightly lower than chondritic. Depletions of the BSE in refractory siderophile elements are due to their sequestration in the core.

8. By using the chondritic reference model for refractory elements in the Earth we can calculate the partitioning of siderophile elements between the core and the mantle. The results may then be used in conjunction with experimental liquid metal–liquid silicate partition coefficients to estimate the conditions of core segregation. This exercise requires the assumption that almost all of the core formed in equilibrium with the mantle. Partition coefficients depend on pressure and temperature, and the depletions of the BSE in siderophile elements can only be matched by very high pressures and temperatures.

9. Results imply segregation of the core at the base of a deep magma ocean which extended to 30–40% of the depth of the mantle in the growing Earth. The bottom of the magma ocean was by definition below the peridotite liquidus temperature and the constraints this imposes leads to the requirement that the Earth became more oxidized during accretion. The BSE composition can only be matched if the oxidation state of the Earth changed as it grew. The resultant core contains about 4% Si, thus indicating that it contains substantial Si and is a major reason for the superchondritic Mg/Si of the BSE.

10. As the Earth accreted it would, when it reached the current-day size of Mars have begun to crystallize silicate perovskites close to the core–mantle boundary. The strong affinity of $(Mg,Fe)SiO_3$ perovskite for ferric iron means that crystallization of this phase forces disproportionation of Fe^{2+} into Fe^{3+} plus metal. This is one mechanism for oxidizing the Earth during accretion since progressive growth through impact would have led to release of Fe^{3+} to the magma ocean and of metal to the core. This mechanism also could explain the observation that the mantle of the Earth is at higher oxygen fugacity than the mantles of Mars and the Moon.

11. The Moon-forming Giant Impact did not result in any significant losses of slightly volatile elements such as Li, Mg, or Fe. The slightly heavy Fe isotopic composition of the Earth and Moon relative to Mars and Vesta may result from the self-oxidation process hypothesized as a consequence of perovskite fractionation. The similarity between the Fe isotopic composition of the Earth and Moon would then have resulted from isotopic equilibration between the material accreting to form the Moon with the proto-Earth as recently proposed as an explanation for the O isotopic similarities.

12. Major losses of volatiles affected the Earth during its accretion, including losses of early-formed atmospheres. Such losses are apparent from Sr, Xe, Tl, and Pb isotopic compositions and time-integrated Rb/Sr, I–Pu/Xe and Pb/Tl ratios. The late U–Pb age for the Earth may reflect this process. However, it is also possible that the effect is caused by changes in partitioning during the final stages of core formation or simply that all estimates of the Pb isotopic composition of the BSE are in error.

13. After the Giant Impact crystallization of the Earth would have led to production of majoritic garnet in the transition zone and of $(Mg,Fe)SiO_3$ and $CaSiO_3$ perovskites in the lower mantle. Crystal–liquid partition coefficients for these phases can be used to estimate the extent of any long-lived 'hidden' reservoir produced in the lower mantle after mantle crystallization. Results indicate a maximum

extent of majorite crystallization of 14% of mantle volume and of perovskite fractionation of 10% could be hidden without perturbing the chondritic refractory lithophile element ratios of the BSE.

14. The 'hidden' reservoir with superchondritic Nd/Sm required to balance the BSE could not be perovskite or majorite since these phases all prefer Sm over Nd. One possibility would be isolation and freezing of dense high-pressure melts (complementary to perovskite) in the lower mantle. Such a low-volume reservoir would have appropriate Nd/Sm to generate superchondritic $^{142}Nd/^{144}Nd$ in the remaining BSE. However, the preservation of such a reservoir is hard to reconcile with the strong tidal melting anticipated for the deep Earth following Moon formation. Furthermore, the ^{142}Nd effects can be explained by small differences between the Earth and chondrites, as exemplified by other lines of evidence.

Acknowledgments

BJW acknowledges the support of a Federation Fellowship from the Australian Research Council. ANH acknowledges the support of PPARC for the cosmochemistry research program at Oxford.

References

Abe Y (1997) Thermal and chemical evolution of the terrestrial magma ocean. *Physics of the Earth and Planetary Interiors* 100: 27–39.

Agnor CB and Ward WR (2002) Damping of terrestrial-planet eccentricities by density-wave interactions with a remnant gas disk. *Astrophysical Journal* 567: 579–586.

Allègre CJ and Lewin E (1989) Chemical structure and history of the Earth: Evidence from global non-linear inversion of isotopic data in a three box model. *Earth and Planetary Science Letters* 96: 61–88.

Allègre CJ, Lewin E, and Dupré B (1988) A coherent crust–mantle model for the uranium–thorium–lead isotopic system. *Chemical Geology* 70: 211–234.

Allègre CJ, Manhès G, and Göpel C (1995a) The age of the Earth. *Geochimica et Cosmochimica Acta* 59: 1445–1456.

Allègre CJ, Poirier J-P, Humler E, and Hofmann AW (1995b) The chemical composition of the Earth. *Earth and Planetary Science Letters* 134: 515–526.

Amelin Y, Krot AN, Hutcheon ID, and Ulyanov AA (2002) Lead isotopic ages of chondrules and calcium–aluminum-rich inclusions. *Science* 297: 1678–1683.

Arculus RJ and Delano JW (1981) Siderophile element abundances in the upper mantle: Evidence for a sulfide signature and equilibrium with the core. *Geochimica et Cosmochimica Acta* 45: 1331–1344.

Arnould M, Paulus G, and Meynet G (1997) Short-lived radionuclide production by non-exploding Wolf–Rayet stars. *Astronomy and Astrophysics* 321: 452–464.

Benz W, Cameron AGW, and Slattery WL (1987) Collisional stripping of Mercury's mantle. *Icarus* 74: 516–528.

Bizzarro M, Baker JA, and Haack H (2004) Mg isotope evidence for contemporaneous formation of chondrules and refractory inclusions. *Nature* 431: 275–278.

Bizzarro M, Baker JA, and Haack H (2005) Mg isotope evidence for contemporaneous formation of chondrules and refractory inclusions – Correction. *Nature* 435: 1280.

Blum J (2000) Laboratory experiments on preplanetary dust aggregation. In: Benz W, Kallenbach R, and Lugmair GW (eds.) *Space Science Reviews: From Dust to Terrestrial Planets*, vol. 92, pp. 265–278. Dordrecht: Kluwer Academic Publishers.

Boss AP (1990) 3D Solar nebula models: Implications for Earth origin. In: Newsom HE and Jones JH (eds.) *Origin of the Earth*, pp. 3–15. Oxford: Oxford University Press.

Boss AP (1997) Giant planet formation by gravitational instability. *Science* 276: 1836–1839.

Boyet M and Carlson RW (2005) ^{142}Nd evidence for early (>4.53 Ga) global differentiation of the silicate Earth. *Science* 309: 576–581.

Brearley AJ and Jones RH (1998) Chondritic meteorites. In: Papike J (ed.) *Reviews in Mineralogy 36: Planetary Materials*, ch. 3, pp. 1–398. Washington, DC: Mineralogical Society of America.

Burbidge GR, Burbidge EM, Fowler WA, and Hoyle F (1957) Synthesis of elements in stars. *Reviews in Modern Physics* 29: 547–650.

Busso M, Gallino R, and Wasserburg GJ (2003) Short-lived nuclei in the early solar system: A low mass stellar source? *Publications of the Astronomical Society of Australia* 20: 356–370.

Cameron AGW and Benz W (1991) Origin of the Moon and the single impact hypothesis IV. *Icarus* 92: 204–216.

Cameron AGW and Truran JW (1977) The supernova trigger for formation of the solar system. *Icarus* 30: 447–461.

Canup RM and Agnor CB (2000) Accretion of the terrestrial planets and the Earth–Moon system. In: Canup RM and Righter K (eds.) *Origin of the Earth and Moon*, pp. 113–132. Tucson, AZ: University of Arizona Press.

Canup RM and Asphaug E (2001) Origin of the moon in a giant impact near the end of the Earth's formation. *Nature* 412: 708–712.

Cassen P (1996) Models for the fractionation of moderately volatile elements in the solar nebula. *Meteoritics and Planetary Science* 31: 793–806.

Chabot NL and Agee CB (2003) Core formation in the Earth and Moon: New experimental constraints from V, Cr, and Mn. *Geochimica et Cosmochimica Acta* 67(11): 2077–2091.

Chambers JE (2004) Planetary accretion in the inner solar system. *Earth and Planetary Science Letters* 223: 241–252.

Chou CL (1978) Fractionation of siderophile elements in the Earth's upper mantle. *Proceedings of the Lunar Science Conference* 9: 219–230.

Ciesla FJ and Charnley SB (2006) The physics and chemistry of nebular evolution. In: Lauretta D, Leshin L, and MacSween H (eds.) *Meteorites and the Early Solar System II*, pp. 209–230. Tucson, AZ: University of Arizona Press.

Clayton RN (1986) High temperature isotope effects in the early solar system. In: Valley JW, Taylor HP, and O'Neil JR (eds.) *Stable Isotopes in High Temperature Geological Processes*, pp. 129–140. Washington, DC: Mineralogical Society of America.

Clayton RN (1993) Oxygen isotopes in meteorites. *Annual Review of Earth and Planetary Sciences* 21: 115–149.

Clayton RN, Grossman L, and Mayeda TK (1973) A component of primitive nuclear composition in carbonaceous meteorites. *Science* 182: 485–487.

Clayton RN and Mayeda TK (1975) Genetic relations between the moon and meteorites. *Proceedings of the Lunar Science Conference* XI: 1761–1769.

Corgne A and Wood BJ (2004) Trace element partitioning between majoritic garnet and silicate melt at 25 GPa. *Physics of the Earth and Planetary Interiors* 143–144: 407–419.

Cuzzi JN, Hogan RC, Paque JM, and Dobrovolskis AR (2001) Size-selective concentration of chondrules and other small particles in protoplanetary nebula turbulence. *Astrophysical Journal* 546: 496–508.

Davies GF (1984) Geophysical and isotopic constraints on mantle convection: An interim synthesis. *Journal of Geophysical Research* 89: 6017–6040.

Doe BR and Zartman RE (1979) Plumbtectonics, the Phanerozoic. In: Barnes HL (ed.) *Geochemistry of Hydrothermal Ore Deposits*, pp. 22–70. New York: Wiley.

Drake MJ and Righter K (2002) Determining the composition of the Earth. *Nature* 416: 39–44.

Feigelson ED, Broos P, Gaffney JA, III, et al. (2002a) X-ray-emitting young stars in the Orion nebula. *The Astrophysical Journal* 574: 258–292.

Feigelson ED, Garmire GP, and Pravdo SH (2002b) Magnetic flaring in the pre-main-sequence Sun and implications for the early solar system. *The Astrophysical Journal* 572: 335–349.

Foley CN, Wadhwa M, Borg LE, Janney PE, Hines R, and Grove TL (2005) The early differentiation history of Mars from [182]W–[142]Nd isotope systematics in the SNC meteorites. *Geochimica et Cosmochimica Acta* 69: 4557–4571.

Foster PN and Boss AP (1997) Injection of radioactive nuclides from stellar source that triggered the collapse of the presolar nebula. *The Astrophysical Journal* 489: 346–357.

Foster PN and Boss AP (1996) Triggering star formation with stellar ejecta. *The Astrophysical Journal* 468: 784–796.

Frost DJ, Liebske C, Langenhorst F, McCammon CA, Trønnes RG, and Rubie DC (2004) Experimental evidence for the existence of iron-rich metal in the Earth's lower mantle. *Nature* 428: 409–412.

Galer SJG and Goldstein SL (1991) Depleted mantle Pb isotopic evolution using conformable ore leads. *Terra Abstracts (EUG VI)* 3: 485–486.

Galer SJG and Goldstein SL (1996) Influence of accretion on lead in the Earth. In: Basu AR and Hart SR (eds.) *Isotopic Studies of Crust–Mantle Evolution*, pp. 75–98. Washington, DC: AGU.

Gast PW (1960) Limitations on the composition of the upper mantle. *Journal of Geophysical Research* 65: 1287–1297.

Gessmann CK and Rubie DC (1998) The effect of temperature on the partitioning of nickel, cobalt, manganese, chromium, and vanadium at 9 GPa and constraints on formation of the Earth's core. *Geochimica et Cosmochimica Acta* 62: 867–882.

Gessmann CK, Wood BJ, Rubie DC, and Kilburn MR (2001) Solubility of silicon in liquid metal at high pressure: Implications for the composition of the Earth's core. *Earth and Planetary Science Letters* 184: 367–376.

Goldreich P and Ward WR (1973) The formation of planetesimals. *Astrophysical Journal* 183: 1051–1060.

Gounelle M, Shu FH, Shang H, Glassgold AE, Rehm KE, and Lee T (2001) Extinct radioactivities and protosolar cosmic-rays: Self-shielding and light elements. *The Astrophysical Journal* 548: 1051.

Halliday AN (2003) The origin and earliest history of the Earth. In: Holland HD and Turekian KK (eds.) *Treatise on Geochemistry: Meteorites, Comets and Planets*, vol. 1, pp. 509–557. Oxford: Elsevier-Pergamon.

Halliday AN (2004) Mixing, volatile loss and compositional change during impact-driven accretion of the Earth. *Nature* 427: 505–509.

Halliday AN (2006) The origin of the Earth – What's new? *Elements* 2: 205–210.

Halliday AN and Kleine T (2006) Meteorites and the timing, mechanisms and conditions of terrestrial planet accretion and early differentiation. In: Lauretta D, Leshin L, and MacSween H (eds.) *Meteorites and the Early Solar System II*, pp. 775–801. Tucson, AZ: University of Arizona Press.

Halliday AN, Lee D-C, Christensen JN, et al. (1998) Applications of multiple collector ICPMS to cosmochemistry, geochemistry and paleoceanography. (The 1997 Geochemical Society Presidential Address.) *Geochimica et Cosmochimica Acta* 62: 919–940.

Halliday AN, Lee D-C, Porcelli D, Wiechert U, Schönbächler M, and Rehkämper M (2001) The rates of accretion, core formation and volatile loss in the early solar system. *Philosophical Transactions of the Royal Society of London* 359: 2111–2135.

Halliday AN and Porcelli D (2001) In search of lost planets – The paleocosmochemistry of the inner solar system. *Earth and Planetary Science Letters* 192: 545–559.

Hanan BB and Tilton GR (1987) 60025: Relict of primitive lunar crust? *Earth and Planetary Science Letters* 84: 15–21.

Hanks TC and Anderson DL (1969) The early thermal history of the Earth. *Physics of the Earth and Planetary Interiors* 2: 19–29.

Hayashi C, Nakazawa K, and Mizuno H (1979) Earth's melting due to the blanketing effect of the primordial dense atmosphere. *Earth and Planetary Science Letters* 43: 22–28.

Hayashi C, Nakazawa K, and Nakagawa Y (1985) Formation of the solar system. In: Black DC and Matthews DC (eds.) *Protostars and Planets II*, pp. 1100–1153. Tucson, AZ: University of Arizona Press.

Herd CDK, Borg LE, Jones JH, and Papike JJ (2002) Oxygen fugacity and geochemical variations in the martian basalts: Implications for martian basalt petrogenesis and the oxidation state of the upper mantle of Mars. *Geochimica et Cosmochimica Acta* 66(11): 2025–2036.

Herzberg C, Feigenson M, Skuba C, and Ohtani E (1988) Majorite fractionation recorded in the geochemistry of peridotites from South Africa. *Nature* 332: 823–826.

Hewins RH and Herzberg C (1996) Nebular turbulence, chondrule formation, and the composition of the Earth. *Earth and Planetary Science Letters* 144: 1–7.

Hofmann AW, Hemond C, Sarbas B, and Jochum KP (2005) Yes, there really is a lead paradox. *EOS Transactions AGU Fall Meeting Supplement* 86: Abstract V23D-05.

Hofmann AW, Jochum KP, Seufert M, and White WM (1986) Nb and Pb in oceanic basalts: New constraints on mantle evolution. *Earth and Planetary Science Letters* 79: 33–45.

Hofmann AW and White WM (1983) Ba, Rb and Cs in the Earth's mantle. Zeitschrift für *Naturforsch* 38a: 256–266.

Humayun M and Cassen P (2000) Processes determining the volatile abundances of the meteorites and terrestrial planets. In: Canup RM and Righter K (eds.) *Origin of the Earth and Moon*, pp. 3–23. Tucson, AZ: University of Arizona Press.

Humayun M and Clayton RN (1995) Potassium isotope cosmochemistry: Genetic implications of volatile element depletion. *Geochimica et Cosmochimica Acta* 59: 2131–2151.

Hunten DM, Pepin RO, and Walker JCG (1987) Mass fractionation in hydrodynamic escape. *Icarus* 69: 532–549.

Ikoma M and Genda H (2006) Constraints on the mass of a habitable planet with water of nebular origin. *The Astrophysical Journal* 648: 696–706.

Jacobsen SB (2005) The Hf–W system and the origin of the Earth and Moon. *Annual Reviews in Earth and Planetary Sciences* 33: 531–570.

Jacobsen SB and Harper CL, Jr. (1996) Accretion and early differentiation history of the Earth based on extinct radionuclides. In: Basu A and Hart S (eds.) *Earth Processes: Reading the Isotope Code*, pp. 47–74. Washington, DC: AGU.

Javoy M (1999) Chemical Earth models. *Comptes Rendus de l'Academie des Sciences, Edition Scientifiques et Médicales Elseviers SAS*, 329: 537–555.

Jochum KP, Hofmann AW, and Seufert M (1993) Tin in mantle-derived rocks – Constraints on Earth evolution. *Geochimica et Cosmochimica Acta* 57: 3585–3595.

Jochum KP, Hofmann AW, Ito E, Seufert M, and White WM (1983) K, U and Th in mid-ocean ridge basalt glasses and heat production, K/U and K/Rb in the mantle. *Nature* 306: 431–436.

Jones JH and Drake MJ (1986) Geochemical constraints on core formation in the Earth. *Nature* 322: 221–228.

Kamber BS and Collerson KD (1999) Origin of ocean-island basalts: A new model based on lead and helium isotope systematics. *Journal of Geophysical Research* 104: 25479–25491.

Karato S-I and Rama Murthy V (1997) Core formation and chemical equilibrium in the Earth. Part I: Physical considerations. *Physics of the Earth and Planetary Interiors* 100(1–4): 61–79.

Karner JJ, Sutton SR, Papike JJ, Shearer CK, Jones JH, and Newville M (2006) The application of a new vanadium valence oxybarometer to basaltic glasses from the Earth Moon and Mars. *American Mineralogist* 91(2–3): 270–277.

Kita NT, Huss GR, Tachibana S, Amelin Y, Nyquist LE, and Hutcheon ID (2006) Constraints on the origin of chondrules and CAIs from short-lived and long-lived radionuclides. In: Krot AN, Scott ERD, and Reipurth B (eds.) *Chondrules and the Protoplanetary Disk*, pp. 558–587. San Francisco, CA: Astronomical Society of the Pacific.

Kleine T, Mezger K, Munker C, Palme H, and Bischoff A (2004a) ^{182}Hf–^{182}W isotope systematics of chondrites, eucrites and Martian meteorites: Chronology of core formation and early mantle differentiation in Vesta and Mars. *Geochim et Cosmochim Acta* 68: 2935–2946.

Kleine T, Mezger K, Palme H, and Münker C (2004b) The W isotope evolution of the bulk silicate Earth: Constraints on the timing and mechanisms of core formation and accretion. *Earth and Planetary Science Letters,* 228: 109.

Kleine T, Mezger K, Palme H, Scherer E, and Münker C (2005a) Early core formation in asteroids and late accretion of chondrite parent bodies: Evidence from ^{182}Hf–^{182}W in CAIs, metal–rich chondrites and iron meteorites. *Geochimica et Cosmochimica Acta.* 69: 5805–5818.

Kleine T, Munker C, Mezger K, and Palme H (2002) Rapid accretion and early core formation on asteroids and the terrestrial planets from Hf–W chronometry. *Nature* 418: 952–955.

Kleine T, Palme H, Mezger K, and Halliday AN (2005b) Hf–W chronometry of lunar metals and the age and early differentiation of the Moon. *Science* 310: 1671–1674.

Kortenkamp SJ, Kokubo E, and Weidenschilling SJ (2000) Formation of planetary embryos. In: Canup RM and Righter K (eds.) *Origin of the Earth and Moon*, pp. 85–100. Tucson, AZ: The Tucson: University of Arizona Press.

Kortenkamp SJ and Wetherill GW (2000) Terrestrial planet and asteroid formation in the presence of giant planets. *Icarus* 143: 60–73.

Kramers JD and Tolstikhin IN (1997) Two terrestrial lead isotope paradoxes, forward transport modeling, core formation and the history of the continental crust. *Chemical Geology* 139: 75–110.

Kwon S-T, Tilton GR, and Grünenfelder MH (1989) Lead isotope relationships in carbonatites and alkalic complexes: an overview. In: Bell K (ed.) *Carbonatites-Genesis and Evolution*, pp. 360–387. London: Unwin-Hyman.

Lee D-C and Halliday AN (1996) Hf–W isotopic evidence for rapid accretion and differentiation in the early solar system. *Science* 274: 1876–1879.

Lee D-C and Halliday AN (1997) Core formation on Mars and differentiated asteroids. *Nature* 388: 854–857.

Lee D-C and Halliday AN (2000b) Hf–W isotopic systematics of ordinary chondrites and the initial ^{182}Hf/^{180}Hf of the solar system. *Chemical Geology* 169: 35–43.

Lee D-C and Halliday AN (2000a) Accretion of primitive planetesimals: Hf–W isotopic evidence from enstatite chondrites. *Science* 288: 1629–1631.

Lee D-C, Halliday AN, Davies GR, Essene EJ, Fitton JG, and Temdjim R (1996) Melt enrichment of shallow depleted mantle: a detailed petrological, trace element and isotopic study of mantle derived xenoliths and megacrysts from the Cameroon line. *Journal of Petrology* 37: 415–441.

Lee D-C, Halliday AN, Leya I, Wieler R, and Wiechert U (2002) Cosmogenic tungsten and the origin and earliest differentiation of the Moon. *Earth and Planetary Science Letters* 198: 267–274.

Lee D-C, Halliday AN, Snyder GA, and Taylor LA (1997) Age and origin of the Moon. *Science* 278: 1098–1103.

Lee T, Shu FH, Glassgold AE, and Rehm KE (1998) Protostellar cosmic rays and extinct radioactivities in meteorites. *Astrophysical Journal* 506: 898–912.

Leya I, Halliday AN, and Wieler R (2003) The predictable collateral consequences of nucleosynthesis by spallation reactions in the early solar system. *Astrophysical Journal* 594: 605–616.

Li J and Agee CB (1996) Geochemistry of mantle–core differentiation at high pressure. *Nature* 381: 686–689.

Li J and Agee CB (2001) The effect of pressure, temperature, oxygen fugacity and composition on partitioning of nickel and cobalt between liquid Fe–Ni–S alloy and liquid silicate: Implications for the Earth's core formation. *Geochimica et Cosmochimica Acta* 65: 1821–1832.

Liebske C, Corgne A, Frost DJ, Rubie DC, and Wood BJ (2005) Compositional effects on element partitioning between Mg-silicate perovskite and silicate melts. *Contributions to Mineralogy and Petrology* 149: 113–128.

Liew TC, Milisenda CC, and Hofmann AW (1991) Isotopic constrasts, chronology of element transfers and high-grade metamorphism: The Sri Lanka highland granulites, and the Lewisian (Scotland) and Nuk (S.W. Greenland) gneisses. *Geologische Rundschau* 80: 279–288.

Lissauer JJ (1987) Time-scales for planetary accretion and the structure of the protoplanetry disk. *Icarus* 69: 249–265.

Lissauer JJ (1993) Planet formation. *Annual Review of Astronomy and Astrophysics* 31: 129–174.

Lodders K (2003) Solar system abundances and condensation temperatures of the elements. *The Astrophysical Journal* 591: 1220–1247.

Lugmair GW and Galer SJG (1992) Age and isotopic relationships between the angrites Lewis Cliff 86010 and Angra dos Reis. *Geochimica et Cosmochimica Acta* 56: 1673–1694.

Magna T, Wiechert U, and Halliday AN (2006) New constraints on the lithium isotope compositions of the Moon and terrestrial planets. *Earth and Planetary Science Letters* 243: 336–353.

Markowski A, Quitté G, Halliday AN, and Kleine T (2006) Tungsten isotopic compositions of iron meteorites: Chronological constraints vs. cosmogenic effects. *Earth and Planetary Science Letters* 242: 1–15.

Marty B, Hashizume K, Chaussidon M, and Wieler R (2003) Nitrogen isotopes on the Moon: Archives of the solar and planetary contributions to the inner solar system. *Space Science Reviews* 106: 175–196.

Marty B and Marti K (2002) Signatures of early differentiation of Mars. *Earth and Planetary Science Letters* 196: 251–263.

McCammon C (1997) Perovskite as a possible sink for ferric iron in the lower mantle. *Nature* 387: 694–696.

McCaughrean MJ and O'Dell CR (1996) Direct imaging of circumstellar disks in the Orion nebula. *Astronomical Journal* 111: 1977–1986.

McDonough WF (2003) Compositional model for the Earth's core. In: Carlson RW (ed.) *The Mantle and Core*, vol 2, pp. 547–568. Oxford: Elsevier-Pergamon.

McDonough WF and Sun S-S (1995) The composition of the Earth. *Chemical Geology* 120: 223–253.

Merrill PW (1952) Technetium in the Staps. *Science* 115: 484.

Meyer BS and Clayton DD (2000) Short-lived radioactivities and the birth of the Sun. *Space Science Reviews* 92: 133–152.

Meyer BS and Zinner E (2006) Nucleosynthesis. In: Lauretta D, Leshin L, and MacSween H (eds.) *Meteorites and the Early Solar System II*, pp. 69–108. Tucson, AZ: University of Arizona Press.

Mittlefehldt DW, McCoy TJ, Goodrich CA, and Kracher A (1998) Non-chondritic meteorites from asteroidal bodies. In: Papike J (ed.) *Reviews in Mineralogy 36: Planetary Materials*, ch. 4, pp. 4-1–4-195. Chantilly, VA: Mineralogical Society of America.

Mostefaoui S, Lugmair GW, Hoppe P, and El Goresy A (2004) Evidence for live Fe-60 in meteorites. *New Astronomy Reviews* 48: 155–159.

Murphy DT, Kamber BS, and Collerson KD (2003) A refined solution to the first terrestrial Pb-isotope paradox. *Journal of Petrology* 44: 39–53.

Newsom HE (1990) Accretion and core formation in the Earth: Evidence from siderophile elements. In: Newsom HE and Jones JH (eds.) *Origin of the Earth*, pp. 273–288. Oxford: Oxford University Press.

Newsom HE (1995) Composition of the solar system, planets, meteorites, and major terrestrial reservoirs. In: *AGU Reference Shelf 1: Global Earth Physics, A Handbook of Physical Constants*, pp. 159–189. Washington, DC: American Geophysical Union.

Newsom HE, Sims KWW, Noll PD, Jr., Jaeger WL, Maehr SA, and Bessera TB (1996) The depletion of W in the bulk silicate Earth. *Geochimica et Cosmochimica Acta* 60: 1155–1169.

Newsom HE, White WM, Jochum KP, and Hofmann AW (1986) Siderophile and chalcophile element abundances in oceanic basalts, Pb isotope evolution and growth of the earth's core. *Earth and Planetary Science Letters* 80: 299–313.

Nielsen SG, Rehkämper M, and Halliday AN (2006) Large thallium isotopic variations in iron meteorites and evidence for lead-205 in the early solar system. *Geochimica et Cosmochimica Acta* 70: 2643–2657.

O'Brien DP, Morbidelli A, and Levison HF (2006) Terrestrial planet formation with strong dynamical friction. *Icarus* 184: 39–58.

O'Neill HSt C (1991b) The origin of the Moon and the early history of the Earth – A chemical model. Part II: The Earth. *Geochimica et Cosmochimica Acta* 55: 1159–1172.

O'Neill HS (1991a) The origin of the Moon and the early history of the Earth – A chemical model. Part 1: The Moon. *Geochimica et Cosmochimica Acta* 55(4): 1135–1157.

Ohtani E, Kato T, and Sawamoto H (1986) Melting of a model chondritic mantle to 20 GPa. *Nature* 322: 352–353.

O'Neill HSC and Palme H (1998) Composition of the silicate Earth: Implications for accretion and core formation. In: Jackson I (ed.) *The Earth's Mantle – Composition, Structure and Evolution*, pp. 3–127. Cambridge: Cambridge University Press.

Ozima M and Podosek FA (1999) Formation age of Earth from $^{129}I/^{127}I$ and $^{244}Pu/^{238}U$ systematics and the missing Xe. *Journal of Geophysical Research* 104(B11): 25493–25499.

Pahlevan K and Stevenson DJ (2005) The oxygen isotope similarity between the Earth and Moon – Source region or formation process? *Lunar and Planetary Sciences* XXXVI: 2382.

Palme H (2000) Are there chemical gradients in the inner solar system?. *Space Science Reviews* 92: 237–262.

Pepin RO and Porcelli D (2006) Xenon isotope systematics, giant impacts, and mantle degassing on the early Earth. *Earth and Planetary Science Letters* 250: 470–485.

Patterson CC (1956) Age of meteorites and the Earth. *Geochimica et Cosmochimica Acta* 10: 230–237.

Poitrasson F, Halliday AN, Lee D-C, Levasseur S, and Teutsch N (2004) Iron isotope differences between Earth, Moon, Mars and Vesta as possible records of contrasted accretion mechanisms. *Earth and Planetary Letters* 223: 253–266.

Porcelli D, Cassen P, and Woolum D (2001) Deep Earth rare gases: Initial inventories, capture from the solar nebula and losses during Moon formation. *Earth and Planetary Science Letters* 193: 237–251.

Porcelli D, Cassen P, Woolum D, and Wasserburg GJ (1998) Acquisition and early losses of rare gases from the deep Earth. In: *Origin of the Earth and Moon, abst. LPI Cont. No. 957*, pp. 35–36. Lunar and Planetary Institute, Houston, USA.

Porcelli D and Halliday AN (2001) The possibility of the core as a source of mantle helium. *Earth and Planetary Science Letters* 192: 45–56.

Porcelli D and Pepin RO (2000) Rare gas constraints on early Earth history. In: Canup RM and Righter K (eds.) *Origin of the Earth and Moon*, pp. 435–458. Tucson, AZ: The University of Arizona Press.

Pritchard MS and Stevenson DJ (2000) Thermal constraints on the Origin of the Moon. In: *Origin of the Earth and Moon*, pp. 179–196. Tucson, AZ: University of Arizona Press.

Righter K and Drake MJ (1997) Metal-silicate equilibrium in a homogeneously accreting Earth: New results for Re. *Earth and Planetary Science Letters* 146: 541–553.

Righter K and Drake MJ (1999) Effect of water on metal-silicate partitioning of siderophile elements: A high pressure and temperature terrestrial magma ocean and core formation. *Earth and Planetary Science Letters* 171: 383–399.

Righter K, Drake MJ, and Yaxley G (1997) Prediction of siderophile element metal/silicate partition coefficients to 20 GPa and 2800° C: The effects of pressure, temperature, oxygen fugacity, and silicate and metallic melt compositions. *Physics of the Earth and Planetary Interiors* 100: 115–142.

Ringwood AE (1979) *Origin of the Earth and Moon*, 295 pp. New York: Springer.

Ringwood AE (1989a) Flaws in the giant impact hypothesis of lunar origin. *Earth and Planetary Science Letters* 95: 208–214.

Ringwood AE (1989b) Significance of the terrestrial Mg/Si ratio. *Earth and Planetary Science Letters* 95: 1–7.

Ryan SG, Aoki W, Norris JE, et al. (2001) Lead (Pb) in the C-rich, s-process rich, metal poor subgiant LP625-44. *Nuclear Physics* A688: 209c–212c.

Rubie DC, Melosh HJ, Reid JE, Liebske C, and Righter K (2003) Mechanisms of metal-silicate equilibration in the terrestrial magma ocean. *Earth and Planetary Science Letters* 205: 239–255.

Russell SS, Srinivasan G, Huss GR, Wasserburg GJ, and MacPherson GJ (1996) Evidence for widespread ^{26}Al in the solar nebula and constraints for nebula time scales. *Science* 273: 757–762.

Safronov VS (1954) On the growth of planets in the protoplanetary cloud. *Astronomicheskij Zhurnal* 31: 499–510.

Sasaki S and Nakazawa K (1986) Metal-silicate fractionation in the growing Earth: Energy source for the terrestrial magma ocean. *Journal of Geophysical Research* 91: B9231–B9238.

Scherstén A, Elliott T, Hawkesworth CJ, Russell S, and Masarik J (2006) Hf–W evidence for rapid differentiation of iron meteorite parent bodies. *Earth and Planetary Science Letters* 241: 530–542.

Seitz H-M, Brey GP, Weyer S, Durali S, Ott U, Münker C, and Mezger K (2006) Lithium isotope compositions of Martian and lunar reservoirs. *Earth and Planetary Science Letters* 245: 6–18.

Schönberg R, Kamber BS, Collerson KD, and Eugster O (2002) New W isotope evidence for rapid terrestrial accretion and very early core formation. *Geochimica et Cosmochimica Acta* 66: 3151–3160.

Shu FH, Shang H, Glassgold AE, and Lee T (1997) X-rays and fluctuating x-winds from protostars. *Science* 277: 1475–1479.

Shukolyukov A and Lugmair GW (1993) Live iron-60 in the early solar system. *Science* 259: 1138–1142.

Shukolyukov A and Lugmair GW (2004) Manganese–chromium isotope systematics of enstatite meteorites. *Geochimica et Cosmochimica Acta* 68: 2875–2888.

Solomatov VS (2000) Fluid dynamics of a terrestrial magma ocean. In: Canup RM and Righter K (eds.) *Origin of the Earth and Moon*, pp. 323–360. Tucson, AZ: University of Arizona Press.

Solomon SC (1979) Formation, history and energetics of cores in the terrestrial planets. *Earth and Planetary Science Letters* 19: 168–182.

Stevenson DJ (1981) Models of the Earth's core. *Science* 214: 611–619.

Stixrude L and Kirko B (2005) Structure and freezing of $MgSiO_3$ liquid in Earth's lower mantle. *Science* 310: 297–299.

Stone JM and Norman ML (1992) The 3-dimensional interaction of a supernova remnant with an interstellar cloud. *Astrophysical Journal* 390: L17.

Suzuki A and Ohtani E (2003) Density of peridotite melts at high pressure. *Physics and Chemistry of Minerals* 30: 449–456.

Takafuji N, Hirose K, Mitome M, and Bando Y (2005) Solubilities of O and Si in liquid iron in equilibrium with $(Mg,Fe)SiO_3$ perovskite and the light elements in the core. *Geophysical Research Letters* 32: L06313 (doi:10.1029/2005GL022773).

Taylor SR and Norman MD (1990) Accretion of differentiated planetesimals to the Earth. In: Newsom HE and Jones JH (eds.) *Origin of the Earth*, pp. 29–43. Oxford: Oxford University Press.

Tolstikhin I and Hofmann AW (2005) Early crust on top of the Earth's core. *Physics of the Earth and Planetary Interiors* 148: 109–130.

Truran JW and Cameron AGW (1978) [26]Al production in explosive carbon burning. *The Astrophysical Journal* 219: 230.

Vanhala HAT and Cameron AGW (1998) Numerical simulations of triggered star formation. Part 1: Collapse of dense molecular cloud cores. *The Astrophysical Journal* 508: 291–307.

Wade J and Wood BJ (2005) Core formation and the oxidation state of the Earth. *Earth and Planetary Science Letters* 236: 78–95.

Walker JCG (1986) Impact erosion of planetary atmospheres. *Icarus* 68: 87–98.

Walter MJ, Nakamura E, Tronnes RG, and Frost DJ (2004) Experimental constraints on crystallization differentiation in a deep magma ocean. *Geochimica et Cosmochimica Acta* 68: 4267–4284.

Wänke H (1981) Constitution of the terrestrial planets. *Philosphical Transactions of the Royal Society of London A* 303: 287–302.

Wänke H and Dreibus G (1986) Geochemical evidence for formation of the Moon by impact induced fission of the proto-Earth. In: Hartmann WK, Phillips RJ, and Taylor GJ (eds.) *Origin of the Moon*, pp. 649–672. Houston, TX: Lunar Planetary Institute.

Ward WR (2000) On planetesimal formation: The role of collective behavior. In: Canup RM and Righter K (eds.) *Origin of the Earth and Moon*, pp. 65–289. Tucson, AZ: The University of Arizona Press.

Wasserburg GJ, Busso M, and Gallino R (1996) Abundances of actinides and short-lived nonactinides in the interstellar medium: Diverse supernova sources for the r-processes. *The Astrophysical Journal* 466: L109–L113.

Wasserburg GJ, Busso M, Gallino R, and Raiteri CM (1994) Asymptotic giant branch stars as a source of short-lived radioactive nuclei in the solar nebula. *The Astrophysical Journal* 424: 412–428.

Wasserburg GJ, Gallino R, and Busso M (1998) A test of the supernova trigger hypothesis with [60]Fe and [26]Al. *The Astrophysical Journal* 500: L189–L193.

Wasserburg GJ, MacDonald F, Hoyle F, and Fowler WA (1964) Relative contributions of uranium, thorium, and potassium to heat production in the Earth. *Science* 143: 465–467.

Wasserburg GJ, Papanastassiou DA, Tera F, and Huneke JC (1977) Outline of a lunar chronology. *Philosophical Transactions of the Royal Society of London* A285: 7–22.

Wasson JT (1985) *Meteorites, Their Record of Early Solar-System History*, 267 pp. New York: W.H. Freeman.

Weidenschilling SJ (2006) Models of particle layers in the midplane of the solar nebula. *Icarus* 181: 572–586.

Wetherill GW (1980) Formation of the terrestrial planets. *Annual Review of Astronomy and Astrophysics* 18: 77–113.

Wetherill GW (1986) Accmulation of the terrestrial planets and implications concerning lunar origin. In: Hartmann WK, Phillips RJ, and Taylor GJ (eds.) *Origin of the Moon*, pp. 519–550. Houston, TX: Lunar Planetary Institute.

Wetherill GW (1994) Provenance of the terrestrial planets. *Geochimica et Cosmochimica Acta* 58: 4513–4520.

Weyer S, Anbar AD, Brey GP, Münker C, Mezger K, and Woodland AB (2005) Iron isotope fractionation during planetary differentiation. *Earth and Planetary Science Letters* 240: 251–264.

Wiechert U and Halliday AN (2007) Non-chondritic magnesium and the origin of the terrestrial planets. *Earth and Planetary Science Letters* 256: 360–371.

Wiechert U, Halliday AN, Lee D-C, Snyder GA, Taylor LA, and Rumble DA (2001) Oxygen isotopes and the Moon-forming giant impact. *Science* 294: 345–348.

Wiechert U, Halliday AN, Palme H, and Rumble D (2004) Oxygen isotopes and the differentiation of planetary embryos. *Earth and Planetary Science Letters* 221: 373–382.

Wilde SA, Valley JW, Peck WH, and Graham CM (2001) Evidence from detrital zircons for the existence of continental crust and oceans on the Earth 4.4 Gyr ago. *Nature* 409: 175–178.

Williams HM, Markowski A, Quitté G, Halliday AN, Teutsch N, and Levasseur S (2006) Iron isotope fractionation in iron meteorites: New insights into metal-sulfide segregation and the sulphur contents of planetary cores. *Earth and Planetary Science Letters* 250: 486–500.

Williams HM, McCammon CA, Peslier AH, *et al.* (2004) Iron isotope fractionation and oxygen fugacity in the mantle. *Science* 304: 1656–1659.

Wood BJ (2000) Phase transformations and partitioning relations in peridotite under lower mantle conditions. *Earth and Planetary Science Letters* 174(3–4): 341–354.

Wood BJ, Bryndzia LT, and Johnson KE (1990) Mantle oxidation state and its relationship to tectonic environment and fluid speciation. *Science* 248: 337–345.

Wood BJ and Halliday AN (2005) Cooling of the Earth and core formation after the giant impact. *Nature* 437: 1345–1348.

Wood BJ and Rubie DC (1996) The effect of alumina on phase transformations at the 660-kilometer discontinuity from Fe–Mg partitioning experiments. *Science* 273(5281): 1522–1524.

Yi W, Halliday AN, Alt JC, *et al.* (2000) Cadmium, indium, tin, tellurium and sulfur in oceanic basalts: Implications for chalcophile element fractionation in the Earth. *Journal of Geophysical Research* 105: 18927–18948.

Yi W, Halliday AN, Lee D-C, and Christensen JN (1995) Indium and tin in basalts, sulfides and the mantle. *Geochimica et Cosmochimica Acta* 59: 5081–5090.

Yin QZ, Jacobsen SB, Yamashita K, Blicher-Toft J, Télouk P, and Albarède F (2002) A short time scale for terrestrial planet formation from Hf–W chronometry of meteorites. *Nature* 418: 949–952.

Yoshino T, Walter MJ, and Katsura T (2003) Core formation in planetesimals triggered by permeable flow. *Nature* 422: 154–157.

Zahnle K, Arndt N, Cockell C, *et al.* (2006) Emergence of a habitable planet. *Space Science Reviews* doi: 10.1007/s11214-007-9225-z.

Zanda B, Hewins RH, Bourot-Denise M, Bland PA, and Albarède F (2006) Formation of solar nebula reservoirs by mixing chondritic components. *Earth and Planetary Science Letters* 248: 650–660.

Zartman RE and Haines SM (1988) The plumbotectonic model for Pb isotopic systematics among major terrestrial reservoirs – A case for bi-directional transport. *Geochimica et Cosmochimica Acta* 52: 1327–1339.

3 Formation of Earth's Core

D. C. Rubie, Universität Bayreuth, Bayreuth, Germany

F. Nimmo, University of California, Santa Cruz, CA, USA

H. J. Melosh, University of Arizona, Tucson, AZ, USA

3.1 Core Formation in the Earth and Terrestrial Planets

3.1.1 Introduction and Present State of Cores in Solar System Bodies

The Earth's metallic core, comprising 32% of the total mass of the planet, lies beneath a silicate mantle, with the core–mantle boundary (CMB) located at a depth of 2891 km. The differentiation of the Earth into a metallic core and silicate mantle occurred during the accretion of the planet and represents the most important differentiation event in its history. Other terrestrial planets (e.g., Mercury, Venus, and Mars) and small asteroid bodies also underwent such differentiation events during the early history of the solar system and are thus important for providing additional information that helps in understanding differentiation of the Earth.

'Core formation' implies a single event, whereas in reality the core of the Earth most likely formed through a long series of events, over an extended time period. In this chapter, we consider the period up to which the core reached roughly 99% of its present-day mass; its later evolution is considered (Chapter 9). Other relevant chapters in this volume are concerned with the early Earth's composition (Chapter 2) and the terrestrial magma ocean (Chapter 4).

Here we first provide a general outline of the physical processes that are likely to be involved in core formation, including a discussion of the various uncertainties that arise. The second part of this chapter is focused on observations and experimental data that place constraints on the processes that operated and the state of the core during its formation.

The ultimate reason that the Earth and planets have cores is a matter of elementary physics: because metallic iron and its alloys are denser than silicates or ices, the most stable arrangement of a rotating, self-gravitating mass of material is an oblate spheroid with the dense iron at the center. Although the physical imperative driving core formation is simple, its realization is complicated by the usual factors of contingency and history. How the cores of the terrestrial planets came to their present configuration depends on what materials were available to make them, at what time and in what condition this material was added, and then how this mass of material evolved with time.

The origin and abundance of the elements of our solar system is now understood as the consequence of (mainly) stellar nucleosynthesis. Nuclear burning in stars created the elements and established their abundances in our solar system. Iron, as the end product of nuclear burning, owes its large abundance to its maximal nuclear stability. However, this review cannot go that far back. Fortunately, there are many good reviews of current ideas about the origin of the elements in our solar system (e.g., Busso *et al.*, 1999; Turan, 1984). But we cannot entirely ignore this aspect of solar system history, because one of the major developments in the past few years follows from the contingent circumstance that the solar system formed in conjunction with one or more nearby supernovas. These catastrophic events produced a substantial abundance of the short-lived radioactive isotopes ^{60}Fe and ^{26}Al, among many others. The consequences of this accident reverberate through our present solar system in ways that are just now becoming clear.

However, before exploring the antecedents of planetary cores in detail, we first consider what is known. All of the terrestrial planets and satellites, except perhaps Earth's moon, are believed to possess a dense, probably metallic, core. This belief is founded mainly on the density and moment of inertia of each body. If the average (uncompressed) density of a planet is much larger than the density of the material at its surface, one must infer that its interior is more dense. A homogeneous body has a moment of inertia ratio C/MR^2 equal to 0.400. Substantially smaller values of this ratio indicate a central mass concentration, one that usually implies a change of state or composition in the interior (we here exclude the slight decrease in this ratio due to self compression).

The size of Earth's core is known accurately from seismic data. The limited operating lifetime of the Apollo lunar seismic array provided only ambiguous evidence for a lunar core, and no seismic data exist for any other solar system object. Mercury presently possesses a small magnetic field and Mars probably had one in the past, suggesting the presence of metallic, fluid, and electrically conducting cores in these planets. Although Venus and Mars do not have magnetic fields at the present time, spacecraft measurements of their k_2 Love number indicate a large deformation in response to solar tides, implying an at least partially liquid core; observations of lunar nutations likewise suggest a liquid lunar core. **Table 1** summarizes our present knowledge of cores in the terrestrial planets and other bodies in the solar system.

3.1.2   The Relevance of Iron Meteorites

During the two centuries after meteorites were first accepted as samples of other bodies in our solar system, approximately 18 000 iron meteorites have been cataloged that, on the basis of trace element groupings, appear to have originated as the cores of approximately 100 separate bodies. These iron-meteorite parent bodies were small: cooling rates estimated from Fe/Ni diffusion profiles across taenite/kamacite crystal contacts suggest parent body diameters ranging from 30 to 100 km (Wasson, 1985: Chabot and Haack, 2006). Until recently, it was believed that the iron-meteorite parent bodies differentiated into an iron–nickel core and mantle sometime after most of the other meteorites, principally chondrites, had formed. However, recent dates using the extinct (9 My half-life) ^{182}Hf–^{182}W radioactive system have demonstrated that magmatic iron meteorites actually formed about 3 My before most chondrites (Scherstén *et al.*, 2006). This is nearly the same age as the heretofore oldest objects in the solar system, the calcium–aluminum inclusions (CAI) found in carbonaceous chondrites. This observation has completely changed our perception of the events in the early solar system and, in particular, the nature of the planetesimals that accumulated to form the Earth and other terrestrial planets. Previous to this revision in the age of the iron meteorites, the source of the heat necessary to differentiate the iron-meteorite parent bodies was obscure: suggestions ran the gamut from electromagnetic induction in solar flares to lightening in the solar nebula. Although the short-lived radioactivities ^{26}Al (half-life 0.74 My) and ^{60}Fe (1.5 My) were known to

Table 1   Planetary and satellite cores

Body	Mean density (Mg m^{-3})	Moment of inertia factor C/MR^2	Love number k$_2$	Mean planet radius, R$_p$ (km)	Core radius (km)	Magnetic moment T R$_p^3$	Core mechanical state	Composition
Mercury	5.427	0.33	?	2440	~1600	$4 \times 10^{-7\,a}$	Liquid?	Fe, Ni, ?
Venus	5.204	0.33	~0.25	6051.8	~3200	None at present	Liquid	Fe, Ni, ?
Earth	5.515	0.3308	0.299	6371.0	3485	6.1×10^{-5}	Liquid outer, solid inner core	Fe, Ni, FeO/ FeS ?
Moon	3.344	0.3935	0.0302	1737.53	400?	None at present	Liquid?	Fe, Ni, ?
Mars	3.933	0.366	~0.14	3389.9	~1700	Only local sources	Liquid?	Fe, Ni, FeO/ FeS ?
Io^b	3.53	0.378	?	1821	~950	None	Liquid?	Fe, Ni, FeS ?
Europa^b	2.99	0.346	?	1565	200–700	None	Liquid?	Fe, Ni FeS ?
Ganymede^b	1.94	0.312	?	2631	650–900	$7.15 \times 10^{-7\,b}$	Liquid?	Fe, Ni, FeS ?
Callisto	1.83	0.355	?	2410	None?	None	—	—

[a]Russell and Luhmann (1997).
[b]Schubert et al. (2004).
Unless otherwise noted, data are from Yoder (1995).
'?' indicates that values are currently unknown.

have been present in the early solar system, it was believed that by the time that the iron-meteorite parent bodies formed these potential heat sources had burned themselves out. However, with the new ages, a new view emerges.

The effectiveness of the heating caused by the decay of a radioactive isotope of a stable element can be gauged from the following formula that gives the maximum temperature rise ΔT occurring in a completely insulated sample of material during the time subsequent to its isolation:

$$\Delta T = \frac{f \; C_m \; E_D}{c_p} \qquad [1]$$

where c_p is the heat capacity of the material, C_m is the concentration of the stable element in the material, f is the fraction of radioactive isotope at the beginning of the isolation interval, and E_D is the nuclear decay energy released into heat (we do not count the energy of neutrinos emitted during beta decay). **Table 2** lists the values of this temperature rise for undifferentiated material of carbonaceous chondrite composition at the time of CAI and iron-meteorite formation, and at the time that the bulk of the chondrites formed, 3 million years later.

The principal implication of this recent finding for core formation in the Earth is that many (if not most)

of the planetesimals that accumulated to form the major planets possessed metallic cores at the time of their accretion. As yet, the consequences of accretional impacts among partially molten planetesimals are not well studied. However, it is clear that if the planetesimals contributing to the growth of planetary embryos had already formed iron cores, chemical equilibration between iron and silicates initially occurred in a low-pressure regime. In addition, the iron that was added to the growing embryos might not have been in the form of small, dispersed droplets.

Just how the iron core of an impacting planetesimal mixes with the existing surface is presently somewhat unclear. Even the largest meteorite impacts on the present Earth seldom preserve more than a trace of the projectile. In most cases the projectile, if it can be identified at all, is only revealed by geochemical or isotopic tracers in the rock melted by the impact. The largest intact remnant of an impactor presently known is a 25-cm-diameter fragment of LL6 chondrite discovered in the melt sheet of the c. 70-km-diameter Morokweng crater (Maier et al., 2006). The impactor that created the 170-km-diameter Chicxulub crater is known only from a few millimeter-size fragments recovered from a deep-sea core (Kyte, 1998). The projectiles that created these craters are at the low end of the size spectrum we

Table 2   Temperature rise of undifferentiated carbonaceous chondrites due to radioactive decay

Radioisotope	Half-life (My)	Fractional abundance of isotope at CAI time, f	Nuclear decay energy, E_D (J kg^{-1})	ΔT at CAI time (K)	ΔT 3 My later (K)
^{26}Al	0.74	$(5–7) \times 10^{-5}$	1.16×10^{13a}	4170	251
^{60}Fe	1.5	4.4×10^{-6b}	4.43×10^{12}	2960	740

[a]Schramm et al. (1970).
[b]Quitté et al. (2005).
Assuming the solar system abundances of Anders and Gervesse (1989) and heat capacity $c_p = 1200$ J kg^{-1} K. Abundances of Fe and Al are assumed to be chondritic, at 18.2 wt.% and 0.865 wt.%, respectively (Lodders and Fegley, 1998).

expect for planetesimals: the diameter of the Morokweng impactor was about 4 km or more, while the Chicxulub impactor was probably about 15 km in diameter. Both objects probably impacted near the average velocity for asteroidal impactors on the Earth, about 17 km s^{-1}. Although this velocity is substantially higher than the encounter velocity of the nearly formed Earth with infalling planetesimals (less than about 10 km s^{-1}, or a factor of more than three times less energy than current asteroidal impacts), it still illustrates the fact that impacts typically disrupt the impactor and whatever the arrangement of materials might have been in the impacting objects; this arrangement is greatly distorted during the impact event.

If we can extrapolate from these observations to the impacts of much larger objects, one would conclude that it makes little difference whether the impacting planetesimal was differentiated or not – in either case the result of the impact is a finely dispersed mixture of melted target and projectile material. On the other hand, computer simulations of the much larger moon-forming impact show that the core of a planet-size impactor remains mostly together in a discrete mass during the impact and appears to merge almost *en masse* with the Earth's core (Canup, 2004). In these simulations, however, each 'particle' representing the material of the Earth and projectile is about 200 km in diameter, so that it is not possible to resolve details of how the iron mass from the projectile core really interacts with the Earth's mantle and core. It thus seems possible that the cores of large planetesimals might remain together in homogeneous masses too large for chemical equilibration (i.e., larger than a few centimeters) with their surroundings, at least at the beginning of their descent into the Earth's mantle. This is an area needing further study from the impact perspective. Later in this review we discuss the probable fate of such large masses of iron in their inexorable fall toward the Earth's core.

3.1.3   History of Ideas on Core Formation

Ideas on how the Earth's core formed have shifted dramatically over the past century. At the beginning of the twentieth century most geophysicists believed, following Lord Kelvin, that the Earth began in a completely molten state and its subsequent history was one of secular cooling and solidification (Thomson and Tait, 1883, p. 482). As late as 1952, Harold Jeffreys found core formation totally unproblematic, because dense iron would inevitably sink through the liquid proto-mantle (Jeffreys, 1952, p. 271). However, about the same time, Urey (1952) was elaborating Chamberlin's (1916) hypothesis that the planets had formed from a swarm of small, cold, mutually gravitating 'planetesimals'. In Urey's analysis, core formation becomes highly problematic. The apparent difficulty posed by Urey's view initiated much of our current thinking about how the Earth's metallic iron core originated.

In his famous book *The Planets*, Urey (1952) presented a model of planet formation that strongly appealed to physicists, although, as we shall soon see, it lacked many aspects of reality. Urey approximated the growing Earth as a spherical, homogeneous, isotropic body of radius r that grew from cold matter in space by the addition of infinitesimally thin shells of thickness dr. In this model he equated the gravitational energy of infalling matter, $-GM(r)/r$ per unit mass, where G is Newton's gravitational constant and $M(r)$ is the mass of the nascent Earth at radius r, with the heat gained after the matter accreted to the Earth. This energy was apportioned between heating of the added mass, heat conduction into the interior, and thermal radiation to space. Because the shell of added matter is very thin, thermal radiation dominates and the planet accretes at very low temperatures.

Models of this kind led Hanks and Anderson (1969) to discover that even if the Earth accreted in as little as 1 My it would not reach the melting temperature of rock anywhere in its interior. They showed that radio-active heating would only warm the Earth's interior to the melting point much later, initiating core formation as late as 1.6 Gy after its formation.

By the 1970s, however, study of lead isotopes in ancient crustal rocks indicated that the core had formed within a few hundred million years of the iron meteorites (Gancarz and Wasserburg, 1977), and that a problem existed with the models for thermal evolution of the Earth. Safronov (1978) and Kaula (1979) independently suggested that Urey's model, in which thin shells of infalling matter radiated most of its energy to space, is too drastic. If the impacting planetesimals were of moderate size, a few kilometers or more in diameter, some substantial, but not easily computed, fraction h of the gravitational energy would be buried in the planet. Although this solution has seemed attractive for many decades, it has one central flaw: because the gravitational energy of the planet is initially rather small, the energy added by each unit of mass, $hG\,M(r)/r$, increases roughly as the square of the radius of the Earth. The center of the Earth thus starts out cold, although a hotter, molten, outer shell eventually develops around it. We thus end up with an Earth that is thermally inside-out: cold in the center, hot on the outside.

This apparent conundrum over the initial thermal structure of the Earth led to a series of clever exam-inations of the stability of a shell of molten, segregated iron in the hot outer portion of the Earth. Elsasser (1963) suggested that a shell of molten iron would push its way to the Earth's center by diapiric instabilities. Later, Stevenson (1981) showed that this instability is even stronger than Elsasser suspected, and that the iron would actually fracture the cold kernel of the Earth on a timescale of hours, supposing that such a global iron layer ever had time to form.

In our modern era, in which a much more cata-strophic view of Earth's formation reigns (Wetherill, 1985), the problematic initial thermal profile of the Earth is ameliorated by the ability of gigantic impacts to implant heat deep into a growing planet (Melosh, 1990). Deep, strong heating and core formation can be initiated by impacts themselves, given only that they are large and late enough (Tonks and Melosh, 1992). Magma oceans are now seen as an inevitable conse-quence of the late accretion of planet-scale protoplanets (Tonks and Melosh, 1993). In this era the problem is not so much how cores form, as to how, and under what circumstances, iron and silicate may have equilibrated chemically, and how the current inventories of chemical elements in the crusts and mantles of the Earth and planets were established.

3.2 Physics of Core Formation

The Earth is the end product of multiple collisions between smaller protoplanets. This process of accre-tion results in increased temperatures and, ultimately, melting on a planetary scale. As discussed below, differentiation is unavoidable once melting begins; thus, the accretion process is intimately con-nected to the manner in which the Earth, and its precursor bodies, underwent differentiation and core formation. In this section, our theoretical under-standing of the accretion process and its consequences for core formation are discussed; in Section 3.3, observational and experimental con-straints on these processes are outlined.

Earlier reviews and discussions of the processes enumerated here may be found in Stevenson (1989, 1990), Rubie *et al.* (2003), Walter and Trønnes (2004), and Wood *et al.* (2006). The collection of papers edited by Canup and Righter (2000) is also highly recommended.

3.2.1 Accretion

The basic physics of planetary accretion are now reasonably well understood, although many details remain obscure (see Wetherill (1990) and Chambers (2003) for useful reviews). Growth of kilometer-sized objects (planetesimals) from the initial dusty, gaseous nebula must have been a rapid process (occurring within approximately 10^3 years), because otherwise the dust grains would have been lost due to gas drag. At sizes >1 km, mutual gravitational interac-tions between planetesimals become important. Furthermore, because the largest bodies experience the greatest gravitational focusing, they tend to grow at the expense of smaller surrounding objects. This 'runaway growth' phase, if uninterrupted, can poten-tially result in the development of tens to hundreds of Mars- to Moon-sized embryos in $\sim 10^5$ years at a distance of around 1 astronomical unit (AU) from the Sun (Wetherill and Stewart, 1993). However, runaway growth slows down as the initial swarm of small bodies becomes exhausted and the velocity dispersion of the remaining larger bodies increases (Kokubo and Ida, 1998). Thus, the development of Moon- to Mars-sized embryos probably took $\sim 10^6$ years at 1 AU (Weidenschilling *et al.*, 1997), and involved collisions both between comparably sized embryos, and between embryos and smaller, left-over planetesimals. Based on astronomical observations of dust disks (Haisch *et al.*, 2001), the dissipation of any remaining nebular gas also takes place after a few million years; the dissipation timescale of gas has implications both for the orbital evolution of the bodies (e.g., Kominami *et al.*, 2005), their volatile inventories (e.g., Porcelli *et al.*, 2001), and their surface temperatures (e.g., Abe, 1997), and is currently a critical unknown parameter. Noble gas isotopes, in particular those of xenon, have been used to argue for a primordial, dense, radiatively opaque terrestrial atmosphere (e.g., Porcelli *et al.*, 2001, Halliday, 2003), but this interpretation remains controversial (*see* Chapter 2).

Collisional growth processes lead to a peculiar size–frequency spectrum of the accumulating bodies. At first, the runaway accretional processes produce a spectrum in which the cumulative number of objects (the number of objects equal to, or greater, than diameter D) is proportional to an inverse power of their diameter, generally of form $N_{cum}(D) \sim D^{-b}$, where b is often approximately 2 (Melosh, 1990). One of the principal characteristics of such a distribution is that although the smallest bodies overwhelmingly dominate in number, most of the mass and energy resides in the very largest objects. Accretional impacts are thus catastrophic in the sense that objects at the largest end of the size spectrum dominate planetary growth. Later, during oligarchic growth at the planetary embryo scale, the large bodies represent an even larger fraction of the size spectrum and giant impacts, that is, impacts between bodies of comparable size dominate planetary growth history.

The subsequent growth of Earth-sized bodies from smaller Mars-sized embryos is slow, because the embryos grow only when mutual gravitational perturbations lead to crossing orbits. Numerical simulations show that Earth-sized bodies take 10–100 My to develop (e.g., Chambers and Wetherill, 1998; Agnor *et al.*, 1999; Morbidelli *et al.*, 2000; Raymond *et al.*, 2004), and do so through a relatively small number of collisions between objects of roughly comparable sizes. A recent result of great importance is that geochemical observations, notably using the hafnium–tungsten (Hf–W) isotopic system, have been used to verify the timescales obtained theoretically through computer simulations of accretion processes (see Section 3.3.1).

It should be noted that an important implicit assumption of most late-stage accretion models is

that collisions result in mergers. In fact, this assumption is unlikely to be correct (Agnor and Asphaug, 2004; Asphaug *et al.*, 2006) and many collisions may involve little net transfer of material, though both transient heating and transfer of angular momentum will occur. In fact nearly 80% of the mantle of Mercury may have been 'lost' by collisional erosion after core formation, thus explaining the huge size of its metallic core (Benz *et al.*, 1988). Such disruptive collisions may also have influenced the evolution of the Earth and could explain an excess of Fe in the Earth's bulk composition relative to C1 chondrites (Palme *et al.*, 2003).

Figure 1(a) shows a schematic example (obtained by splicing together two different accretion simulations) of how a roughly Earth-mass ($1M_e$) body might grow. Here the initial mass distribution consists of 11 lunar-mass embryos ($\approx 0.01 M_e$) and 900 smaller ($\approx 0.001 M_e$) noninteracting planetesimals centered around 1 AU. The solid line shows the increase in mass, and the crosses show the impactor:target mass ratio γ (both in log units). The early stage of growth is characterized by steady collision with small planetesimals, and occasional collisions with other, comparably sized embryos (e.g., at 0.068 and 1.9 My). Because the planetesimals do not grow, the impactor:target mass ratio γ of colliding planetesimals declines with time; embryo–embryo collisions show up clearly, having $\gamma \sim 1$. At 2 My, the growing object has a mass of $0.2M_e$ and roughly half of this mass has been delivered by large impacts. The late stage of growth consists entirely of large impacts, between embryos of comparable masses ($\gamma \sim 0.5$). This final stage takes place over a more extended timescale – in this case, the last significant collision occurs at 14 My, resulting in a final mass of $0.73M_e$.

One of the most important outstanding questions regarding this late-stage accretion is the amount of water that was delivered to the Earth. The presence of large quantities of water in the early mantle would have profound implications for the oxidation state and composition of the core (see Williams and Hemley (2001)); furthermore, a byproduct would be a thick steam atmosphere, which would be sufficiently insulating to ensure a magma ocean (Matsui and Abe, 1986). Although the Earth formed inside the 'snow line', where water ice becomes unstable, some of its constituent planetesimals may have been derived from greater heliocentric distances and thus contained more water. Simulations (Morbidelli *et al.*, 2000; Raymond *et al.*, 2004) suggest that a water-rich Earth is quite likely, but the stochastic nature of the outcomes precludes a firm conclusion. Radial mixing of planetesimals is clearly not completely efficient because of the differing oxygen-isotope characteristics of Earth and Mars (e.g., Clayton and Mayeda, 1996).

3.2.2   Thermal Evolution

As discussed below in Section 3.2.3, the actual mechanisms of core formation (metal–silicate separation) that operate are dependent on the thermal state of a planetary body and at least some degree of partial melting is required. However, in addition to understanding the thermal state of the Earth during core formation, it is also important to understand the thermal histories of small bodies (planetesimals and asteroids) because these determine whether or not the material that accreted to form the Earth had already undergone core–mantle differentiation.

There are three main sources of energy that can produce the melting that is required for core formation. First, the decay of short-lived radioactive nuclides (^{26}Al and ^{60}Fe) is an important source of energy when accretion occurs very soon after the formation of the solar system (**Figure 1(b)**). (These isotopes have half-lives of 0.73×10^6 and 1.5×10^6 years, respectively.) Second, the kinetic energy delivered by impacts can be sufficient to generate local or global melting, especially during the late stages of Earth accretion (**Figures 1(b)** and **3**).

Finally, as discussed below, the process of differentiation itself, by reducing the gravitational potential energy of the body, also releases heat and may lead to runaway differentiation.

3.2.2.1   Decay of radioactive nuclides

Thermal models show that the decay of ^{26}Al and ^{60}Fe in a body with a minimum radius of 30 km can result in maximum temperatures that range from below the Fe–FeS eutectic temperature to above silicate melting temperatures, depending on the initial concentrations of these isotopes (**Figure 2**). In contrast, the energy released through collisions between bodies less than a few hundred kilometer radius is insufficient to cause global melting (Keil et al., 1997; see also Section 3.2.2.2). This means that the melting required for core–mantle differentiation in a small body could only occur at a very early stage during the history of the solar system – for example, within the first 1 My (Baker et al., 2005). In support of the thermal models, there is geochemical evidence for large-scale melting and magma ocean formation on at least some small bodies (Greenwood et al., 2005). In addition, the parent body of the HED meteorites (which is likely Asteriod-4 Vesta, 530 km in diameter) underwent early core–mantle differentiation. These considerations support the view that planetesimals that accreted to form the Earth were already differentiated (e.g., Taylor and Norman, 1990).

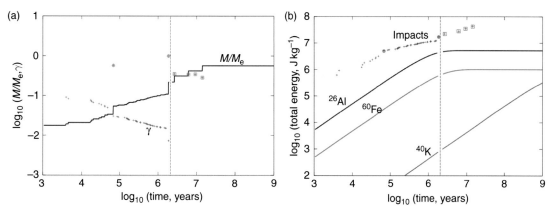

Figure 1   (a) Schematic growth of a proto-Earth, obtained by splicing two accretion simulations together. Early growth is from Agnor (unpublished) where the initial mass distribution consists of 11 embryos ($\approx 0.01M_e$) and 900 noninteracting planetesimals ($\approx 0.001M_e$) centred around 1 AU. Late growth is from particle 12 in run 3 of Agnor et al. (1999). The vertical dashed line denotes the splicing time. The solid line shows the mass evolution of the body, and the crosses denote the impactor:target mass ratio γ. Circles denote embryo–embryo collisions; squares late-stage giant impacts. The general reduction in γ prior to 2 My is a result of the fact that the planetesimals cannot merge with each other, but only with embryos. (b) Corresponding energy production (J kg^{-1}). The cumulative energy due to impacts (crosses) is calculated using eqn [2] for each impact. The solid lines show the cumulative energy associated with the decay of radioactive elements ^{26}Al, ^{60}Fe, and ^{40}K. Half-lives are 0.73 My, 1.5 My, and 1.25 Gy, respectively; initial bulk concentrations are 5×10^{-7}, 2×10^{-7}, and 4.5×10^{-7}, respectively (Ghosh and McSween, 1998; Tachibana et al., 2006; Turcotte and Schubert, 2002).

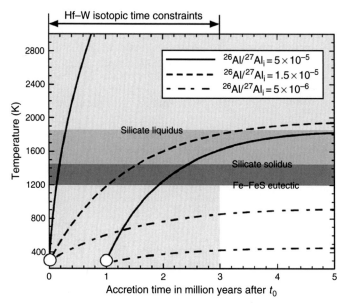

Figure 2 Models of the thermal evolution of a small body (>30 km radius) as a function of the initial concentration of ^{26}Al. Temperature is calculated as a function of time after the start of the solar system (t_0). Depending on the concentrations of ^{26}Al and ^{60}Fe and when accretion starts, maximum temperatures range from below the Fe–FeS eutectic to above the silicate liquidus (see also Yoshino et al., 2003, figure 2). Reproduced from Walter MJ and Trønnes RG (2004) Early Earth differentiation. *Earth and Planetary Science Letters* 225: 253–269, with permission from Elsevier.

However, if small bodies accreted somewhat later, melting and differentiation may not have occurred because the concentrations of ^{26}Al and ^{60}Fe would have been too low (**Figure 2**). The density of the solar nebula must have decreased with increasing distance from the Sun. Because the rate of a body's accretion depends on the density of the solar nebula, bodies in the outer regions would have accreted slower and therefore later, than those close to the Sun (Grimm and McSween, 1993). Thus not all planetesimals would have differentiated. Ceres is an example of an asteroid that may not have differentiated to form a metallic core (Thomas *et al.*, 2005). It is therefore currently not clear whether the Earth accreted from only differentiated bodies or a mixture of differentiated and undifferentiated bodies. The answer depends on the extent of the Earth's feeding zone during accretion.

3.2.2.2 *Heating due to the energy of impacts*

Figure 1(a) shows that the bulk of late-stage Earth accretion involves large impacts well separated in time. The energetic consequences of such impacts have been discussed elsewhere (Melosh, 1990; Benz and Cameron, 1990; Tonks and Melosh, 1993) and strongly suggest that, even in the absence of a thick primordial (insulating) atmosphere, the final stages of Earth's growth must have involved of one or more global magma oceans. This conclusion has important implications for the mode of core formation, and may be understood using the following simple analysis.

For an impact between a target of mass M and an impactor of mass γM, the mean change in energy per unit mass of the merged object due to kinetic and gravitational potential energy is

$$\Delta E = \frac{1}{1+\gamma}\left[-\frac{3}{5}\left(\frac{4\pi\rho}{3}\right)^{1/3} \right.$$
$$\left. \times GM^{2/3}\left(1+\gamma^{5/3}-(1+\gamma)^{5/3}\right) + \frac{1}{2}\gamma V_\infty^2 \right] \quad [2]$$

Here ρ is the mean density of the merged object, G is the universal gravitational constant, V_∞ is the velocity of the impactor at a large distance from the target, and the factor of 3/5 comes from considering the binding energy of the bodies (assumed uniform) prior to and after the collision. Neglecting V_∞ and taking γ to be small, the global average temperature rise associated with one such impact is given by

$$\Delta T \approx 6000K\left(\frac{\gamma}{0.1}\right)\left(\frac{M}{M_e}\right)^{2/3} \quad [3]$$

where we have assumed that $\rho = 5000 \, \text{kg m}^{-3}$ and a heat capacity of $1 \, \text{kJ kg}^{-1} \, \text{K}^{-1}$. Note that this temperature change is a globally averaged value; for small impacts in particular, real temperatures will vary significantly with distance from the impact site.

Equation [2] is based on several simplifying assumptions. It assumes that the energy is deposited uniformly, which is unlikely to be correct. More importantly, it assumes that all the kinetic energy is converted into heat and retained. Such an assumption is unlikely to be correct for small impacts, where most of the energy is deposited at shallow depths where it can be radiated back to space (Stevenson, 1989). For larger impacts, however, the energy will be deposited at greater depths, and thus the only major energy loss mechanism is the ejection of hot material. The amount and temperature of material ejected depends strongly on the geometry of the impact, but is in general rather small compared to the target mass (Canup et al., 2001). Since we are primarily concerned with large impacts ($\gamma > 0.1$), the assumption that the majority of the energy is retained as heat energy is a reasonable one. Thus, impactors with a size similar to the one that is believed to have formed the Earth's Moon (Cameron, 2000; Canup and Asphaug, 2001) probably resulted in the bulk of the Earth being melted.

Although a Mars-sized ($0.1 M_e$) proto-Earth has a smaller mass, it experiences collisions with bodies comparable in size to itself ($\gamma \approx 1$; see Figure 1). In this case, eqn [2] shows that $\Delta T \approx 4500 \, \text{K}$. Thus, it seems likely that Mars-sized embryos were also molten and thus differentiated. Although there is currently little direct evidence for an ancient Martian magma ocean (see Elkins-Tanton et al., 2003), Blichert-Toft et al. (1999) and Borg and Draper (2003) have used Lu–Hf systematics and incompatible element abundances, respectively, to argue for such an ocean. Conversely, Righter (2003) argued that the temperatures and pressures inferred from siderophile element abundances do not necessarily require a magma ocean.

In considering the thermal effects of impacts, it may also be useful to consider the cumulative energy delivered for comparison with other sources of energy. Figure 1(b) shows the cumulative impact energy in J kg^{-1}. The bulk of the energy is delivered by the few largest impacts, as expected. For comparison, the radioactive heat production due to one long-lived ([40]K) and two short-lived isotopes ([26]Al and [60]Fe) are shown. Long-lived isotopes have no effect at all on the thermal evolution of the Earth over its first 10 My.

The total energy associated with [26]Al depends very strongly on the accretion time (Figure 2) and in this case is roughly one order of magnitude smaller than that due to the impacts. Figure 1(b) shows that the thermal evolution of the Earth naturally divides into two stages: the early stage (up to ~ 10 My) when heating due to impacts and short-lived isotopes dominate; and the later stage, when long-lived isotopes and secular cooling are important.

Figure 3 summarizes the expected mean global temperature change due to impacts and short-lived radionuclides as a function of planetary size. The effect of a single impact (solid lines) is calculated using eqn [2] and demonstrates the strong dependence on both body mass and the impactor:target mass ratio γ. It should be re-emphasized that, particularly for small impacts, the energy will not be distributed evenly and that the calculated temperature rise is only a mean global value. The effect of [26]Al decay (dashed lines) does not depend on the body mass, but only on the accretion time relative to CAI formation. For small bodies, only radioactive decay contributes substantially to warming; for large

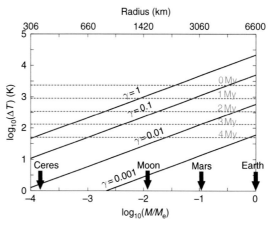

Figure 3 Mean global temperature change ΔT as a function of planetary mass (in units of M_e = one Earth mass). Solid lines show gravitational energy due to a single collision where γ denotes the impactor:target mass ratio and ΔT is calculated from eqn [2] assuming that $V_\infty = 0$, $C_p = 1000 \, \text{J kg}^{-1} \, \text{K}^{-1}$ and $\rho = 5000 \, \text{kg m}^{-3}$. Dashed lines show temperature change due to [26]Al decay as a function of (instantaneous) planet formation time (in million years) after solar system formation. Total energy release by [26]Al is $6.2 \times 10^6 \, \text{J kg}^{-1}$ (assuming 1 wt.% Al and a fractional abundance of [26]Al of 5×10^{-5}) and half-life is 0.73 My. Planetary melting is expected to be widespread for $\Delta T > 1000 \, \text{K}$. Note that heat losses by conduction or convection are neglected when calculating the temperature rise.

bodies, gravitational energy release will tend to dominate. **Figure 3** also emphasizes that the population of
impactors a growing planet encounters has a very
important effect on its ultimate temperature structure. If all the impactors are small ($\gamma \ll 1$) then the
temperature changes are quite modest. Conversely,
the later stages of planetary accretion ensure that
growing bodies will encounter other bodies of similar
size (i.e., $\gamma \sim 1$; Section 3.2.1), and thus melting is
expected for bodies of roughly Moon-size and larger.

Although the average temperature jumps in large
planets struck by comparable-sized objects are
impressive, it must be kept in mind that these are
averages only. Impacts are not good at homogenizing
temperatures. An impact initially deposits a large
fraction of its kinetic energy in a region comparable
in size to the projectile itself. Shortly after a time
measured by D/v_i, the projectile diameter D divided
by its impact velocity v_i, the shock waves generated
by the impact expand away from the impact site,
accelerating the target material and depositing heat
as they spread out and weaken. The heat is therefore
mostly deposited in a region called the 'isobaric core'
whose shape is initially approximately that of an
immersed sphere tangent to the surface of the target.
Outside of this region the shock pressure, and therefore the amount of heat deposited, falls off rapidly,
generally as more than the cube of the distance from
the impact (Pierazzo et al., 1997). A velocity field that
is established by the spreading shock wave accompanies this heat deposition. The moving material
generally opens a crater whose size and form depends
on the relative importance of gravity and strength in
the crater excavation. In general, about half of the
hot, isobaric core is ejected from the impact site, to
spread out beyond the crater rim, while the other half
remains in the distorted target beneath the crater
floor (i.e., for impacts that do not vaporize a large
volume of the target, which is the case for all accretional impacts on Earth-sized planets). The major
part of the kinetic energy of the impact is thus converted to heat in a roughly hemispherical region
centered on the crater. This hot zone extends to a
depth of only a few projectile diameters.

For small impacts, generally those of objects less
than a kilometer in diameter, the heat is deposited so
close to the surface that a major fraction is radiated to
space before the next similar-size impact occurs
(Melosh, 1990). Larger impactors deposit more of
their energy deeper in the planet, up to the truly
gigantic impacts of objects comparable in size to the
growing Earth, which may deposit their energy over
an entire hemisphere. Nevertheless, detailed computations show that this energy is not homogeneously
distributed over the target Earth. Later processes,
such as the re-impact of large fractions of the projectile (seen in some of the Moon-forming scenarios:
Canup, 2004), or thermal convection of the mantle
driven by a suddenly heated core, are needed to
homogenize the heat input. Impact heating is thus
characterized by large initial temperature variations:
the part of the Earth near the impact site becomes
intensely hot while more distant regions remain cold.
Averages, therefore, tell only a small part of the overall story: the aftermath of heating by a large impact is
characterized by strong temperature gradients and
the evolution of the planet may be dominated by
contrasts between very hot and cold, mostly unaffected, portions of the growing Earth.

3.2.2.3 Heating through the reduction of gravitational potential energy

Although the energies involved in late-stage impacts
imply the formation of a magma ocean, such an ocean
may not reach the CMB, because the solidus temperature of mantle material is a strong function of
pressure (see **Figure 9(b)** below). The mechanical
properties of the magma ocean, which control the
rate of iron transport across the ocean, change dramatically when the melt fraction drops below $\approx 60\%$
(Solomatov, 2000; see also Chapter 4). The effective
base of the magma ocean occurs at this rheological
transition. Descending iron droplets will tend to pond
at this interface. However, the resulting iron layer is
still denser than the underlying (mantle) material and
will therefore tend to undergo further transport
towards the center of the planet. The transport
mechanism might be percolation through a partially
molten mantle, or the motion of larger iron bodies via
brittle fractures (dyking) or through a viscously
deformable mantle (diapirism). These different
mechanisms are discussed in Stevenson (1990) and
will be addressed in more detail in Section 3.2.3.

The redistribution of mass involved with the descent of iron toward the center of the planet results in
a release of gravitational energy. Because the gravitational energy change only depends on the initial
and final density distributions, the mechanism by
which the iron is transported is of only secondary
importance. The extent to which the iron, rather than
the surrounding mantle material, is heated depends
on the rate of transport (rapid for diapirs or dykes,
slower for percolation) but the total energy released

would be the same. The magnitude of the heating can be large, and may be calculated as follows.

Consider a uniform, thin layer of iron at the base of the magma ocean, overlying a mantle and core (**Figure 4**). The top and bottom of the iron layer and the underlying core are at radii R_o, $R_m = (1 - \varepsilon)R_o$ and $R_c = \beta R_o$, respectively, where ε is a measure of the thickness of the iron layer ($\varepsilon \lll 1$) and β is a measure of the initial core radius. After removal of iron to the centre, whether by diapirism, dyking, or percolation, the core will have grown and the situation will have a lower potential energy. The difference in potential energy may be calculated and used to infer the mean temperature change in the final core, assuming that all the potential energy is converted to core heat (e.g., Solomon, 1979). For the specific case of a constant core density twice that of the mantle, it may be shown that the mean temperature change of the entire post-impact core is given by

$$\Delta T = \frac{\pi G \rho_c R_o^2}{\beta^3 C_p (1 + 3\varepsilon\beta^{-3})}$$
$$\times \left[\frac{1}{10} \beta^5 \left((1 + 3\varepsilon\beta^{-3})^{5/3} - 1 \right) + \frac{1}{2}\varepsilon - \beta^3\varepsilon \right] \quad [4]$$

Here C_p is the specific heat capacity of the core and ρ_c is the core density, and it is assumed that the core is well-mixed (isothermal). As before, the temperature change is a strong function of planetary size, specifically the radial distance to the base of the magma ocean R_o. The temperature change also depends on ε, which controls the mass of iron being delivered to the

core, and the initial core radius R_c when $\beta > 0$. The temperature change goes to zero when $\beta = 1$, as expected, while when $\beta = 0$ (i.e., no initial core) the temperature change is essentially independent of the mass of iron delivered.

Figure 4 shows the expected temperature change as a function of R_o plotted for different values of β and ε. It is clear that for Earth-sized planets, the addition of iron to the core by individual impacts can lead to core temperature increases of several hundred to a few thousand kelvin.

For example, consider two cases, appropriate to the Moon-forming impact (Canup and Asphaug, 2001): a $0.9M_e$ planet hit by a $0.1M_e$ impactor, and a $0.8M_e$ planet hit by a $0.2M_e$ impactor, all bodies having a core of density 10^4 kg m^{-3} and radius half the body radius. For magma oceans of depths 500 and 1000 km, respectively ($R_o = 5900$ and 5400 km), we obtain $\varepsilon = 0.0054$ and 0.0139, and $\beta = 0.52$ and 0.55. The core adiabat is roughly 1 K km^{-1}, giving temperature increases of 2400 and 1900 K. The further increases from gravitational heating (eqn [4]) are 1250 and 2000 K, respectively. Thus, the post-impact core temperature is likely to have increased by 3500–4000 K from the temperature it attained at the base of the magma ocean. Because the base of the magma ocean is estimated to be in the range 2500–4000 K (Section 3.3.2), the initial core temperature was probably at least 6000 K, sufficient to cause substantial lower-mantle melting.

This estimate is only approximate, because of the assumptions made (e.g., no transfer of heat to the mantle) and the fact that the Earth probably suffered several comparably sized impacts. However, the result is important because the initial temperature contrast between the core and the lowermost mantle determines the initial CMB heat flux, and thus the ability of the core to generate a dynamo.

3.2.3   Differentiation Mechanisms

A crucial question is at what stage during the accretion history did core–mantle differentiation actually occur and how long did the process take? These questions depend on the thermal history of the accreting body which, in turn, determines the physical mechanisms by which metal and silicate separate. The physics of differentiation have been reviewed previously by Stevenson (1990) and Rushmer *et al.* (2000) and here we provide an updated account. Differentiation occurs because of the large density contrast between silicates and metal, but the rate at

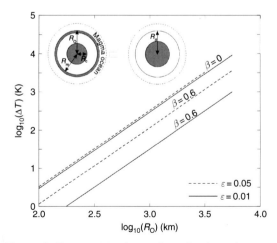

Figure 4 Temperature change due to iron layer descent, from eqn [4]. Here ε is a measure of the thickness of the iron layer being added to the core and is given by $1 - (R_m/R_o)$, while β is a measure of the initial core radius and is given by $\beta = R_c/R_o$.

which it occurs depends on the length scales and properties of the phases involved. Considering the current size of the Earth, the segregation process involved the transport of metal through the proto-mantle over length scales of up to almost 3000 km. As reviewed by Stevenson (1990), the process involved the transport of liquid metal through either solid (crystalline) silicates or through partially or fully molten silicate (i.e., in a magma ocean). The former mechanism is possible because iron (plus alloying elements) has a lower melting temperature than mantle silicates, so that liquid iron can coexist with solid silicates. In contrast, the separation of solid metal from solid silicate is too sluggish to have been a significant process during core formation (Stevenson, 1990). Identifying whether metal separated from solid or molten silicates during core formation clearly provides information about the early thermal state of the planet. In addition, the separation mechanism may have affected the conditions and extent of chemical equilibration between metal and silicate and therefore affects mantle geochemistry, as discussed further below. Here we review recent results pertaining to core formation by (1) grain-scale percolation of liquid metal through crystalline silicates, (2) separation of molten metal from molten silicate in a magma ocean, and (3) descent of large (kilometer scale) diapirs of molten metal through crystalline silicate and/or transport by a fracture/dyking mechanism.

As discussed above, it seems likely that the final stages of Earth's accretion involved large impacts between previously differentiated objects which were at least partly molten. What happens in detail during these impacts is poorly understood. Hydrocode simulations of impacts (Cameron, 2000; Canup and Asphaug, 2001) show that the cores of the target and impactor merge rapidly, within a few free-fall timescales (hours), although a small fraction (typically <1%) of core material may be spun out into a disk. Unfortunately, the resolution of these simulations is on the order of 100 km, while the extent to which chemical re-equilibration occurs depends on length scales that are probably on the order of centimeters (Stevenson, 1990). Another approach to understanding this important question is presented in Section 3.2.3.2.

3.2.3.1 Percolation

Liquid metal can percolate through a matrix of poly-crystalline silicates by porous flow provided the liquid is interconnected and does not form isolated pockets

(Stevenson, 1990; Rushmer *et al.*, 2000). The theory is well developed for the ideal case of a monomineralic system consisting of crystals with isotropic surface energy (i.e., no dependence on crystallographic orientation). Whether the liquid is interconnected depends on the value of the wetting or dihedral angle (θ) between two solid–liquid boundaries that are intersected at a triple junction by a solid–solid boundary (**Figure 5**) (von Bargen and Waff, 1986; Stevenson, 1990). For $\theta < 60°$, the liquid is fully interconnected and can percolate through the solid irrespective of its volume fraction; under such conditions, complete metal–silicate segregation can occur efficiently by porous flow. In the case that $\theta > 60°$, the liquid forms isolated pockets when the melt fraction is low ($\leq 0.8\%$) and connectivity exists only when the melt fraction exceeds a critical value. This critical melt fraction is known as the connection boundary and ranges from 2% to 6% for dihedral angles in the range 60–85°. If the melt fraction lies above the connection boundary, the melt is interconnected and can percolate. However, as the melt fraction decreases due to percolation, a pinch-off melt fraction is reached below which interconnectivity is broken. At this point the remaining melt is stranded in the crystalline matrix. The pinch-off melt fraction lies slightly below

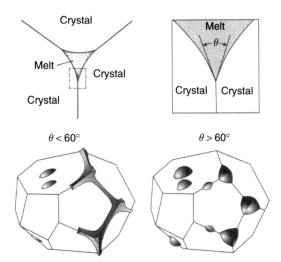

Figure 5 Relation between melt connectivity and dihedral angle in a polycrystalline aggregate containing a small amount of dispersed melt. The dihedral angle is defined in the top diagrams and the dependence of melt connectivity on the dihedral angle is shown in the lower diagrams. Note that an ideal case is shown here in which the crystals have an isotropic surface energy. Adapted from Stevenson DJ (1990) Fluid dynamics of core formation. In: Newsom HE and Jones JH (eds.) *Origin of the Earth*, pp. 231–250. New York: Oxford University Press.

the connection boundary and ranges from 0% to 4.5% for dihedral angles in the range 60–85°. The dependence of the pinch-off boundary on dihedral angle has been formulated theoretically as $0.009(\theta - 60)^{0.5}$ for values of θ in the range 60–100° (von Bargen and Waff, 1986). Yoshino *et al.* (2003) have shown, through electrical conductivity measurements, that the percolation threshold for Fe–S melt in an olivine matrix is approximately 5 vol.%. This result suggests that core formation could have occurred by percolation under hydrostatic conditions (see below) but $\sim$5 vol.% metal would have been stranded in the mantle. In fact, 5 vol.% may be a lower bound on the amount of stranded metal because the percolation threshold theory was developed assuming that the liquid is uniformly and finely dispersed throughout the crystalline matrix (von Bargen and Waff, 1986). However, due to the minimization of surface energy, textures might evolve over long time periods (relative to normal experimental timescales of hours to days) such that the liquid becomes concentrated in pools that are widely dispersed and relatively large, in which case the percolation threshold could be much greater than 5 vol.% (Stevenson, 1990; Walte *et al.*, 2007).

The effects of crystal anisotropy and crystal faceting are not taken into account by the above theory (which is based on surface energies being isotropic). It has been argued that the effects of crystal anisotropy and crystal faceting reduce permeability (Faul, 1997; Laporte and Watson, 1995; Yoshino *et al.*, 2006); however, there is no experimental evidence to suggest that such effects are significant for liquid metal–silicate systems relevant to core formation.

The dihedral angle θ depends on the energies of the respective interfaces that intersect at a triple junction that is occupied by a melt pocket (**Figure 5**):

$$\theta = 2\cos^{-1}\left(\frac{\gamma_{ss}}{2\gamma_{sl}}\right) \qquad [5]$$

where γ_{sl} is the solid–liquid interfacial energy and γ_{ss} is the solid–solid interfacial energy. Note that, as emphasized above, this expression is based on the assumption that interfacial energies are independent of crystal orientation and that stress is hydrostatic. When considering metal–silicate systems that are applicable to core formation, the interfacial energies, and therefore the dihedral angle, can be affected by (1) the structure and composition of the crystalline phase, (2) the structure and composition of the liquid metal alloy, and (3) temperature and pressure. Dihedral angles in metal–silicate systems relevant to core formation and the effects of the above variables have been

investigated experimentally in recent years (Ballhaus and Ellis, 1996; Minarik *et al.*, 1996; Shannon and Agee, 1996, 1998; Gaetani and Grove, 1999; Holzheid *et al.*, 2000a; Rose and Brenan, 2001; Takafuji *et al.*, 2004; Terasaki *et al.*, 2005, 2007a, 2007b). These studies, performed on a range of different starting materials at pressure–temperature conditions up to those of the lower mantle, now enable the most important factors that control dihedral angles in metal–silicate systems to be identified.

The effects of pressure, temperature, and the nature of the silicate crystalline phase appear to be relatively unimportant, at least up to $\sim$23 GPa. For example, through experiments on the Homestead meteorite, Shannon and Agee (1996) found that dihedral angles have an average value of 108° and remain essentially constant over the pressure range 2–20 GPa, irrespective of the dominant silicate mineral (e.g., olivine or ringwoodite). However, their subsequent study of the Homestead meteorite under lower-mantle conditions (25 GPa), suggests that dihedral angles decrease to $\sim$71° when silicate perovskite is the dominant silicate phase (Shannon and Agee, 1998).

The most important parameter controlling dihedral angles is evidently the anion (oxygen and/or sulfur) content of the liquid metal phase. The reason is that dissolved O and S act as 'surface-active' elements in the metallic melt and thus reduce the solid–liquid interfacial energy (e.g., Iida and Guthrie, 1988). Dissolved oxygen, in particular, makes the structure of the metal more compatible with that of the adjacent silicate, thus reducing the interfacial energy and promoting wetting.

In a recent review, Rushmer *et al.* (2000, figure 4) showed that dihedral angles decrease from 100–125° to 50–60° as the anion to cation ratio, defined as $(O + S)/(Fe + Ni + Co + Mn + Cr)$, increases from $\sim$0.3 to $\sim$1.2. Based on their data compilation, only metallic melts with the highest contents of O and S are wetting ($\theta < 60°$). Recently, the effects of the anion content of the metallic liquid were investigated systematically in the stability fields of olivine and ringwoodite by Terasaki *et al.* (2005) through a study of dihedral angles in the system Fe–S–$(Mg, Fe)_2SiO_4$ in the pressure range 2–20 GPa. By varying the FeO content of the silicate phase ($Fe\# = FeO/(FeO + MgO) = 0.01–0.44$), the oxygen fugacity, which controls the O content of the metal, could be varied over a wide range. They confirmed the importance of the anion content of the liquid phase and showed that the effect of dissolved oxygen is greater than that of dissolved S (**Figure 6**). Because

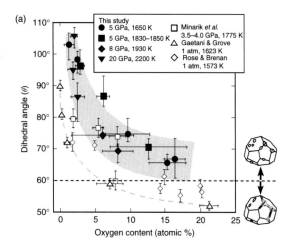

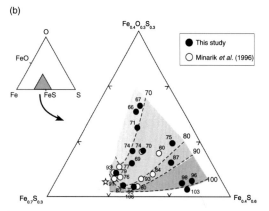

Figure 6 Dihedral angles in aggregates of olivine ($\leq$8 GPa) and ringwoodite (20 GPa) containing several vol.% Fe–FeS melt as a function of the melt composition. (a) Dihedral angle as a function of the oxygen content of the Fe-alloy liquid. As the oxygen content increases, the structures of the silicate and liquid Fe alloy become more similar with the result that both the interfacial energy and the dihedral angle decrease. (b) Effects of O and S contents of the Fe-alloy liquid on dihedral angles. The data points are plotted on a triangular section of the Fe–O–S system and each data point is labeled with the dihedral angle. The dashed lines are contours of constant dihedral angle. These results suggest that the effect of dissolved O on the dihedral angle is greater than that of dissolved S. Reproduced from Terasaki H, Frost DJ, Rubie DC, and Langenhorst F (2005) The effect of oxygen and sulphur on the dihedral angle between Fe–O–S melt and silicate minerals at high pressure: Implications for Martian core formation. *Earth and Planetary Science Letters* 232: 379–392 (doi:10.1016/j.epsl.2005.01.030), with permission from Elsevier.

the oxygen content of the metallic liquid decreases with pressure up to 10–15 GPa (Rubie *et al.*, 2004; Asahara *et al.*, 2007), dihedral angles increase with increasing pressure in this range. Terasaki *et al.* (2007a) showed that for metallic liquids with a high

O + S content, dihedral angles lie below 60° at pressures below 2–3 GPa, depending on the FeO content of the silicate phase. Percolation could therefore contribute significantly to core formation in planetesimals during the early stages of heating before the temperatures reach the silicate solidus.

Based on a recent study of liquid Fe + silicate perovskite at 27–47 GPa and 2400–3000 K, Takafuji *et al.* (2004) have suggested that dihedral angles decrease to ~51° under deep-mantle conditions. This result may be consistent with recent results which show that the solubility of oxygen in liquid Fe increases strongly with temperature and weakly with pressure above 10–15 GPa (Asahara *et al.*, 2007). However, the study of Takafuji *et al.* (2004) was performed using a laser-heated diamond anvil cell (LH-DAC), in which samples are exceedingly small and temperature gradients are very high. Furthermore, dihedral angles had to be measured using transmission electron microscopy – which is far from ideal for obtaining good statistics. This preliminary result there awaits confirmation from further studies. In addition, in a study of wetting in a similar system at the lower pressures of 25 GPa, Terasaki *et al.* (2007b) found that the dihedral angle increases with the FeSiO$_3$ component of silicate perovskite but failed to find any obvious correlation with the oxygen content of the metal liquid.

In summary, experimental results for systems under hydrostatic stress show that dihedral angles significantly exceed the critical angle of 60° at pressures of 3–25 GPa in chemical systems that are relevant for core formation in terrestrial planets. This means that, for percolation under static conditions, at least several vol.% metal would have been stranded in the mantle – which is inconsistent with the current concentration of siderophile elements in the mantle (see Section 3.3.2). Efficient percolation can occur at low pressures (<3 GPa) when the S and O contents of the metal are very high (i.e., close to the Fe–S eutectic) but such conditions are not applicable to core formation in a large planet such as the Earth. There is a preliminary indication that efficient percolation ($\theta < 60°$) may be a feasible mechanism under deep lower-mantle conditions but this result awaits confirmation.

When dihedral angles significantly exceed 60°, experimental evidence suggests that liquid metal can separate from crystalline silicates when the material is undergoing shear deformation due to nonhydrostatic stress (Bruhn *et al.*, 2000; Rushmer *et al.*, 2000; Groebner and Kohlstedt, 2006; Hustoft and Kohlstedt, 2006). The most recent results indicate that ~1 vol.% liquid metal remains stranded in the silicate matrix so that reasonably efficient percolation of liquid metal, with percolation velocities on the order of $150 \, km \, yr^{-1}$, might occur in crystalline mantle that is undergoing solid-state convection (Hustoft and Kohlstedt, 2006). There are, however, two potential problems with this shear-induced percolation mechanism. First, experiments have been performed at high strain rates (10^{-2} to $10^{-5} \, s^{-1}$) and it is uncertain if the mechanism would also be effective at much lower strain rates. This is a consequence of the fact that at high strain rates the solid crystals do not deform appreciably and so can generate pore space by dilatant expansion of the crystal–melt mixture. At much lower strain rates the crystals can accommodate shear by deformation and so may greatly reduce the interconnections available for liquid-phase percolation. Second, the conversion of potential energy to heat during percolation of iron liquid in a planet the size of Earth might be sufficient to melt the silicates and thus change the mechanism to the one discussed in the next section (see Section 3.2.2.3).

3.2.3.2   Metal–silicate separation in a magma ocean

According to the results of calculations of the energy released by giant impacts, it is clear that the collision of a Mars-sized body with the proto-Earth would have resulted in the melting of a large part of, or even the entire planet (Section 3.2.2.2). In the case of partial melting, although the distribution of melt may initially have been concentrated on the side affected by the impact, isostatic readjustment would have led rapidly to the formation of a global magma ocean of approximately uniform depth (Tonks and Melosh, 1993). An important, and presently unresolved, question is whether this isostatic adjustment takes place on a timescale longer or shorter than that of iron separation. The mechanics of separation and chemical equilibration is quite different if the isostatic adjustment is much slower than iron separation, because then the melt region is not a global ocean of approximately uniform depth, but a restricted 'sea' that is both hotter and deeper than the ocean that eventually develops. This question is further raised below in Section 3.3.2.3.

Because of the large density difference between liquid iron and liquid silicate, magma ocean formation provides a rapid and efficient mechanism for the separation of these two phases. Here we examine the

physics of metal–silicate separation in some detail because of the consequences for interpreting mantle geochemistry, as discussed below in Section 3.3.2.

A fundamental property that controls the dynamic behavior of a deep magma ocean is the viscosity of ultramafic silicate liquid. The viscosity of peridotite liquid has been determined at 2043–2523 K and 2.8–13.0 GPa by Liebske *et al.* (2005). Their results (**Figure 7**) show that viscosity increases with pressure up to 9–10 GPa and then decreases to at least 13 GPa (Reid *et al.*, 2003, found a similar trend for $CaMgSi_2O_6$ liquid). Based on a study of the self-diffusion of O and Si in silicate liquid, viscosity is expected to increase again at pressures above 18 GPa (Schmickler *et al.*, 2005). The transient decrease in viscosity in the pressure range 9–18 GPa may be caused by pressure-induced coordination changes (e.g., formation of fivefold and sixfold coordinated Si) in the melt structure (Liebske *et al.* 2005).

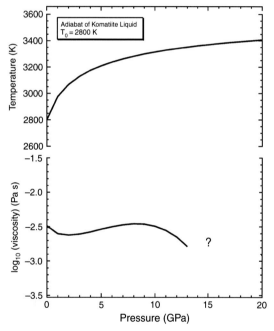

Figure 7   The viscosity of the upper part of a peridotitic magma ocean that has a total depth of ~1800 km (bottom). The viscosity is shown only to a depth of ~400 km because the existing experimental data cannot be extrapolated reliably to higher pressures. Based on results of Schmickler *et al.* (2005), the viscosity likely increases again above ~18 GPa. The viscosity profile is based on the adiabat shown in the top part of the figure. Reproduced from Liebske C, Schmickler B, Terasaki H, *et al.* (2005) Viscosity of peridotite liquid up to 13 GPa: Implications for magma ocean viscosities. *Earth and Planetary Science Letters* 240: 589–604, with permission from Elsevier.

Unfortunately, it is currently not possible to extrapolate these experimental results reliably in order to make predictions of melt viscosities at pressures significantly higher than 18 GPa; the large uncertainties involved in doing this are illustrated by Liebske *et al.* (2005, figure 7).

On the basis of the experimental data of Liebske *et al.* (2005), the viscosity of a magma ocean, at least to a depth of ~500 km, is estimated to lie in the range 0.01–0.003 Pa s, which is extremely low (for comparison, the viscosity of water at ambient conditions is 0.001 Pa s). Consequently, the Rayleigh number (which provides a measure of the vigor of convection) is extraordinarily high, on the order of 10^{27}–10^{32} (e.g., Solomatov 2000; Rubie *et al.*, 2003; *see* Chapter 4). This means that a deep magma ocean undergoes vigorous turbulent convection, with convection velocities on the order of at least a few meters per second. Using a simple parametrized convection model, the rate of heat loss can also be estimated, which leads to the conclusion that, in the absence of other effects (see below), the life time of a deep magma ocean on Earth is only a few thousand years (Solomatov, 2000).

What is the physical state of molten iron in a vigorously convecting magma ocean? Initially, iron metal may be present in states that range from finely dispersed submillimeter particles (as in undifferentiated chondritic material) to large masses that originated as cores of previously differentiated bodies (ranging in size from planetesimals to Mars-sized planets) that impacted the accreting Earth. In a molten system, very small particles tend to grow in size by coalescing with each other in order to reduce surface energy. Large molten bodies, on the other hand, are unstable as they settle and tend to break up into smaller bodies. A crucial question concerns the extent to which an impacted core breaks up and becomes emulsified as it travels through the target's molten mantle (see Stevenson (1990), Karato and Murthy (1997), and Rubie *et al.* (2003) for discussions of this issue). Hallworth *et al.* (1993) noted that laboratory-scale turbidity currents travel only a few times their initial dimension before being dispersed by turbulent instabilities). Such a core will experience both shear (Kelvin–Helmholtz) and buoyancy (Rayleigh–Taylor) instabilities. These processes operate at different length scales (R–T instabilities are small-scale features; see Dalziel *et al.*, 1999), but both processes will tend to break the body up until a stable droplet size is reached at which surface tension inhibits further break-up. The stable droplet size can

be predicted using the dimensionless Weber number, which is defined as

$$W_e = \frac{(\rho_m - \rho_s)dv_s^2}{\sigma} \qquad [6]$$

where ρ_m and ρ_s are the densities of metal and silicate respectively, d is the diameter of metal bodies, v_s is the settling velocity, and σ is the surface energy of the metal–silicate interface (Young, 1965). A balance between coalescence and breakup is reached when the value of W_e is approximately 10: when the value is larger than 10, instabilities cause further breakup to occur and when it is less than 10, coalescence occurs. The settling velocity v_s is determined using Stokes' law when the flow regime is lamellar or an equation that incorporates a drag coefficient when the flow around the falling droplet is turbulent (Rubie *et al.*, 2003). Both the settling velocity and the droplet size depend on silicate melt viscosity. For likely magma ocean viscosities and assuming a metal–silicate surface energy of $1\,N\,m^{-1}$ (which is not well constrained – see Stevenson 1990), the droplet diameter is estimated to be ~1 cm and the settling velocity ~0.5 m s^{-1} (**Figure 8**; Rubie *et al.*, 2003, see also Stevenson (1990) and Karato and Murthy (1997)).

Having estimated the stable size of metal droplets, the next question is how quickly does a large mass of metal (e.g., 50–500 km in diameter) become emulsified and break up into a 'rain' of small droplets of stable diameter? Although the process is currently not well understood, Rubie *et al.* (2003) argued that emulsification should occur within a falling distance equal to a few times the original diameter of the body. Thus, the cores of all, except perhaps the largest, impacting bodies probably experienced a very large degree of emulsification.

The primary importance of emulsification is that it determines the degree to which chemical and thermal re-equilibration occurs (Karato and Murthy, 1997; Rubie *et al.*, 2003). Because thermal diffusivities are higher than chemical diffusivities, thermal equilibrium is always reached first. For typical settling velocities, iron blobs on the order of ~0.01 m in diameter will remain in chemical equilibration with the surrounding silicate liquid as they fall (Rubie *et al.*, 2003). Droplets that are much larger (e.g., on the order of meters or more in diameter) will fall rapidly and will not remain in equilibrium with the adjacent silicate because diffusion distances are too large. The physical arguments for emulsification summarized here are supported by evidence for chemical equilibration from both Hf–W observations

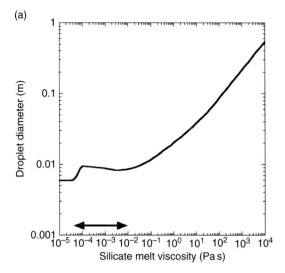

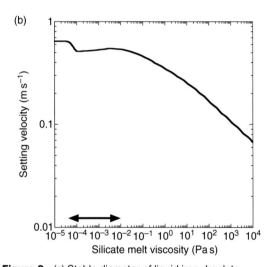

Figure 8 (a) Stable diameter of liquid iron droplets dispersed in a magma ocean as a function of silicate melt viscosity. The droplet diameter is calculated using the Weber number, as explained in the text. (b) Terminal settling velocity of liquid Fe droplets (of stable droplet diameter) as a function of silicate melt viscosity. The nonlinear trend arises from a transition from Stokes to turbulent flow at low viscosities. In both (a) and (b), the arrowed lines indicate the range of likely magma ocean viscosities. Reproduced from Rubie DC, Melosh HJ, Reid JE, Liebske C, and Righter K (2003) Mechanisms of metal-silicate equilibration in the terrestrial magma ocean. *Earth and Planetary Science Letters* 205: 239–255, with permission from Elsevier.

(Section 3.3.1) and siderophile element abundances in the mantle (Section 3.3.2). Since chemical equilibration of macroscale iron bodies is very slow, this apparent equilibration strongly suggests that the iron

was present, at least to a large extent, as small dispersed droplets (Righter and Drake, 2003).

Because likely settling velocities ($\sim 0.5\,\mathrm{m\,s^{-1}}$) of small iron droplets are much lower than typical convection velocities ($\sim 10\,\mathrm{m\,s^{-1}}$), iron droplets may remained entrained for a significant time in the magma ocean and accumulation through sedimentation at the base of the ocean will be a slow and gradual process. The dynamics of the settling and accumulation processes are important because they determine the chemical consequences of core formation (e.g., siderophile element geochemistry), as discussed below in Section 3.3.2.3.

The time taken for a magma ocean to start to crystallize is an important parameter when evaluating metal–silicate equilibrium models, as discussed below (Section 3.3.2.3). The existence of an early, thick atmosphere has little effect on large incoming impactors, but may be sufficiently insulating that, by itself, it ensures a magma ocean (e.g., Matsui and Abe, 1986). The depth to the (rheologically determined) base of the magma ocean is determined by the point at which the adiabat crosses the geotherm defining a melt fraction of roughly 60% (Solomatov, 2000). The survival time of the magma ocean depends on both the atmosphere and whether or not an insulating lid can develop. In the absence of these two effects, the lifetimes are very short, of order 10^3 years (e.g., Solomatov, 2000; Pritchard and Stevenson, 2000). However, if a conductive lid develops, the lifetime may be much longer, of order 10^8 years (Spohn and Schubert, 1991), and similar lifetimes can arise due to a thick atmosphere (Abe, 1997). Thus the lifetime of magma oceans is currently very unclear. The Moon evidently developed a chemically buoyant, insulating crust on top of its magma ocean (Warren, 1985). However, it did so because at low pressures aluminium partitions into low-density phases, especially plagioclase. At higher pressures, Al instead partitions into dense garnet, in which case a chemically buoyant crust will not develop (e.g., Elkins-Tanton *et al.*, 2003). In the absence of chemical buoyancy, a solid crust will still develop, but will be vulnerable to disruption by impacts or foundering (Stevenson, 1989). The latter process in particular is currently very poorly understood, and thus the lifetime of magma oceans remains an open question. Fortunately, even the short-lived magma oceans persist for timescales long compared to most other processes of interest.

3.2.3.3 Diapirs and dyking

If percolation is not an effective mechanism, then differentiation may occur either by downwards migration of large iron blobs (diapirism) or by propagation of iron-filled fractures (dyking).

The transport of iron though crystalline mantle as large diapirs, 1–10 km in diameter or larger, has been discussed in detail by Karato and Murthy (1997). When liquid iron ponds as a layer at the base of a magma ocean (**Figure 9**), gravitational instabilities develop due to the density contrast with the underlying silicate-rich material and cause diapir formation. Their size and rate of descent through the mantle depend on the initial thickness of the metal layer and the viscosity of the silicate mantle. Clearly gravitational heating will be important and will facilitate diapir descent by reducing the viscosity of the adjacent mantle. In contrast to magma ocean segregation, there will be no significant chemical exchange between metal and silicate, chemical disequilibrium will result and siderophile element abundances in the mantle cannot be a consequence of this mechanism (Karato and Murthy, 1997).

Liquid iron ponded at the base of the magma ocean may also, under the right conditions, sink rapidly toward the Earth's core by dyking. Although it may be supposed that the hot, but nevertheless crystalline, mantle underlying the magma ocean

cannot support brittle cracks, numerical studies summarized in Rubin (1995) indicate that dykes can still form, so long as the contrast in viscosity between the fluid in the dyke and the surrounding host rocks is greater than 10^{11}–10^{14}. With a viscosity in the neighborhood of 10^{-2} Pa s, liquid iron is thus expected to form dykes if the viscosity of the host rock exceeds 10^{9}–10^{12} Pa s. Given that the viscosity of the asthenosphere today is around 10^{19} Pa s, it is not unreasonable to expect the iron to reach the core via narrow dikes rather than as diapirs. In this case even less time is required for the rapidly descending iron to reach the core and thus less time for the iron to chemically equilibrate with the surrounding mantle. Indeed, even in the present Earth, Stevenson (2003) has proposed that masses of molten iron as small as 10^{8} kg (which would fill a cube about 25 m on a side) could travel from the Earth's surface to the core in about 1 week.

3.2.3.4 Summary and implications for chemical equilibration

A schematic illustration of how the various differentiation mechanisms might operate together is shown in **Figure 9**. Liquid metal separates rapidly from liquid silicate in a deep magma ocean and accumulates as ponded layers at the rheological base of the magma ocean. The ponded iron then migrates through the largely crystalline underlying mantle

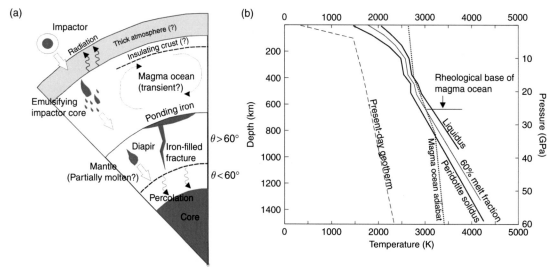

Figure 9 (a) Schematic diagram showing the various processes of metal–silicate segregation that could have operated during Earth accretion and core formation. The rheological base of the magma ocean is defined as the point at which the melt fraction drops below ~60% (Solomatov, 2000). (b) Temperature structure in the early Earth. The solidus curve in the lower mantle is uncertain (see Boehler, 2000), and the 60% melt fraction line is schematic, (b) Adapted from Walter MJ and Trønnes RG (2004) Early Earth differentiation. *Earth and Planetary Science Letters* 225: 253–269.

towards the proto-core by either percolation, diapirism, or dyking. According to experimental results summarized above, percolation is unlikely to be a completely efficient mechanism, even when catalyzed by shear deformation, but the mechanism may at least contribute to core formation. In addition, R–T instabilities develop at the base of the ponded layer and result in diapir formation. Given a sufficiently large viscosity contrast between liquid iron and the early silicate mantle, the dyking mechanism could also be important.

Despite the rapid transit times associated with falling iron drops and the somewhat longer timescales of percolative flow, the inferred length scales are small enough that complete chemical equilibration is expected. Conversely, descending iron diapirs are sufficiently large that chemical equilibrium is expected to be negligible (Karato and Murthy, 1997). Thus, the different differentiation mechanisms have very different chemical consequences. However, the magnitude of these chemical effects also depends on the relative abundances: a late passage of 1% core material through the mantle may well have a strong effect on mantle siderophile element abundances (e.g., W or Pt), but will have little effect on major element concentrations (e.g., oxygen) simply because the core material will become saturated and thus transport insignificant amounts of these more-abundant elements.

3.3 Observational and Experimental Constraints

Having discussed our theoretical expectations of accretion and core-formation processes, we will now go on to discuss the extent to which observations and experimental data may be used to differentiate between the various theoretical possibilities. Excellent summaries of many of these observations and experimental results may be found in Halliday (2003) and in the volume edited by Canup and Righter (2000).

3.3.1 Core-Formation Timescales

An extremely important development in recent years has been the recognition that some isotopic systems provide an observational constraint on core-formation timescales, and thus planetary accretion rates. The most useful isotopic system is Hf–W (Harper and Jacobsen, 1996; Kleine *et al.*, 2002; Schoenberg

et al., 2002; Yin *et al.*, 2002; Halliday, 2004; Jacobsen, 2005; note that Hf–W measurements published prior to 2002 were erroneous and led to conclusions that are now considered to be incorrect). The U–Pb system is more problematic, but generates results which can be reconciled with the more robust Hf–W technique (Wood and Halliday, 2005).

The Hf–W chronometer works as follows. W is siderophile (i.e., 'metal loving'), while Hf is lithophile (it remains in silicates). Furthermore, ^{182}Hf decays to stable ^{182}W with a half-life of 9 My. If an initially undifferentiated object suddenly forms a core after all the ^{182}Hf has decayed, the W will be extracted into the core and the mantle will be strongly depleted in all tungsten isotopes. However, if core formation occurs early, while ^{182}Hf is live, then the subsequent decay of ^{182}Hf to ^{182}W will enrich the mantle in radiogenic tungsten compared with nonradiogenic tungsten. Thus, a radiogenic tungsten excess, or tungsten anomaly, in the mantle is a sign of early core formation. Furthermore, if the silicate:iron mass ratio and the mantle concentrations of Hf and W compared with undifferentiated materials (chondrites) are known, then the observed tungsten anomaly can be used to infer a single-stage core-formation age. In the case of the Earth, this single-stage age is roughly 30 My (Jacobsen, 2005), while the Hf–W age of the Moon suggests that the last giant impact experienced by the Earth occurred at 30–50 My (Halliday, 2004; Kleine *et al.*, 2005).

There are three characteristics of the Hf–W system which makes it especially suitable for examining core formation. First, the half-life is comparable to the timescale over which planets are expected to form. Second, there are few other processes likely to lead to tungsten fractionation and perturbation of the isotopic system, though very early crustal formation or neutron capture (Kleine *et al.*, 2005) can have effects. Finally, both Hf and W are refractory; certain other isotopic systems suffer from the fact that one or more elements (e.g., lead) are volatile and can be easily lost during accretion.

The fact that tungsten isotope anomalies exist in the terrestrial mantle imply that core formation was essentially complete before about five half-lives (50 My) had elapsed. Mars and Vesta have larger tungsten anomalies, indicating that core formation ended earlier on these smaller bodies (Kleine *et al.*, 2002). The timescales implied are compatible with the theoretical picture of planetary accretion described in Section 3.2.1. The simple model of a single core-formation event is of course a simplification of the real picture, in which the bulk of the core mass is added during stochastic giant impacts. However, more complicated models, in which the mass is added in a series of discrete events, do not substantially alter the overall timescale derived.

The observed tungsten anomaly depends mainly on the timescale over which the core forms, the relative affinities of Hf and W for silicates, and the extent to which the cores of the impactors re-equilibrate with the target mantle. The relative affinities of Hf and W can be determined, in a time-averaged sense, by measuring the present-day concentrations of these elements in the mantle. These affinities (i.e., the partition coefficients) may have varied with time, due to changing conditions (P, T, oxygen fugacity f_{O_2}) in the Earth. Although the dependence of the partition coefficients on these variables is known (e.g., Righter, 2003), how conditions actually evolved as the Earth grew is very poorly understood (e.g., Halliday, 2004). This caveat aside, if one accepts that the accretion timescales determined by numerical accretion models are reasonable, these models can then be used to investigate the extent to which re-equilibration must have occurred.

Figure 10 shows examples of the tungsten anomalies generated from numerical models of late-stage accretion (Nimmo and Agnor 2006). **Figure 10(b)** assumes that undifferentiated bodies undergo differentiation on impact, and that the bodies' cores then merge without any re-equilibration. In this case, the

tungsten anomaly of a body is set by the mass-weighted average of the anomalies generated when each constituent planetesimal differentiated. Bodies made up of early colliding planetesimals tend to be bigger, and also have higher tungsten anomalies. The tungsten anomalies generated for Earth-mass bodies are much larger than those actually observed. **Figure 10(a)** shows results from the same accretion simulation, but now assuming that during each impact the core of the impactor re-equilibrates with the mantle of the target. This re-equilibration drives down the tungsten anomaly during each impact, and results in lower tungsten anomalies for large bodies than for small ones. The measured tungsten anomalies of Earth, Mars, and the HED parent body (probably Vesta) are all compatible with this mantle re-equilibration scenario. Thus, assuming that the accretion timescales generated by the simulations are correct, the Hf–W data suggest that even the largest impacts result in complete or near-complete equilibration of the impactor's core with the target's mantle. This conclusion in turn places constraints on the physics of these very large impacts and, in particular, suggests that emulsification of the impacting core occurs as it travels through the magma ocean (see also Section 3.2.3.2).

Although **Figure 10** only models the late stages of accretion, this is the stage when most of the Earth's mass is added. If re-equilibration occurs, earlier isotopic signatures will be overprinted. However, if core merging takes place, the overall signature will be set

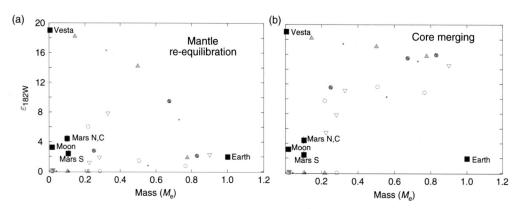

Figure 10 Isotopic outcomes of core formation based on N-body accretion codes. M_e is the mass of the final bodies in Earth masses, ε_{182W} is the final tungsten anomaly. Black squares are the observed values, tabulated in Jacobsen (2005) (Mars has two values because different meteorite classes give different answers). Colored symbols are the model results. Differentiation is assumed to occur when a body first collides with another object. (a) Outcome assuming that the impactor core re-equilibrates with the mantle of the target (a scenario favored if emulsification occurs). (b) Outcome assuming that the impactor core merges with the target core without any re-equilibration. Reproduced from Nimmo F and Agnor CB (2006) Isotopic outcomes of N-body accretion simulations: Constraints on equilibration processes during large impacts from Hf-W observations. *Earth and Planetary Science Letters* 243: 26–43, with permission from Elsevier.

by the time that the accreting objects differentiate. Since it appears that at least some bodies have differentiated very early and, in the case of Vesta, have correspondingly high tungsten anomalies (Kleine *et al.*, 2002), **Figure 10(b)** probably underestimates the tungsten anomalies that would result if re-equilibration did not occur.

Finally, because the early history of the core and the mantle are intimately coupled, there are some isotopic systems governed by mantle processes which are also relevant to core-formation timescales. In particular, the nonchondritic ^{142}Nd isotope signature of the Earth's upper mantle has been used to argue for global melting of the mantle within 30 My of solar system formation (Boyet and Carlson, 2005). This constraint is entirely consistent with the core-formation timescale derived above; however, Ba isotope measurements have been used to argue that the Earth as a whole is not chondritic (Ranen and Jacobsen, 2006), thus making the ^{142}Nd measurements more difficult to interpret. Sm–Nd and Lu–Hf chronometers are also consistent with the solidification of a magma ocean with the first ~100 Myr of Earth's history (Caro *et al.*, 2005), and U–Pb dates have been interpreted as resulting from the final stage of magma ocean crystallization at about 80 Myr after solar system formation (Wood and Halliday, 2005). Xe-isotope data give a comparable time for loss of xenon from the mantle (e.g., Porcelli *et al.*, 2001; Halliday, 2003).

In summary, the Hf–W system is important for two reasons: it constrains the timescale over which core formation occurred, and it also constrains the extent of re-equilibration between core and mantle material. To obtain the pressure–temperature conditions under which this re-equilibration took place, it is necessary to look at other siderophile elements as well. Inferring the equilibration conditions is important both because constraints are placed on the early thermal state of both core and mantle, and because these conditions strongly influence the ultimate composition of the core.

3.3.2 Constraints from Siderophile Element Geochemistry

3.3.2.1 Introduction to siderophile element geochemistry

The primary geochemical evidence for core formation in the Earth is provided by a comparison of the composition of the Earth's silicate mantle with its bulk composition. Estimates of the composition of

the mantle are based on numerous geochemical studies of mantle peridotites and are well established (e.g., McDonough and Sun, 1995; Palme and O'Neill, 2003), assuming of course that the mantle is homogeneous. Although the bulk composition of the Earth is not known precisely (O'Neill and Palme, 1998; Drake and Righter, 2002), it is often approximated by the composition of C1 carbonaceous chondrites, which are the most pristine (undifferentiated) relicts known from the early solar system. As seen in **Figure 11**, refractory lithophile elements (e.g., Al, Ca, Ti, Ta, Zr, and the rare earth elements, REE's) are present in the mantle in C1 chondritic concentrations, and their concentrations have therefore been unaffected by accretion or differentiation processes. Compared with the bulk composition of the Earth, the mantle is strongly depleted in (1) siderophile (metal-loving) elements that have partitioned into iron-rich metal during formation of the Earth's core (Walter *et al.*, 2000) and (2) volatile elements that are considered to have been partly lost during accretion. Note that some of the volatile elements are also siderophile (e.g., sulfur) so that current mantle concentrations can be the result of both core formation and volatility (for a full classification of depleted elements, see Walter *et al.*, 2000, **Table 1**). Siderophile elements that are unaffected by volatility are most valuable for understanding core formation and these include the moderately siderophile elements (MSEs) (e.g., Fe, Ni, Co, W, and Mo) and the highly-siderophile elements (HSEs), which include the platinum group elements (PGE's, e.g. Re, Ru, Rh, Os, Ir, Pd, Pt, and Au).

The degree of siderophile behavior is described for element M by the metal–silicate partition coefficient $D_M^{met-sil}$ which is defined as

$$D_M^{met-sil} = \frac{C_M^{met}}{C_M^{sil}} \qquad [7]$$

where C_M^{met} and C_M^{sil} are the wt.% concentrations of M in metal and silicate, respectively. The MSEs are defined as having values (determined experimentally at 1 bar) of $D_M^{met-sil} < 10^4$, whereas HSEs have 1-bar values that are greater than 10^4 and can be, in the case of Ir for example, as high as 10^{10}. The boundary between siderophile and lithophile behavior is defined as $D_M^{met-sil} = 1$.

As discussed below, partition coefficients are a function of pressure and temperature. An additional controlling parameter, which is critical in the context of core formation, is oxygen fugacity, f_{O_2}. The effect

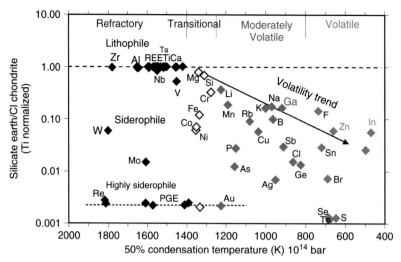

Figure 11 Element abundances of the Earth's mantle normalized to CI chondrite and the refractory element Ti (data after Palme and O'Neill, 2003) plotted against their 50% condensation temperatures as given by Wasson (1985). Note that the errors on these data points may be large (e.g. at least an order of magnitude for some volatile elements). Courtesy of Ute Mann.

of f_{O_2} on the partitioning of a siderophile element depends on its valence state when it is dissolved in the silicate phase, as can be understood by considering the following metal–silicate equilibrium:

$$M + (n/4)O_2 = MO_{n/2}$$

metal   silicate liquid

(e.g., Righter and Drake, 2003). Here M is the element of interest and n is its valence when it is dissolved in silicate liquid. Increasing the oxygen fugacity drives the equilibrium toward the right-hand side, thus increasing the concentration of M in the silicate liquid and reducing the value of $D_M^{met-sil}$. Typically, the dependence of $D_M^{met-sil}$ on f_{O_2}, pressure (P), and temperature (T) is expressed by a relationship of the form

$$\ln D_M^{met-sil} = a \ln fO_2 + b/T + cP/T + g \quad [8]$$

where a is related to the valence state of M in the silicate liquid and b, c, and g are related to thermodynamic free energy terms which have generally assumed to be constants even when the P–T range of extrapolation is large (see Righter and Drake (2003)). In reality, these parameters are not likely to be constant over large ranges of pressure, temperature, and f_{O_2}. For example, the valence state of an element can change as a function of f_{O_2}. Pressure-induced changes in silicate melt structure may also cause a strong nonlinear pressure dependence at certain conditions (Keppler and Rubie, 1993; Kegler et al., 2005). In addition to P, T, and

f_{O_2}, the effects of additional factors, such as the composition/structure of the silicate melt and sulfur and carbon contents of the metal, may also need to be included empirically (Righter and Drake, 2003; Walter et al., 2000).

For core formation, the value of f_{O_2} is constrained by the partitioning of Fe between the mantle and core and is estimated to have been 1 to 2 log units below the oxygen fugacity defined by the iron–wüstite (Fe-FeO or 'IW') buffer (which is abbreviated as IW-1 to IW-2).

Values of $D_M^{met-sil}$ for the core–mantle system lie in the range 13–30 for MSEs and 600–1000 for HSEs (for compilations of values, see Wade and Wood, 2005; Wood et al., 2006). These values are much lower than experimentally determined 1-bar metal–silicate partition coefficients, in some cases by a few orders of magnitude, and show that the mantle contains an apparent 'overabundance' of siderophile elements. Possible explanations of this anomaly, the so-called 'excess siderophile element problem', form the basis of theories of how and under which conditions the Earth's core formed (e.g., Newsom, 1990; Righter, 2003), as described further in Section 3.3.2.2. An important additional constraint is provided by the observation that the HSEs are present in the mantle in approximately chondritic relative abundances (**Figure 12**), even though the experimentally determined 1-bar $D_M^{met-sil}$ values vary by orders of magnitude.

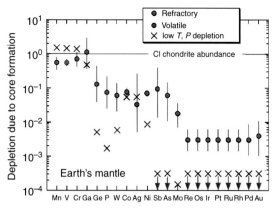

Figure 12 Depletion of siderophile elements in the Earth's mantle as the result of core formation. The siderophile elements are distinguished as being either refractory or volatile. The crosses show depletion values that would be expected on the basis of partitioning experiments performed at one bar and moderate temperatures (e.g., 1300°C); in the case of the highly-siderophile elements, the predicted values plot far below the bottom of the graph and are highly variable. The large discrepancies between the observed and calculated values shown here form the basis for the 'siderophile element anomaly'. For further details see Walter et al., (2000). After Walter MJ, Newsom HE, Ertel W, and Holzheid A (2000). Siderophile elements in the Earth and Moon: Metal/silicate partitioning and implications for core formation. Reprinted from Canup RM and Righter K (eds.) Origin of the Earth and Moon, pp. 265–290. Tucson, AZ: University of Arizona Press; Courtesy of Michael Walter, with permission of the University of Arizona Press.

Important constraints on how the Earth's core formed are potentially provided by also considering the geochemical consequences of core formation in other planetary bodies. If the sizes of such bodies differ significantly from that of the Earth, pressure–temperature conditions of metal–silicate equilibration will also be different, thus enabling some of the theories detailed below (Section 3.3.2.2) to be tested. Estimates of the geochemical effects of core formation in Mars, Vesta, and the Earth's Moon are also based on comparisons of siderophile element concentrations in the mantles of these bodies with their likely bulk compositions (e.g., Treiman et al., 1987; Righter and Drake, 1996; Walter et al., 2000). Both bulk and mantle compositions are poorly constrained compared with the Earth. Mantle compositions are inferred from studies of SNC meteorites for Mars, the eucritic meteorites for Vesta and samples collected from the Moon's surface (Warren 1993; McSween, 1999). In all cases, the samples (e.g., lavas and cumulates) are the products of crustal differentiation processes and therefore mantle compositions

have to be inferred by taking the effects of differentiation into account. In the case of the Moon, there is an additional problem because the petrogenesis of samples is poorly constrained because of their extremely small size (Warren, 1993).

Compared to the Earth's mantle, current assessments indicate that the Martian mantle is relatively depleted in Ni and Co, whereas concentrations of HSE's are similar. As in the Earth's mantle, the HSEs appear to be present in chondritic relative abundances (Kong et al., 1999; Warren et al., 1999; McSween, 1999; Jones et al., 2003). The mantles of both the Moon and Vesta show relatively large depletions in most siderophile elements (Righter and Drake, 1996; Walter et al., 2000). It is not possible in any of these cases to explain mantle siderophile element abundances by simple metal–silicate equilibrium at moderate pressures and temperatures, that is, based on 1-bar experimental data.

3.3.2.2 Core formation/accretion models

As described in the previous section, siderophile element concentrations are depleted in the Earth's mantle relative to chondritic compositions as a consequence of core formation (**Figures 11** and **12**). However, compared with predicted depletions based on element partitioning studies at 1 bar and moderate temperatures (e.g., 1300–1400°C) the concentrations of siderophile elements are too high (**Figure 12**; Wood et al., 2006, Table 1). In the case of MSEs, the discrepancies are around 1–2 orders of magnitude. In the case of the HSEs, the discrepancies are on the order of 5–10 orders of magnitude.

The apparent overabundance of siderophile elements in the mantle has led to a number of core-formation hypotheses, most notably:

- metal–silicate equilibration at high pressures and temperatures,
- the late-veneer hypothesis,
- inefficient core formation, and
- addition of core material to the lower mantle.

We briefly review each of these hypotheses in turn.

3.3.2.2.(i) Metal–silicate equilibration at high pressure and temperature Murthy (1991) proposed that mantle siderophile element abundances could be explained if temperatures during core formation were extremely high. Although there were significant problems with the thermodynamic arguments on which this suggestion was based

(Jones *et al.*, 1992; O'Neill, 1992), metal–silicate equilibration at combined high temperatures and pressures may provide the explanation for at least the MSE abundances. This topic has been reviewed recently in detail by Walter *et al.* (2000), Righter and Drake (2003), and Wood *et al.* (2006).

Preliminary high-pressure studies produced results that were inconclusive in determining whether metal–silicate equilibration can explain the siderophile element anomaly (Walker *et al.*, 1993; Hillgren *et al.*, 1994; Thibault and Walter, 1995; Walter and Thibault, 1995). However, an early study of the partitioning of Ni and Co between liquid Fe alloy and silicate melt suggested that the metal–silicate partition coefficients for these elements reach values that are consistent with mantle abundances at a pressure of 25–30 GPa (**Figure 13(a)**; Li and Agee, 1996). This important result led to the idea (which has been disputed recently by Kegler *et al.*, 2005) that metal–silicate equilibration at the base of a magma ocean ~800 km deep can explain the mantle abundances of at least the MSEs (Li and Agee, 1996; Righter *et al.*, 1997). Since the late 1990s, there have been numerous subsequent studies of the partitioning of the MSEs (e.g., Li and Agee, 2001; O'Neill *et al.*, 1998; Gessmann and Rubie, 1998, 2000; Tschauner *et al.*, 1999; Righter and Drake, 1999, 2000, 2001; Chabot and Agee, 2003; Bouhifd and Jephcoat, 2003; Chabot *et al.*, 2005; Wade and Wood, 2005; Kegler *et al.*, 2005). A growing consensus emerging from such studies is that the pressures and temperatures required for metal–silicate equilibration may have been considerably higher than originally suggested. Estimated conditions are quite variable and range up to >4000 K and 60 GPa (**Table 3**; **Figure 14**). One of the reasons for the large scatter of P–T estimates is that, based on the current experimental data set and the associated uncertainties, a wide range of P–T–f_{O_2} conditions can satisfy mantle abundances of the MSEs (**Figure 14**). In addition, the difficulty of identifying a unique set of conditions is hardly surprising considering that core formation occurred over a protracted time period during accretion as the likely consequence of multiple melting events under a range of conditions.

The solubilities of HSEs in silicate liquid have been investigated extensively at 1 bar and moderate temperatures of 1300–1400°C (e.g., Ertel *et al.*, 1999, 2001; Fortenfant *et al.*, 2003a, 2006), whereas there have been only a few studies of the solubility or partitioning of HSE's at high pressure. All such studies are beset by a serious technical problem. The quenched samples of silicate liquid inevitably contain numerous metal micronuggets consisting of or rich in the element of interest, especially at low f_{O_2}, which make it very difficult to obtain reliable chemical analyses. Although it is normally considered that such nuggets were present in the melt at high temperature (e.g., Lindstrom and Jones, 1996; Holzheid *et al.*, 2000b), it has also been suggested that they form by exsolution from the silicate liquid during quenching (Cottrell and Walker, 2006). Depending on the interpretation adopted, greatly different results are obtained. So far, the solubilities of Pt and Pd in silicate liquid (Holzheid *et al.*, 2000b; Ertel *et al.* 2006; Cottrell and Walker, 2006), the metal–silicate partitioning of Re (Ohtani and Yurimoto, 1996) and the partitioning of Re and Os between magnesiowüstite and liquid Fe (Fortenfant *et al.*, 2003b) have been investigated at high pressure. The concentrations of Re and Pt in the mantle may possibly be explained by metal–silicate equilibration (Righter and Drake, 1997; Cottrell and Walker, 2006). However, the conclusion of the majority of these studies is that metal–silicate equilibration at high pressure is very unlikely to explain the concentrations of all HSEs in the mantle and, in particular, their chondritic ratios. In addition, based on studies of Martian meteorites, HSE abundances are similar in the mantles of both Earth and Mars (Warren *et al.*, 1999). This is difficult to explain by metal–silicate equilibration because, in planets of dissimilar sizes, P–T conditions of core formation should be quite different. Therefore, metal–silicate HSE partitioning results are generally interpreted to support the late-veneer hypothesis (see next section) for both Earth and Mars (e.g., Righter, 2005).

There are several problems with the simple concept of metal–silicate equilibration at the base of a magma ocean. First, the complete process of core formation in the Earth cannot be accomplished through a single event involving a magma ocean of limited depth (e.g., 800–1500 km). Several large impacts are likely to have occurred during accretion, each generating a magma ocean with different characteristics, while there may also have been a steady background flux of smaller impactors. Second, many of the proposed P–T conditions (e.g., 3750 K and 40 GPa) lie far (e.g. ~650 K) above the peridotite liquidus temperature (Wade and Wood, 2005; see **Figure 15**). The temperature at the base of a magma ocean should lie between the solidus and liquidus temperatures (**Figure 9(b)**); therefore, this observation appears to require that the metal ceased equilibrating with the silicate liquid far above the bottom of the magma ocean. Based on the arguments given above concerning emulsification (Section 3.2.3.2), this possibility is very unlikely. Finally, the

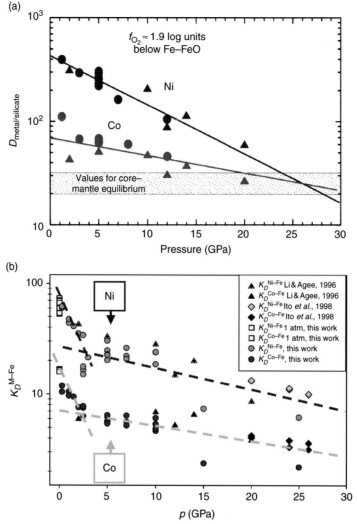

Figure 13 (a) Early experimental results on the effects of pressure on the partitioning of Ni (blue symbols) and Co (red symbols) between liquid Fe alloy and liquid silicate melt, at 2123–2750 K and $f_{O_2} = IW$-1.9, that led to the hypothesis of metal and silicate equilibrating during core formation at the bottom of a deep magma ocean. Triangles show data from Li and Agee (1996) and filled circles show data from Thibault and Walter (1995). (b) More recent results on the partitioning of Ni and Co show that the pressure dependence undergoes a pronounced change at ~3 GPa, probably because of pressure-induced structural changes in the silicate liquid (see Keppler and Rubie, 1993); above 3 GPa the pressure dependences for both Ni and Co are considerably weaker than the trends shown in (a). Here $K_D^{M\text{-}Fe}$ is a distribution coefficient in which the partition coefficient has been normalized to Fe partitioning and is thus independent of f_{O_2}; in addition, the data have all been normalized to 2273 K by extrapolation. (a) Reprinted from Wood BJ, Walter MJ, and Wade J (2006) Accretion of the Earth and segregation of its core. *Nature* 441: 825–833, doi: 10.1038/nature 04763, with permission from Macmillan Publishers Ltd. (b) From Kegler P, Holzheid A, Rubie DC, Frost DJ, and Palme H (2005) New results of metal/silicate partitioning of Ni and Co at elevated pressures and temperatures. *XXXVI Lunar and Planetary Science Conference*, Abstract #2030.

assumption is made that a layer of ponded liquid iron can equilibrate chemically with the overlying convecting magma ocean (see **Figure 9(a)**). As discussed further below (Section 3.3.2.3) this is unlikely to be realistic and more complex metal–silicate fractionation models are required.

In order to overcome the first two problems listed above, Wade and Wood (2005) presented a model in which core formation occurs continuously during accretion. This is based on the metal–silicate partitioning of a range of elements, including Fe, Ni, Co, V, W, Nb, Cr, and Mn, assuming equilibration at the

Table 3   Metal–silicate equilibration conditions during core formation inferred from experimental studies of siderophile element partitioning

P(GPa)	T(K)	Ref.	Notes
28	2400–2700*	1	Ni,Co; *T fixed by peridotite liquidus; $f_{O_2} \approx$ IW-0.5 (expt.)
27	2200	2	Ni,Co,P,Mo,W; f_{O_2} = IW-0.15 (inf.)
37	2300	3	Ni,Co,Fe; Cr requires 3400K
>35	>3600	4	V,Cr,Mn; f_{O_2} = IW-2.3 (inf.)
43–59	2400–4200	5	Ni,Co; f_{O_2} = IW to IW-2 (expt.)
25*	3350	6	Si; *P fixed by Ni/Co data; temp exceeds peridotite liquidus
27	2250	7	P,W,Co,Ni,Mo,Re,Ga,Sn,Cu; f_{O_2} = IW-0.4 (inf.)
40	2800	8	Ni,Co
40	3750*	9	V,Ni,Co,Mn,Si; *T fixed by peridotite liquidus; evolving f_{O_2}?
30–60	>2000	10	Ni,Co; f_{O_2} = IW-2.2 (inf.); demonstrates solution tradeoffs

f_{O_2} is the oxygen fugacity, IW indicates the iron–wustite buffer, 'inf.' means inferred f_{O_2} for the magma ocean and 'expt.' indicates the experimental value. 1, Li and Agee (1996); 2, Righter et al. (1997); 3, O'Neill et al., (1998); 4, Gessmann and Rubie, (2000); 5, Li and Agee (2001); 6, Gessmann et al. (2001); 7, Righter and Drake (2003); 8, Walter and Trønnes (2004); 9, Wade and Wood (2005); 10, Chabot et al. (2005).

base of a magma ocean that deepens as the Earth grows in size. The temperature at the base of the magma ocean is constrained to lie on the peridotite liquidus. In order to satisfy the observed mantle abundances of all elements considered (and especially V), it is necessary that the oxygen fugacity is initially very low but increases during accretion to satisfy the current FeO content of the mantle (**Figure 15**). The explanation for the increase in f_{O_2} involves the crystallization of silicate perovskite from the magma ocean, which can only occur once the Earth has reached a critical size such that the pressure at the base of the magma ocean reaches ~24 GPa. Because of its crystal chemistry, the crystallization of silicate perovskite causes ferrous iron to dissociate to ferric iron + metallic iron by the reaction:

$$Fe^{2+} \rightarrow Fe^{3+} + Fe$$

silicate liquid   perovskite   metal

(Frost et al., 2004). If some or all of the metal produced by this reaction is transferred from the mantle to the core, the Fe^{3+} content and the f_{O_2} of the mantle both increase. Although this model can explain the abundances of MSEs in the mantle, ~0.5% of chondritic material has to be added to the Earth at a late stage of accretion in order to generate the observed chondritic ratios of the HSEs (see Section 3.3.2.2.(ii)).

There is at least one potential problem with the model of Wade and Wood (2005). At very low oxygen fugacities (≤IW-3), Ta becomes siderophile and would therefore be extracted from the mantle during core formation (Mann et al., 2006). However, the abundance of Ta in the mantle is chondritic which may exclude the possibility of the initially low oxygen fugacity conditions proposed by Wade and Wood (2005).

3.3.2.2.(ii)   The   'late-veneer'   hypothesis

According to this hypothesis, the main stage of core formation involved the almost complete extraction of HSEs from the mantle into the metallic core, under reducing oxygen-fugacity conditions (e.g., Kimura et al., 1974; Morgan, 1986; O'Neill and Palme, 1998; Righter, 2005). At a late stage of accretion, a thin veneer of chondritic material was added to the Earth under relatively oxidizing conditions, such that the HSEs were retained in the mantle in approximately chondritic ratios. The mass of material added at this late stage is considered to be <1% of the entire mantle. This is currently the most widely accepted 'heterogeneous' core-formation model. However, the likelihood that Ta would be extracted into the core under reducing conditions (Mann et al., 2006) is a problem, as discussed in the previous section, because the mantle has not been depleted in this element.

3.3.2.2.(iii)   Inefficient   core   formation

Originally suggested by Jones and Drake (1986), this hypothesis proposes that a small quantity of metallic Fe was trapped in the mantle during core formation. The current HSE budget of the mantle was supplied by this stranded metal. One objection to this hypothesis is that Earth is likely to have been largely molten during core formation, with the result that metal segregation should have been very efficient (Righter, 2005). However, based on the study of Frost et al. (2004), it is likely that not all metal

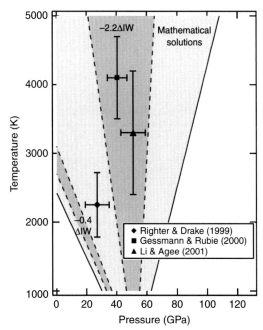

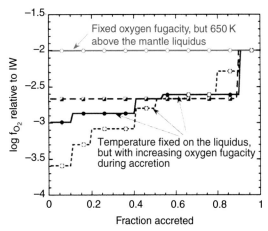

Figure 14 Pressure–temperature conditions of metal–silicate equilibration during core formation that are consistent with the concentrations of Ni and Co in the Earth's mantle, based on experimental studies of the metal–silicate partitioning of these elements. The two dark-gray shaded regions show results calculated for oxygen fugacities of IW-0.4 and IW-2.2, respectively, whereas the light-shaded region shows solutions for all oxygen fugacities. *P–T* estimates from three previous studies are shown by the symbols with error bars. Temperature is poorly constrained because the partition coefficients for Ni and Co depend only weakly on this variable. In order to better constrain the *P–T* conditions of equilibration, it is necessary to consider additional siderophile elements for which partitioning is more strongly temperature dependent (e.g., Wade and Wood, 2005). Reproduced from Chabot NL, Draper DS, and Agee CB (2005) Conditions of core formation in the Earth: Constraints from nickel and cobalt partitioning. *Geochemica et Cosmochemica Acta* 69: 2141–2151, with permission from Elsevier.

Figure 15 Models involving continuous core formation during accretion of the Earth based on metal–silicate equilibration at the base of a deepening magma ocean. Oxygen fugacity is plotted as a function of fraction accreted. The temperature at the base of the magma ocean is constrained to lie on the peridotite liquidus for the variable oxygen fugacity models. In order to satisfy concentrations of a range of siderophile elements in the mantle, the oxygen fugacity has to be low initially and then increase during accretion for reasons described in the text. Reprinted from Wood BJ, Walter MJ, and Wade J (2006) Accretion of the Earth and segregation of its core. *Nature* 441: 825–833 (doi:10.1038/nature04763), with permission of Macmillan Publishers Ltd.

segregated to the core during crystallization of the molten Earth. The argument runs as follows: As described above (Section 3.3.2.2.(i)), the crystallization of silicate perovskite, the dominant phase of the lower mantle, involved the disproportionation reaction $Fe^{2+} \rightarrow Fe^{3+} + Fe$ metal. The subsequent loss of some of this metal to the core resulted in an increase in the oxidation state of the mantle – which explains why the Earth's mantle has been oxidized during or after core formation (Frost *et al.*, 2004; Wood and Halliday, 2005; Wood *et al.*, 2006). However, if all the metal produced by perovskite crystallization had been extracted to the core, the mantle would now be

much more oxidized than is observed. In the Martian mantle, there is, at most, only a thin lower mantle containing silicate perovskite. The paucity (or absence) of silicate perovskite thus explains, based on the observations of Frost *et al.* (2004), why the Martian mantle is more reduced than Earth's mantle (Wadhwa, 2001; Herd *et al.*, 2002). However, the possibility of a small fraction of metal being trapped during core formation on Mars still exists and could explain the similarities between HSE abundances in the mantles of Earth and Mars (thus making the more complex late-veneer hypothesis redundant). Recently, inefficient core formation has also been suggested to result from deformation-enhanced percolation (Hustoft and Kohlstedt, 2006), as discussed above in Section 3.2.3.1.

3.3.2.2.(iv) Addition of outer-core material to the lower mantle
It has been suggested that HSE abundances in the Earth's mantle have resulted from core–mantle interaction (Brandon and Walker, 2005), for example, by the addition of a small amount of core metal to the mantle (Snow and Schmidt, 1998). This could potentially occur by capillary

action (considered highly unlikely by Poirier et al. (1998)) or by dilatancy caused by volume strain (Rushmer et al., 2005; Kanda and Stevenson, 2006). Alternatively, siderophile elements could be added to the base of the mantle by crystallization of oxides or silicates due to the growth of the inner core (Walker, 2000, 2005) and chemical exchange could be facilitated by the (possible) presence of partial melt in the lower few kilometers of the mantle. In this case, the metal–silicate partition coefficients at CMB conditions would have to be orders of magnitude lower than at low $P–T$ conditions and would need to result in the addition of the HSEs in approximately chondritic proportions.

3.3.2.3 Metal–silicate fractionation models

The simplest model of metal–silicate fractionation during core formation involves the ponding of liquid iron at the base of a convecting magma ocean with chemical equilibration occurring at the metal–silicate interface (Section 3.3.2.2.(i), **Figure 9(a)**). An appealing feature of this model is that siderophile element abundances in the mantle can be interpreted directly in terms of magma ocean depth (e.g., Li and Agee, 1996; Righter et al., 1997). The timescale required for chemical equilibration across the metal–silicate interface has been investigated by Rubie et al. (2003) assuming that mass transport occurs by chemical diffusion across boundary layers that exist above and below the interface. As described earlier, a magma ocean is expected to crystallize from the bottom up which means that the initial crystallization of silicate minerals at the base of the magma ocean will terminate equilibration between the ponded iron and the overlying magma ocean. Thus, Rubie et al. (2003) also calculated the timescale required for the initial crystallization of the base of the magma ocean. Results are dependent upon the depth of the magma ocean and suggest that equilibration times are almost three orders of magnitude greater than cooling times (**Figure 16**). Such a result is also predicted by considering that rates of conductive heat transfer are much faster than rates of chemical diffusion (Rubie et al., 2003). Therefore, these results appear to rule out simple chemical equilibration at the base of the magma ocean.

The equilibration model of Rubie et al. (2003) is based on the assumption that a dense atmosphere was not present during magma ocean crystallization. The effect of such an insulating atmosphere would be to reduce the rate of heat loss and therefore prolong the lifetime of the magma ocean. However, the rate of convection would also be reduced, which would slow the rate of chemical exchange across the metal–silicate interface. Thus, the presence of an insulating atmosphere is unlikely to affect the conclusions of Rubie et al. (2003). However, there has been a recent suggestion that a magma ocean adiabat is much steeper than formerly believed with the consequence that terrestrial magma oceans might crystallize from the top down rather than from the bottom up (Mosenfelder et al., 2007; cf. Miller et al., 1991). In this case, magma ocean crystallization would be much slower and the possibility of simple metal–silicate equilibration at its base might need to be reconsidered.

It is currently assumed in core-formation models that metal–silicate separation takes place in a magma ocean of global extent and of more or less constant depth (**Figure 9**). As discussed above in Section 3.2.3.2, this assumption requires reconsideration. **Figure 17** shows an alternative model in which the magma ocean that is generated by a large impact is initially a hemispherical body of limited lateral extent. The attainment of isostatic equilibrium eventually results in the formation of a global magma ocean but metal segregation could already have taken place before this developed. Pressure at the base of the initial hemispherical magma ocean is clearly much higher than at the base of the final global magma ocean, which means that the partitioning of siderophile elements could depend critically on the timing of metal segregation.

As discussed above, liquid metal is likely to be present in a magma ocean in the form of small droplets ~1 cm in diameter. Such droplets remain in chemical equilibrium with the magma as the they settle out and $P–T$ conditions change because diffusion distances are short (Karato and Murthy, 1997; Rubie et al., 2003). It is necessary to understand the chemical consequences of the settling out of such droplets in order to interpret siderophile element geochemistry in terms of magma ocean depths. Rubie et al. (2003) considered two end-member models for the partitioning of Ni. The models are based on the parametrized Ni partitioning formulation of Righter and Drake (1999) which requires a magma ocean depth of ~800 km for metal–silicate equilibration at its base to produce the estimated core–mantle partition coefficient of ~28. In model 1 of Rubie et al. (2003), the magma ocean is static (i.e., no convection or mixing) and droplets that are initially uniformly dispersed settle out and re-equilibrate progressively as they

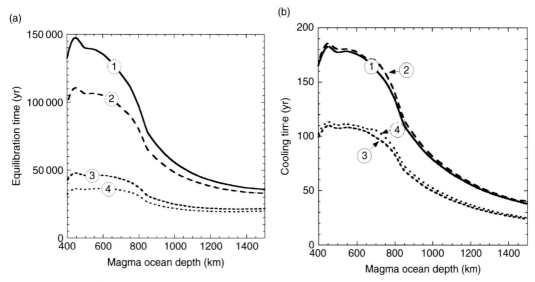

Figure 16 Results of a model in which a layer of segregated (ponded) liquid iron equilibrates chemically with an overlying convecting magma ocean. (a) Time to reach 99% equilibration between metal and silicate as a function of magma ocean depth. (b) Time required for initial crystallization at the base of the magma ocean (thus effectively terminating the equilibration process). The four curves (1–4) are results for a wide range of plausible model parameters. Reproduced from Rubie DC, Melosh HJ, Reid JE, Liebske C, and Righter K (2003) Mechanisms of metal-silicate equilibration in the terrestrial magma ocean. *Earth and Planetary Science Letters* 205: 239–255, with permission from Elsevier.

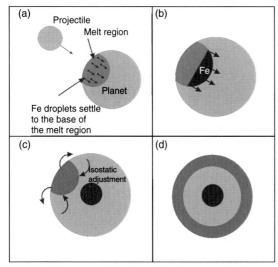

Figure 17 Possible evolution of a deep magma ocean that forms as a consequence of a giant impact. (a) Initially a hemispherical magma ocean forms that is of limited lateral extent. There is the possibility that iron, emulsified in the form of small dispersed droplets, settles and segregates rapidly to form a protocore at the bottom of this magma ocean (b and c). Subsequent isostatic adjustment causes the magma ocean to evolve into a layer of global extent (d). The relative timing of the processes of metal segregation and isostatic adjustment determines the pressure (and probably also temperature) conditions of metal–silicate equilibration. Reproduced from Tonks WB and Melosh HJ (1992) Core formation by giant impacts. *Icarus* 100: 326–346, with permission from Elsevier.

fall. This polybaric equilibration model requires a magma ocean ~1400 km deep (**Figure 18**) because much of the silicate equilibrates at relatively low pressures where $D_{Ni}^{met-sil}$ values are high. Model 2 of Rubie *et al.* (2003) is based on an assumption of vigorous convection keeping the magma ocean fully mixed and chemically homogeneous. The iron droplets equilibrate finally with silicate liquid at the base of the magma ocean just before segregating into a ponded layer. Because the mass fraction of metal that is available to equilibrate with silicate progressively decreases with time, the effectiveness of metal to remove siderophile elements from the magma ocean also decreases. This model requires a magma ocean ~550 km deep to produce the desired core–mantle partition coefficient of 28 (**Figure 18**).

The two end-member models 1 and 2 are clearly both physically unrealistic and a more realistic result must lie somewhere between the two extremes shown in **Figure 18**. In an attempt to investigate the chemical consequences of metal–silicate segregation in a deep magma ocean more rigorously, Höink *et al.* (2006) have combined two- and three-dimensional numerical convection models with a tracer-based sedimentation method. They found that metal droplets stabilize the magma ocean against convection and that convection only develops in the upper layer of a magma ocean after it has become depleted

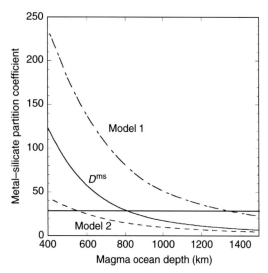

Figure 18 Results of different metal–silicate fractionation models for the siderophile element Ni as a function of magma ocean depth. The core–mantle partition coefficient for this element is estimated to be ~28 (as indicated by the horizontal line). The curve labeled D^{ms} indicates results for simple equilibration between a ponded liquid metal layer and the overlying magma ocean and requires a magma ocean ~800 km deep to obtain the core–mantle value. However, this model is unlikely to be realistic (**Figure 16**). Models 1 and 2 represent extreme end-member cases (see text for details) and require magma ocean depths of ~1400 and ~550 km, respectively. Reproduced from Rubie DC, Melosh HJ, Reid JE, Liebske C, and Righter K (2003) Mechanisms of metal-silicate equilibration in the terrestrial magma ocean. *Earth and Planetary Science Letters* 205: 239–255, with permission from Elsevier.

in iron droplets. The timescale for droplet separation was found to be identical to the Stokes' settling time. Because of the lack of convection in the droplet-dominated layer, the chemical consequences are similar to those of model 1 of Rubie *et al.* (2003), although the results of the two- and three-dimensional models of Höink *et al.* (2006) appear to be very different from each other (see their figure 20). An alternative approach has been adopted by Melosh and Rubie (2007) using a two-dimensional model for computing the flow of two interpenetrating fluids (liquid silicate and liquid iron). Their preliminary results also suggest that droplets settle out in approximately the Stokes' settling time but show that strong density currents develop in the droplet-bearing region of the magma ocean due to density perturbations. The velocities of these density currents range up to $50\,\mathrm{m\,s}^{-1}$, which is much greater than velocities induced by thermal convection. In addition, the gravitational energy of sinking droplets

is converted into heat which raises the temperature at the base of the magma ocean by at least several hundred degrees. Based on these preliminary results, it is not yet clear how the partitioning of siderophile elements compares with the predictions of the Höink *et al.* models. This is a rapidly developing area of research and controversy at the moment.

3.3.2.4 Concluding remarks

The consistency of MSE abundances with metal–silicate equilibration at high P,T conditions, implies that abundances produced by earlier equilibration in smaller bodies at lower P,T conditions must have been overprinted. This conclusion supports the idea of impactor cores undergoing re-equilibration in the magma ocean, presumably as a result of emulsification (Rubie *et al.*, 2003). Later events which did not involve re-equilibration (e.g., the descent of large iron diapirs through the lower mantle) would not leave any signature in the siderophile element abundances.

3.3.3 Light Elements in the Core

The density of the Earth's core is too low, by 5–10%, for it to consist only of Fe and Ni (e.g., Birch, 1952; Anderson and Isaak, 2002). It has therefore been postulated that the core must contain up to ~10 wt.% of one or more light elements, with the most likely candidates being S, O, Si, C, P, and H (Poirier, 1994). Knowledge of the identity of the light element(s) is important for constraining the bulk composition of the Earth, for understanding processes occurring at the CMB and how such processes are affected by crystallization of the inner core (e.g., Buffett *et al.* (2000); Helffrich and Kaneshima, 2004). The main constraints on the identity of the light elements present in the core are based on cosmochemical arguments (McDonough, 2003), experimental data (Hillgren *et al.*, 2000; Li and Fei, 2003), and computational simulations (e.g., Alfe *et al.*, 2002). The topic has been reviewed recently by Hillgren *et al.* (2000), McDonough (2003), and Li and Fei (2003) and here we provide only a brief summary of the main arguments and recent experimental results.

The sulfur content of the core, based on the relative volatility of this element, is likely to be no more than 1.5–2 wt.%, (McDonough and Sun, 1995; Dreibus and Palme, 1996; McDonough, 2003). Similarly, cosmochemical constraints suggest that only very small amounts (e.g., ≤0.2 wt.%) of C and P are present in the core (McDonough, 2003);

therefore these elements are unlikely to contribute significantly to the density deficit.

One of the main controversies concerning the identity of the principle light element(s) in the core involves silicon and oxygen. A high oxygen content is favored by high oxygen fugacity, whereas a high Si content is favored by low oxygen fugacity (Kilburn and Wood, 1997; Hillgren *et al.*, 2000; Li and Fei, 2003; Malavergne *et al.*, 2004). It has therefore been proposed that these two elements are almost mutually exclusive (**Figure 19**; O' Neill *et al.*, 1998). Some models of core composition have been based on this exclusivity; that is, it is assumed that the core contains either Si or O but not both of these elements (McDonough, 2003; table 7). However, the effects of high pressure and temperature are also critical.

Based on experimental results of Gessmann *et al.* (2001), obtained up to 23 GPa and 2473 K, the solubility of Si in liquid iron increases with both P and T. An extrapolation of their experimental data shows that ~7 wt.% Si can be dissolved in liquid Fe at 25–30 GPa, 3100–3300 K, and an f_{O_2} (IW-2) that is expected for core formation (**Figure 20(a)**). This result is in accordance with the geochemical model of the core of Allegre *et al.* (1995) in which ~7 wt.% Si in the core was proposed based on the Mg/Si ratio of the mantle being high compared with CI chondrites. However, according to thermodynamic arguments of Gessmann *et al.* (2001), the solubility

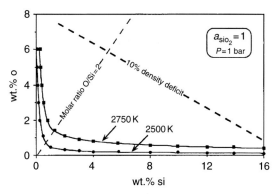

Figure 19 Mutual solubilities of O and Si in liquid iron at 1 bar. Because high O concentrations are favored by high f_{O_2} and high Si solubilities require a low f_{O_2}, the solubilities of these elements are mutually exclusive. The dashed line indicates solubilities required to account for the 10% density deficit of the core. Reproduced from O'Neill HSC, Canil D, and Rubie DC (1998) Oxide-metal equilibria to 2500°C and 25 GPa: Implications for core formation and the light component in the Earth's core. *Journal of Geophysical Research* 103: 12239–12260, with permission of American geophysical Union.

of Si in liquid Fe is predicted to decrease with pressure above ~30 GPa in the mantle and to approach zero at conditions of the CMB (**Figure 20(b)**). The reason for the postulated reversal in pressure dependence is that the coordination of Si^{4+} in mantle silicates changes from fourfold to sixfold at ~25 GPa which reverses the sign of the volume change (ΔV) of the exchange reaction of Si between silicates and liquid Fe.

Oxygen was first proposed as the principal light element in the core about 30 years ago (e.g., Ringwood, 1977; Ohtani and Ringwood, 1984; Ohtani *et al.*, 1984). Although it is clear that the solubility of this element in liquid Fe increases with temperature, the effect of pressure has been controversial. According to studies of phase relations in the Fe–FeO system, solubility increases with pressure (Ringwood, 1977; Kato and Ringwood, 1989). However, investigations of the partitioning of FeO between magnesiowüstite and liquid Fe have indicated that solubility decreases with increasing pressure and, based on a very large extrapolation, is essentially zero at conditions of the CMB (O'Neill *et al.*, 1998; Rubie *et al.*, 2004). Asahara *et al.* (2007) may have resolved the controversy by showing that the partitioning of FeO into liquid Fe decreases weakly with pressure up to ~15 GPa and then increases again at high pressures. The reason for the change in pressure dependence is that the FeO component dissolved in liquid Fe is more compressible than FeO dissolved in an oxide or silicate phase. Thus, the sign of the volume change (ΔV) of the exchange reaction of oxygen between silicates/oxides and liquid Fe reverses at 10–15 GPa (see also Ohtani *et al.*, 1984; Walker, 2005). As discussed below, the partitioning of oxygen into liquid iron appears to be high enough at core-formation conditions for this element to be the most abundant light element in the Earth's core (Rubie *et al.*, 2004; Asahara *et al.*, 2007).

Recent studies using the LH-DAC suggest that concentrations of both oxygen and silicon could be significant in the core and could, in combination, account for the density deficit. Takafuji *et al.* (2005) found 3 wt.% Si and 5 wt.% O in liquid Fe in equilibrium with $(Mg,Fe)SiO_3$ perovskite at 97 GPa and 3150 K. In similar experiments on liquid Fe coexisting with the post-perovskite phase, Sakai *et al.* (2006) found up to 6.3 wt.% O and 4 wt.% Si in liquid Fe at 139 GPa and 3000 K. These results should be regarded as preliminary because of the huge temperature gradients and large temperature uncertainties that are characteristic of LH-DAC experiments; in addition,

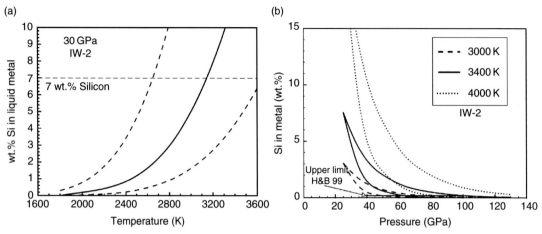

Figure 20 (a) Solubility of Si in liquid Fe alloy at 30 GPa and an oxygen fugacity of IW-2 based on an extrapolation of experimental data. The dashed lines show the extent of the uncertainty envelope based on error propagation. (b) Extrapolation of the experimental data to core conditions at an oxygen fugacity of IW-2. Here the pressure dependence is predicted to be opposite to that at low pressures due to Si being incorporated in silicates in six-fold coordination at pressures above 25–30 GPa; this structural change causes the volume change of the exchange reaction of Si between metal and silicate to reverse. Also shown are diamond anvil cell results (H&B99) of Hillgren and Boehler (1999) that were obtained at 45–100 GPa and an unknown oxygen fugacity. Reproduced from Gessmann CK, Wood BJ, Rubie DC, and Kilburn MR (2001) Solubility of silicon in liquid metal at high pressure: Implications for the composition of the Earth's core. *Earth Planetary Science Letters* 184: 367–376, with permission from Elsevier.

the oxygen fugacity was probably not buffered. In the case of oxygen, the LH-DAC results are mostly consistent with the lower pressure data of Asahara *et al.* (2007). However, the silicon results indicate considerably higher solubilities than those predicted by Gessmann *et al.* (2001) which may indicate that the volume change of the exchange reaction of Si between silicates and Fe metal assumed in the latter study is too large. The high mutual solubilities of O and Si, compared with low *P–T* results (O'Neill *et al.*, 1998), could be largely the consequence of high temperatures.

An additional important observation is that the outer core has a larger density deficit and therefore appears to contain a higher concentration of light element(s) than the inner core (Jephcoat and Olson, 1987; Alfe *et al.*, 2002). This implies that the light element(s) partitions strongly into liquid iron during freezing, which is potentially diagnostic behavior. For instance, Alfe *et al.* (2002), using molecular dynamics simulations, found that oxygen, due to its small atomic radius, tends to be expelled during the freezing of liquid iron. Conversely, S and Si have atomic radii similar to that of iron at core pressures, and thus substitute freely for iron in the solid inner core. Based on the observed density difference between the inner and outer cores, Alfe *et al.* (2002) predicted ~8 wt.% oxygen in the outer core. These

results support the case for O being the main light element and are in agreement with the predictions of Asahara *et al.* (2007). Such molecular dynamics calculations have not yet been performed for either H or C, which might also behave in a similar manner to O.

The Earth's core is considered not to be in chemical equilibrium with the mantle (Stevenson, 1981) and light-element solubilities at core conditions are therefore not necessarily indicative of the actual concentrations of these elements in the core. Instead, the light-element content of the core is likely to have been set during core formation, as was the case for the siderophile elements. As discussed above, studies of the metal–silicate partitioning of the MSEs indicate that core–mantle partitioning is consistent with metal–silicate equilibration at conditions of 30–60 GPa and ≤4000 K (**Table 3**). Thus metal–silicate partitioning of light elements (e.g., O and Si) at such conditions may have determined the light-element content of the core. Rubie *et al.* (2004) investigated this possibility by modeling the partitioning of oxygen (actually the FeO component) between a silicate magma ocean and liquid Fe alloy during core formation as a function of magma ocean depth. Because oxygen solubility in liquid Fe increases strongly with temperature, FeO partitions increasingly into the Fe alloy as the magma ocean depth is increased beyond 1000 km, leaving the silicate depleted in FeO (**Figure 21(a)**). The results of this model are consistent with equilibration in a magma ocean 1200–2000 km deep and with the Earth's core containing 7–8 wt.% oxygen.

The Earth's mantle contains ~8 wt.% FeO, whereas, based on studies of Martian meteorites, the Martian mantle is considered to contain ~18 wt.% FeO. This difference can be explained by oxygen partitioning during core formation so that the possibility that the two planets have similar (or even identical) bulk compositions cannot be excluded (Rubie *et al.*, 2004). Because Mars is a much smaller planet than Earth, temperatures and pressures in a Martian magma ocean are expected to have been relatively low, so that little FeO was extracted from the mantle during core formation (**Figure 21(b)**). This model also explains why the mass fraction of the Martian core is smaller than that of the Earth's core.

One advantage of modeling the metal–silicate partitioning of oxygen during core formation, in addition to the siderophile elements, is that assumptions concerning oxygen fugacity are not required. This is because the partitioning of oxygen determines the oxygen fugacity. Instead it is necessary to estimate the bulk composition (i.e., oxygen content) of the metal–silicate system, for example, on the basis of the chemistry of chondritic meteorites (Rubie *et al.*, 2004).

3.4 Summary

Theoretical arguments and geochemical observations suggest that the bulk of the mass of the Earth accreted through a few, large impacts within about 50 My of solar system formation, and that each of these impacts generated a global, if transient, magma ocean. Although the impacting bodies were undoubtedly differentiated, pre-existing chemical signals appear to have been overprinted by the impact process. Siderophile element concentrations are consistent with magma oceans extending at least to mid-mantle depths (800–1500 km, 2500–4000 K). The impactor cores likely underwent emulsification

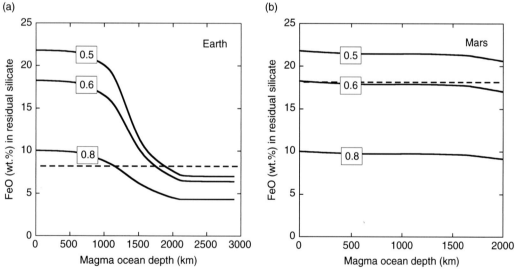

Figure 21   Results of core-formation models for Earth (a) and Mars (b), based on metal–silicate partitioning of FeO in a magma ocean. The models are based on a chondritic bulk composition that consists initially of a mixture of metal and silicate components. The bulk oxygen content determines the fraction of metal that is present. The three curves, in each case, show results for different initial bulk oxygen contents and the labels indicate the weight fraction of Fe that is initially present as metal. The horizontal dashed lines show the current FeO contents of the respective mantles. On Earth, the FeO content of the residual silicate decreases when the magma ocean depth exceeds 1000 km, because, at high temperatures, FeO partitions into the metal phase. On Mars, such an effect is almost absent because magma ocean temperatures are relatively low because of the small size of the planet. These results show that the bulk compositions of Earth and Mars could be similar (e.g., curve labeled '0.6') and that the current FeO content of the Earth's mantle resulted from core formation in a magma ocean ~1800 km deep: in this case, the Earth's core could contain 7–8 wt.% oxygen. Reprinted from Rubie DC, Gessmann CK, and, Frost DJ (2004) Partitioning of oxygen during core formation on the Earth and Mars. *Nature* 429: 58–61, with permission from Macmillan Publishers Ltd.

as they sank through the magma ocean, resulting in a chemical re-equilibration that is also suggested by both siderophile and Hf–W isotopic observations. This re-equilibration ceased as the metal pooled at the base of the magma ocean; subsequent transport of the resulting large-scale iron masses to the growing core was rapid and will have resulted in a significant increase in core temperatures. Following the Moon-forming impact, the initial core temperature was probably at least 6000 K, suggesting that extensive melting occurred in the lowermost mantle.

Unusually, there is broad agreement between the geochemical constraints and geophysical expectations of the core's early history. Nonetheless, several important outstanding questions remain:

1. The physics of exactly what happens during giant impacts is poorly understood (Sections 3.2.2.2 and 3.2.3.2). In particular, although there are physical and geochemical arguments for impactor emulsification, this process has not yet been investigated by numerical or laboratory models.

2. The lifetime of magma oceans is also poorly known (Section 3.2.3.2). This is in part because complicating factors such as the possible presence of an insulating atmosphere or a foundering crust have a large effect on the outcome. It may be that this is an issue which can only be resolved using radiogenic isotopes with appropriate half-lives, rather than geophysical modeling. In addition there is a possibility that terrestrial magma oceans may crystallize from the top down rather than from the bottom up (Mosenfelder et al., 2007) – which could have major implications by greatly extending magma ocean lifetimes.

3. Geochemical models of core formation are currently hampered by a lack of partitioning data, especially for the HSEs at high pressures and temperatures. Even partitioning of MSEs is poorly known at pressures significantly above 25 GPa. Rectifying the latter deficiency requires laser-heated diamond anvil cell experiments (e.g., Tschauner et al., 1999; Bouhifd and Jephcoat, 2003), which are difficult to perform successfully because of large temperature gradients (that can drive chemical diffusion), temperature uncertainties, and difficulties in sample analysis. An additional problem is that models of chemical fractionation during metal–silicate separation are not yet fully developed (Rubie et al., 2003; Höink et al., 2006, Melosh and Rubie, 2007).

4. Uncertainties remain concerning the identity of the light element(s) in the core. Based on cosmochemical arguments and recent high-pressure studies, oxygen may be the main light element together with $\sim$2 wt.% S and a small amount of Si (see also Badro et al. 2007).

5. Most of the models of accretion and core formation to date have assumed single-stage processes. In practice, of course, accretion and core formation occurs as a series of discrete events, under evolving conditions. The effect of these changing conditions on the behavior and chemistry of the core and mantle is just beginning to be addressed (e.g., Halliday, 2004; Wade and Wood, 2005; Wood et al., 2006). Unfortunately, although parameters like f_{O_2} likely evolved with time, all observations (except those of unstable isotopes) constrain only some time-weighted mean value of the parameter. Thus, resolving the time evolution of parameters such as f_{O_2} will be challenging and it may be difficult to identify unique solutions.

Acknowledgments

Portions of this work were supported by NSF-EAR, NASA-Origins, the German Science Foundation Priority Programme 'Mars and the Terrestrial Planets' (Grant Ru437) and an Alexander von Humboldt Senior Research Award to HJM.

References

Abe Y (1997) Thermal and chemical evolution of the terrestrial magma ocean. *Physics of the Earth and Planetary Interiors* 100: 27–39.

Agnor C and Asphaug E (2004) Accretion efficiency during planetary collisions. *Astrophysical Journal* 613: L157–L160.

Agnor CB, Canup RM, and Levison HF (1999) On the character and consequences of large impacts in the late stage of terrestrial planet formation. *Icarus* 142: 219–237.

Alfe D, Gillan MJ, and Price GD (2002) Composition and temperature of the Earth's core constrained by combining *ab initio* calculations and seismic data. *Earth and Planetary Science Letters* 195: 91–98.

Allegre CJ, Poirier J-P, Humler E, and Hofmann AW (1995) The chemical composition of the Earth. *Earth and Planetary Science Letters* 134: 515–526.

Anders E and Grevesse N (1989) Abundances of the elements: Meteoritic and solar. *Geochemica et Cosmochemica Acta* 53: 197–214.

Anderson OL and Isaak DG (2002) Another look at the core density deficit of Earth's outer core. *Physics of the Earth and Planetary Interiors* 131: 19–27.

Asahara Y, Frost DJ, and Rubie DC (2007) Partitioning of FeO between magnesiowüstite and liquid iron at high pressures and temperatures: Implications for the composition of the

Earth's outer core. *Earth and Planetary Science Letters* doi: 10.1016/j.epsl.2007.03.006.

Asphaug E, Agnor CB, and Williams Q (2006) Hit-and-run planetary collisions. *Nature* 439: 155–160.

Badro J, Fiquet G, Guyot F, *et al.* (2007) Effect of light elements on the sound velocities in solid iron: Implications for the composition of Earth's core. *Earth and Planetary Science Letters* 254: 233–238.

Baker J, Bizzarro M, Wittig N, Connelly J, and Haack H (2005) Early planetesimal melting from an age of 4.5662 Gyr for differentiated meteorites. *Nature* 436: 1127–1131.

Ballhaus C and Ellis DJ (1996) Mobility of core melts during Earth's accretion. *Earth and Planetary Science Letters* 143: 137–145.

Benz W and Cameron AGW (1990) Terrestrial effects of the giant impact. In: Newsom HE and Jones JH (eds.) *Origin of the Earth*, pp. 61–67. New York: Oxford University Press.

Benz W, Slattery WL, and Cameron AGW (1988) Collisional stripping of Mercury's mantle. *Icarus* 74: 516–528.

Birch F (1952) Elasticity and constitution of the Earth's interior. *Journal of Geophysical Research* 69: 227–286.

Blichert-Toft J, Gleason JD, Télouk P, and Albarède F (1999) The Lu–Hf isotope geochemistry of shergottites and the evolution of the Martian mantle–crust system. *Earth and Planetary Science Letters* 173: 25–39.

Boehler R (2000) High-pressure experiments and the phase diagram of lower mantle and core materials. *Reviews of Geophysics* 38: 221–245.

Borg LE and Draper DS (2003) A petrogenetic model for the origin and compositional variation of the martian basaltic meteorites. *Meteoritics and Planetary Science* 38: 1713–1731.

Bouhifd MA and Jephcoat AP (2003) The effect of pressure on the partitioning of Ni and Co between silicate and iron-rich metal liquids: A diamond-anvil cell study. *Earth and Planetary Science Letters* 209: 245–255.

Boyet M and Carlson RW (2005) Nd-142 evidence for early (> 4.53 Ga) global differentiation of the silicate Earth. *Science* 309: 576–581.

Brandon AD and Walker RJ (2005) The debate over core–mantle interaction. *Earth and Planetary Science Letters* 232: 211–225.

Bruhn D, Groebner N, and Kohlstedt DL (2000) An interconnected network of core-forming melts produced by shear deformation. *Nature* 403: 883–886.

Buffett BA, Garnero EJ, and Jeanloz R (2000) Sediments at the top of Earth's core. *Science* 290: 1338–1342.

Busso M, Gallino R, and Wasserburg GJ (1999) Nucleosynthesis in asymptotic giant branch stars: Relevance for galactic enrichment and solar system formation. *Annual Reviews of Astronomy and Astrophysics* 37: 239–309.

Cameron AGW (2000) High-resolution simulations of the giant impact. In: Canup RM and Righter K (eds.) *Origin of the Earth and Moon*, pp. 133–144. Tucson, AZ: University of Arizona Press.

Canup RM (2004) Simulations of a late lunar-forming impact. *Icarus* 168: 433–456.

Canup RM and Asphaug E (2001) Origin of the Moon in a giant impact near the end of the Earth's formation. *Nature* 412: 708–712.

Canup RM and Righter K (eds.) (2000) *Origin of the Earth and Moon*. Tucson, AZ: University of Arizona Press.

Canup RM, Ward WR, and Cameron AGW (2001) A scaling relationship for satellite-forming impacts. *Icarus* 150: 288–296.

Caro G, Bourdon B, Wood BJ, and Corgne A (2005) Trace-element fractionation in Hadean mantle generated by melt segregation from a magma ocean. *Nature* 436: 246–249.

Chabot NL and Agee CB (2003) Core formation in the Earth and Moon: New experimental constraints from V, Cr, and Mn. *Geochemica et Cosmochemica Acta* 67: 2077–2091.

Chabot NL, Draper DS, and Agee CB (2005) Conditions of core formation in the Earth: Constraints from Nickel and Cobalt partitioning. *Geochemica et Cosmochemica Acta* 69: 2141–2151.

Chabot NL and Haack H (2006) Evolution of asteroidal cores. In: Lauretta DS and McSween HY (eds.) *Meteorites and the Early Solar System*, pp. 747–771. Tucson, AZ: University of Arizona Press.

Chamberlin TC (1916) *The Origin of the Earth*, 271p. Chicago: University of Chicago Press.

Chambers JE (2003) Planet formation. In: Davis AM (ed.) *Treatise on Geochemistry, Vol. 1: Meteorites, Planets and Comets*, pp. 461–475. Oxford: Elsevier-Pergamon.

Chambers JE and Wetherill GW (1998) Making the terrestrial planets: N-body integrations of planetary embryos in three dimensions. *Icarus* 136: 304–327.

Clayton RN and Mayeda TK (1996) Oxygen isotope studies of achondrites. *Geochemica et Cosmochemica Acta* 60: 1999–2017.

Cottrell E and Walker D (2006) Constraints on core formation from Pt partitioning in mafic silicate liquids at high temperatures. *Geochemica et Cosmochemica Acta* 70: 1565–1580.

Dalziel SB, Linden PF, and Youngs DL (1999) Self-similarity and internal structure of turbulence induced by Rayleigh–Taylor instability. *Journal of Fluid Mechanics* 399: 1–48.

Drake MJ (2001) The eucrite/Vesta story. *Meteoritics and Planetary Science* 36: 501–513.

Drake MJ and Righter K (2002) Determining the composition of the Earth. *Nature* 416: 39–44.

Dreibus G and Palme H (1996) Cosmochemical constraints on the sulfur content in the Earth's core. *Geochemica et Cosmochemica Acta* 60: 1125–1130.

Elkins-Tanton LT, Parmentier EM, and Hess PC (2003) Magma ocean fractional crystallization and cumulate overturn in terrestrial planets: Implications for Mars. *Meteoritics and Planetary Science* 38: 1753–1771.

Elsasser WM (1963) Early history of the earth. In: Geiss J and Goldberg ED (eds.) *Earth Science and Meteoritics*, pp. 1–30. Amsterdam: North-Holland.

Ertel W, O'Neill HSC, Sylvester PJ, and Dingwell DB (1999) Solubilities of Pt and Rh in a haplobasaltic silicate melt at 1300°C. *Geochemica et Cosmochemica Acta* 63: 2439–2449.

Ertel W, O'Neill HSC, Sylvester PJ, Dingwell DB, and Spettel B (2001) The solubility of rhenium in silicate melts. Implications for the geochemical properties of rhenium at high temperatures. *Geochemica et Cosmochemica Acta* 65: 2161–2170.

Ertel W, Walter MJ, Drake MJ, and Sylvester PJ (2006) Experimental study of platinum solubility in silicate melt to 14 GPa and 2273 K: Implications for accretion and core formation in the Earth. *Geochemica et Cosmochemica Acta* 70: 2591–2602.

Faul UH (1997) Permeability of partially molten upper mantle rocks from experiments and percolation theory. *Journal of Geophysical Research* 102: 10299–10312 (doi:10.1029/96JB03460).

Fortenfant SS, Dingwell DB, Ertel-Ingrisch W, Capmas F, Birck JL, and Dalpé C (2006) Oxygen fugacity dependence of Os solubility in haplobasaltic melt. *Geochemica et Cosmochemica Acta* 70: 742–756.

Fortenfant SS, Günther D, Dingwell DB, and Rubie DC (2003a) Temperature dependence of Pt and Rh solubilities in a haplobasaltic melt. *Geochemica et Cosmochemica Acta* 67: 123–131.

Fortenfant SS, Rubie DC, Reid J, Dalpé C, and Gessmann CK (2003b) Partitioning of Re and Os between liquid metal and magnesiowüstite at high pressure. *Physics of the Earth and Planetary Interiors* 139: 77–91.

Frost DJ, Liebske C, Langenhorst F, McCammon CA, Trønnes RG, and Rubie DC (2004) Experimental evidence for the existence of iron-rich metal in the Earth's lower mantle. *Nature* 428: 409–412.

Gaetani GA and Grove TL (1999) Wetting of mantle olivine by sulfide melt: Implications for Re/Os ratios in mantle peridotite and late-stage core formation. *Earth and Planetary Science Letters* 169: 147–163.

Gancarz AJ and Wasserburg GJ (1977) Initial Pb of the Amîtsoq gneiss, west Greenland, and implications for the age of the Earth. *Geochemica et Cosmochemica Acta* 41(9): 1283–1301.

Gessmann CK and Rubie DC (1998) The effect of temperature on the partitioning of Ni, Co, Mn, Cr and V at 9 GPa and constraints on formation of the Earth's core. *Geochemica et Cosmochemica Acta* 62: 867–882.

Gessmann CK and Rubie DC (2000) The origin of the depletions of V, Cr and Mn in the mantles of the Earth and Moon. *Earth and Planetary Science Letters* 184: 95–107.

Gessmann CK, Wood BJ, Rubie DC, and Kilburn MR (2001) Solubility of silicon in liquid metal at high pressure: Implications for the composition of the Earth's core. *Earth and Planetary Science Letters* 184: 367–376.

Ghosh A and McSween HY (1998) A thermal model for the differentiation of Asteroid 4 Vesta, based on radiogenic heating. *Icarus* 134: 187–206.

Greenwood RC, Franchi IA, Jambon A, and Buchanan PC (2005) Widespread magma oceans on asteroid bodies in the early solar system. *Nature* 435: 916–918 (doi:10.1038/nature03612).

Grimm RE and McSween HY (1993) Heliocentric zoning of the asteroid belt by aluminum-26 heating. *Science* 259: 653–655.

Groebner N and Kohlstedt DL (2006) Deformation-induced metal melt networks in silicates: Implications for core–mantle interactions in planetary bodies. *Earth and Planetary Science Letters* 245: 571–580.

Haisch KE, Lada EA, and Lada CJ (2001) Disk frequencies and lifetimes in young clusters. *Astronomical Journal* 553: L153–L156.

Halliday AN (2003) The origin and earliest history of the Earth. In: Holland HD and Turekian KK (eds.) *Treatise on Geochemistry*, vol. 1, pp. 509–557. Amsterdam: Elsevier.

Halliday AN (2004) Mixing, volatile loss and compositional change during impact-driven accretion of the Earth. *Nature* 427: 505–509.

Hallworth MA, Phillips JC, Hubbert HE, and Sparks RSJ (1993) Entrainment in turbulent gravity currents. *Nature* 362: 829–831.

Hanks TC and Anderson DL (1969) The early thermal history of the Earth. *Physics of the Earth and Planetary Interiors* 2: 19–29.

Harper CL and Jacobsen SB (1996) Evidence for Hf-182 in the early solar system and constraints on the timescale of terrestrial accretion and core formation. *Geochemica et Cosmochemica Acta* 60: 1131–1153.

Helffrich G and Kaneshima S (2004) Seismological constraints on core composition from Fe-O-S liquid immiscibility. *Science* 306: 2239–2242.

Herd CDK, Borg LE, Jones JH, and Papike JJ (2002) Oxygen fugacity and geochemical variations in the martian basalts: Implications for Martian basalt petrogenesis and the oxidation state of the upper mantle of Mars. *Geochemica et Cosmochemica Acta* 66: 2025–2036.

Hillgren VJ, Drake MJ, and Rubie DC (1994) High-pressure and high-temperature experiments on core-mantle segregation in the accreting Earth. *Science* 264: 1442–1445.

Hillgren VJ, Gessmann CK, and Li J (2000) An experimental perspective on the light element in Earth's core. In: Canup RM and Righter K (eds.) *Origin of the Earth and Moon*, pp. 245–264. Tucson, AZ: University of Arizona Press.

Höink T, Schmalzl J, and Hansen U (2006) Dynamics of metal-silicate separation in a terrestrial magma ocean. *Geochemistry Geophysics Geosystems* 7: Q09008 (doi:10.1029/2006GC001268).

Holzheid A, Schmitz MD, and Grove TL (2000a) Textural equilibria of iron sulphide liquids in partly molten silicate aggregates and their relevance to core formation scenarios. *Journal of Geophysical Research* 105: 13555–13567.

Holzheid A, Sylvester P, O'Neill HSC, Rubie DC, and Palme H (2000b) Evidence for a late chondritic veneer in the Earth's mantle from high-pressure partitioning of palladium and platinum. *Nature* 406: 396–399.

Hustoft JW and Kohlstedt DL (2006) Metal-silicate segregation in deforming dunitic rocks. *Geochemistry Geophysics Geosystems* 7: Q02001 (doi:10.1029/2005GC001048).

Iida T and Guthrie RL (1988) *The Physical Properties of Liquid Metals*, pp. 109–146. Oxford: Clarendon.

Jacobsen SB (2005) The Hf-W isotopic system and the origin of the Earth and Moon. *Annual Review of Earth and Planetary Sciences* 33: 531–570.

Jeffreys H (1952) *The Earth*, 3rd edn. Cambridge: Cambridge University Press.

Jephcoat A and Olson P (1987) Is the inner core of the Earth pure iron? *Nature* 325: 332–335.

Jones JH, Capobianco CJ, and Drake MJ (1992) Siderophile elements and the Earth's formation. *Science* 257: 1281–1282.

Jones JH and Drake MJ (1986) Geochemical constraints on core formation in the Earth. *Nature* 322: 221–228.

Jones JH, Neal CR, and Ely JC (2003) Signatures of the siderophile elements in the SNC meteorites and Mars: A review and petrologic synthesis. *Chemical Geology* 196: 21–41.

Kanda R and Stevenson DJ (2006) Suction mechanism for iron entrainment into the lower mantle. *Geophysical Research Letters* 33: L02310 (doi:10.1029/2005GL025009).

Karato SI and Murthy VR (1997) Core formation and chemical equilibrium in the Earth. Part I: Physical considerations. *Physics of the Earth and Planetary Interiors* 100: 61–79.

Kato T and Ringwood AE (1989) Melting relationships in the system Fe–FeO at high pressure: Implications for the composition and formation of the Earth's core. *Physics and Chemistry of Minerals* 16: 524–538.

Kaula WM (1979) Thermal evolution of the Earth and Moon growing by planetesimal impacts. *Journal of Geophysical Research* 84: 999–1008.

Kegler P, Holzheid A, Rubie DC, Frost DJ, and Palme H (2005) New results of metal/silicate partitioning of Ni and Co at elevated pressures and temperatures. *XXXVI Lunar and Planetary Science Conference*, Abstract #2030.

Keil K, Stoeffler D, Love SG, and Scott ERD (1997) Constraints on the role of impact heating and melting in asteroids. *Meteoritics and Planetary Science* 32: 349–363.

Keppler H and Rubie DC (1993) Pressure-induced coordination changes of transition-metal ions in silicate melts. *Nature* 364: 54–55.

Kilburn MR and Wood BJ (1997) Metal-silicate partitioning and the incompatibility of S and Si during core formation. *Earth and Planetary Science Letters* 152: 139–148.

Kimura K, Lewis RS, and Anders E (1974) Distribution of gold between nickel–iron and silicate melts: Implications for the abundances of siderophile elements on the Earth and Moon. *Geochemica et Cosmochemica Acta* 38: 683–701.

Kleine T, Palme H, Mezger M, and Halliday AN (2005) Hf-W chronometry of lunar metals and the age and early differentiation of the Moon. *Science* 310: 1671–1673.

Kleine T, Munker C, Mezger K, and Palme H (2002) Rapid accretion and early core formation on asteroids and the terrestrial planets from Hf-W chronometry. *Nature* 418: 952–955.

Kokubo E and Ida S (1998) Oligarchic growth of protoplanets. *Icarus* 131: 171–178.

Kominami J, Tanaka H, and Ida S (2005) Orbital evolution and accretion of protoplanets tidally interacting with a gas disk. Part I: Effects of interaction with planetesimals and other protoplanets. *Icarus* 178: 540–552.

Kong P, Ebihara M, and Palme H (1999) Siderophile elements in Martian meteorites and implications for core formation in Mars. *Geochemica et Cosmochemica Acta* 63: 1865–1875.

Kyte FT (1998) A meteorite from the Cretaceous/Tertiary boundary. *Nature* 396: 237–239.

Laporte D and Watson EB (1995) Experimental and theoretical constraints on melt distribution in crustal sources: The effect of crystalline anisotropy on melt interconnectivity. *Chemical Geology* 124: 161–184.

Li J and Agee CB (1996) Geochemistry of mantle–core differentiation at high pressure. *Nature* 381: 686–689.

Li J and Agee CB (2001) The effect of pressure, temperature, oxygen fugacity and composition on partitioning of nickel and cobalt between liquid Fe–Ni–S alloy and liquid silicate: Implications for the Earth's core formation. *Geochemica et Cosmochemica Acta* 65: 1821–1832.

Li J and Fei Y (2003) Experimental constraints on core composition. In: Holland HD and Turekian KK (eds.) *Treatise on Geochemistry, vol. 2: The Mantle and Core*, pp. 521–546. Oxford: Elsevier-Pergamon.

Liebske C, Schmickler B, Terasaki H, *et al.* (2005) Viscosity of peridotite liquid up to 13 GPa: Implications for magma ocean viscosities. *Earth and Planetary Science Letters* 240: 589–604.

Lindstrom DJ and Jones JH (1996) Neutron activation analysis of multiple 10–100 μg glass samples from siderophile element partitioning experiments. *Geochemica et Cosmochemica Acta* 60: 1195–1203.

Lodders K and Fegley B (1998) *The Planetary Scientist's Companion*. New York: Oxford University Press.

Maier WD, Andreoli MAG, McDonald I, *et al.* (2006) Discovery of a 25-cm asteroid clast in the giant Morokweng impact crater, South Africa. *Nature* 441: 203–206.

Malavergne V, Siebert J, Guyot F, *et al.* (2004) Si in the core? New high-pressure and high-temperature experimental data. *Geochimica et Cosmochemica Acta* 68: 4201–4211.

Mann U, Frost DJ, Rubie DC, Shearer CK, and Agee CB (2006) Is silicon a light element in Earth's core? – Constraints from liquid metal – liquid silicate partitioning of some lithophile elements. *XXXVII Lunar and Planetary Science Conference*, Abstract #1161.

Matsui T and Abe Y (1986) Impact-induced atmospheres and oceans on Earth and Venus. *Nature* 322: 526–528.

McDonough WF (2003) Compositional model for the Earth's core. In: Holland HD and Turekian KK (eds.) *Treatise on Geochemistry*, pp. 517–568. Oxford: Elsevier.

McDonough WF and Sun S-s (1995) The composition of the Earth. *Chemical Geology* 120: 223–253.

McSween HY (1999) *Meteorites and Their Parent Planets*. Cambridge, UK: Cambridge University Press.

Melosh HJ (1990) Giant impacts and the thermal state of the early Earth. In: Newsom NE and Jones JE (eds.) *Origin of the Earth*, pp. 69–83. Oxford: Oxford University Press.

Melosh HJ and Rubie DC (2007) Ni partitioning in the terrestrial magma ocean: A polybaric numerical model. *Lunar and Planetary Science XXXVIII* Abstract #1593.

Miller GH, Stolper EM, and Ahrens TJ (1991) The equation of state of a molten komatiite. Part 2: Application to komatiite petrogenesis and the Hadean mantle. *Journal of Geophysical Research* 96: 11849–11864.

Minarik WG, Ryerson FJ, and Watson EB (1996) Textural entrapment of core-forming melts. *Science* 272: 530–533.

Morbidelli A, Chambers J, Lunine JI, *et al.* (2000) Source regions and timescales for the delivery of water to the Earth. *Meteoritics and Planetary Science* 35: 1309–1320.

Morgan JW (1986) Ultramafic xenoliths: Clues to Earth's late accretionary history. *Journal of Geophysical Research* 91: 12375–12387.

Mosenfelder JL, Asimow PD, and Ahrens TJ (2007) Thermodynamic properties of Mg_2SiO_4 liquid at ultra-high pressures from shock measurements to 200 GPa on forsterite and wadsleyite. *Journal of Geophysical Research* (in press).

Murthy VR (1991) Early differentiation of the Earth and the problem of mantle siderophile elements – A new approach. *Science* 253: 303–306.

Nimmo F and Agnor CB (2006) Isotopic outcomes of N-body accretion simulations: Constraints on equilibration processes during large impacts from Hf-W observations. *Earth and Planetary Science Letters* 243: 26–43.

Ohtani E and Ringwood AE (1984) Composition of the core. Part I: Solubility of oxygen in molten iron at high temperatures. *Earth and Planetary Science Letters* 71: 85–93.

Ohtani E, Ringwood AE, and Hibberson W (1984) Composition of the core. Part II: Effect of high pressure on solubility of FeO in molten iron. *Earth and Planetary Science Letters* 71: 94–103.

Ohtani E and Yurimoto H (1996) Element partitioning between metallic liquid, magnesiowüstite, and silicate liquid at 20 GPa and 2500°C: A secondary ion mass spectrometric study. *Geophysical Research Letters* 23: 1993–1996.

O'Neill HSC (1992) Siderophile elements and the Earth's formation. *Science* 257: 1282–1284.

O'Neill HSC, Canil D, and Rubie DC (1998) Oxide-metal equilibria to 2500°C and 25 GPa: Implications for core formation and the light component in the Earth's core. *Journal of Geophysical Research* 103: 12239–12260.

O'Neill HSC and Palme H (1998) Composition of the silicate Earth: Implications for accretion and core formation. In: Jackson I (ed.) *The Earth's Mantle*, pp. 3–126. Cambridge, UK: Cambridge University Press.

Newsom HE (1990) Accretion and core formation in the Earth: Evidence from siderophile elements. In: Newsom HE and Jones JH (eds.) *Origin of the Earth*, pp. 273–288. New York: Oxford University Press.

Palme H and O'Neill HSC (2003) Cosmochemical estimates of mantle composition. In: Holland HD and Turekian KK (eds.) *Treatise on Geochemistry, vol. 2: The Mantle and Core*, pp. 1–38. Oxford: Elsevier-Pergamon.

Palme H, O'Neill H, St. C, and Benz W (2003) Evidence for collisional erosion of the Earth. *XXXIV Lunar and Planetary Science Conference*, Abstract #1741.

Pierazzo E, Vickery AM, and Melosh HJ (1997) A re-evaluation of impact melt production. *Icarus* 127: 408–432.

Poirier JP (1994) Light elements in the Earth's outer core: A critical review. *Physics of the Earth and Planetary Interiors* 85: 319–337.

Poirier JP, Malavergne V, and Le Mouël JL (1998) Is there a thin electrically conducting layer at the base of the mantle? In: Gurnis M, Wysession ME, Knittle E, and Buffett BA (eds.) *Geodynamics 28: The Core–Mantle Boundary Region*, pp. 131–137. Washington, DC: American Geophysical Union.

Porcelli D, Woolum D, and Cassen P (2001) Deep Earth rare gases: Initial inventories, capture from the solar nebula, and losses during Moon formation. *Earth and Planetary Science Letters* 193: 237–251.

Pritchard ME and Stevenson DJ (2000) Thermal aspects of a lunar origin by giant impact. In: Canum RM and Righter K (eds.) *Origin of the Earth and Moon*, pp. 179–196. Tucson, AZ: University of Arizona Press.

Quitté G, Latkoczy C, Halliday AN, Schönbächler M, and Günther D (2005) Iron-60 in the Eucrite parent body and the initial $^{60}Fe/^{56}Fe$ of the solar system. *LPSC XXXVI*, Abstract 1827.

Ranen MC and Jacobsen SB (2006) Barium isotopes in chondritic meteorites: Implications for planetary reservoir models. *Science* 314: 809–812.

Raymond SN, Quinn T, and Lunine JI (2004) Making other Earths: Dynamical simulations of terrestrial planet formation and water delivery. *Icarus* 168: 1–17.

Reid JE, Suzuki A, Funakoshi K, et al. (2003) The viscosity of $CaMgSi_2O_6$ liquid at pressures up to 13 GPa. *Physics of the Earth and Planetary Interiors* 139: 45–54.

Righter K (2003) Metal-silicate partitioning of siderophile elements and core formation in the early Earth. *Annual Review of Earth and Planetary Sciences* 31: 135–174.

Righter K (2005) Highly siderophile elements: Constraints on Earth accretion and early differentiation. In: van der Hilst RD, Bass JD, Matas J, and Trampert J (eds.) *Geophysical Monograph 160: Earth's Deep Mantle: Structure, Composition and Evolution*, pp. 201–218. Washington, DC: American Geophysical Union.

Righter K and Drake MJ (1996) Core Formation in Earth's Moon, Mars, and Vesta. *Icarus* 124: 513–529.

Righter K and Drake MJ (1997) Metal-silicate equilibrium in a homogeneously accreting Earth: New results for Re. *Earth and Planetary Science Letters* 146: 541–553.

Righter K and Drake MJ (1999) Effect of water on metal-silicate partitioning of siderophile elements: A high pressure and temperature terrestrial magma ocean and core formation. *Earth and Planetary Science Letters* 171: 383–399.

Righter K and Drake MJ (2000) Metal-silicate equilibration in the early Earth: New constraints from volatile moderately siderophile elements Ga, Sn, Cu and P. *Geochemica et Cosmochemica Acta* 64: 3581–3597.

Righter K and Drake MJ (2001) Constraints on the depth of an early terrestrial magma ocean. *Meteoritics and Planetary Science* 36: A173.

Righter K and Drake MJ (2003) Partition coefficients at high pressure and temperature. In: Carlson RW (ed.) *Treatise on Geochemistry*, vol. 2, pp. 425–449. Oxford: Elsevier.

Righter K, Drake MJ, and Yaxley G (1997) Prediction of siderophile element metal-silicate partition coefficients to 20 GPa and 2800 degrees C: The effects of pressure, temperature, oxygen fugacity, and silicate and metallic melt compositions. *Physics of the Earth and Planetary Interiors* 100: 115–134.

Ringwood AE (1977) Composition of the core and implications for the origin of the Earth. *Geochemical Journal* 11: 111–135.

Rose LA and Brenan JM (2001) Wetting properties of Fe–Ni–Co–Cu–O–S melts against olivine: Implications for sulfide melt mobility. *Economic Geology and Bulletin of the Society of Economics Geologists* 96: 145–157.

Rubie DC, Gessmann CK, and Frost DJ (2004) Partitioning of oxygen during core formation on the Earth and Mars. *Nature* 429: 58–61.

Rubie DC, Melosh HJ, Reid JE, Liebske C, and Righter K (2003) Mechanisms of metal-silicate equilibration in the terrestrial magma ocean. *Earth and Planetary Science Letters* 205: 239–255.

Rubin AM (1995) Propagation of magma-filled cracks. *Annual Review of Earth and Planetary Sciences* 23: 287–336.

Rushmer T, Minarik WG, and Taylor GJ (2000) Physical processes of core formation. In: Canup RM and Righter K (eds.) *Origin of the Earth and Moon*, pp. 227–244. Tucson, AZ: University of Arizona Press.

Rushmer T, Petford N, Humayun M, and Campbell AJ (2005) Fe-liquid segregation in deforming planetesimals: Coupling core-forming compositions with transport phenomena. *Earth and Planetary Science Letters* 239: 185–202.

Russell CT and Luhmann JG (1997) Mercury: Magnetic field and magnetosphere. In: Shirley JH and Fairbridge RW (eds.) *Encyclopedia of Planetary Sciences*, pp. 476–478. New York: Chapman and Hall.

Ruzicka A, Snyder GA, and Taylor LA (1997) Vesta as the howardite, eucrite and diogenite parent body: Implications for the size of a core and for large-scale differentiation. *Meteoritics and Planetary Science* 32: 825–840.

Safronov VS (1978) The heating of the Earth during its formation. *Icarus* 33: 1–12.

Sakai T, Kanto T, Ohtani E, et al. (2006) Interaction between iron and post-perovskite at core–mantle boundary and core signature in plume source region. *Geophysical Research Letters* 33: L15317 (doi 10.1029/2006GL026868).

Scherstén A, Elliott T, Hawkesworth C, Russell S, and Massarik J (2006) Hf-W evidence for rapid differentiation of iron meteorite parent bodies. *Earth and Planetary Science Letters* 241: 530–542.

Schmickler B, Liebske C, Holzapfel C, and Rubie DC (2005) Viscosity of peridotite liquid up to 24 GPa: Predictions from self-diffusion coefficients. *EOS Transactions American Geophysical Union* 86(52): Fall Meeting Supplement, Abstract MR11A-06.

Schoenberg R, Kamber BS, Collerson KD, and Eugster O (2002) New W-isotope evidence for rapid terrestrial accretion and very early core formation. *Geochemica et Cosmochemica Acta* 66: 3151–3160.

Schramm DN, Tera F, and Wasserburg GJ (1970) The isotopic abundance of ^{26}Mg and limits on ^{26}Al in the early solar system. *Earth and Planetary Science Letters* 10: 44–59.

Schubert G, Anderson JD, Spohn T, and McKinnon WB (2004) Interior composition, structure and dynamics of the Galilean satellites. In: Bagenal F, Dowling T, and Mckinnon WB (eds.) *Jupiter: The Planet, Satellites and Magnetosphere*, pp. 281–306. Cambridge, UK: Cambridge University Press.

Shannon MC and Agee CB (1996) High pressure constraints on percolative core formation. *Geophysical Research Letters* 23: 2717–2720.

Shannon MC and Agee CB (1998) Percolation of core melts at lower mantle conditions. *Science* 280: 1059–1061.

Snow JE and Schmidt G (1998) Constraints on Earth accretion deduced from noble metals in the oceanic mantle. *Nature* 391: 166–169.

Solomatov VS (2000) Fluid dynamics of a terrestrial magma ocean. In: Canup RM and Righter K (eds.) *Origin of the Earth and Moon*, pp. 323–338. Tucson, AZ: University of Arizona Press.

Solomon S (1979) Formation, history and energetics of cores in the terrestrial planets. *Physics of the Earth and Planetary Interiors* 19: 168–182.

Spohn T and Schubert G (1991) Thermal equilibration of the Earth following a giant impact. *Geophysical Journal International* 107: 163–170.

Stevenson DJ (1981) Models of the Earth's core. *Science* 214: 611–619.

Stevenson DJ (1989) Formation and early evolution of the Earth. In: Peltier WR (ed.) *Mantle Convection plate tectonics and global dynamics*, pp. 817–873. New York: Gordon and Breach.

Stevenson DJ (1990) Fluid dynamics of core formation. In: Newsom HE and Jones JH (eds.) *Origin of the Earth*, pp. 231–250. New York: Oxford University Press.

Stevenson DJ (2003) Mission to the Earth's core – A modest proposal. *Nature* 423: 239–240.

Tachibana S, Huss GR, Kita NT, Shimoda G, and Morishita Y (2006) Fe-60 in chondrites: Debris from a nearby supernova in the early solar system? *Astrophysical Journal* 639: L87–L90.

Taylor SR and Norman MD (1990) Accretion of differentiated planetesimals to the Earth. In: Newsom HE and Jones JH (eds.) *Origin of the Earth*, pp. 29–43. New York: Oxford University Press.

Takafuji N, Hirose K, Mitome M, and Bando Y (2005) Solubilities of O and Si in liquid iron in equilibrium with (Mg,Fe), SiO_3 perovskite and the light elements in the core. *Geophysical Research Letters* 32: L06313 (doi: 10.1029/2005GLO22773).

Takafuji N, Hirose K, Ono S, Xu F, Mitome M, and Bando Y (2004) Segregation of core melts by permeable flow in the lower mantle. *Earth and Planetary Science Letters* 224: 249–257.

Terasaki H, Frost DJ, Rubie DC, and Langenhorst F (2005) The effect of oxygen and sulphur on the dihedral angle between Fe–O–S melt and silicate minerals at high pressure: Implications for Martian core formation. *Earth and Planetary Science Letters* 232: 379–392 (doi:10.1016/j.epsl.2005.01.030).

Terasaki H, Frost DJ, Rubie DC, and Langenhorst F (2007a) Percolative core formation in planetesimals. *Earth and Planetary Science Letters* (in press).

Terasaki H, Frost DJ, Rubie DC, and Langenhorst F (2007b) The interconnectivity of Fe–O–S liquid in polycrystalline silicate perovskite at lower mantle conditions. *Physics of the Earth and Planetary Interiors* (in press).

Thibault Y and Walter MJ (1995) The influence of pressure and temperature on the metal-silicate partition coefficients of nickel and cobalt in a model C1 chondrite and implications for metal segregation in a deep magma ocean. *Geochemica et Cosmochemica Acta* 59: 991–1002.

Thomas PC, Parker JW, McFadden LA, et al. (2005) Differentiation of the asteroid Ceres as revealed by its shape. *Nature* 437: 224–226.

Thomson W and Tait PG (1883) *Treatise on Natural Philosophy*, 527p. Cambridge: Cambridge University Press.

Tonks WB and Melosh HJ (1992) Core formation by giant impacts. *Icarus* 100: 326–346.

Tonks WB and Melosh HJ (1993) Magma ocean formation due to giant impacts. *Journal of Geophysical Research* 98: 5319–5333.

Treiman AH, Jones JH, and Drake MJ (1987) Core formation in the Shergottite parent body and comparison with the Earth. *Journal of Geophysical Research* 92: E627–E632.

Tschauner O, Zerr A, Specht S, Rocholl A, Boehler R, and Palme H (1999) Partitioning of nickel and cobalt between silicate perovskite and metal at pressures up to 80 GPa. *Nature* 398: 604–607 (doi:10.1038/19287).

Turan JW (1984) Nucleosynthesis. *Annual Review of Nuclear and Particle Science* 34: 53–97.

Turcotte DL and Schubert G (2002) *Geodynamics*. Cambridge, UK: Cambridge University Press.

Urey HC (1952) *The Planets: Their Origin and Development*, 245p. New Haven, CT: Yale University Press.

von Bargen N and Waff HS (1986) Permeabilities, interfacial areas and curvatures of partially molten systems: Results of numerical computations of equilibrium microstructures. *Journal of Geophysical Research* 91: 9261–9276.

Wade J and Wood BJ (2005) Core formation and the oxidation state of the Earth. *Earth and Planetary Science Letters* 236: 78–95.

Wadhwa M (2001) Redox state of Mars' upper mantle and crust from Eu anomalies in Shergottite pyroxenes. *Science* 291: 1527–1530.

Walker D (2000) Core participation in mantle geochemistry: Geochemical Society Ingerson Lecture. *Geochemica et Cosmochemica Acta* 64: 2897–2911.

Walker D (2005) Core–mantle chemical issues. *Canadean Mineralogist* 43: 1553–1564.

Walker D, Norby L, and Jones JH (1993) Superheating effects on metal-silicate partitioning of siderophile elements. *Science* 262: 1858–1861.

Walte NP, Becker JK, Bons PD, Rubie DC, and Frost DJ (2007) Liquid distribution and attainment of textural equilibrium in a partially-molten crystalline system with a high-dihedral-angle liquid phase. *Earth and Planetary Science Letters* (in press).

Walter MJ, Newsom HE, Ertel W, and Holzheid A (2000) Siderophile elements in the Earth and Moon: Metal/silicate partitioning and implications for core formation. In: Canup RM and Righter K (eds.) *Origin of the Earth and Moon*, pp. 265–290. Tucson, AZ: University of Arizona Press.

Walter MJ and Thibault Y (1995) Partitioning of tungsten and molybdenum between metallic liquid and silicate melt. *Science* 270: 1186–1189.

Walter MJ and Trønnes RG (2004) Early Earth differentiation. *Earth and Planetary Science Letters* 225: 253–269.

Warren PH (1985) The magma ocean concept and lunar evolution. *Annual Review of Earth and Planetary Sciences* 13: 201–240.

Warren PH (1993) A concise compilation of petrologic information on possible pristine nonmare Moon rocks. *American Mineralogist* 78: 360–376.

Warren PH, Kallemeyn GW, and Kyte FT (1999) Origin of planetary cores: Evidence from highly siderophile elements in Martian meteorities. *Geochemica et Cosmochemica Acta* 63: 2105–2122.

Wasson JT (1985) *Meteorites: Their Record of Early Solar-System History*, 267p. New York: W. H. Freeman.

Weidenschilling SJ, Spaute D, Davis DR, Marzari F, and Ohtsuki K (1997) Accretional evolution of a planetesimal swarm. Part 2: The terrestrial zone. *Icarus* 128: 429–455.

Wetherill GW (1985) Occurrence of giant impacts during the growth of the terrestrial planets. *Science* 228: 877–879.

Wetherill GW (1990) Formation of the Earth. *Annual Review of Earth and Planetary Sciences* 18: 205–256.

Wetherill GW and Stewart GR (1993) Formation of planetary embryos – Effects of fragmentation, low relative velocity, and independent variation of eccentricity and inclination. *Icarus* 106: 190–209.

Williams J-P and Nimmo F (2004) Thermal evolution of the Martian core: Implications for an early dynamo. *Geology* 32: 97–100.

Williams Q and Hemley RJ (2001) Hydrogen in the deep Earth. *Annual Review of Earth and Planetary Sciences* 29: 365–418.

Wood BJ and Halliday AN (2005) Cooling of the Earth and core formation after the giant impact. *Nature* 437: 1345–1348.

Wood BJ, Walter MJ, and Wade J (2006) Accretion of the Earth and segregation of its core. *Nature* 441: 825–833 (doi:10.1038/nature04763).

Yin QZ, Jacobsen SB, Yamashita K, Blichert-Toft J, Telouk P, and Albarede F (2002) A short timescale for terrestrial planet formation from Hf-W chronometry of meteorites. *Nature* 418: 949–952.

Yoder CF (1995) Astrometric and geodetic properties of Earth and the solar system. In: Ahrens TJ (ed.) *Global Earth Physics*, pp. 1–31. Washington, DC: American Geophysical Union.

Yoshino T, Price JD, Wark DA, and Watson EB (2006) Effect of faceting on pore geometry in texturally equilibrated rocks: Implications for low permeability at low porosity. *Contributions to Mineralogy and Petrology* 152: 169–186 (doi:10.1007/s00410-006-0099-y).

Yoshino T, Walter MJ, and Katsura T (2003) Core formation in planetesimals triggered by permeable flow. *Nature* 422: 154–157.

Young GA (1965) *The Physics of the Base Surge*. US White Oak, MD: Naval Ordnance Laboratory.

4 Magma Oceans and Primordial Mantle Differentiation

V. Solomatov, Washington University, St. Louis, MO, USA

Nomenclature

a	prefactor in nucleation function
a_u	prefactor in scaling law for velocity
c_p	isobaric specific heat
c_p'	apparent isobaric specific heat in two-phase regions
d	crystal diameter
d_{crit}	crystal diameter separating equilibrium and fractional crystallization
d_e	critical crystal diameter for equilibrium crystallization
d_f	critical crystal diameter for fractional crystallization
d_{nucl}	crystal diameter after nucleation
d_{ost}	crystal diameter controlled by Ostwald ripening
d_{ost}'	crystal diameter in shallow magma ocean
f_ϕ	hindering settling function
g	acceleration due to gravity
k_B	Boltzmann's constant
n_i	mole fraction of component i
n_i^l	mole fraction of liquid component i
n_i^s	mole fraction of solid component i
n_{mw}	mole fraction of magnesiowüstite
t	time

t_{conv}	crystallization time due to solid-state convection	V	volume of magma ocean
t_{ost}	Ostwald ripening time in deep magma ocean	W	mechanical work per unit time
t'_{ost}	Ostwald ripening time in shallow magma ocean	α	thermal expansion
		α'	apparent thermal expansion in two-phase regions
t_{RT}	Rayleigh-Taylor instability time	α_η	coefficient in melt fraction-dependent viscosity
t_s	settling time		
u_D	Darcy velocity	δ	thermal boundary layer thickness
u_0	amplitude of convective velocity	ϵ	efficiency factor
u_{perc}	percolation velocity of melt	ζ	ratio of supersaturation to supercooling
u_s	settling velocity of crystals	η	viscosity
x^*	velocity coefficient for turbulent boundary layer	η_l	viscosity of melt
		η_s	viscosity of solid
z	depth	θ	parameter in analytical solution for nucleation
A	surface area of magma ocean		
B	exponential coefficient in nucleation function	κ	coefficient of thermal diffusivity
		λ	aspect ratio of mean flow
D	diffusion coefficient	μ_i^l	chemical potential of component i in liquid
E	activation energy of solid	μ_i^s	chemical potential of component i in solid
E_l	activation energy of melt	$\mu_i^l(0)$	chemical potential of pure component i in liquid state
F	heat flux		
G	Gibbs energy	$\mu_i^s(0)$	chemical potential of pure component i in solid state
I	mixing length		
L	depth of magma ocean	ν	kinematic viscosity
L'	depth of partially molten upper mantle	ρ	density
M	mass of magma ocean	σ	surface energy
P	pressure	σ_{app}	apparent surface energy
Pr	Prandtl number	σ_{SB}	Stefan-Boltzmann constant
R	gas constant	τ_T	convective stress scale
Ra	Rayleigh number	τ^*	skin friction
Ra_*	critical Rayleigh number for transition to no-rotation regime	ϕ	crystal fraction
		ϕ_l	melt fraction
T	temperature	ϕ_m	maximum packing fraction
$\dot{T}$	cooling rate	ΔH	enthalpy change upon melting
T'	supercooling	ΔS	entropy change upon melting
T_{ad}	adiabatic temperature	ΔS_{per}	entropy change upon melting of perovskite
T_e	effective temperature		
T_{liq}	liquidus temperature	ΔT_{rh}	rheological temperature scale
T_m	magma ocean temperature	$\Delta \rho$	crystal/melt density difference
T_{per}	melting temperature of perovskite	$\Delta \rho_{RT}$	density difference for Rayleigh-Taylor instability
T_s	surface temperature		
T_{sol}	solidus temperature	Φ	energy release per unit time
Ta	Taylor number	Ω	angular velocity

4.1 Earth Accretion and the Giant Impact Hypothesis

The idea that the early Earth could have been substantially molten is not new. Lord Kelvin's famous estimate of the age of the Earth was based on the assumption that the Earth was once completely molten and that the present-day surface heat flux resulted from the cooling of the molten Earth (Thomson, 1864). Later, isotopic dating showed that his estimates of the Earth's age were far too short. Radiogenic heating and mantle convection proved to play a major role in the thermal history of the Earth. The assumption of initially molten Earth did not find much observational or theoretical support and was not accepted either. The theory of Earth formation, which dominated in the middle of the twentieth century, was quite the opposite to Lord Kelvin's theory, and can be characterized as soft accretion: Earth accreted from a relatively uniform influx of small particles which gently settled on Earth without causing any 'harm'. Melting, necessary for core formation, was thought to be caused by radiogenic heating later in Earth's history (Elsasser, 1963).

The first observational evidence that accretion was not so soft came from the Moon. The analysis of samples returned by the Apollo missions suggested a large-scale differentiation of the Moon which was most easily explained by the crystallization of a lunar magma ocean (Wood *et al.*, 1970; Wood, 1975; Ringwood and Kesson, 1976; Longhi, 1980; Warren, 1985). This led Hostetler and Drake to propose as early as in 1980 that, perhaps, the Earth and other planets were molten in their early histories as well (Hostetler and Drake, 1980).

That Earth accretion was very energetic became obvious upon the realization that the Earth-forming planetesimals also grew with time, like the Earth itself. It was shown that at the latest stages of planetary accretion the population of Earth-forming planetesimals was very likely to include large bodies. Initially, the largest size was estimated at 10^{-3} of the Earth's mass or about 1000 km in diameter, based on coagulation theory and the observed planetary obiliquities which were assumed to be affected by large impacts (Safronov, 1964, 1978). Numerical investigations showed that the largest impactors could have reached about 0.1 of the Earth's mass which is equivalent to the size of Mars (Safronov, 1964, 1978; Wetherill, 1975, 1985, 1990; Weidenschilling *et al*, 1997). Mars-size impactors had enough energy to melt and partially vaporize the Earth (Safronov, 1964, 1978; Kaula, 1979; Benz and Cameron, 1990; Melosh, 1990).

The giant impact hypothesis suggested a successful explanation for the origin of the Moon, its composition, and the angular momentum of the Earth–Moon system (Benz *et al.*, 1986, 1987, 1989; Stevenson, 1987; Newsom and Taylor, 1989; Canup and Esposito, 1996; Ida *et al.*, 1997; Cameron, 1997; Canup and Agnor, 2000; Canup and Asphaug, 2001; Canup, 2004). The conclusion that seems inevitable is that the Earth was born as result of rather 'violent accretion' during which it was substantially melted. An interesting description of this paradigm shift can be found in a short review by Drake (2000).

Although the kinetic energy of giant impacts was undoubtedly the largest energy source, there are other factors which contributed to Earth heating and melting. Numerical models showed that thermal blanketing effects of the atmosphere on the growing Earth helped to retain the heat during the accretion and could have maintained the surface temperature above the solidus (Hayashi *et al.*, 1979; Nakazawa *et al.*, 1985; Abe and Matsui, 1986; Matsui and Abe, 1986; Zahnle *et al.*, 1988). Core formation released energy corresponding to about 2000 K heating of the entire Earth (Flasar and Birch, 1973). Some energy sources were active on a timescale of $\sim 10^6$ years, which is much shorter than the timescale of planetary formation, $\sim 10^7 - 10^8$ years (Wetherill, 1990; Chambers and Wetherill, 1998; Canup and Agnor, 2000). These include radiogenic heating by short-lived isotopes such as ^{26}Al and ^{60}Fe (Urey, 1955; Shukolyukov and Lugmair, 1993; Srinivasan *et al.*, 1999; Yoshino *et al*, 2003; Mostefaoui *et al.*, 2005) and electromagnetic induction heating (Sonett *et al.*, 1968). These energy sources contributed to heating and differentiation in both the proto-Earth and the Earth-forming planetesimals during the early stages of planetary formation.

Although the accretion was a stochastic process and the Earth was melted to some degree many times by bodies of varying sizes, the major event was probably the Moon-forming impact at the end of accretion as suggested by numerical simulations (Canup, 2004). Such an impact seems to explain best the properties of the Earth–Moon system, including its angular momentum, the masses of the Earth and Moon, and the iron depletion of the Moon. After the impact, the mantle temperature varies significantly – from 2000 K to 10^4 K, with some parts of the mantle being completely molten and others remaining solid. These estimates are consistent with earlier simulations (Melosh, 1990; Tonks and Melosh, 1993; Pierazzo *et al.*, 1997). Unfortunately, the numerical simulations can look only at the first hours of evolution (1 day at most). It is hard to predict the 'final' temperature distribution – after all major compositional and thermal heterogeneities settle down. This must be a very fast process because the stresses associated with these heterogeneities are huge and might even exceed the ultimate strength of solid rocks, which is of the order of 1–2 GPa (Davies, 1990). The gravitational energy release due to the redistribution of thermal heterogeneities in the mantle and segregation of the remaining iron increases the temperature further by probably another several hundred degrees (Tonks and Melosh, 1990, 1992). Thus, at the end of the large-scale redistribution of density heterogeneities, the hottest mantle material is completely molten and a small fraction of the coldest material remains solid and accumulate at the bottom of the magma ocean (the solidus temperature at the base of the mantle is around 4000–5000 K; Holland and Ahrens, 1997; Zerr *et al.*, 1998).

Interestingly, Canup's (2004) simulations show that the material which is heated most ($>10^4$ K) is the iron core of the impactor. It quickly sinks into the core of the proto-Earth, thus suggesting that the Earth's core was likely to be very hot. A substantial fraction of the energy of the superheated core is expected to be quickly transferred to the mantle (Ke and Solomatov, 2006). This may cause an additional melting of the Earth's mantle.

Another important and yet poorly quantified energy source is tidal heating. Given the fact that the Moon was very close to the Earth right after the Moon formation, it is reasonable to expect that tidal heating contributed significantly to sustaining the magma ocean after the Moon-forming event (Sears 1993).

The term 'magma ocean' has been widely used to describe the initial molten state of the Earth. Originally, this term was applied to the lunar magma ocean and was defined as "a global, near-surface shell of magma, tens or hundreds of kilometers thick" (Warren, 1985). Warren warned about the shortcomings of this term (he proposed a different term, 'magmasphere' but it is not used as widely as 'magma ocean'): "Among other things, 'ocean' implies the system is virtually 100% liquid, with a mainly gas/liquid upper surface and a waterlike viscosity of the order of 10^{-2} poise". It has been realized that the structure of the early molten mantle of the Earth and other planets can be quite complex: the molten Earth may have no stable crust in the early times but may develop a thin crust later on, while still remaining molten over hundreds of kilometers below the crust; the molten layer may form not only near the surface but also in the middle of the mantle and even at the base of the mantle; and the degree of melting can vary both radially and laterally. Here, we will use the term 'magma ocean' in a broad sense – to describe a partially or a completely molten layer within the Earth's mantle whose motions are controlled by liquid viscosity. This means that it can be underground and it can have a relatively large viscosity.

4.2 Geochemical Evidence for Magma Ocean

An early global magma ocean would undoubtedly have left some record of its existence in geochemical data. Indeed, early studies showed that segregation of crystals in a deep magma ocean seemed to produce the major element composition of the upper mantle of the Earth (defined by the spinel–perovskite phase transition around 23 GPa), and in particular could explain the elevated Mg/Si ratios in the upper mantle (Ohtani, 1985; Ohtani and Sawamoto, 1987; Agee and Walker, 1988; Herzberg and Gasparik, 1991). However, geochemical models of differentiation of a terrestrial magma ocean showed that any substantial segregation of perovskite crystals in the lower mantle would drive the ratios of minor and trace elements well outside of their observed range (Kato *et al.*, 1988a, 1988b; Ringwood, 1990; McFarlane and Drake, 1990; McFarlane *et al.*, 1994). On the other hand, it was pointed out that the partition coefficients were not well constrained at realistic temperatures, pressures, and compositions of magma oceans and it was premature to draw any conclusions (Presnall *et al.*, 1998). Later studies showed that fractionation of small amounts of Ca-perovskite in addition to Mg-perovskite can increase the allowed amount of fractionation of perovskite. Possible amounts of perovskite fractionation are in the range from 5% to 15% (Corgne and Wood, 2002; Walter *et al.*, 2004; Walter and Trønnes, 2004; Corgne *et al.*, 2005; Liebske *et al.*, 2005a). Ito *et al.* (2004) suggest that it can be as high as 40%. The third phase, magnesiowüstite (or ferroperi- clase), does not accommodate trace elements well so that its fractionation is not expected to have any significant effect on elemental ratios (Walter *et al.*, 2004).

Analysis of isotopic data for $^{176}Hf–^{176}Lu$ and $^{147}Sm–^{143}Nd$ systems indicates that continental crust formed very early, perhaps in the first 200 million years of evolution, which might require a magma ocean (Collerson *et al.*, 1991; Bennett *et al.*, 1993; Bowring and Housh, 1995; Albarède *et al.*, 2000; Amelin *et al.*, 1999; Bizzarro *et al.*, 2003, Harrison *et al.*, 2005). Isotopic data on compositions of ^{142}Nd produced by short-lived ^{146}Sm (Caro *et al.*, 2003; Boyet and Carlson, 2005) indicate even shorter timescales, within 30 million years, for mantle differentiation.

A simultaneous modeling of several isotopic systems can potentially provide a very tight constraint on mantle differentiation (e.g., Halliday, 2004). Caro *et al.* (2005) argued that superchondritic Sm/Nd ratios of early Archaean rocks can be reconciled with a nearly chondritic Lu/Hf ratio observed in zircons from the same place and of the same age by extraction of about 0.3 wt.% of melt from the upper mantle. Differentiation at small degrees of melting was also proposed earlier to explain the ratio of major and minor elements (Gasparik and Drake, 1995).

Another argument for an early molten state of the Earth comes from the fact that core formation on Earth as well as other planetary bodies happened fast, perhaps within 30–50 My of the solar system history, which is inferred from the analysis of the short-lived [182]Hf–[182]W system (Lee and Halliday, 1995; Halliday et al., 1996; Kleine et al., 2002; Yin et al., 2002; Foley et al., 2005; Halliday, 2004; Jacobsen, 2005). This requires some degree of melting of the Earth's mantle very early in planetary history (Stevenson, 1990). To what extent the Earth was molten during core formation is not constrained by Hf–W data. The depth of the molten layer of the Earth can be inferred from the analysis of siderophile elements. Laboratory experiments on partition coefficients of siderophile elements suggest that metal/silicate chemical equilibrium was established around 28 GPa and 2200 K which can be interpreted as the bottom of a magma ocean (Li and Agee, 1996; Righter et al., 1997; Righter and Drake, 1997; Righter, 2003). According to this model, iron delivered by impacts first accumulates at the bottom of a magma ocean and then sinks through the mostly solid layers below the magma ocean (Elsasser, 1963; Stevenson, 1981, 1990; Karato and Murthy, 1997a, 1997b; Solomatov, 2000). The pressure at the bottom of the magma ocean might be noticeably different from the apparent equilibration pressure because metal/silicate equilibrium was established during a 'rainfall' of iron droplets rather than in the iron layer at the base of magma ocean (Rubie et al., 2003). Besides, the bottom of the magma ocean may not be well described as a simple boundary but rather as an extensive partially molten region with variable melt fraction and a complicated structure and dynamics.

The geochemical constraints can be summarized as follows. Only a small amount of perovskite could have been fractionated via crystal–melt segregation (~10%). Chemical differentiation was likely to occur in the upper parts of the mantle via percolation of melt in a partially molten mantle at small melt fractions. The timescales for core formation and mantle differentiation appear to be very close to each other and both are close to the accretion timescale constrained by numerical simulations (Wetherill, 1990; Chambers and Wetherill, 1998; Canup and Agnor, 2000). This suggests that core

formation, mantle differentiation, and accretion might have been occurring simultaneously. The discussion below is based largely on Slamotov and Stevenson (1993a, 1993b, 1993c), Solomatov et al. (1993), and Solomatov (2000).

4.3 Thermal Structure of a Convecting Magma Ocean

4.3.1 Adiabats

In vigorously convecting systems such as magma oceans, the temperature distribution is nearly adiabatic and isentropic. In one-phase systems, such as a completely molten or a completely solid layer, the equation for an adiabat is

$$\frac{dT}{dP} = \frac{\alpha T}{\rho c_p} \qquad [1]$$

where T is the temperature, P is the pressure, α is the coefficient of thermal expansion, c_p is the isobaric specific heat, and ρ is the density.

In two-phase systems, the effects of phase changes need to be considered (Abe, 1993, 1995, 1997; Miller et al., 1991b; Solomatov and Stevenson, 1993b). In the absence of chemical differentiation and assuming thermodynamic equilibrium, the volume fractions as well as compositions of each mineral phase depend only on T and p. Thus, the system can be described with the help of two thermodynamic variables, T and p. All thermodynamic parameters such as the coefficient of thermal expansion and the isobaric specific heat can be obtained the same way as for a one-phase system. In terms of two-phase parameters, α' and c_p', the equation for the adiabat in the temperature range between liquidus and solidus can be written as

$$\frac{dT}{dP} = \frac{\alpha' T}{\rho c_p'} \qquad [2]$$

The order-of-magnitude values of α' and c_p' are

$$\alpha' \sim \alpha + \frac{\Delta\rho}{\rho(T_{liq} - T_{sol})} \qquad [3]$$

and

$$c_p' \sim c_p + \frac{\Delta H}{T_{liq} - T_{sol}} \qquad [4]$$

where ΔH is the specific enthalpy change upon melting, $\Delta\rho$ is the density difference between solid and liquid, T_{liq} is the liquidus temperature, and T_{sol} the solidus temperature. Usually, the second term dominates,

$\alpha' \gg \alpha$ and $c_p' \gg c_p$. Also, α' is affected more strongly than c_p' so that the two-phase adiabat is usually steeper than either one-phase adiabat (see Solomatov and Stevenson (1993b) for the examples when this is not the case). Note that the situation is different for the atmosphere where the so-called 'wet' adiabat (due to condensation of water) is always less steep than the 'dry' adiabat (e.g., Goody, 1995). Also, because $\alpha' \gg \alpha$ and $c_p' \gg c_p$, eqn [2] is approximately equal to the Clapeyron slope of the dominant crystallizing phase and the adiabats tend to align with the curves of constant crystal fraction (e.g., solidus and liquidus).

There are several approaches to calculate the adiabats in the magma ocean. The simplest approach is to parametrize the melt fraction as a function of temperature and pressure without considering what phases crystallize and what composition they are (Miller *et al.*, 1991b; Abe, 1997). The most rigorous approach would involve building a self-consistent thermodynamic model which would use the available database for both solid and liquid silicates (e.g., Ghiorso, 1997; Asimow *et al.*, 1997). An intermediate approach is to choose some simple system consisting of only a few components and treat it self-consistently (Solomatov and Stevenson, 1993b).

4.3.2 Upper Mantle

Solomatov and Stevenson (1993b) consider two types of three-component systems to describe thermodynamics of the upper mantle. Here we only consider the eutectic-like system which seems to be most accurate at low pressures ($P < 10\,\mathrm{GPa}$). In this system, the three components, olivine, orthopyroxene, and clinopyroxene, are completely insoluble in each other in the solid state. The Gibbs free energy of the system (solid plus melt) is written as

$$G = \sum_{i=1}^{3}(n_i^s \mu_i^s + n_i^l \mu_i^l) \qquad [5]$$

where μ_i^s and μ_i^l are the chemical potentials of the component i ($i = 1, 2, 3$) in the solid and liquid, respectively, n_i^s and n_i^l are the mole fractions of the the solid and liquid component i, respectively,

$$n_i^s + n_i^l = n_i \qquad [6]$$

and n_i is a total mole fraction of the component i in the system,

$$\sum_{i=1}^{3} n_i = 1 \qquad [7]$$

The solid phases are assumed to be insoluable in each other. Thus, the chemical potentials of the solid phases are the same as the chemical potentials $\mu_i^{s(0)}$ of the pure components:

$$\mu_i^s = \mu_i^{s(0)} \qquad [8]$$

The liquid phases are assumed to form ideal solutions:

$$\mu_i^l = \mu_i^{l(0)} + RT \ln N_i^l \qquad [9]$$

where $\mu_i^{l(0)}$ is the chemical potential of a pure component i in liquid state, and N_i^l is the mole fraction of the component i in the liquid,

$$N_i^l = \frac{n_i^l}{\sum_{i=1}^{3} n_i^l} \qquad [10]$$

The requirement of thermodynamic equilibrium

$$\mu_i^l = \mu_i^s \qquad [11]$$

allows us to uniquely determine the composition of liquid and solid (n_i^s and n_i^l) and the melt fraction (the mole melt fraction is simply $\sum_{i=1}^{3} n_i^l$) at any given pressure and temperature. All thermodynamic parameters of the combined melt/crystal system, such as the specific heat and the coefficient of thermal expansion can be calculated from the Gibbs energy using standard thermodynamic relationships.

Figure 1 shows adiabats for this system at $P < 10\,\mathrm{GPa}$. Variation of melt fraction between liquidus and solidus is given in **Figure 2** along with the experimental data from McKenzie and Bickle (1988). The melting temperatures, their gradients, and the fractions of the three components are slightly adjusted to fit the experimental data. **Figure 2** shows that the data follow the expected general pattern of an ideal three-component eutectic system. Near the solidus all three solid phases undergo eutectic melting characterized by a jump in the melt fraction and one component (clinopyroxene) is completely liquid at the end of eutectic melting. At some melt fraction (around 30%), the slope changes because the second component is completely molten (orthopyroxene). Olivine remains the only solid phase all the way to liquidus.

4.3.3 Lower Mantle

The three components in the lower mantle are periclase, MgO, wüstite, FeO, and perovskite, MgSiO_3. Perovskite has only a small amount of iron and is considered to be a pure $MgSiO_3$. The system is very similar to the three-component system described

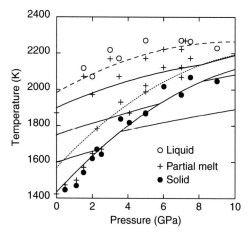

Figure 1 Adiabats in the upper mantle (thick solid lines). Liquidus (dashed line), solidus (solid line) and the beginning of crystallization of orthopyroxene (dotted line) are shown together with experimental data for peridotites (McKenzie and Bickle, 1988; Scarfe and Takahashi, 1986; Ito and Takahashi, 1987). Reproduced from Solomatov VS (2000) Fluid dynamics of a terrestrial magma ocean. In: Canup RM and Righter K (eds.) *Origin of the Earth and Moon*, pp. 323–338. Tucson, AZ: University of Arizona Press.

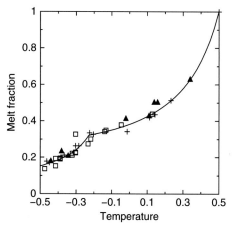

Figure 2 Variation of melt fraction between liquidus and solidus. The temperature is normalized in such a way that at the solidus $T = -0.5$ and at the liquidus $T = 0.5$. The experimental data are from McKenzie and Bickle (1988): $0 \leq P \leq 0.5\,GPa$ (crosses), $0.5 < P \leq 1.5\,GPa$ (squares), $P > 1.5\,GPa$ (triangles). The theoretical curve is calculated for an idealized eutectic system consisting of three immiscible components: olivine, orthopyroxene, and clinopyroxene. It shows a typical behavior of this type of systems: a step-like melting at the solidus during which one component – clinopyroxene – melts completely and an abrupt slope change around ~30% melt fraction where the second component – orthopyroxene – melts completely so that only olivine crystals are present until the temperature reaches liquidus. Reproduced from Solomatov VS (2000) Fluid dynamics of a terrestrial magma ocean. In: Canup RM and Righter K (eds.) *Origin of the Earth and Moon*, pp. 323–338. Tucson, AZ: University of Arizona Press.

above for the upper mantle. The main difference is that the components 1 and 2 (periclase and wüstite) are assumed to form an ideal solid solution with each other (magnesiowüstite). Thus, the chemical potentials of the solid phases are described as follows:

$$\mu_i^s = \mu_i^{s(0)} + RT \ln N_i^s, \quad i = 1, 2 \qquad [12]$$

$$\mu_3^s = \mu_3^{s(0)} \qquad [13]$$

where N_i^s is the mole fraction of the component i in the solid:

$$N_i^s = \frac{n_i^s}{\sum_{i=1}^{3} n_i^s} \qquad [14]$$

Figure 3 shows an example of calculations for the lower mantle. The melting temperatures of $MgSiO_3$, MgO, and FeO are based on laboratory experiments and extrapolation to high pressures by Boehler (1992) and Zerr and Boehler (1993, 1994).

The liquidus and solidus curves predicted by this model for the lower mantle are steeper than single-phase adiabats. The solidus is very close to one estimated by Holland and Ahrens (1997) and Zerr *et al.* (1998). The liquidus T_{liq} approximately follows the melting temperature T_{prv} of pure perovskite $MgSiO_3$ lowered by the presence of MgO and FeO:

$$T_{prv} - T_{liq} \approx \frac{k_B}{\Delta S_{prv}} n_{mw} T_{prv} \qquad [15]$$

where ΔS_{prv} is the entropy change per atom upon melting of pure perovskite, and n_{mw} is the mole fraction of magnesiowüstite. For $n_{mw} \approx 0.3$, $\Delta S_{prv} \approx 5k_B$ (assuming that the entropy change is approximately k_B per atom, where k_B is Boltzman's constant; Stishov (1988)) and $T_{prv} \approx 7500\,K$ (at the base of the mantle) we obtain $T_{prv} - T_{liq} \approx 450\,K$. This shows that the liquidus strongly depends on the assumed melting temperature of perovskite. For comparison, Miller *et al.* (1991b) and Solomatov and Stevenson (1993b) assumed a lower melting temperature of perovskite based on the data by Knittle and Jeanloz (1989). The liquidus also depends on the details of the phase diagram. Abe's (1997) lower liquidus is related to his assumption that $MgSiO_3$ and MgO form an ideal solid solution while in our calculations, perovskite forms eutectic-like subsystems with periclase and wüstite.

Agee (1990) and Zhang and Herzberg (1994) found that magnesiowüstite is the liquidus phase in both

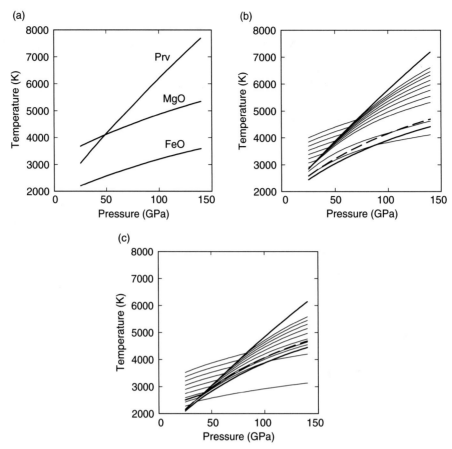

Figure 3 (a) Melting curves of perovskite, MgO, and FeO (Boehler, 1992; Zerr and Boehler, 1993, 1994). (b) Adiabats in the convective magma ocean. Liquidus and solidus are shown with heavy solid line. The beginning of magnesiowüstite crystallization is shown with a dashed line. (c) An example of calculations where magnesiowüstite is the first liquidus phase at the top of the lower mantle. Reproduced from Solomatov VS (2000) Fluid dynamics of a terrestrial magma ocean. In: Canup RM and Righter K (eds.) *Origin of the Earth and Moon*, pp. 323–338. Tucson, AZ: University of Arizona Press.

chondrites and peridotites just slightly below the upper mantle/lower mantle boundary. A small fraction of iron in the solid perovskite and the presence of other components (Ca-perovskite) in melt would decrease the liquidus temperature. The estimates of the entropy change are also not very accurate. As a result, magnesiowüstite can be the first liquidus phase at low pressures, although at higher pressures it would still be substituted by perovskite (**Figure 3**).

4.4 Viscosity of the Magma Ocean

Viscosity is an important parameter which controls virtually all dynamic processes in the magma ocean and affects the outcome of crystallization of the magma ocean. The viscosity of the magma ocean varies in a very broad range.

At low degrees of melting the viscosity of peridotitic magmas can be quite high, on the order of 100 Pa s (Kushiro, 1980, 1986). This is related to relatively low temperature as well as relatively felsic composition – the large concentration of silicon in the initial melts tends to increase the viscosity. The viscosity decreases with temperature very rapidly. Experimental and theoretical studies suggest that the viscosity of many near-liquidus ultramafic silicates at low pressures is around $\eta_l \sim 0.1$ Pa s (Bottinga and Weill, 1972; Shaw, 1972; Persikov *et al.*, 1997; Bottinga *et al.*, 1995; Dingwell *et al.*, 2004; Liebske *et al.*, 2005b).

There are no experimental data for the viscosity of the lower mantle. The viscosity behaves nonmonotonically with pressure because of the structural changes in the polymerized melts (e.g., Kushiro, 1980; Poe *et al.*, 1997). The viscosity of a completely depolymerized melt weakly increases with pressure (Andrade, 1952;

Gans, 1972). An increase of less than one order of magnitude can be expected along the liquidus throughout the lower mantle. Thus, the viscosity of magma oceans near the liquidus is probably around 10^{-1} Pa s. In the region between solidus and liquidus, the viscosity of pure melt (without crystals) is somewhat higher because of the lower temperatures. However, the temperature effect is small for low viscosity liquids (<1 Pa s), which exhibit a power law rather than Arrhenius behavior (Bottinga et al., 1995; Liebske et al., 2005b). The value of 10^{-1} Pa s with the uncertainty of a factor of 10 can be assumed for the upper mantle as well as for the lower mantle (**Table 1**).

Many authors pointed out that magma oceans can contain substantial amounts of water (Ahrens, 1992; Righter et al., 1997; Abe et al., 2000). Although water reduces the viscosity of magmas, in the limit of high temperatures the viscosities and diffusivities of many liquids including water are very similar (Persikov et al., 1997). Therefore, the viscosity of completely depolymerized, high-temperature hydrous magma is unlikely to be much different from that of anhydrous magma.

The most important effect on magma viscosity is the presence of crystals. The viscosity of a melt/crystal mixture change relatively little with the solid fraction. At small volume fraction ϕ of solids, the viscosity η of the mixture can approximately be described with the help of Einstein's linear formula for diluted suspensions of perfectly spherical particles (Einstein 1906):

$$\eta = \eta_l(1 + 2.5\phi) \qquad [16]$$

Various formulas have been proposed to describe the viscosity of suspensions at large crystal fractions (Mooney, 1951; Roscoe, 1952; Brinkman, 1952; Krieger

Table 1 Physical parameters for a deep magma ocean

Thermal expansion, α	5×10^{-5}	K^{-1}
Thermal capacity, c_p	10^3	$J\,kg^{-1}K^{-1}$
Gravity, g	10	$m\,s^{-2}$
Crystal/melt density difference, $\Delta\rho$	300	$kg\,m^{-3}$
Viscosity of melt at liquidus, η_l	0.1	Pa s
Viscosity of solids at solidus, η_s	10^{18}	Pa s
Diffusion coefficient, D	10^{-9}	$m^2\,s^{-1}$
Enthalpy change upon melting, ΔH	10^6	$J\,kg^{-1}$
Surface energy, σ	0.5	$J\,m^{-2}$
Apparent surface energy, σ_{app}	0.02	$J\,m^{-2}$
Heat flux, F	10^6	$J\,m^{-2}s^{-1}$
Magma ocean depth, L	3×10^6	m
Remaining motten layer depth, L'	5×10^5	m
Convective velocity, u_0	10	$m\,s^{-1}$

Adapted from Solomatov VS (2000) Fluid dynamics of a terrestrial magma ocean. In: Canup RM Righter K (eds.) Origin of the Earth and Moon, pp. 323–338. Tucson, AZ: University of Arizona Press.

and Dougherty, 1959; Murray, 1965; Frankel and Acrivos, 1967; McBirney and Murase, 1984; Campbell and Forgacs, 1990; Costa, 2005). For example, one of the most popular expressions is (Roscoe 1952)

$$\eta = \frac{\eta_l}{(1-\phi/\phi_m)^{2.5}} \qquad [17]$$

where ϕ_m is the maximum packing crystal fraction.

Although the above equation does not describe the viscosity accurately near $\phi = \phi_m$, it does show an important physical aspect of solid/liquid mixtures, that is, the viscosity becomes infinite when the crystal fraction reaches ϕ_m. The deformation beyond $\phi = \phi_m$ is only possible if crystals deform via solid-state creep, which is many orders of magnitude slower.

The value of ϕ_m for ideal spheres is around 0.64. For a binary mixture in which the small spheres have just the right size to fit the space in between large spheres (about 30% smaller than the large spheres) $\phi_m = 0.86$ (McGeary, 1961). In reality, the crystal distribution is far from any hypothetical one. Moreover, the shape of crystals is not spherical. Sarr et al. (2001) showed that crystal network can form at crystal fraction as low as 10% or 20% depending on the elongation (the smaller number is for a rather large ratio of the longest side of the crystal to the shortest one – over a factor of 10). In the presence of shear, the elongation can affect ϕ_m in the opposite direction. Elongated crystals can align with the flow and develop a hydrodense suspension with as much as 80% crystal fraction (Nicolas and Ildefonse, 1996).

The experiments on partial melts provide the most direct constraint on magma viscosity and ϕ_m, although the melt composition is still different from that of the magma ocean. These experiments show a rapid increase of the viscosity at high crystal fractions and $\phi_m \approx 60\%$ (Arzi, 1978; van der Molen and Paterson, 1979; Lejeune and Richet, 1995). The value of ϕ_m in the experiments is determined from to theoretical fit using equations similar to eqn [17].

Complications arise due to crystal–crystal interactions in the vicinity of ϕ_m. They are most likely to be responsible for the non-Newtonian behavior of partial melts characterized by substantial yield strength (reported values range from 10^2 to 10^6 Pa) and nonlinear relationship between stress and strain rate (Robson, 1967; Shaw et al., 1968; Shaw, 1969; Pinkerton and Sparks, 1978; McBirney and Murase, 1984; Ryerson et al., 1988; Lejeune and Richet, 1995).

The transition from fluid-like behavior to solid-like behavior occurs in a rather narrow range of crystal fraction. Thus, it can be called the 'rheological transition'. At ϕ only slightly smaller than ϕ_m, the viscosity is not much different from that of a pure melt except for corrections described by eqns [16] and [17]. At $\phi > \phi_m$ the deformation rate is controlled by the viscosity of the solid matrix reduced by the presence of melt (van der Molen and Paterson, 1979; Cooper and Kohlstedt, 1986; Jin et al., 1994; Rutter and Neumann, 1995; Hirth and Kohlstedt, 1995a, 1995b; Kohlstedt and Zimmerman, 1996; Mei and Kohlstedt 2000a, 2000b; Mei et al., 2002). The dependence of the solid-state viscosity on the melt fraction is parametrized with the help of simple exponential function:

$$\eta = \eta_s \exp(-\alpha_\eta \phi) \qquad [18]$$

where η_s is the viscosity of melt-free rock and α_η is a constant which depends on the creep mechanism. Mei et al. (2002) give $\alpha_\eta = 26$ for diffusion creep and $\alpha_\eta = 31$ for dislocation creep.

4.5 Convection in the Magma Ocean

4.5.1 Convective Heat Flux

During the early stages of crystallization of the magma ocean, the viscosity is small and convection is extremely turbulent. The convective heat flux is usually calculated as

$$F_{soft} = 0.089 \frac{k(T_m - T_s)}{L} Ra^{1/3} \qquad [19]$$

where

$$Ra = \frac{\alpha g (T_m - T_s) L^3}{\kappa \nu} \qquad [20]$$

is the Rayleigh number, T_m is the potential temperature of the magma ocean, T_s is the surface temperature, and k is the coefficient of thermal conductivity, $\kappa = k/\rho c_p$ is the coefficient of thermal diffusivity, and $\nu = \eta/\rho$ is the kinematic viscosity (Kraichnan, 1962; Siggia, 1994).

However, at very high Rayleigh numbers such as those in the magma ocean, convection changes to a regime sometimes called hard turbulence (Castaing et al., 1989; Shraiman and Siggia, 1990; Grossmann and Lohse, 1992; Siggia, 1994). By contrast, the ordinary turbulence is called soft turbulence. One of the important features of hard-turbulent convection is the existence of a large-scale circulation. This coherent motion within the highly turbulent fluid is an example of self-organization in complex systems (e.g., Nicolis

and Prigogine, 1977), that is, systems which have many degrees of freedom interacting with each other. Turbulence is caused by numerous plumes originating from the upper and lower boundaries of the convective layer. The coherent circulation of the fluid emerges from random actions of these plumes. In the magma ocean this mainly includes cold plumes originating from the surface but may also include hot plumes originating from the core–mantle boundary.

Several scaling laws have been proposed for the hard turbulence regime (Siggia, 1994; Kadanoff, 2001). Shraiman and Siggia (1990) obtained the following equation:

$$F_{hard} = 0.22 \frac{k(T_m - T_s)}{L} Ra^{2/7} Pr^{-1/7} \lambda^{-3/7} \qquad [21]$$

where $Pr = v/\kappa$ is the Prandtl number and λ is the aspect ratio for the mean flow.

Another effect to consider is the rotation of the Earth. Although the rotation of the Earth does not affect subsolidus mantle dynamics (Coriolis forces are negligible compared to viscous forces), the dynamics of low viscosity magma ocean may be affected by rotation. How large is the effect of rotation on the heat flux? At low Rayleigh numbers (soft turbulence), the heat flux depends on the rotation as follows:

$$F \sim \frac{k(T_m - T_s)}{L} Ra^3 Ta^{-2} \qquad [22]$$

where

$$Ta = \frac{4\Omega^2 L^4}{v^2} \qquad [23]$$

is the Taylor number (Boubnov and Golitsyn, 1986, 1990; Canuto and Dubovikov, 1998).

At high Rayleigh numbers, the effect of rotation becomes negligible and the heat flux is the same as in the absence of rotation. The critical Rayleigh number for the transition to the no-rotation regime is approximately the one at which the two scaling laws (with and without rotation) give the same heat flux. In the soft turbulence regime, the requirement that the heat flux [19] is the same as [22] gives the Rayleigh number above which rotation does not affect the heat flux (see also Canuto and Dubovikov (1998)):

$$Ra_* \sim Ta^{3/4} \qquad [24]$$

For the magma ocean $Ta \sim 10^{25}$. This gives $Ra_* \sim 10^{19}$ which is much smaller than the typical values $Ra \sim 10^{28}$–10^{29}. Although these estimates are for soft turbulence, the difference between soft turbulence and hard turbulence is relatively small. Thus, it is reasonable to assume that rotation does not have any significant effect on the heat flux in the hard turbulence regime.

Application of the hard turbulence regime to magma oceans requires substantial extrapolation: laboratory experiments have only been performed on helium ($Pr \sim 1$) to about $Ra \sim 10^{17}$ while in the magma ocean $Ra \sim 10^{28}$–10^{29} and $Pr \sim 10^2$–10^3. At very high Rayleigh numbers, convection was expected to enter a new regime of turbulent convection (Kraichnan, 1962; Siggia, 1994). So far experiments did not show any evidence of such an 'ultrahard' regime (Glazier, 1999; Niemela et al, 2000). Thus, hard turbulence is probably applicable to the extreme conditions of magma oceans.

The spherical geometry of the magma ocean is different from the geometries in laboratory experiments. However, like in other convection problems the changes in the scaling laws due to spherical geometry are expected to be minor. A major uncertainty associated with spherical geometry is probably the aspect ratio λ of the mean flow. The simplest assumption is that $\lambda \sim 1$, that is, the horizontal scale of the mean flow is of the order of the depth of the magma ocean. This parameter can be effected by rotation.

Temperature-dependent viscosity does not seem to play a significant role in controlling the heat flux. When the potential temperature is close to liquidus, the temperature contrast in the thermal boundary layer is 200–600 K depending on the scaling law. According to Liebske et al. (2005b), the corresponding viscosity contrast is at most one order of magnitude. This is too small to have a significant effect on the heat transport (Solomatov, 1995a).

Does radiative heat transport play a role in the thermal boundary layer near the surface? The thickness $\delta = k(T_m - T_s)/F$ of the thermal boundary layer is of the order of 1 cm. The free path of infrared photons in silicates is smaller, of the order of 1 mm (Clark, 1957; Shankland et al, 1979). This means that the thermal boundary layer is not thin enough to be transparent to infrared radiation. Also, the thermal conductivity of magma does not seem to be significantly affected by the radiative heat transport. The data collected by Murase and McBirney (1973) suggest that for solid lherzolite the thermal conductivity is not very different from that at low temperatures: $k \approx 4 \, \text{W m}^{-1} \text{K}^{-1}$ around 1500°C.

The heat flux depends on the surface temperature. Since the heat flux from the magma ocean is much higher than the incoming solar radiation, the surface temperature is established not by the radiative equilibrium between the incoming and outgoing radiation but by the equilibrium between the heat flux transported by convection in the magma ocean to the surface and the heat flux radiated from the surface of the magma ocean. The latter depends on the type of the atmosphere covering the magma ocean (Hayashi et al., 1979; Nakazawa et al., 1985; Abe and Matsui, 1986; Matsui and Abe, 1986; Zahnle et al., 1988). In the early stages of crystallization, the surface temperature is very high, more than 2000 K. In this case the atmosphere is a 'silicate' one (Hayashi et al., 1979; Nakazawa et al., 1985; Thompson and Stevenson, 1988). The heat flux can be written as

$$F = \sigma_{SB} T_e^4 \qquad [25]$$

where T_e is the effective temperature (the temperature that corresponds to a blackbody emitting the heat flux F) and $\sigma_{SB} = 5.67 \times 10^{-8} \, \text{J m}^{-2} \text{K}^{-4}$ is the Stefan–Boltzmann constant. The magnitude of T_e depends on the mass of the silicate atmosphere. It can be smaller than the surface temperature T_s by several hundred degrees kelvin (Thompson and Stevenson, 1988). Fortunately, the dependence on the mass of the atmosphere is weak. The simplest assumption one can make without considering the heat transfer in the atmosphere is that $T_e \approx T_s$, that is, magma ocean radiates like a blackbody with temperature T_s. With this assumption, the surface heat flux can be overestimated by a factor of $(T_s/T_e)^4$. For a moderately opaque atmosphere with $T_e \approx 1500$ K and $T_s \approx 2000$ K, $(T_s/T_e)^4 \sim 3$.

Figure 4 shows the heat flux and the surface temperature as functions of the potential temperature of the magma ocean during the initial period of crystallization. The scaling law for hard-turbulent convection predicts that the heat flux is close to $10^6 \, \text{W m}^{-2}$. The formula for soft-turbulent convection gives lower values of the heat flux, by a factor of ~ 3.

4.5.2 Convective Velocities

In the soft turbulence regime, the convective velocities can be estimated in terms of the mixing length l as (Priestly, 1959; Kraichnan, 1962)

$$u_0 \approx 0.6 \left(\frac{\alpha g l F}{\rho c_p} \right)^{1/3} \qquad [26]$$

where F is the surface heat flux. The coefficient is constrained by laboratory experiments (Deardorff, 1970; Willis and Deardorff, 1974) and atmospheric measurements (Caughey and Palmer, 1979). The simplest assumption is that the mixing length is

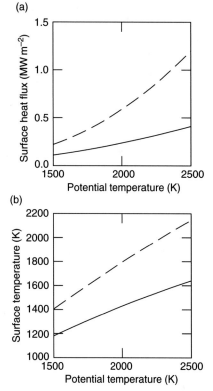

Figure 4 (a) The surface heat flux and (b) the surface temperature are shown as functions of the potential temperature calculated using scaling laws for soft turbulence convection (solid line) and hard turbulence convection (dashed line) The viscosity is $\eta = 0.1\,\mathrm{Pa\,s}$. The uncertainties associated with the viscosity are smaller than those associated with the scaling law.

approximately equal to the depth of the magma ocean $l \sim L$ (if the magma ocean has partially crystallized beyond the rheological transition, then l should be the thickness of the remaining low viscosity region). For $F \approx 10^6\,\mathrm{W\,m^{-2}}$ (**Figure 4**) and other parameters from **Table 1**, eqn [26] gives $u_0 \approx 4\,\mathrm{m\,s^{-1}}$ (note that the velocity is slightly larger in two-phase regions where α and c_p need to be replaced by eqns [3] and [4].

The ratio F/ρ is almost the same for the magma ocean and the atmosphere. This explains why the above estimate is very similar to the observed velocities in the convective boundary layer in the atmosphere (Caughey and Palmer, 1979).

The equations suggested by Shraiman and Siggia (1990) can be rewritten in the form similar to eqn [26]

$$u_0 \approx a_u \left(\frac{\alpha g l F}{\rho c_p}\right)^{1/3} \qquad [27]$$

where the coefficient $a_u \approx 0.086 x^*$ and x^* is related to u_0 through the following equation:

$$x^* = 2.5 \ln\left[\frac{\rho u_0 l}{\eta}\frac{1}{x^*}\right] + 6 \qquad [28]$$

The solution to the above transcendental equations gives $a_u \approx 5.9$ which varies weakly with $\rho u_0 l/\eta$. In the hard turbulence regime, the velocity increases by a factor of 10. This brings the estimate of the convective velocity in the magma ocean up to $40\,\mathrm{m\,s^{-1}}$. The effect of Earth's rotation can be estimated in the same way as it was done for soft turbulence: replacing the thickness of the convective layer by a reduced length scale

$$l \sim \frac{u_0}{\Omega} \qquad [29]$$

imposed by rotation explains rather well the experimentally observed reduction in the convective velocity (Golitsyn, 1980, 1981; Hopfinger *et al.*, 1982; Hopfinger, 1989; Boubnov and Golitsyn, 1986, 1990; Chen *et al.*, 1989; Fernando *et al.*, 1991; Solomatov and Stevenson, 1993a).

Assuming that this length scale works for hard turbulence, Solomatov (2000) obtained

$$u_0 \approx 14\left(\frac{\alpha g F}{\rho c_p \Omega}\right)^{1/2} \qquad [30]$$

This gives velocities around $16\,\mathrm{m\,s^{-1}}$ with the uncertainty of a factor of 3. The value of $10\,\mathrm{m\,s^{-1}}$ can be assumed for the magma ocean (**Table 1**).

4.6 Fractional versus Equilibrium Crystallization

4.6.1 How Are Crystals Suspended by Convection?

Early studies of convective suspensions used several approaches. Some looked at the trajectories of particles in steady flows and assumed that the particles whose trajectories are closed can be suspended indefinitely (Marsh and Maxey, 1985; Weinstein *et al.*, 1988; Rudman, 1992). This approach cannot work for turbulent flows where closed trajectories do not exist.

Others used a phenomenological description of the balance between the donward flux of particles due to settling and the upward flux due to convection (Bartlett, 1969; Huppert and Sparks, 1980). This approach predicted that crystals stay in suspension if the settling velocity is smaller than the convective velocity.

However, this approach did not consider the fact that the convective velocities decrease near the bottom. The first systematic laboratory experiments on convective suspensions by Martin and Nokes (1988, 1989) showed that even when the settling velocity is much smaller than the convective velocity, the particles eventually settle down and they do so nearly as fast as in the absence of convection. They argued that because the convective velocity at the lower boundary vanishes, the particles cannot be re-entrained and remain at the boundary. Yet, in one experiment the amount of suspended particles remained constant after some partial initial sedimentation.

Tonks and Melosh (1990) addressed the problem of suspension in convective systems using the analogy with the entrainment of particles in shear flows. They argued that turbulence is the key factor for suspension for both shear flows and convection. According to their arguments, turbulence was not strong enough in the Martin and Nokes (1988, 1989) experiments. In particular, they argued that sedimentation in convective systems does not occur if the settling velocity is smaller than the effective friction velocity associated with turbulent fluctuations. However, Solomatov and Stevenson (1993a) pointed out that the Reynolds number in the only experiment where Martin and Nokes reported suspension was about 0.03 which is below the critical value for the transition to turbulence by two orders of magnitude. Thus, it remained unclear how particles are re-entrained from the bottom.

Solomatov et al. (1993) investigated the mechanisms of entrainment experimentally using polystyrene spheres in an aqueous $CaCl_2$ solution. They showed that for either laminar or turbulent convection, the particles are moved at the bottom by the stresses generated by thermal plumes. These stresses are much larger than the stresses generated by turbulence. They also showed that the particles are not re-entrained right away but they are piled together in 'dunes'. If the flow is strong enough, it picks up the particles at the crests of the dunes and re-entrain them back into the interior region of the convective layer. The critical condition for re-entrainment is

$$\frac{\tau_T}{\Delta \rho g d} \sim 0.1 \qquad [31]$$

where

$$\tau_T = \left(\frac{\eta \alpha g F}{c_p} \right)^{1/2} \qquad [32]$$

is the convection stress scale and d is the diameter of the particles.

Application of these criteria to magma oceans is still not straightforward because the conditions in magma oceans are quite different from the laboratory experiments. Although the stresses [32] dominate at low Rayleigh numbers, at high Rayleigh numbers, the situation can reverse. In fact, in the hard turbulence regime the stresses associated with turbulence are larger than [32] (Shraiman and Siggia 1990). The former are

$$\tau^* = \rho u^{*2} \qquad [33]$$

where $u^* = u_0/x^*$ (eqn [28]). With typical values of $x^* \sim 60$ and $u_0 \sim 10\,\mathrm{m\,s}^{-1}$ we obtain $\tau^* \sim 50\,\mathrm{Pa}$. Assuming that eqn [34] can be applied to the hard turbulence regime, that is,

$$\frac{\tau^*}{\Delta \rho g d} \sim 0.1 \qquad [34]$$

we find that crystals up to several centimeters in diameter can be suspended in magma oceans. The criterion [32] gives much smaller critical crystal size, by almost two orders of magnitude.

Perhaps a more important factor is that unlike noninteracting particles in the laboratory experiments, crystals can form solid bonds as observed in experiments on magmas (e.g., Lejeune and Richet, 1995). Formation and growth of solid bonds is a well-studied phenomenon in metallurgy, which plays a key role in liquid phase sintering (e.g., German, 1985). The bonds between crystals and the underlying bed can easily prevent re-entrainment.

4.6.2 Energetics of Convective Suspension

Even though re-entrainment of crystals from the bottom of the convective layer is the key process which keeps particles suspended in laboratory experiments, this is not the most important factor that controls sedimentation in magma oceans. The crucial factor for magma oceans is that the presence of crystals can suppress convection as a result of viscous heating and density stratification associated with crystal settling (Solomatov and Stevenson, 1993a).

The total amount of energy released per unit time due to crystal settling is

$$\Phi = u_s g \phi \Delta \rho V \qquad [35]$$

where u_s is the settling velocity and V is the volume of the magma ocean. Thus, convection has to do work to suspend crystals.

The total amount of energy per unit time, which is available for mechanical work, is

$$W \approx \frac{\alpha g L}{c_p} FA \approx FA \qquad [36]$$

where A is the surface area of the Earth, and $\alpha g L / c_p \sim 1$. Most of this work is spent to overcome viscous friction associated with convection, that is just to keep convection going.

Experiments show that only a small fraction $\epsilon \sim 0.1$–1% of W is available for re-entrainment (Solomatov et al., 1993) If re-entrainment is possible, the equilibrium volume fraction of crystals is determined by the condition that

$$\Phi = \epsilon W \qquad [37]$$

When $\Phi > \epsilon W$, the sedimentation rate exceeds the re-entrainment rate and vice versa.

When viscous dissipation due to sedimentation exceeds the available mechanical power, that is,

$$\Phi > W \qquad [38]$$

convection is suppressed and suspension ceases to exist (Solomatov and Stevenson, 1993a). Although this theoretical prediction has not been tested experimentally, Solomatov and Stevenson (1993a) showed that this condition approximately coincides with the condition that the turbulence collapses due to density stratification caused by crystal settling (Solomatov and Stevenson, 1993a). The latter problem is analogous to the problem of turbulence in stratified fluids and has been well studied in laboratory experiments (Hopfinger, 1987).

When viscous dissipation exceeds the total heat loss rate from the magma ocean, that is,

$$\Phi > FA \qquad [39]$$

then cooling can continue only after fractionation of some amount of crystals reduces Φ to the level where $\Phi < FA$.

4.6.3 Conditions for Fractional Crystallization

Fractional crystallization is expected to happen when the condition [39] is satisfied (cooling is impossible without sedimentation) and even when a less severe condition [38] is satisfied (convection is suppressed) although for deep magma oceans these two conditions almost coincide. In either case, crystal–melt segregation occurs independently of whether or not there is any re-entrainment of crystals at the bottom of the magma ocean.

To estimate the crystal size at which this happens, we need the equation for the settling velocity

$$u_s = f_\phi \frac{\Delta \rho g d^2}{18 \eta_1} \qquad [40]$$

where f_ϕ is a hindered settling function such that $f_\phi = 1$ at $\phi = 0$ (Davis and Acrivos, 1985). If the crystal fraction in the magma ocean varies from 0 to the maximum packing fraction $\phi_m = 0.6$, then the average crystal fraction is about $\phi \sim 30\%$ at which $f_\phi \sim 0.15$ (Davis and Acrivos, 1985).

The critical crystal size above which fractional crystallization occurs is then found from eqn [38]:

$$d_f = \left(\frac{18 \alpha \eta_1 FAL}{f_\phi g c_p \Delta \rho^2 \phi V} \right)^{1/2} \qquad [41]$$

or

$$d_f \approx 10^{-3} \left(\frac{\eta_1}{0.1 \, \text{Pa s}} \right)^{1/2} \left(\frac{F}{10^6 \, \text{W m}^{-2}} \right)^{1/2} \, \text{m} \qquad [42]$$

4.6.4 Conditions for Equilibrium Crystallization

A simple condition for the magma ocean to crystallize without any substantial chemical differentiation is one which requires that the sedimentation time is much smaller than the crystallization time (Solomatov, 2000).

The crystallization time is

$$t_c \approx \frac{(\Delta H \phi + c_p \Delta T) M}{FA} \approx 400 \, \text{years} \qquad [43]$$

where $\Delta T \sim 1000 \, \text{K}$ is the average temperature drop upon crystallization of the magma ocean up to the crystal fraction $\phi \sim 60\%$, and M is the mass of the magma ocean.

The sedimentation time even in the presence of turbulent convection is (Martin and Nokes 1988)

$$t_s \approx \frac{L}{u_s} \quad [44]$$

Crystallization is faster than sedimentation, that is, $t_c < t_s$, provided the crystal size is smaller than

$$d_e = \left(\frac{18L\eta_l}{f_\phi g \Delta \rho t_c}\right)^{1/2} \quad [45]$$

or

$$d_e \approx 10^{-3}\left(\frac{\eta_l}{0.1\,\mathrm{Pa\,s}}\right)^{1/2}\left(\frac{F}{10^6\,\mathrm{W\,m^{-2}}}\right)^{1/2}\mathrm{m} \quad [46]$$

The difference between d_f and d_e is negligible compared to the uncertainties in the crystal size. Therefore, the critical crystal size which separates equilibrium and fractional crystallization is

$$d_{crit} \approx 10^{-3}\left(\frac{\eta_l}{0.1\,\mathrm{Pa\,s}}\right)^{1/2}\left(\frac{F}{10^6\,\mathrm{W\,m^{-2}}}\right)^{1/2}\mathrm{m} \quad [47]$$

The fact that these two estimates are very close to each other is not surprising – their ratio scales approximately as

$$\frac{d_e}{d_f} \sim \left(\frac{\Delta\rho/\rho}{\alpha T}\frac{c_p}{\Delta S}\right)^{1/2} \quad [48]$$

where we used $M \sim \rho A L$ and $\Delta H \phi + c_p \Delta T \sim \Delta H \phi \sim T \Delta S \phi$.

The entropy change upon melting ΔS is of the order of several k_B per atom (Stishov, 1988) while $c_p = 3k_B$ per atom. Therefore, $\Delta S \sim c_p$. Also, $\Delta\rho/\rho \sim \alpha T \sim 0.1$. Thus, $d_e/d_f \sim 1$.

4.7 Crystal Size in the Magma Ocean

The estimate of the critical crystal size ~1 mm separating equilibrium and factional crystallization is significantly smaller than the early estimates which pedicted that the critical crystal size is of the order of tens of meters (Tonks and Melosh, 1990; Miller et al., 1991b). If the cirtical crystal size were indeed this big, no physical mechanism could allow the crystals to reach this size during the lifetime of the magma ocean. The new estimate is in the range of typical crystal sizes observed in magmas (e.g., Cashman and Marsh, 1988), thus bringing up the question of

whether the crystal size in the magma ocean was larger or smaller than ~1 mm.

4.7.1 What Processes Control the Crystal Size in the Magma Ocean

In thermodynamic equilibrium, the crystal diameter d is realted to the crystal fraction ϕ and the number of crystals N per unit volume as (assuming spherical shape)

$$d = \left(\frac{6\phi}{\pi N}\right)^{1/3} \quad [49]$$

The number of crystals is controlled by nuecleation in the decending flow. When crystal-free magma reaches the pressure where the temperature of the convecting magma ocean (adiabat) drops below the liquidus, crystals nucleate and grow until the equilibrium crystal fraction is reached.

If the number of crystals does not change with time, then the crystals grow simply because the equilibrium crystal fraction ϕ changes with depth along the adiabat – increases on the way down and decreases on the way up. However, dissolution of smaller crystals and growth of larger crystals decreases the number of crystals per unit volume and thus increases the average crystal size according to eqn [49]. This process is called Ostwald ripening.

4.7.2 Nucleation

When the temperature in the downwelling convective flow drops below the liquidus, crystals form via nucleation and growth mechanism (**Figures 5** and **6**). During nucleation, tiny crystals (tens of atoms in diameter, depending on the supercooling) precipitate from the supercooled melt. The nucleation rate (the number of nuclei produced per unit volume per unit time) is an extremely sensitive function of the supercooling $\Delta T = T_{liq} - T$ – the difference between the equilibrium value of the liquidus and the actual temperature of the fluid. It can approximately be described as

$$\mathcal{J}(\Delta T) = \alpha \exp\left(-\frac{B}{\Delta T^2}\right) \quad [50]$$

$$B = \frac{16\pi\sigma_{app}^3 T_0^2}{3k_B T \rho^2 \Delta H^2} \quad [51]$$

where a is a constant, T_0 is the melting temperature of the crystallizing phase, and σ_{app} is the apparent surface

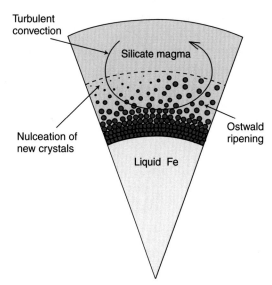

Figure 5   Magma ocean in the early stages of crystallization. Nucleation occurs in the downgoing convective flow when it enters the two-phase region. Crystals continue to growth because of Ostwald ripening – larger crystals grow at the expense of smaller ones. The crystals which have resided longer in the two-phase region are schematically shown bigger.

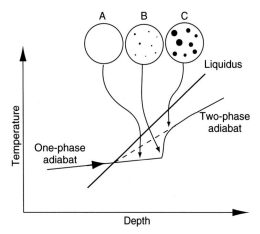

Figure 6   Evolution of a fluid parcel (large circle) in the downwelling flow. When the temperature of a fluid parcel drops below the liquidus the temperature of the parcel initially follows a metastable one-phase adiabat (solid line) rather than the two-phase adiabat (dashed line), without any crystallization (stage A – pure liquid). When the supercooling reaches the nucleation barrier, an avalanche-like nucleation produces tiny crystals (stage B – liquid with nuclei). The subsequent crystal growth brings the temperature of the fluid up to the equilibrium two-phase adiabat (stage C – nearly equilibrium liquid-crystal mixture).

energy. Once a crystal has nucleated, it grows at a rate which can be controlled by either interface kinetics or diffusion. In the former case, the bottleneck in crystal growth is the attachment of atoms to the crystal (e.g., in some cases the atom needs to find an imperfection on the crystal surface to be able to attach to it) while in the latter case the attachment is relatively easy and the bottleneck is the long-range transport of atoms in the melt surrounding the growing crystal.

Although nucleation and crystal growth in the magma ocean might seem to be very difficult to constrain, fortunately, this process is mathematically very similar to bulk crystallization of continuously cooling liquids in laboratory conditions (Lofgren et al., 1974; Flemings et al., 1976; Grove and Walker, 1977; Walker et al., 1978; Ichikawa et al., 1985; Grove, 1990; Smith et al., 1991; Cashman, 1993). In both cases the controlling parameter is the cooling rate $\dot{T}$. The only difference is that the effective cooling rate of an adiabatically descending fluid parcel is the rate of change of the difference between the liquidus temperature T_{liq} and the adiabatic temperature T_{ad}:

$$\dot{T} = u_0 \left(\frac{\mathrm{d}T_{liq}}{\mathrm{d}z} - \frac{\mathrm{d}T_{ad}}{\mathrm{d}z} \right) \quad [52]$$

where $\mathrm{d}T_{liq}/\mathrm{d}z$ is the liquidus gradient and $\mathrm{d}T_{ad}/\mathrm{d}z$ is the adiabatic gradient.

Solomatov and Stevenson (1993c) solved the problem of nucleation in continuously cooling multicomponent liquids for interface kinetics controlled growth. Their solution is somewhat similar to one obtained by Raizer (1960) (see also Zel'dovich and Raizer (2002)) for condensation in an adiabatically expanding cloud of vapor except that it uses a more general mathematical method developed by Buyevich and Mansurov. Solomatov (1995b) solved this problem for the case when crystal growth is controlled by diffusion rather than interface kinetics and showed that this solution is more consistent with various experimental data on silicates and alloys.

According to Solomatov (1995b), nucleation in continuously cooling liquids operates during a very short period of time (an avalanche-like event). The abrupt start of nucleation is due to the fact that the liquid needs to reach a critical supersaturation (the nucleation barrier). The short duration and abrupt cessation of the nucleation is due to the fact that the precipitated crystals quickly change the composition of the fluid, which reduces the supersaturation below the nucleation barrier. Although the duration of nucleation is short, this is the key process which determines the number of crystals nucleated per unit volume and thus, the size of these crystals after the system reaches an equilibrium.

The number of nuclei produced per unit volume during the nucleation period is calculated as follows:

$$N = 0.2 \frac{\zeta^{3/2}\, \dot{T}^{3/2} \theta^{3/2}}{B|\mathrm{d}T_{\mathrm{liq}}/\mathrm{d}\phi|D^{3/2}} \qquad [53]$$

where D is the diffusion coefficient and $\zeta = \Delta T/\Delta C$ determines the relationship between the supercooling ΔT and the supersaturation ΔC. It is of the order of the difference between liquidus and solidus temperatures, $\zeta \sim T_{\mathrm{liq}} - T_{\mathrm{sol}}$.

The parameter θ in eqn [53] is nearly constant ($\sim$30). It can be found from the transcendental equation

$$\theta = \ln\left[\frac{6.05|\mathrm{d}T_{\mathrm{liq}}/\mathrm{d}\phi|aB^{3/2}D^{3/2}}{\zeta^{3/2}\, \dot{T}^{5/2}\theta^3}\right] \qquad [54]$$

From eqns [49] and [53] with $\phi \sim 60\%$, we estimate

$$\begin{aligned} d_{\mathrm{nucl}} &\approx 10^{-3} \left(\frac{\sigma_{\mathrm{app}}}{0.02\,\mathrm{J\,m^{-2}}}\right) \left(\frac{D}{10^{-9}\,\mathrm{m^2\,s^{-1}}}\right)^{1/2} \\ &\quad \times \left(\frac{u_0}{10\,\mathrm{m\,s^{-1}}}\right)^{-1/2}\mathrm{m} \end{aligned} \qquad [55]$$

This relationship can be interpreted as follows. The velocity controls the cooling rate of fluid parcels: the faster the cooling rate, the further the system is driven into the metastable state, the higher the nucleation rate and the more crystals are nucleated. This corresponds to the well-known fact that faster cooling produces smaller crystals. The diffusion coefficient controls the crystal size somewhat indirectly: during the nucleation and growth stage, the thermodynamic equilibrium is achieved via both the nucleation of new crystals and the growth of already nucleated crystals. The larger the diffusion coefficient, the more this balance is shifted toward growth and the fewer crystals are nucleated. The surface energy controls nucleation: the larger the surface energy, the more difficult it is to nucleate crystals and the fewer crystals are produced.

Note that the apparent cooling rates $\dot{T}$ in the magma ocean are not that much different from those in the laboratory conditions (**Figure 7**). This means that there is no extrapolation in this parameter. The parameter which can significantly affect the estimate of the crystal size is the apparent surface energy, σ_{app}. It was argued that it is not the normal

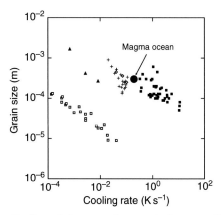

Figure 7 Crystal size vs cooling rate for five crystallizing phases (from Solomatov 1995b): Sn (Sn–Pb system, solid boxes), Al (Al–Cu, pluses), Si (Al–Si, diamonds), diopside (diopside–plagioclase, solid triangles), and plagioclase (diopside–plagioclase, open boxes). The data are from Lofgren et al. (1974), Flemings et al. (1976), Grove and Walker (1977), Walker et al. (1978), Ichikawa et al. (1985), Grove (1990), Smith et al. (1991), and Cashman (1993). The data approximately follow the theoretical 1/2 slope. The location of the magma ocean is shown.

surface energy but rather an apparent one for nucleation (Dowty, 1980; Solomatov, 1995b). The value of $0.02\,\mathrm{J\,m^{-2}}$ (**Table 1**) is based on the analogy with other cases that suggest that it is almost one order of magnitude smaller compared to the usual one. Higher values of σ_{app} in the magma ocean cannot be excluded. This means that it can increase the crystal size by as much as one order of magnitude.

It is also worth noting that because of a large value of the logarithm in eqn [54], the value of θ and thus the crystal size, is insensitive to the uncertainties in the parameters under the logarithm. In particular, the most uncertain parameter is the prefactor a (the uncertainties can be as large as 10 orders of magnitude). However, this uncertainty effects the results only by a factor of 2 or so.

4.7.3 Ostwald Ripening

For diffusion-controlled Ostwald ripening, crystal size is calculated as follows (Lifshitz and Slyozov, 1961; Voorhees, 1992):

$$d_{\mathrm{ost}}^3 - d_{\mathrm{nucl}}^3 = \frac{32}{9}b_\phi\alpha_0 Dt_{\mathrm{ost}} \qquad [56]$$

where d_{nucl} is the initial crystal size after nucleation, $\alpha_0 = 2\sigma c_\infty v_{\mathrm{m}}/RT$, σ is the surface energy, c_∞ is the

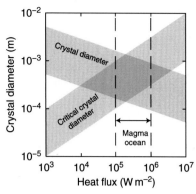

Figure 8 The crystal size during the early stages of crystallization of the magma ocean (assuming that it is controlled by nucleation in the downwelling flow) and the critical size for suspension are shown as a function of the heat flux. The width of the curves (gray bands) represents the uncertainty range of one order of magnitude.

equilibrium concentration of the crystallizing mineral in the melt, v_m is the molar volume of the crystallizing mineral, t_{ost} is the time available for Ostwald ripening, and b_ϕ is a function of ϕ such that $b_\phi = 1$ in the limit $\phi \ll 1$. The time available for Ostwald ripening is roughly the characteristic residence time in the two-phase region, that is, $t_{ost} \sim H/u_0$. Neglecting the initial crystal size d_{nucl}, we obtain that

$$d_{ost} \approx 10^{-3}\left(\frac{D}{10^{-9}\,\text{m}^2\,\text{s}^{-1}}\right)^{1/3}\left(\frac{u_0}{10\,\text{m}\,\text{s}^{-1}}\right)^{-1/3}\text{m} \quad [57]$$

This is similar to estimate [55]. Thus, Ostwald ripening does not increase the crystal size substantially.

The above estimates show that the crystal size during the early crystallization of the magma ocean is very close to the critical crystal size separating fractional and equilibrium crystallization of the magma ocean (**Figure 8**) which is about 1 mm. This means that both equilibrium and fractional crystallization (up to 60% crystal fraction) are equally acceptable within the uncertainties of the physical parameters.

4.8 Crystallization beyond the Rheological Transition

As was mentioned earlier, at the crystal fraction around $\phi_m \sim 60\%$, melt/crystal mixture undergoes a rheological transition to a solid-like behavior and the deformation is controlled by the solid-state creep of crystals. Solomatov and Stevenson (1993b) and

Solomatov (2000) showed that rapid cooling and crystallization may continue even at $\phi > \phi_m$.

This can be explained as follows. When the viscosity of magma abruptly increases from liquid-like viscosity to the solid-like viscosity around $\phi \sim \phi_m$, convection stops at the bottom of the magma ocean or somewhere near it – the hot core may keep the base of the mantle molten even after crystallization of the magma ocean. A nonconvecting layer with the crystal fraction $\phi \sim \phi_m$ starts growing from the bottom of the magma ocean (**Figure 9**). If solid-state convection does not start, then in about 400 years, eqn [43], the temperature in the entire magma ocean would approximately follow the curve $\phi = \phi_m = \text{const}$. However, the temperature gradient of the magma ocean which is partially crystallized to $\phi = \phi_m$ is roughly parallel to solidus and thus is steeper than the adiabat. Since the adiabat is the temperature profile of a neutrally stable fluid, a superadiabatic temperature profile would make the mantle gravitationally unstable.

The instability is so fast that one can ignore thermal diffusion and treat it like a Rayleigh–Taylor instability. The driving effective density contrast is of the order of

$$\triangle_{\rho RT} \approx \frac{1}{2}\alpha\rho L\left(\frac{dT_{Sol}}{dz} - \frac{dT_{ad}}{dz}\right) \approx 100\,\text{kg}\,\text{m}^{-3} \quad [58]$$

for the whole mantle ($L \approx 3 \times 10^6$ m).

The timescale for the overturn can then be estimated as (Turcotte and Schubert 1982)

$$t_{RT} \approx 26\frac{\eta_S}{\triangle_{\rho RT}gL} \quad [59]$$

where η_s is the viscosity of the solid mantle and g is the gravity.

The viscosity η_s of the lower mantle is poorly constrained. The present-day lower mantle viscosity is around 10^{22} Pa s (e.g., King, 1995). The temperature of the mantle is much below solidus, by $\sim 1000-1500$ K at the base of the mantle (Boehler, 1996, 2000; Zerr et al., 1998). Thus, the viscosity near the solidus must be substantially lower than 10^{22} Pa s. Based on experimental and theoretical constraints on the lower mantle viscosity (Karato and Li, 1992; Wright and Price, 1993; Li et al., 1996; Ita and Cohen, 1998; Yamazaki et al., 2000; Yamazaki and Karato, 2001; Béjina et al., 2003), we can assume that the viscosity near the solidus is about 10^{18} Pa s for the grain size of 10^{-3} m.

The presence of melt reduces the viscosity further. Equation [18] suggests that the viscosity is $\sim 10^4$–10^5 times smaller at $\phi \sim \phi_m$ and one order of magnitude

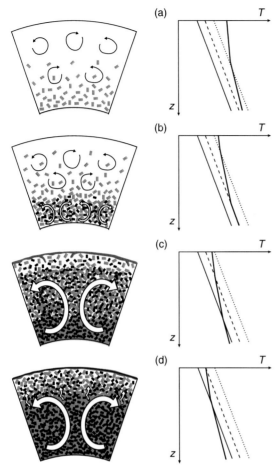

Figure 9 Crystallization of the magma ocean: (a) the lower part of the magma ocean is below liquidus (dotted line), convection is controlled by melt viscosity and the temperature (heavy solid line) is adiabatic; (b) high viscosity, gravitationally unstable region with the maximum packing crystal fraction forms near the bottom of the magma ocean ($\phi = \phi_m$ curve is shown with a dashed line); (c) cooling beyond $\phi = \phi_m$ proceeds via solid state convection which is still fast at this stage. The temperature in the high viscosity region below solidus (solid line) can be superadiabatic; (d) $\phi > \phi_m$ everywhere and the rate of cooling and crystallization of the remaining melt is controlled by solid-state convection and melt percolation. The gray and black crystals schemically illustrate different mineral phases.

smaller when only 10% melt remains in the mantle. These estimates are very approximate because eqn [18] is a phenomenological equation whose underlying physical mechanisms are poorly understood and thus, the errors due to extrapolation to the magma ocean conditions are unknown. Nevertheless, eqns [18] and [59] predict that for a near-solidus mantle, the overturn is of the order of 100 years and for $\phi \sim \phi_m$ it can be as short as a few days.

Thus, the instability starts developing soon after the maximum packing fraction is reached at the bottom of the magma ocean and takes the form of solid-state convection which is the primary mechanism of cooling and crystallization in the range $\phi_m < \phi < 1$. A complete solidification front would follow the rheological front toward the surface (**Figure 9**). When the adiabatic temperature in the solidified regions becomes so low that the viscosity exceeds about 10^{19} Pa s, then Rayleigh–Taylor will be unable to catch up with the solidification front. It is likely that the temperatures in the high viscosity regions will be superadiabatic. In this case, a substantial superadiabatic temperature gradient can be created upon crystallization of the magma ocean. The instability would continue after crystallization of the magma ocean. It may take a form of overturn as suggested for Mars by Elkins-Tanton *et al.* (2005), which would create a subadiabatic and gravitationally stable mantle. It should be emphasized that this remains one of the most difficult and uncertain aspects of magma ocean crystallization.

4.9 The Last Stages of Crystallization

4.9.1 Cessation of Suspension

The first important transition occurs when the molten layer disappears, that is, the potential temperature drops below liquidus everywhere. The crystal size increases because the sequence 'nucleation–growth–dissolution' changes to just 'growth'. During the nucleation–growth–dissolution cycle, the time available for Ostwald ripening is controlled by the residence time of crystals in the partially molten region, which is of the order of $t_{ost} \sim 3H/u_0 \sim 10^6$ s or roughly 1 week. When the completely molten layer disappears, crystals never exit the two-phase region and the characteristic time for crystal growth is much larger. It can approximately be estimated as the time it takes for the potential temperature to drop from the liquidus temperature to the critical temperature for the rheological transition (i.e., while liquid-state convection still occurs). Since the difference between liquidus T_{liq} and solidus T_{sol} is about 600 K near the surface (**Figure 1**) and $\phi \approx 60\%$ is reached approximately in the middle between liquidus and solidus (**Figure 2**), then $t'_{ost} \sim 3 \times 10^9$ s (100 years) according to eqn [43]. The crystal size increases by $(t'_{ost}/t_{ost})^{1/3} \approx 14$ times, that is, $d'_{ost} \sim 10^{-2}$ m. If crystal/melt segregation did not start earlier, at this stage it is impossible to avoid (**Figure 8**).

In addition, the surface temperature would eventually drop below 1500 K (**Figure 4**). If substantial amounts of water were present at that time, the atmosphere would change from silicate to steam one. The blanketing effect of the steam atmosphere can reduce the heat flux to $10^2-10^3\,W\,m^{-2}$ (Abe and Matsui, 1986; Zahnle *et al.*, 1988; Kasting, 1988). This would increase the crystal size even further and at the same time substantially reduce the ability of convection to suspend crystals (**Figure 8**).

The pressure at the bottom of the partially molten layer (the complete crystallization front) at the time when the completely molten layer disappears can be estimated from an adiabat which starts at the liquidus temperature at the surface. The bottom of the partially molten layer is the pressure where this adiabat intersects with the solidus. It depends strongly on the details of the thermodynamics of the partially molten layer. Calculations by Miller *et al.* (1991a, 1991b), Solomatov and Stevenson (1993b), Abe (1997), and Solomatov (2000) (**Figures 1 and 3**) suggest that the bottom of the partially molten layer is around 30–40 GPa.

When crystal/melt segregation begins, convection can become quite complex. This regime is poorly understood. It may develop layers and instabilities as experiments on convective suspension show (e.g., Koyaguchi *et al.*, 1993; Sparks *et al.*, 1993). Although convection is suppressed by sedimentation, it may not necessarily cease. Crystal fraction near the surface is below 60%; thus, the viscosity in a thin region below the surface is still small. This might be sufficient to maintain convection at least in this layer. Note that convection is driven by the instabilities of the 1 cm thick surface thermal boundary layer and the heat flux does not depend much on the processes at large depth, such as crystal/melt segregation. This means that as long as a low viscosity layer exists below the surface, one might expect that this layer would continue to convect vigorously generating the high heat flux predicted by eqns [19] and [21].

is about 60% and can even reach the solidus (**Figures 1, 2**, and **4**). First, it seems that it is unlikely that a solid or partially solid crust can form on the timescale of $\delta^2/4\kappa \sim 30s$ – this is too short for crystals to nucleate or grow. Second, if a rigid crust does form on top of the thermal boundary layer, its effect on the heat flux is not as big as one might expect. This can be estimated as follows. The viscosity decreases within the thermal boundary layer due to both temperature and crystal fraction. As a result, while the whole boundary layer cannot participate in convection because of the rigid crust, the bottom part which has the lowest viscosity can. This convection regime is called stagnant lid convection.

The heat flux in the stagnant lid convection regime can be approximately described by eqns [19] and [21], provided the temperature difference $T_m - T_s$ is replaced by the rheological temperature scale (Morris and Canright, 1984; Fowler, 1985; Davaille and Jaupart, 1993a, 1993b; Solomatov, 1995a; Solomatov and Moresi, 2000). For magma, it is calculated as follows (Davaille and Jaupart, 1993b; Reese and Solomatov, 2006):

$$\triangle T_{rh} \approx 2\left|\frac{d\ln\eta}{dT}\right|^{-1}_{T=T_m} = 2\left|\frac{\partial\ln\eta}{\partial T} + \frac{\partial\ln\eta}{\partial\phi}\frac{d\phi}{dT}\right|^{-1}_{T=T_m} \quad [60]$$

Assuming an Arrhenius dependence of melt viscosity on temperature, $\eta_l \sim \exp(E_l/RT)$, where the activation energy $E_l = 190\,kJ\,mol^{-1}$ (Liebske *et al.*, 2005a, 2005b) and using eqn [17] and **Figure 2**, we obtain that near liquidus, $\triangle T_{rh} \approx 100$ K. This scale is mostly determined by the variation in the crystal fraction within the boundary layer.

Thus, the driving temperature difference for convection can be reduced by a factor of 2–6 compared to the temperature difference without solid crust (**Figure 4**) and the heat flux can drop by one order of magnitude (eqns [19] and [2]).

4.9.2   Formation of Thin Crust within the Thermal Boundary Layer

Even though a low viscosity region may exist below the surface, a stable crust can form within the thin thermal boundary layer, which would reduce the heat flux. Indeed, when the potential temperature drops below the liquidus, the surface temperature can reach the critical one where the crystal fraction

4.9.3   Cessation of Liquid-State Convection

When the crystal fraction increases to 60% all the way to the surface, the effective viscosity of magma is controlled by solid-state deformation of crystals. Convection becomes much slower, the heat flux drops, and a thick crust forms at the surface. The location of the bottom of the remaining partially molten layer can be obtained from an adiabat starting at the critical temperature for the rheological transition. **Figure 1** suggests that such an adiabat intersects the solidus around 10 GPa or 300 km. This estimate is rather uncertain since even a 100 K error in the estimate of the rheological transition increases this pressure by a factor of 2. A similar factor can be caused by variations in the solidus and in the thermodynamics of the partial melt. Also, this gives only a lower limit because the temperatures can be somewhat superadiabatic due to the strong variation of the viscosity with depth and because of the compositional stratification at shallow depth. After solid-state convection replaces liquid-state convection, percolation of melt through the solid matrix becomes an important process.

4.9.4 Percolation

Melt extraction from the remaining partially molten layer can be controlled by either the percolation of melt through the solid matrix or by the compaction of the solid matrix. (Bercovici *et al.*, 2001). The compaction length is of the order of meters, which is much smaller than the depth, L, of the remaining partially molten layer. This implies that compaction does not effect melt extraction and the rate limiting process is melt percolation. Although the melt fraction and the viscosities of solid and liquid vary in the partially molten region, an order-of-magnitude melt percolation time is

$$t_{\text{diff}} \sim \frac{L'}{u_{\text{perc}}} \qquad [61]$$

where

$$u_{\text{perc}} = \frac{g\Delta\rho d^2 \phi_l^2}{150\eta_l(1-\phi_l)} \qquad [62]$$

is the percolation velocity (Soo, 1967; Dullien, 1979), which is related to Darcy velocity, $u_D \approx \phi_l u_{\text{perc}}$, and $\phi_l = 1 - \phi$ is the melt fraction. Melt is likely to migrate very quickly at the initial stages of percolation (near the rheological transition), when the melt fraction is large and the melt viscosity is low, and significantly slow down near the solidus when only a few percent melt is left. For $\phi_l \sim 2\%$, $d \sim 1$ mm, and $\eta_l \sim 100$ Pa s (the value of the viscosity of a low pressure, high silica, polymerized magma just above the solidus; Kushiro (1980, 1986)):

$$t_{\text{diff}} \sim 10^8 \left(\frac{\phi_l}{0.02}\right)^2 \left(\frac{d}{10^{-3}\text{m}}\right)^2 \left(\frac{\eta_l}{100\,\text{Pa s}}\right) \text{years} \qquad [63]$$

Melt/crystal density inversions (Agee, 1998) might result in the formation of more than one molten layer. In particular, a molten layer could be formed at the bottom of the upper mantle in addition to the shallow magma ocean beneath the crust. Melt/crystal density inversions in the mantle are very important for chemical differentiation in the magma ocean and yet we have very little understanding of where these inversions happen (Ohtani and Maeda, 2001; Akins *et al.*, 2004; Matsukage *et al.*, 2005; Stixrude and Karki, 2005).

4.9.5 Remelting due to Melt Extraction

The extraction of melt is accompanied by gravitational energy release which is converted to heat via viscous dissipation associated with melt migration. Assuming that most of the energy goes to melting, Solomatov and Stevenson (1993b) showed that the additional amount of melt generated is proportional to

$$R_m = \frac{\Delta\rho g L'}{\rho\Delta H} \qquad [64]$$

The value of R_m is only about 0.1 for 50 km deep magma chambers but close to unity for a layer of $L' \sim 500$ km depth. Thus, the total degree of melting can roughly be twice as much as the initial degree of melting before melt extraction.

4.9.6 Solid-State Convection

The timescale for complete crystallization of the mantle depends on the style of convection in the remaining partially molten layers. When the deformation is controlled by solid-state creep, the heat flux is determined by the equations which are similar to those describing plate tectonics. In its simplest form, it can be written as (Davies, 1990; Turcotte and Schubert, 2002)

$$F = 0.3k(T_m - T_s)^{4/3}\left(\frac{\alpha\rho g}{k\eta_s}\right)^{1/3} \qquad [65]$$

For the viscosity $\eta_s \sim 10^{18}$ Pa s, and the temperature difference across the surface thermal boundary layer $T_m - T_s \sim 1500$ K (i.e., the surface temperature is around 300 K), the above equation gives $F \sim 1 \, W \, m^{-2}$. The thickness of the thermal boundary layer is about 1 km. The crystallization of a partially molten upper layer of thickness $L' \sim 500 \, km$ takes

$$t_{conv} \sim \frac{\Delta H (1 - \phi_m) \rho L'}{F} \sim 10 \, My \qquad [66]$$

If one assumes that the surface layer is rigid and convection occurs in the stagnant lid regime, then the driving temperature $T_m - T_s \approx 1500 \, K$ in eqn [65] needs to be replaced by the rheological temperature scale (eqn [60]). Equations [60] and [18] give $\Delta T_{rh} \approx 100 \, K$, where we assumed $d\phi / dT \approx 5 \times 10^{-4} \, K^{-1}$ near the rheological transition (the melt fraction is about 40%; **Figure 2**) and $E \approx 240 \, kJ \, mol^{-1}$ which is the activation energy for diffusion creep in water-saturated olivine (Karato and Wu, 1993). The heat flux can drop by a factor of ~ 40, thus increasing the crystallization time to $\sim 400 \, My$. If the viscosity η_s and/or $d\phi / dT$ is larger (the latter depends on the details of the phase diagram), the crystallization time can increase further, perhaps, to 2 billions years or so.

It is also possible that density stratification due to chemical differentiation completely suppresses convection. In the stagnant lid convection regime, where the driving temperature differences are small, this does not take much (Zaranek and Parmentier, 2004). If this happens, cooling and crystallization would be controlled by thermal diffusion whose timescale $\sim L'^2 / 4\kappa$ is about 2 billion years. These order-of-magnitude estimates show that the crystallization time can be comparable with the age of the Earth. This means that it is possible that the Earth may never had a chance to crystallize completely.

4.10 Summary

Crystallization of the molten Earth was undoubtedly very complex. A variety of physical and chemical processes determined the final product of crystallization during the first 10^8 years, before the long-term solid-state convection took over. This includes convection in liquid, solid and partially molten states, crystal settling during the early stages of crystallization, and extraction of residual melt from the partially molten mantle during the later stages.

A scenario which seems to agree with the geochemical constraints and is physically feasible is as follows. The latest and the largest impact melts a significant part of the Earth. Gravitational instabilities quickly redistribute the material so that denser materials (due to temperature and composition) accumulate at the bottom and lighter ones accumulate at the top. A small fraction of the mantle which survived the impact and remained solid sink to the bottom of the magma ocean.

In the beginning the viscosity of the magma ocean is very low ($\sim 0.1 \, Pa \, s$), the magma ocean is vigorously convecting, and the heat flux $\sim 10^6 \, W \, m^{-2}$. The temperature profile is approximately adiabatic. Since the melting curve is likely to be steeper than the adiabat, crystallization proceeds from the bottom up. It takes about $\sim 10^3$ years to crystallize the lower part of the mantle. Highly turbulent convection helps to cool and crystallize the magma ocean to about 60% crystal fraction (the rheological transition from a low viscosity suspension to a high viscosity partially molten solid).

Whether or not fractional crystallization is prevented during the early stages of crystallization depends on the crystal size. It is controlled by nucleation in the downgoing convective flow and by Ostwald ripening. The crystal diameter established by these processes can be sufficiently small ($\sim 1 \, mm$) to allow equilibrium crystallization. The lower part of the mantle underwent only a small degree of crystal-melt segregation, which can be reconciled with the geochemical constrained on the fractionation of minor and trace elements.

Crystallization beyond the crystal fraction of 60% is accomplished by solid-state creep convection. Although the viscosity of solids is very high ($\sim 10^{18} \, Pa \, s$ at the solidus and $\sim 10^{14} \, Pa \, s$ near 60% crystal fraction), convection remains sufficiently fast and is able to help crystallize the lower mantle completely.

Several important changes happen during the last stages of crystallization of the magma ocean. The purely molten region disappears and the crystals undergo continuous growth rather than cycle between nucleation, growth, and dissolution. The early silicate atmosphere changes to a steam atmosphere whose blanketing effect can reduce the heat flux by several orders of magnitude. Convection becomes weaker while the crystals grow larger and cannot be suspended by convection. Crystal settling or flotation (depending on pressure) creates a compositional and density stratification which tends to suppress convection, although vigorous convection and cooling are likely to continue near the surface. When the crystal fraction reaches the maximum packing fraction of about $\sim 60\%$ all the way to the surface, the heat flux drops to ~1 W m^{-2} because of the viscosity jump from liquid-like to solid-like viscosity (the rheological transition). Formation of the solid crust can reduce the heat flux further by one order of magnitude. All these changes contribute to the onset of crystal-melt segregation in the magma ocean. The depth of the differentiated region is from a few hundred kilometers to 1000 km, with the melt fraction varying from zero at the bottom to ~60% at the top.

Crystallization of this remaining partially molten layer takes $10^7–10^9$ years, depending on the regime of mantle convection (surface recycling or stagnant lid convection). This is comparable with the time it takes the melt to escape via percolation. Radiogenic heating and subsolidus convection become the main factors determining the subsequent evolution of the planet.

The two very different timescales, 10^3 years for crystallization of the lower mantle and $10^7–10^9$ years for crystallization of the upper mantle, suggest that whenever the Earth was melted by a giant impact the lower mantle healed very quickly while the upper mantle never had enough time to crystallize completely. If so, then iron delivered by impacts accumulated at the base of the partially molten upper mantle before it sank into the Earth's core. Thus, chemical equilibrium between iron and silicates was established at high pressures near the bottom of the partially molten upper mantle, in agreement with models of abundances of siderophile elements.

Although the model of equilibrium crystallization provides a simple conceptual framework which can help to reconcile various geochemical constraints, some form of fractional crystallization cannot be excluded within the uncertainties of physical parameters. Also, fractional crystallization is inevitable at some stage of crystallization even within this model. The dynamics of fractional crystallization is undoubtedly very complex and involves a variety of dynamic processes, including crystal settling/flotation, convective mixing in the melt as well as in the partially molten solid, melt percolation, gravitational instabilities due to unstable density stratification and compositional and convective layering. This is an important problem which needs to be addressed in the future. Finally, future studies will need to consider primordial differentiation of the Earth's mantle in combination with planetary accretion, core formation, and atmospheric evolution.

Acknowledgment

This work was supported by NASA.

References

Abe Y (1993) Physical state of the very early Earth. *Lithos* 30: 223–235.

Abe Y (1995) Early evolution of the terrestrial planets. *Journal of Physics of the Earth* 43: 515–532.

Abe Y (1997) Thermal and chemical evolution of the terrestrial magma ocean. *Physics of the Earth and Planetary Interiors* 100: 27–39.

Abe Y and Matsui T (1986) Early evolution of the Earth: Accretion, atmosphere formation, and thermal history. *Journal of Geophysical Research* 91: E291–E302.

Abe Y, Ohtani E, Okuchi T, Righter K, and Drake M (2000) Water in the early Earth. In: Canup RM and Righter K (eds.) *Origin of the Earth and Moon*, pp. 413–433. Tucson, AZ: University of Arizona Press.

Agee CB (1990) A new look at differentiation of the Earth from melting experiments on the Allende meteorite. *Nature* 346: 834–837.

Agee CB (1998) Crystal-liquid density inversions in terrestrial and lunar magmas. *Physics of the Earth and Planetary Interiors* 107: 63–74.

Agee CB and Walker D (1988) Mass balance and phase density constraints on early differentiation of chondritic mantle. *Earth and Planetary Science Letters* 90: 144–156.

Ahrens TJ (1990) Earth accretion. In: Newsom HE and Jones JH (eds.) *Origin of the Earth*, pp. 211–227. New York: Oxford University Press.

Ahrens TJ (1992) A magma ocean and the Earth's internal water budget. In: Agee CB and Longhi J (eds.) *Workshop on the Physics and Chemistry of Magma Oceans from 1 bar to 4 Mbar, LPI Tech. Rep. 92–03*, pp. 13–14. Houston: Lunar and Planetary Institute.

Akins JA, Luo SN, Asimow PD, and Ahrens TJ (2004) Shock-induced melting of MgSiO$_3$ perovskite and implications for melts in Earth's lowermost mantle. *Geophysical Research Letters* 31: (doi:10.1029/2004GL020237).

Albarède F, Blichert-Toft J, Vervoort JD, Gleason JD, and Rosing M (2000) Hf–Nd isotope evidence for a transient dynamic regime in the early terrestrial mantle. *Nature* 404: 488–490.

Amelin Y, Lee DC, Halliday AN, and Pidgeon RT (1999) Nature of the Earth's earliest crust from hafnium isotopes in single detrital zircons. *Nature* 399: 252–255.

Andrade ENC (1952) Viscosity of liquids. *Proceedings of the Royal Society of London Series. A* 215: 36–43.

Arzi AA (1978) Critical phenomena in the rheology of partially melted rocks. *Tectonophys* 44: 173–184.

Asimow PD, Hirschmann MM, and Stolper EM (1997) An analysis of variations in isentropic melt productivity. *Philosophical Transactions of the Royal Society of London Series A* 355: 255–281.

Bartlett RB (1969) Magma convection, temperature distribution, and differentiation. *American Journal of Science* 267: 1067–1082.

Béjina F, Jaoul O, and Liebermann RC (2003) Diffusion in minerals at high pressure: A review. *Physics of the Earth and Planetary Interiors* 139: 3–20.

Bennett VC (2003) Compositional evolution of the mantle. In: Holland HD and Turekian KK (eds.) *Treatise on Geochemistry*, vol. 2, ch. 13, pp. 493–519. New York: Elsevier.

Bennett VC, Nutman AP, and McCulloch MT (1993) Nd isotopic evidence for transient highly-depleted mantle reservoirs in the early history of the Earth. *Earth and Planetary Science Lettes* 119: 299–317.

Benz W and Cameron AGW (1990) Terrestrial effects of the giant impact. In: Newsom HE and Jones, JH (eds.) *Origin of the Earth*, pp. 61–67. New York: Oxford University Press.

Benz W, Cameron AGW, and Melosh HJ (1989) The origin of the Moon and the single impact hypothesis, III. *Icarus* 81: 113–131.

Benz W, Slattery WL, and Cameron AGW (1986) The origin of the Moon and the single impact hypothesis, I. *Icarus* 66: 515–535.

Benz W, Slattery WL, and Cameron AGW (1987) The origin of the Moon and the single impact hypothesis, II. *Icarus* 71: 30–45.

Bercovici D, Richard Y, and Schubert G (2001) A two-phase model for compaction and damage 1. General theory. *Journal of Geophysical Research* 106: 8887–8906.

Bizzarro M, Baker JA, Haack H, Ulfbeck D, and Rosing MG (2003) Early history of Earth's crust–mantle system inferred from hafnium isotopes in chondrites. *Nature* 421: 931–933.

Boehler R (1992) Melting of the Fe–FeO and the Fe–FeS systems at high pressure: Constraints on core temperatures. *Earth and Planetary Science Letters* 111: 217–227.

Boehler R (1996) Melting temperature of the Earth's mantle and core: Earth's thermal structure. *Annual Review of Earth and Planetary Sciences* 24: 15–40.

Boehler R (2000) High-pressure experiments and the phase diagram of lower mantle and core materials. *Reviews of Geophysics* 38: 221–245.

Bottinga Y, Richet P, and Sipp A (1995) Viscosity regimes of homogeneous silicate melts. *American Mineralogist* 80: 305–318.

Bottinga Y and Weill DF (1972) The viscosity of magmatic silicate liquids: A model for calculation. *American Journal of Science* 272: 438–475.

Boubnov BM and Golitsyn GS (1986) Experimental study of convective structures in rotating fluids. *Journal of Fluid Mechanics* 167: 503–531.

Boubnov BM and Golitsyn GS (1990) Temperature and velocity field regimes of convective motions in a rotating plane fluid layer. *Journal of Fluid Mechanics* 219: 215–239.

Bowring SA and Housh T (1995) The Earth's early evolution. *Science* 269: 1535–1540.

Boyet M and Carlson RW (2005) [142]Nd evidence for early (>4.53 Ga) global differentiation of the silicate Earth. *Science* 309: 576–581.

Brinkman HC (1952) The viscosity of concentrated suspensions and solutions. *Journal of Chemical Physics* 20: 571–581.

Buyevich YA and Mansurov VV (1991) Kinetics of the intermediate stage of phase transition in batch crystallization. *Journal of Crystal Growth* 104: 861–867.

Campbell GA and Forgacs G (1990) Viscosity of concentrated suspensions: An approach based on percolation theory. *Physical Review A* 41: 4570–4573.

Cameron AGW (1997) The origin of the Moon and the single impact hypothesis V. *Icarus* 126: 126–137.

Canup RM (2004) Simulations of a late lunar-forming impact. *Icarus* 168: 433–456.

Canup RM and Agnor CB (2000) Accretion of the terrestrial planets and the Earth–Moon system. In: Canup RM and Righter K (eds.) *Origin of the Earth and Moon*, pp. 113–129. Tucson, AZ: University of Arizona Press.

Canup RM and Asphaug E (2001) Origin of the Moon in a giant impact near the end of the Earth's formation. *Nature* 412: 708–712.

Canuto VM and Dubovikov MS (1998) Two scaling regimes for rotating Rayleigh–Bènard convection. *Physical Review Letters* 80: 281–284.

Canup RM and Esposito LW (1996) Accretion of the Moon from an impact-generated disk. *Icarus* 119: 427–446.

Caro G, Bourdon B, Birck JL, and Moorbath S (2003) [146]Sm–[142]Nd evidence from Isua metamorphosed sediments for early differentiation of the Earth's mantle. *Nature* 423: 428–432.

Caro G, Bourdon B, Wood BJ, and Corgne A (2005) Trace-element fractionation in Hadean Mantle generated by melt segregation from a magma ocean. *Nature* 436: 246–249.

Cashman KV (1993) Relationship between plagioclase crystallization and cooling rate in basaltic melts. *Contribution to Mineralogy and Petrology* 113: 126–142.

Cashman KV and Marsh BD (1988) Crystal size distribution (CSD) in rocks and the kinetics and dynamics of crystallization II. Makaopuhi lava lake. *Contribution to Mineralogy and Petrology* 99: 292–305.

Castaing B, Gunaratne G, Heslot F, *et al.* (1989) Scaling of hard thermal turbulence in Rayleigh–Benard convection. *Journal of Fluid Mechanics* 204: 1–30.

Caughey SJ and Palmer SG (1979) Some aspects of turbulence structure through the depth of the convective boundary layer. *Quarterly journal of the Royal Meteorological Society* 105: 811–827.

Chambers JE and Wetherill GW (1998) Making the terrestrial planets: N-body integrations of planetary embryos in three dimensions. *Icarus* 136: 304–327.

Chen R, Fernando HJS, and Boyer DL (1989) Formation of isolated vortices in a rotating convecting fluid. *Journal of Geophysical Research* 94: 18445–18453.

Clark SP (1957) Radiative transfer in the Earth's mantle. *Transition American Geophysical Union* 38: 931–938.

Collerson KD, Campbell LM, Weaver BL, and Palacz ZA (1991) Evidence for extreme mantle fractionation in early Archaean untramafic rocks from northern Labrador. *Nature* 349: 209–214.

Cooper RF and Kohlstedt DL (1986) Rheology and structure of olivine-basalt partial melts. *Journal of Geophysical Research* 91: 9315–9323.

Corgne A, Liebske C, Wood BJ, Rubie DC, and Frost DJ (2005) Silicate perovskite-melt partitioning of trace elements and geochemical signature of a deep perovskitic reservoir. *Geochem et Cosmochim Acta* 69: 485–496.

Corgne A and Wood BJ (2002) $CaSiO_3$ and $CaTiO_3$ perovskite-melt partitioning of trace elements: Implications for gross mantle differentiation. *Geophysical Research Letters* 29, (doi:10.1029/2001GL014398).

Costa A (2005) Viscosity of high crystal content melts: Dependence on solid fraction. *Geophysical Research Letters* 32: (doi: 10.1029/2005GL024303).

Davaille A and Jaupart C (1993a) Transient high Rayleigh number thermal convection with large viscosity variations. *Journal of Fluid Mechanics* 253: 141–166.

Davaille A and Jaupart C (1993b) Thermal convection in lava lakes. *Geophysical Research Letters* 20: 1827–1830.

Davies GF (1982) Ultimate strength of solids and formation of planetary cores. *Physical Review Letters* 9: 1267–1270.

Davies GF (1990) Heat and mass transport in the early Earth. In: Newsom HE and Jones JH (eds.) *Origin of the Earth*, pp. 175–194. New York: Oxford University Press.

Davis RH and Acrivos A (1985) Sedimentation of noncolloidal particles at low Reynolds numbers. *Annual Review of Fluid Mechanics* 17: 91–118.

Deardorff JW (1970) Convective velocity and temperature scales for the unstabe planetary boundary layer and for Rayleigh convection. *Journal of Atmospheric Sciences* 17: 1211–1213.

Dingwell DB, courtial P, Giordano D, and Nicholas ARL (2004) Viscosity of peridotite liquid. *Earth and Planetary Science Letters* 226: 127–138.

Dowty E (1980) Crystal growth and nucleation theory and the numerical simulation of igneous crystallization.

In: Hargraves RV (ed.) *Physics of Magmatic Processes*, pp. 419–485. Princeton: Princeton University Press.

Drake MJ (2000) Accretion and primary differentiation of the Earth: A personal journey. *Geochem et Cosmochim Acta* 64: 2363–2370.

Dullien FAL (1979) *Porous Media: Fluid Transport and Pore Structure*. San Diego, CA: Academic Press.

Einstein A (1906) Eine neue Bestimmung der Moleküldimensionen. *Annals of Physics* 19: 289–306.

Elkins-Tanton LT, Zaranek SE, Parmentier EM, and Hess PC (2005) Early magnetic field and crust on Mars from magma ocean cumulate overturn. *Earth and Planetary Science Letters* 236: 1–12.

Elsasser WM (1963) Early history of the Earth; dedicated to F. G. Houtermans on his sixtieth birthday. In: Geiss J and Goldberg ED (eds.) *Earth Science and Meteoritics*, pp. 1–30. Amsterdam: North-Holland.

Fernando HJS, Chen RR, and Boyer DL (1991) Effects of rotation on convective turbulence. *Journal of Fluid Mechanics* 228: 513–547.

Flasar FM and Birch F (1973) Energetics of core formation: A correction. *Journal of Geophysical Research* 78: 6101–6103.

Flemings MC, Riek RG, and Young KP (1976) Rheocasting. *Mater. Sci. Eng* 25: 103–117.

Foley CN, Wadhwa M, Borg LE, Janney PE, Hines R, and Grove TL (2005) The early differentiation history of Mars from W-182–Nd-142 isotope systematics in the SNC meteorites. *Geochem et Cosmochim Acta* 69: 4557–4571.

Fowler AC (1985) Fast thermoviscous convection. *Studies in Applied Mathematics* 72: 189–219.

Frankel NA and Acrivos A (1967) On the viscosity of a concentrated suspension of solid spheres. *Chemical Engineering Science* 22: 847–853.

Gans RF (1972) Viscosity of the Earth's core. *Journal of Geophysical Research* 77: 360–366.

Gasparik T and Drake MJ (1995) Partitioning of elements among two silicate perovskites, superphase B, and volatile-bearing melt at 23 GPa and 1500–1600°C. *Earth and Planetary Science Letters* 134: 307–318.

German RM (1985) *Liquid Phase Sintering*. New York: Plenum Press.

Ghiorso MS (1997) Thermodynamic models of igneous processes. *Annual Review of Earth and Planetary Sciences* 25: 221–241.

Glazier JA (1999) Evidence against 'ultrahard' thermal turbulence at very high Rayleigh numbers. *Nature* 398: 307–310.

Golitsyn GS (1980) Geostrophic convection. *Doklady Akademii Nauk SSSR* 251: 1356–1360.

Golitsyn GS (1981) Structure of convection in rapid rotation. *Doklady Akademii Nauk SSSR* 261: 317–320.

Goody RM (1995) *Principles of Atmospheric Physics and Chemistry*. New York: Oxford University Press.

Grossmann S and Lohse D (1992) Scaling in hard turbulent Rayleigh–Bénard flow. *Physical Review A* 46: 903–917.

Grove TL (1990) Cooling histories of lavas from Scrocki volcano. *Proceedings of the Ocean Driling Programme* 106/109: 3–8.

Grove TL and Walker D (1977) Cooling histories of Apollo 15 quartz-normative basalts. *Proceedings of the 8th Lunar Planetary Science Conference* 1501–1520.

Halliday AN (2004) Mixing, volatile loss and compositional change during impact-driven accretion of the Earth. *Nature* 427: 505–509.

Halliday A, Rehkämper M, Lee D-C, and Yi W (1996) Early evolution of the Earth and Moon: New constraints from Hf–W isotope geochemistry. *Earth and Planetary Science Letters* 142: 75–89.

Harrison TM, Blichert-Toft J, Müller W, Albarede F, Holden P, and Mojzsis SJ (2005) Heterogeneous Hadean hafnium: Evidence of continental crust at 4.4 to 4.5 Ga. *Science* 310: 1947–1950.

Hayashi C, Nakazawa K, and Mizuno H (1979) Earth's melting due to the blanketing effect of the primordial dense atmosphere. *Earth and Planetary Science Letters* 43: 22–28.

Herbert F, Drake MJ, and Sonett CP (1978) Geophysical and geochemical evolution of the lunar magma ocean. *Proceedings of the 9th Lunar and Planetary Science Conference* 249–262.

Herzberg C and Gasparik T (1991) Garnet and pyroxenes in the mantle: A test of the majorite fractionation hypothesis. *Journal of Geophysical Research* 96: 16263–16274.

Hirose K, Shimizu N, van Westrenen W, and Fei Y (2004) Trace element partitioning in Earth's lower mantle and implications for geochemical consequences of partial melting at the core–mantle boundary. *Physics of the Earth and Planetary Interiors* 146: 249–260.

Hirth G and Kohlstedt DL (1995a) Experimental constraints on the dynamics of the partially molten upper mantle: Deformation in the diffusion creep regime. *Journal of Geophysical Research* 100: 1981–2001.

Hirth G and Kohlstedt DL (1995b) Experimental constraints on the dynamics of the partially molten upper mantle 2. Deformation in the dislocation creep regime. *Journal of Geophysical Research* 100: 15441–15449.

Holland KG and Ahrens TJ (1997) Melting of (Mg, Fe)$_2$SiO$_4$ at the core–mantle boundary of the Earth. *Science* 275: 1623–1625.

Hopfinger EJ (1987) Turbulence in stratified fluids: A review. *Journal of Geophysical Research* 92: 5297–5303.

Hopfinger EJ (1989) Turbulence and vortices in rotating fluids. In: Germain P, Piau M, and Caillerie D (eds.) *Theoretical and Applied Mechanics*, pp. 117–138. New York: Elsevier.

Hopfinger EJ, Browand FK, and Gagne Y (1982) Turbulence and waves in a rotating tank. *Journal of Fluid Mechanics* 125: 505–534.

Hostetler CJ and Drake MJ (1980) On the early global melting of the terrestrial planets. *Proceeding of the 11th Lunar and Planetary Science Conference*, pp.1915–1929.

Huppert HE and Sparks RSJ (1980) The fluid dynamics of a basaltic magma chamber replenished by influx of hot, dense ultrabasic magma. *Contributions to Mineralogy and Petrology* 75: 279–289.

Ichikawa K, Kinoshita Y, and Shimamura S (1985) Grain refinement in Al–Cu binary alloys by rheocasting. *Transactions of the Japan Institute of Metals* 26: 513–522.

Ida S, Canup RM, and Stewart GR (1997) Lunar accretion from an impact-generated disk. *Nature* 389: 353–357.

Ita J and Cohen RE (1998) Diffusion in MgO at high pressure: Implications for lower mantle rheology. *Geophysical Research Letters* 25: 1095–1098.

Ito E and Takahashi E (1987) Melting of peridotite at uppermost lower-mantle conditions. *Nature* 328: 514–517.

Ito E, Kubo A, Katsura T, and Walter MJ (2004) Melting experiments of mantle materials under lower mantle conditions with implications for magma ocean differentiation. *Physics of the Earth and Planetary Interiors* 143: 397–406.

Jacobsen SB (2005) The Hf–W isotopic system and the origin of the Earth and Moon. *Annual Review of Earth and Planetary Sciences* 33: 531–570.

Jin Z-M, Green HW, and Zhou Y (1994) Melt topology in partially molten mantle peridotite during ductile deformation. *Nature* 372: 164–167.

Julien K, Legg S, McWilliams J, and Werne J (1996a) Hard turbulence in rotating Rayleigh–Bénard convection. *Physical Review E* 53: 5557–5560.

Julien K, Legg S, McWilliams J, and Werne J (1996b) Rapidly rotating turbulent Rayleigh–Benard convection. *Journal of Fluid Mechanics* 322: 243–273.

Kadanoff LP (2001) Turbulent heat flow: Structures and scalings. *Physics Today* 54: 34–39.

Karato SI and Li P (1992) Diffusion creep in perovskite: Implications for the rheology of the lower mantle. *Science* 255: 1238–1240.

Karato S-I and Wu P (1993) Rheology of the upper mantle: A synthesis. *Science* 260: 771–778.

Karato S-I and Murthy VR (1997a) Core formation and chemical equilibrium in the Earth - I. Physical considerations. *Physics of the Earth and Planetary Interiors* 100: 61–79.

Karato S-I and Murthy VR (1997b) Core formation and chemical equilibrium in the Earth - II. Chemical consequences for the mantle and core. *Physics of the Earth and Planetary Interiors* 100: 81–95.

Kasting JF (1988) Runaway and moist greenhouse atmosphere and the evolution of Earth and Venus. *Icarus* 74: 472–494.

Kato T, Ringwood AE, and Irifune T (1988a) Experimental determination of element partitioning between silicate perovskites, garnets and liquids: Constraints on early differentiation of the mantle. *Earth and Planetary Science Letters* 89: 123–145.

Kato T, Ringwood AE, and Irifune T (1988b) Constraints on element partition coefficients between MgSiO$_3$ perovskite and liquid determined by direct measurements. *Earth and Planetary Science Letters* 90: 65–68.

Kaula WM (1979) Thermal evolution of Earth and Moon growing by planetesimals impacts. *Journal of Geophysical Research* 84: 999–1008.

Ke Y and Solomatov VS (2006) Early transient superplumes and the origin of the Martian crustal dichotomy. *Journal of Geophysical Research* 111 (doi:10.1029/2005JE002631).

King SD (1995) Models of mantle viscosity. In: Ahrens TJ (ed.) *Mineral Physics and Crystallography: A Handbook of Physical Constants*, pp. 227–236. Washington, DC: American Geophysical Union.

Kleine T, Münker C, Mezger K, and Palme H (2002) Rapid accretion and early core formation on asteroids and the terrestrial planets from Hf-W chronometry. *Nature* 418: 952–955.

Knittle E and Jeanloz R (1989) Melting curve of (Mg, Fe)SiO$_3$ perovskite to 96 GPa: Evidence for a structural transition in lower mantle melts. *Geophysical Research Letters* 16: 421–424.

Kohlstedt DL and Zimmerman ME (1996) Rheology of partially molten mantle rocks. *Annual Review of Earth and Planetary Sciences* 24: 41–62.

Koyaguchi T, Hallworth MA, and Huppert HE (1993) An experimental study of the effects of phenocrysts on convection in magmas. *Journal of Volcanology and Geothermal Research* 55: 15–32.

Kraichnan RH (1962) Turbulent thermal convection at arbitrary Prandtl number. *Physical of Fluids* 5: 1374–1389.

Krieger IM and Dougherty TJ (1959) A mechanism for non-Newtonian flow in suspensions of rigid spheres. *Transactions of the Society of Rheology* 3: 137–152.

Kushiro I (1980) Viscosity, density, and structure of silicate melts at high pressures, and their petrological applications. In: Hargraves RB (ed.) *Physics of Magmatic Processes*, pp. 93–120. Princeton: Princeton University Press.

Kushiro I (1986) Viscosity of partial melts in the upper mantle. *Journal of Geophysical Research* 91: 9343–9350.

Lee D-C and Halliday AN (1995) Hafnium–tungsten chronometry and the timing of terrestrial core formation. *Nature* 378: 771–774.

Lejeune A-M and Richet P (1995) Rheology of crystal-bearing silicate melts: An experimental study at high viscosities. *Journal of Geophysical Research* 100: 4215–4229.

Li J and Agee CB (1996) Geochemistry of mantle–core differentiation at high pressure. *Nature* 381: 686–689.

Li P, Karato S, and Wang Z (1996) high-temperature creep in fine-grained polycrystalline CaTiO$_3$, an analogue material of (Mg,Fe)SiO$_3$. *Physics of the Earth and Planetary Interiors* 95: 19–36.

Liebske C, Corgne A, Frost DJ, Rubie DC, and Wood BJ (2005a) Compositional effects on element partitioning between Mg-silicate perovskite and silicate melts. *Contributions to Mineralogy and Petrology* 149: 113–128.

Liebske C, Schmickler B, Terasaki H, et al. (2005b) Viscosity of peridotite liquid up to 13 GPa: Implications for magma ocean viscosities. *Earth and Planetary Science Letters* 240: 589–604.

Lifshitz IM and Slyozov VV (1961) The kinetics of precipitation from supersaturated solid solution. *Journal of Physics and Chemistry of Solids* 19: 35–50.

Lofgren G, Donaldson CH, Williams RJ, Mullins O, and Usselman TM (1974) Experimentally reproduced textures and mineral chemistry of Apollo 15 quartz-normative basalts. *Proceedings of the 5th Lunar Science Conference* pp. 549–567.

Longhi J (1980) A model of early lunar differentiation. *Proceeding of the 11th Lunar Science Conference* 289–315.

Marsh BD and Maxey MR (1985) On the distribution and separation of crystals in convecting magma. *Journal of Volcanology and Geothermal Research* 24: 95–150.

Martin D and Nokes R (1988) crystal settling in a vigorously convecting magma chamber. *Nature* 332: 534–536.

Martin D and Nokes R (1989) A fluid dynamical study of crystal settling in convecting magmas. *Journal of Petrology* 30: 1471–1500.

Matsui T and Abe Y (1986) Formation of a "magma ocean" on the terrestrial planets due to the blanketing effect of an impact-induced atmosphere. *Earth Moon Planets* 34: 223–230.

Matsukage KN, Jing Z, and Karato S-I (2005) Density of hydrous silicate melt at the conditions of Earth's deep upper mantle. *Nature* 438: 488–491.

McBirney AR and Murase T (1984) Rheological properties of magmas. *Annual Review of Earth and Planetary Sciences* 12: 337–357.

McFarlane EA and Drake MJ (1990) Element partitioning and the early thermal history of the Earth. In: Newsom HE and Jones JH (eds.) *Origin of the Earth*, pp. 135–150. New York: Oxford University Press.

McFarlane EA, Drake MJ, and Rubie DC (1994) Element partitioning between Mg-perovskite, magnesiowüstite, and silicate melt at conditions of the Earth's mantle. *Geochem et Cosmochim Acta* 58: 5161–5172.

McGeary RK (1961) Mechanical packing of spherical particles. *Journal of the American Ceramic Society* 44: 513–522.

McKenzie D and Bickle MJ (1988) The volume and composition of melt generated by extension of the lithosphere. *Journal of Petrology* 29: 625–679.

Mei S, Bai W, Hiraga T, and Kohlstedt DL (2002) Influence of melt on the creep behavior of olivine–basalt aggregates under hydrous conditions. *Earth and Planetary Science Letters* 201: 491–507.

Mei S and Kohlstedt DL (2000a) Influence of water on plastic deformation of olivine aggregates 1. Diffusion creep regime. *Journal of Geophysical Research* 105: 21457–21469.

Mei S and Kohlstedt DL (2000b) Influence of water on plastic deformation of olivine aggregates 2. Dislocation creep regime. *Journal of Geophysical Research* 105: 21471–21481.

Melosh HJ (1990) Giant impacts and the thermal state of the early Earth. In: Newsom HE and Jones JH (eds.) *Origin of the Earth*, pp. 69–83. New York: Oxford University Press.

Miller GH, Stolper EM, and Ahrens TJ (1991a) The equation of state of a molten komatiite, 1, Shock wave compression to 36 GPa. *Journal of Geophysical Research* 96: 11831–11848.

Miller GH, Stolper EM, and Ahrens TJ (1991b) The equation of state of a molten komatiite, 2, Application to komatiite petrogenesis and the Hadean mantle. *Journal of Geophysical Research* 96: 11849–11864.

Mooney M (1951) The viscosity of a concentrated suspension of spherical particles. *Journal of Colloid Science* 6: 162–170.

Morris S and Canright D (1984) A boundary layer analysis of Bénard convection in a fluid of strongly temperature dependent viscosity. *Physics of the Earth and Planetary Interiors* 36: 355–373.

Mostefaoui S, Lugmair GW, and Hoppe P (2005) Fe-60: A heat source for planetary differentiation from a nearby supernova explosion. *Journal of Geomagnetism and Geoelectricity* 625: 271–277.

Murase T and McBirney AR (1973) Properties of some common igneous rocks and their melts at high temperatures. *Geological Society of America Bulletin* 84: 3563–3593.

Murray JD (1965) On the mathematics of fluidization. Part I: Fundamental equations and wave propagation. *Journal of Fluid Mechanics* 21: 465–493.

Nakazawa K, Mizuno H, Sekiya M, and Hayashi C (1985) Structure of the primordial atmosphere surrounding the early Earth. *Journal of Geomagnetism and Geoelectricity* 37: 781–799.

Newsom HE and Taylor SR (1989) Geochemical implications of the formation of the Moon by a single giant impact. *Nature* 338: 29–34.

Nicolas A and Ildefonse B (1996) Flow mechanism and viscosity in basaltic magma chambers. *Geophysical Research Letters* 23: 2013–2016.

Nicolis G and Prigogine I (1977) *Self-Organization in Non-Equilibrium Systems*. New York: Wiley.

Niemela JJ, Skrbek L, Sreenivasan KR, and Donnelly RJ (2000) turbulent convection at very high Rayleigh numbers. *Nature* 404: 837–840.

Ohtani E (1985) The primordial terrestrial magma ocean and its implication for statification of the mantle. *Physics of the Earth and Planetary Interiors* 38: 70–80.

Ohtani E and Maeda M (2001) Density of basaltic melt at high pressure and stability of the melt at the base of the lower mantle. *Physics of the Earth and Planetary Interiors* 193: 69–75.

Ohtani E and Sawamoto H (1987) Melting experiment on a model chondritic mantle composition at 25 GPa. *Geophysical Research Letters* 14: 733–736.

Persikov ES, Zharikov VA, Bukhtiyarov PG, and Polskoy SF (1997) The effect of volatiles on the properties of magmatic melts. *European Journal of Mineralogy* 2: 621–642.

Pierazzo E, Vickery AM, and Melosh HJ (1997) A reevaluation of impact melt production. *Icarus* 127: 408–423.

Pinkerton H and Sparks RSJ (1978) Field measurements of the rheology of lava. *Nature* 276: 383–385.

Priestly CHB (1959) *Turbulent Transfer in the Lower Atmosphere*. Chicago, IL: University of Chicago Press.

Poe BT, McMillan PF, Rubie DC, Chakraborty S, Yarger J, and Diefenbacher J (1997) Silicon and oxygen self-diffusivities in silicate liquids measured to 15 gigapascals and 2800 kelvin. *Science* 276: 1245–1248.

Presnall DC, Weng Y-H, Milholland CS, and Walter MJ (1998) Liquidus phase relations in the system MgO–MgSiO$_3$ at pressures up to 25 GPa – constraints on crystallization of a molten Hadean mantle. *Physics of the Earth and Planetary Interiors* 107: 83–95.

Raizer YuP (1960) Condensation of a cloud of vaporized matter expanding in vacuum. *Soviet Phys. JETP (English Transl.)* 10: 1229–1235.

Reese CC and Solomatov VS (2006) Fluid dynamics of local martian magma oceans. *Icarus* 184: 102–120.

Righter K (2003) Metal-silicate partitioning of siderophile elements and core formation in the early Earth. *Annual Review of Earth and Planetary Sciences* 31: 135–174.

Righter K and Drake MJ (1997) Metal-silicate equilibrium in a homogeneously accreting Earth: New results for Re. *Earth and Planetary Science Letters* 146: 541–553.

Righter K and Drake MJ (2003) Partition coefficients at high pressure and temperature. In: Holland HD and Turekian KK (eds.) *Treatise on Geochemistry*, vol. 2, ch. 10, pp. 425–449. New York: Elsevier.

Righter K, Drake MJ, and Yaxley G (1997) Prediction of siderophile element metal-silicate partition coefficients to 20 GPa and 2800°C: The effects of pressure, temperature, oxygen fugacity, and silicate and metallic melt compositions. *Physics of the Earth and Planetary Interiors* 100: 115–134.

Ringwood AE (1990) Earliest history of the Earth–Moon system. In: Newsom HE and Jones JH (eds.) *Origin of the Earth*, pp. 101–134. New York: Oxford University Press.

Ringwood AE and Kesson SE (1976) A dynamic model for mare basalt petrogenesis. *Proceeding of the 7th Lunar Plantary Science Conference* 1697–1722.

Robson GR (1967) Thickness of Etnean lavas. *Nature* 216: 251–252.

Roscoe R (1952) The viscosity of suspensions of rigid spheres. *British Journal of Applied Physics* 3: 267–269.

Rubie DC, Melosh HJ, Reid JE, Liebske C, and Righter K (2003) Mechanisms of metal-silicate equilibration in the terrestrial magma ocean. *Earth and Planetary Science Letters* 205: 239–255.

Rudman M (1992) Two-phase natural convection: Implications for crystal settling in magma chambers. *Physics of the Earth and Planetary Interiors* 72: 153–172.

Rutter EH and Neumann DHK (1995) Experimental deformation of partially molten Westerly granite under fluid-absent conditions, with implications for the extraction of granitic magmas. *Journal of Geophysical Research* 100: 15697–15715.

Ryerson FJ, Weed HC, and Piwinskii AJ (1988) Rheology of subliquidus magmas. Part I: Picritic compositions. *Journal of Geophysical Research* 93: 3421–3436.

Safronov VS (1964) The primary inhomogeneities of the Earth's mantle. *Tectonophys* 1: 217–221.

Safronov VS (1978) The heating of the Earth during its formation. *Icarus* 33: 3–12.

Sasaki S and Nakazawa K (1986) Metal-silicate fractionation in the growing Earth: Energy source for the terrestrial magma ocean. *Journal of Geophysical Research* 91: 9231–9238.

Scarfe CM and Takahashi E (1986) Melting of garnet peridotite to 13 GPa and the early history of the upper mantle. *Nature* 322: 354–356.

Scherer E, Münker C, and Mezger K (2001) Calibration of the lutetium–hafnium clock. *Science* 293: 683–687.

Schnetzler CC and Philpotts JA (1971) Alkali, alkaline earth and rare earth element concentrations in some Apollo 12 solids, rocks and separated phases. *Proceedings of the 2nd Lunar and Planetary Science Conference* 1101–1122.

Sears WD (1993) Tidal dissipation and the giant impact origin for the Moon. *Lunar and Planetary Science Conference* 23: 1255–1256.

Shankland TJ, Nitsan U, and Duba AG (1979) Optical absorption and radiative heat transport in olivine at high temperatures. *Journal of Geophysical Research* 84: 1603–1610.

Shaw HR (1969) Rheology of basalt in the melting range. *Journal of Petrology* 10: 510–535.

Shaw HR (1972) Viscosities of magmatic silicate liquids: An empirical method of prediction. *Americal Journal of Science* 272: 870–893.

Shaw HR, Wright TL, Peck DL, and Okamura R (1968) The viscosity of basaltic magma: An analysis of field measurements in Makaopuhi lava lake, Hawaii. *Americal Journal of Science* 266: 225–264.

Shraiman BI and Siggia ED (1990) Heat transport in high-Rayleigh-number convection. *Physical Review A* 42: 3650–3653.

Schubert G, Solomatov VS, Tackley PJ, and Turcotte DL (1997) Mantle convection and the thermal evolution of Venus. In: Bougher SW, Hunten DM, and Philliphs RJ (eds.) *Venus II – Geology, Geophysics, Atmosphere, and Solar Wind Environment*, pp. 1245–1287. Tucson, AZ: University of Arizona Press.

Shukolyukov A and Lugmair GW (1993) Live iron-60 in the early solar system. *Science* 259: 1138–1142.

Siggia ED (1994) High Rayleigh number convection. *Annual Review of Fluid Mechanics* 26: 137–168.

Smith DM, Eady JA, Hogan LM, and Irwin DW (1991) Crystallization of a faceted primary phase in a stirred slurry. *Metallurgical Transactions* 22A: 575–584.

Solomatov VS (1995a) Scaling of temperature- and stress-dependent viscosity convection. *Physics of Fluids* 7: 266–274.

Solomatov VS (1995b) Batch crystallization under continuous cooling: Analytical solution for diffusion limited crystal growth. *Journal of Crystal Growth* 148: 421–431.

Solomatov VS (2000) Fluid dynamics of a terrestrial magma ocean. In: Canup RM and Righter K (eds.) *Origin of the Earth and Moon*, pp. 323–338. Tucson, AZ: University of Arizona Press.

Solomatov VS and Moresi LN (2000) Scaling of time-dependent stagnant lid convection: Application to small-scale convection on the Earth and other terrestrial planets. *Journal of Geophysical Research* 105: 21795–21818.

Solomatov VS and Moresi L-M (1996a) Stagnant lid convection on Venus. *Journal of Geophysical Research* 101: 4737–4753.

Solomatov VS and Moresi L-M (1996b) Scaling of time-dependent stagnant lid convection: Application to small-scale convection on the Earth and other terrestrial planets. *Journal of Geophysical Research* 105: 21795–21818.

Solomatov VS and Stevenson DJ (1993a) Suspension in convective layers and style of differentiation of a terrestrial magma ocean. *Journal of Geophysical Research* 98: 5375–5390.

Solomatov VS and Stevenson DJ (1993b) Nonfractional crystallization of a terrestrial magma ocean. *Journal of Geophysical Research* 98: 5391–5406.

Solomatov VS and Stevenson DJ (1993c) Kinetics of crystal growth in a terrestrial magma ocean. *Journal of Geophysical Research* 98: 5407–5418.

Solomatov VS, Olson P, and Stevenson DJ (1993) Entrainment from a bed of particles by thermal convection. *Earth and Planetary Science Letters* 120: 387–393.

Sonett CP, Colburn DS, and Schwartz K (1968) Electrical heating of meteorite parent bodies and planets by dynamo induction from a pre-main sequence T Tauri solar wind. *Nature* 219: 924–926.

Soo SL (1967) *Fluid Dynamics of Multiphase Systems*. Waltham: Blaisdell.

Sparks RS, Huppert HE, Koyaguchi T, and Hallworth MA (1993) Origin of modal and rhythmic igneous layering by sedimentation in a convecting magma chamber. *Nature* 361: 246–249.

Spera FJ (1992) Lunar magma transport phenomena. *Geochem et Cosmochim Acta* 56: 2253–2265.

Srinivasan G, Goswami JN, and Bhandari N (1999) [26]Al in eucrite Piplia Kalan: Plausible heat source and formation chronology. *Science* 284: 1348–1350.

Stevenson DJ (1981) Models of the Earth's core. *Science* 214: 611–619.

Stevenson DJ (1987) Origin of the Moon – the collision hypothesis. *Annual Review of Earth and Planetary Sciences* 15: 271–315.

Stevenson DJ (1990) Fluid dynamics of core formation. In: Newsom HE and Jones JH (eds.) *Origin of the Earth*, pp. 231–249. New York: Oxford University Press.

Stishov SM (1988) Entropy, disorder, melting. *Soviet Physics Uspekhi* 31: 52–67.

Stixrude L and Karki B (2005) Structure and freezing of $MgSiO_3$ liquid in Earth's lower mantle. *Science* 14: 297–299.

Taborek P, Kleiman RN, and Bishop DJ (1986) Power-law behavior in the viscosity of super cooled liquids. *Physical Review B* 34: 1835–1840.

Thompson C and Stevenson DJ (1988) Gravitational instabilities in 2-phase disks and the origin of the Moon. *Astrophysical Journal* 333: 452–481.

Thomson W (1864) On the secular cooling of the Earth. *Transactions of the Royal Society of Edinburgh* 23: 157–169.

Tonks WB and Melosh HJ (1990) The physics of crystal settling and suspension in a turbulent magma ocean. In: Newsom HE and Jones JH (eds.) *Origin of the Earth*, pp. 151–174. New York: Oxford University Press.

Tonks WB and Melosh HJ (1992) Core formation by giant impacts. *Icarus* 100: 326–346.

Tonks WB and Melosh HJ (1993) Magma ocean formation due to giant impacts. *Journal of Geophysical Research* 98: 5319–5333.

Tschauner O, Zerr A, Specht S, Rocholl A, Boehler R, and Palme H (1999) Partitioning of nickel and cobalt between silicate perovskite and metal at pressures up to 80 GPa. *Nature* 398: 604–607.

Turcotte DL and Schubert G (2002) *Geodynamics*, 2nd edn. New York: Cambridge University Press.

Urey HC (1955) The cosmic abundances of potassium, uranium, and thorium and the heat balances of the Earth, the Moon, and Mars. *National Academy of Sciences USA* 41: 127–144.

van der Molen I and Paterson MS (1979) Experimental deformation of partially-melted granite. *Contributions to Mineralogy and Petrology* 70: 299–318.

Voorhees PW (1992) Ostwald ripening of two-phase mixtures. *Annual Review of Materials Science* 22: 197–215.

Walker D, Powell MA, Lofgren GE, and Hays JF (1978) Dynamic crystallization of a eucrite basalt. *Proceedings of the 9th Lunar and Planetary Science Conference*, pp.1369–1391.

Walter MJ, Nakamura E, Trønnes RG, and Frost DJ (2004) Experimental constraints on crystallization differentiation in a deep magma ocean. *Geochem et Cosmochim Acta* 68: 4267–4284.

Walter MJ and Trønnes RG (2004) Early Earth differentiation. *Earth and Planetary Science Letters* 225: 253–269.

Warren PH (1985) The magma ocean concept and lunar evolution. *Annual Review of Earth and Planetary Sciences* 13: 201–240.

Wetherill GW (1975) Radiometric chronology of the early solar system. *Annual Review of Nuclear Science* 25: 283–328.

Wetherill GW (1985) Occurrence of giant impacts during the growth of the terrestrial planets. *Science* 228: 877–879.

Wetherill GW (1990) Formation of the Earth. *Annual Review of Earth and Planetary Science* 18: 205–256.

Weidenschilling SJ, Spaute D, Davis DR, Marzari F, and Ohtsuki K (1997) Accretional evolution of a planetesimal swarm. 2. The terrestrial zone. *Icarus* 128: 429–455.

Weinstein SA, Yuen DA, and Olson PL (1988) Evolution of crystal-settling in magma-chamber convection. *Earth and Planetary Science Letters* 87: 237–248.

Willis GE and Deardorff JW (1974) A laboratory model of the unstable planetary boundary layer. *Journal of Atmospheric Sciences* 31: 1297–1307.

Wood JA (1975) Lunar petrogenesis in a well-stirred magma ocean. *Proceedings of the 6th Lunar and Planetary Science Conference*, pp.1087–1102.

Wood JA, Dickey JS, Marvin UB, and Powell BN (1970) Lunar anorthosites and a geophysical model of the moon. In: Levinson AA (ed.) *Proceedings of the Apollo 11 Lunar Science Conference*, pp. 965–988. New York: Pergamon.

Wright K and Price GD (1993) Computer simulation of defects and diffusion in perovskites. *Journal of Geophysical Research* 98: 22245–22253.

Yamazaki D and Karato S-I (2001) Some mineral physics constraints on the rheology and geothermal structure of Earth's lower mantle. *American Mineralogist* 86: 385–391.

Yamazaki D, Kato T, Yurimoto H, Ohtani E, and Toriumi M (2000) Silicon self-diffusion in $MgSiO_3$ perovskite at 25 GPa. *Physics of the Earth and Planetary Interiors* 119: 299–309.

Yan XD (2004) On limits to convective heat transport at infinite Prandtl number with or without rotation. *Journal of Mathematical Physics* 45: 2718–2743.

Yin Q, Jacobsen SB, Yamashita K, Blichert-Toft J, Telouk P, and Albarède F (2002) A short timescale for terrestrial planet formation from Hf–W chronometry of meteorites. *Nature* 418: 949–952.

Yoshino T, Walter MJ, and Katsura T (2003) Core formation in planetesimals triggered by permeable flow. *Nature* 422: 154–157.

Zahnle KJ, Kasting JF, and Pollack JB (1988) Evolution of a steam atmosphere during Earth's accretion. *Icarus* 74: 62–97.

Zaranek SE and Parmentier EM (2004) Convective cooling of an initially stably stratified fluid with temperature-dependent viscosity: Implications for the role of solid-state convection in planetary evolution. *Journal of Geophysical Research* 109 (doi:10.1029/2003JB002462).

Zel'dovich YaB and Raizer YuP (2002) *Physics of Shock Waves and High-Temperature Hydrodynamic Phenomena*. Mineola, NY: Dover.

Zerr A and Boehler R (1993) Melting of (Mg,Fe) SiO_3-perovskite to 625 kilobars: Indication of a high melting temperature in the lower mantle. *Science* 262: 553–555.

Zerr A and Boehler R (1994) Constraints on the melting temperature of the lower mantle from high-pressure experiments on MgO and magnesiowüstite. *Nature* 371: 506–508.

Zerr A, Diegeler A, and Boehler R (1998) Solidus of Earth's deep mantle. *Science* 281: 243–246.

Zhang J and Herzberg CT (1994) Melting experiments on anhydrous peridotite KLB-1 from 5.0 to 22.5 GPa. *Journal of Geophysical Research* 99: 17729–17742.

5 Water, the Solid Earth, and the Atmosphere: The Genesis and Effects of a Wet Surface on a Mostly Dry Planet

Q. Williams, University of California, Santa Cruz, CA, USA

5.1 Introduction

The current surface water content of planet Earth results from a fortuitous and poorly constrained sequence of events. It is fortuitous in the sense that our present near-surface hydrosphere is near 1.4×10^{21} kg, or a little over 0.02% of the mass of the planet. While liquid water is a critical ingredient for the genesis of life (e.g., Kasting and Catling, 2003; Gaidos *et al.*, 2005), it is also a component of which one can certainly have too much or too little on a planet. Had the Earth formed or evolved in such a manner that the surface water reservoir was either an order of magnitude larger or smaller, the planet's surface environment would be vastly different – and, in either instance, likely incompatible with our local form of intelligent life. Indeed, variations in water content that are far greater than an order of magnitude or two are observed in our solar system among the rocky bodies of Mercury's size or larger – from the Jovian moon Ganymede with its approximately 40 wt.% water content to Venus which may be nearly dry. Thus, from the perspective of its current sentient life, our planet seems to have a most fortunate amount of surface water.

How, and when did the Earth's oceans originate? Loosely, two long-standing end-member ideas exist on the evolution of Earth's oceans. As early as 1914, the geologist John Joly, famous for his ingenious but erroneous estimate of Earth's age based on the salinity of seawater and the rate of riverine inputs of salt to the ocean, wrote "Geological history opened with the condensation of an atmosphere of immense extent which, after long fluctuations between the states of steam and water, finally settled to the surface..." (Joly, 1914). This viewpoint is, by its nature, catastrophic, with an ocean of more-or-less its current volume forming abruptly at the start of Earth history and evolving only through second-order effects, such as the erosional addition of salts. In contrast, Rubey (1951) adopted a more uniformitarian view, in which the oceans (and other 'excess' volatiles, such as carbon, nitrogen, and sulfur) were largely generated by progressive degassing of the solid Earth, with ongoing volcanic exhalations as the primary source for the atmosphere and hydrosphere. Almost synchronously, Brown (1952) made similar arguments, focusing on deriving the atmosphere itself from the solid Earth. In these types of ongoing-degassing scenarios, the history of the oceans and atmosphere is a tandem history of hydrogen and carbon, and their interactions with the solid planet. These two venerable extremes can still be recognized at the present day, although modern viewpoints on both the interplay between volatiles and the solid Earth and on the properties and persistence of a primordial thick atmosphere have led to substantially more nuanced, and often intermediate viewpoints on the evolution of the oceans.

Why is there such uncertainty in one of the most fundamental of Earth's properties – the amount of water at its surface as a function of time? In the face of an ocean floor that is geologically transient, a continental crust that may be growing with time, and ongoing vertical and horizontal motions of the continents, there are no direct and robust geologic constraints on the volume of ocean water. Indeed, with the exception of the single observation that continental freeboard (the mean elevation of continents above sea level) seems to have varied by of the order of a few hundred meters or less since the Archean (Galer, 1991), there is little control on whether the volume of ocean water has varied by tens of percent or a few percent (or less/not at all) since the initiation of the rock record on Earth. Additionally, there are petrologic/geochemical indications, from the oxygen isotopic composition of 4.3-Gy-old zircons, that there was likely water entrained to depth during this part of Earth history (Mojsis et al., 2001). Yet, the volume of the oceans at this stage of Earth's history remains largely conjectural.

Thus, constraints on the volume of ocean water through Earth history have been largely arrived at through a combination of inferences and modeling. This is simply because the processes and events that have controlled the present surface volume of water are both complex and depend on an understanding of both the planet's starting conditions and its behavior integrated over the subsequent 4.5 Gy of geologic evolution. Effects that play principal roles in the planetary and/or surface water content include the size and water content of objects that accreted to form the planet, the impact-induced loss of protohydrospheres and atmospheres during the accretion process, the amount of water that reacted with iron during Earth's formation and whose hydrogen was lost to space or sequestered in the core, the amount of water degassed from (and retained by) a terrestrial magma ocean, Earth's climate throughout history and the amount of water present in the stratosphere and subject to photodissociation and loss, the rate of subduction into, and degassing of water out of, Earth's interior via tectonic processes, and the rate of non-tectonic (at present, principally hot-spot-related) degassing from the planet's interior.

The intent of this chapter is to examine the likely sources and sinks that have given rise to the present terrestrial hydrosphere, and to a lesser extent, the planet's far less massive atmosphere. The particular focus is on how phenomena that have driven macro-scale planetary evolution – accretion and large impacts, core formation, mantle convection, and plate tectonics – may have affected the presence and/or amount of water and atmospheric components at Earth's surface. This chapter does not examine the chemistry of seawater through time in detail – the evolution of its salinity, carbon content, and pH, as this is the subject of prior reviews (Holland, 1984; Morse and Mackenzie, 1998).

5.2   Accretion

Where and how did water reside during the transitory stage between small solid object formation in the earliest solar system and final planet formation? Carbonaceous chondrites are well known to contain significant quantities ($O \sim 10$ wt.%) of water of hydration – typically present as serpentines or even clays. By the same token, comets are composed of $\sim 50\%$ water ice. However, comets and carbonaceous chondrites need not be the only important carriers of water incorporated into the accreting planets: they simply happen to be heavily hydrated objects that occasionally cross Earth's orbit and have persisted to the present day. Therefore, the location in the solar system where hydrated objects formed and how efficiently – and at what stage of Earth accretion – they were incorporated into the nascent Earth is a pivotal issue in the evolution and genesis of Earth's hydrosphere. Loosely, the questions here can be rephrased as: how much of Earth's water was derived from chondritic sources, how much from cometary sources, and how much from sources that may have no known present-day analog? And, was Earth's water incorporated more-or-less uniformly during its accretion, or did it arrive in a 'late veneer'? If the former is the case, hydration of material at depth early in Earth history is anticipated, and the history of the ocean volume on Earth may well reflect progressive degassing of the interior. The latter possibility implies that ocean formation may have occurred essentially synchronously with the last stages of accretion, and that the bulk of the deep Earth is dry.

A range of dynamical simulations of planetary accretion have recently been conducted to examine the probable provenance and arrival time of Earth's water (Morbidelli *et al.*, 2000; Raymond *et al.*, 2004, 2006). Such simulations indicate that a significant fraction of Earth's accreting material (varying between about 10% and 50% of the planet's mass) was formed beyond 2.5 AU from the sun. Using estimates of probable water contents of such objects, these accretionary models resulted in water amounts varying from near 0 to 350 hydrospheres delivered into Earth-like planets, with most containing substantially more than 1 hydrosphere. The delivery mechanisms of water to Earth appear to be via a size range of different objects, with a number of simulations producing water influx from comparatively large objects (of order 10% Earth mass) relatively late in Earth accretion (Morbidelli *et al.*, 2000; Raymond *et al.*, 2004). Such simulations are particularly sensitive to the location and eccentricity of Jupiter in the early solar system: more eccentric Jovian orbits disrupt the asteroid belt, with wet material being both ejected from the solar system and perturbed into the Sun, resulting in drier planets (Morbidelli *et al.*, 2000; Raymond *et al.*, 2004, 2006).

The derivation of material from beyond a distance of 2.5 AU distance is of particular interest, as this is thought to represent a distance from the early Sun where meter-sized (or larger) chunks of water ice could have migrated from near the 'snow line' (perhaps near 5 AU – Jupiter's present orbit) and been either accreted, or formed a radial maximum in water vapor pressure (Cyr *et al.*, 1998). The snow line is simply the distance from the Sun where temperatures in the protoplanetary disk were sufficiently low that water-ice could form. Yet, the precise location of the snow-line is subject to a broad range of variability and uncertainties, including those associated with vertical temperature gradients in the disk, cooler regions induced by opacity variations, and its time-dependent migration inward as the early solar system cooled (Jones *et al.*, 1990; Drouart *et al.*, 1999). The underpinning importance of the snow line for the incorporation of water in the terrestrial planets lies in the not-particularly-well-constrained idea that hydrous silicates are kinetically difficult to form from the vapor phase (Fegley, 2000). Therefore, the prevailing viewpoint is that the hydrous silicates that occur within a broad suite of chondrites formed via ice-ball accretion into, and subsequent hydrothermal metamorphism of a larger parent body. If this view is correct, then the formation of hydrous silicates in meteorites fundamentally hinges on the existence

of icy material in the zone in which their parent bodies initially formed, and this zone is defined to be in a region where sublimation rates of ice balls was small.

The principal result that the accreting terrestrial planets incorporate a distribution of material with points of origin in the protoplanetary disk that span well into the present-day asteroid belt provides a remarkably simple syllogism for water incorporation into the nascent Earth (Morbidelli *et al.*, 2000; Raymond *et al.*, 2004, 2006). Put briefly, wet objects (hydrated rocks) are generated beyond a certain distance from the Sun; Earth accretion samples material formed over a broad range of distances from the Sun; therefore, some of the material incorporated into the early Earth is wet. Two questions thus emerge: (1) If this chain of reasoning is correct, what are the implications for how water was incorporated into the early Earth? and (2) Does this syllogism accurately describe the means by which Earth's hydrosphere was generated? The answers to the first query are broad-reaching. First, this model effectively could involve incorporation of water-bearing bodies into Earth throughout its accretion (or, worse yet, randomly during its accretion), although there may be some bias in the delivery of water-rich bodies toward the later stages of accretion (Morbidelli *et al.*, 2000). Therefore, water should have been introduced into the interior of the growing planet, through either early incorporation or through retention during giant impacts of hydrated objects. The obvious implication of this model is that the conceptual picture of the interior of the planet degassing to form the hydrosphere is likely correct. Second, the amount of water incorporated into the planet could have been large – far greater than our current hydrosphere. As such, the planet may have either lost a large complement of its original water (through loss processes described in subsequent sections) and/or a significant quantity of water may have been sequestered into Earth's core. The answer to the second query, which concerns whether or not this scenario is the likely source of Earth's water hinges, to some extent, on the viability of possible alternatives, of which two proposals are: (1) that Earth's hydrosphere (and indeed, many of its volatile elements) accreted in a late veneer toward the end of Earth's accretion (e.g., Chyba, 1987); or (2) whether Earth's water may have been generated through an entirely different mechanism, such as early capture of nebular gas and oxidation of hydrogen through interaction with molten silicate. The former mechanism is discussed in the following sections. The latter scenario follows on a suggestion that an early thick atmosphere derived from the solar nebula could have been present on the accreting Earth (Hayashi *et al.*, 1979), with possible production of water from the nebular gas through oxidation (Sasaki, 1990). More recently, Ikoma and Genda (2006) have conducted simulations of this possible process for Earth-like planets, and have noted the incompatibilities between a gas of nebular composition and the atmospheric noble-gas abundances and D/H ratio of the Earth.

5.3 Noble Gases: Evidence for Early, Rapid Degassing

The abundance and isotopic characteristics of noble gases both residing in Earth's atmosphere and being presently degassed from the planet's interior provide much of the evidence that constrain the rates of early atmospheric formation and ongoing planetary degassing. The advantages of the noble gases in inferring atmospheric history include their general incompatibility (with the possible exception of xenon: Lee and Steinle-Neumann (2006)) in most Earth materials, and the ability to isotopically distinguish likely primordial noble gases (such as ^{3}He, ^{20}Ne, ^{36}Ar, and ^{130}Xe) from those that are radiogenically produced. The disadvantages are the possible role of hydrodynamic escape, particularly following events such as large impacts, in mass fractionating the isotopes of these elements (e.g., Pepin, 1997; Kramers, 2003). However, it is the differences between the atmospheric rare-gas isotopic content and those of the various possible storage reservoirs of rare gases within the solid Earth that produce the most significant constraints on the degree to, and rate at which mantle degassing contributed to the atmospheric rare-gas signature. Specifically, the atmosphere is depleted in isotopes such as ^{129}Xe, ^{136}Xe, and ^{40}Ar relative to the degassing mantle: radiogenic isotopes produced principally by the decay of ^{129}I (half-life of 15.7 My), ^{244}Pu (82 My), ^{238}U (4.45 Gy), and ^{40}K (1.25 Gy), respectively.

The models utilized to construct evolutionary histories of the isotopes of the rare gases incorporate at a minimum the atmosphere, a largely degassed upper mantle, and an undegassed deep-mantle reservoir (Allegre *et al.*, 1986/87; Porcelli and Wasserburg, 1995; Kamijo *et al.*, 1998). The rationales for including an undegassed deep-mantle reservoir include the

different (and more enriched in primordial, nonradiogenic ^{3}He and ^{36}Ar relative to mid-ocean ridge-associated degassed upper mantle) isotopic compositions of rare gases from hot spots/ocean-island basalts. However, the effects on noble-gas contents and isotopic ratios of potentially important processes such as vesiculation and magma chamber processes in hot-spot-related magmas relative to mid-ocean ridge basalts (MORBs) remains unclear (Moreira and Sarda, 2000), and the possible presence of a sizable modern air component in mantle-derived samples remains a topic of active inquiry (Ballentine and Barfod, 2000; Sarda, 2004). If present, the nature of the inferred undegassed reservoir remains controversial: it may simply represent regions that have not undergone extensive convective mixing over a multibillion year period (Class and Goldstein, 2005) or, for some of the rare-gas isotopes, might be associated with predominantly upper-mantle processes (Meibom *et al.*, 2003). Indeed, the maintenance of such undepleted rare-gas enriched zones in the face of billions of years of mantle convection has proved to be a challenging problem for geodynamic models (Ballentine *et al.*, 2002). And, the possibility that seawater subduction may control the heavy rare-gas content of the mantle has attracted recent attention (Holland and Ballentine, 2006). Yet, irrespective of the spatial distribution or history of primordial rare gases within the mantle, the mere persistence to the present day of one of the most readily degassed and unlikely-to-be-recycled isotopes, ^{3}He, within the mantle implies that primordial volatiles are retained to the present day within Earth's mantle. It is the retention of more geochemically compatible volatile elements such as hydrogen and carbon, and their rate of degassing that is of critical importance for the genesis of the vast bulk of the atmosphere, and essentially the entirety of the oceans.

With respect to the timing of atmospheric evolution, three scenarios have been designed to explain the atmospheric paucity of noble gases produced by short-lived isotopic decay processes. The net goal of each is to generate the atmospheric rare-gas reservoir either sufficiently rapidly that short-lived decay products are under represented in Earth's atmosphere, or to explain the isotopic difference between Earth's atmosphere and mantle by deriving the Earth's atmosphere by different, non-mantle-associated sources or processes. First, there is the possibility of very early, large-scale Earth degassing, with much of Earth's rare-gas degassing occurring within the first ~50 My of Earth history (Fanale, 1971; Sarda *et al.*,

1985; Allegre *et al.*, 1986/87). In detail, these types of scenarios are often two-staged, with an initial brief period of rapid degassing followed by a more gradual degassing spanning hundred of millions of years (and likely extending to the present) with of order 20% of the stable rare gases being degassed in the slower stage, and almost all of the rare gases produced by decay of long-lived nuclides (Sarda *et al.*, 1985; Allegre *et al.*, 1986/87). The second scenario involves rare-gas introduction into that atmosphere through the accretion of a late veneer of rare-gas enriched material (perhaps cometary in origin): such an origin takes advantage of the considerable variations in rare-gas isotopic compositions of protoplanetary materials (e.g., Ott, 2002), and the perhaps probable loss of earlier, possibly internally derived, atmospheres through giant impacts (Porcelli and Wasserburg, 1995). This late-accreting mechanism largely decouples the noble gases in the atmosphere from those in the underlying mantle: the present atmosphere would represent a mixture between those gases degassed from the interior, and those introduced in the final stages of planetary accretion. Third, the distribution of rare-gas isotopes in Earth's present-day atmosphere may be produced by a combination of two episodes of enhanced hydrodynamic escape from Earth's atmosphere: the first of these might be associated with the heating associated with a lunar-forming impact event, while the second might be associated with a period of enhanced ultraviolet flux from the Sun (Pepin, 1997). The former episode could produce mass fractionation in the heavy xenon isotopes, while the latter event could produce the neon signature of Earth's atmosphere. This escape-modified atmospheric rare-gas content is likely subsequently altered by addition of a component degassed from Earth's mantle. Indeed, the probable occurrence of a lunar-forming impact has expanded the range and severity of processes that might have affected the planet's noble-gas abundances: it has been suggested that 97–99% of the rare gases initially present in the atmosphere and mantle were lost in the first 100–200 My of Earth history, possibly as a consequence of one or more giant impacts (Porcelli *et al.*, 2001; Tolstikhin and Marty, 1998).

While each evolutionary scenario for the noble gases places a large fraction of the rare gases in Earth's atmosphere very early in Earth history, they each have profoundly different implications for how (or indeed, whether) volatiles have interacted with the interior, and for the relevance of Earth's atmospheric noble-gas content for the degassing history of the planet. With respect to Earth's hydrosphere and atmosphere, the principal question raised by these results is how closely correlated either the degassing or accretionary history of rare gases is with that of water and carbon dioxide. Moreover, the degree to which atmospheric noble-gas isotopic abundances might have been modulated by mass fractionation processes associated with hydrodynamic escape (Pepin, 1997), and the inferred noble-gas abundances of the solid Earth shifted by magmatic fractionation processes (Moreira and Sarda, 2000) and atmospheric contamination (Ballentine and Barfod, 2000; Sarda, 2004) each represent principal uncertainties in detailed modeling of the evolution of the rare-gas content of Earth's atmosphere. Moreover, the degree to which noble-gas signatures may have been inherited from evolutionary processes occurring on smaller precursory bodies (as opposed to solely on the nascent Earth) remains unclear. From the perspective of possible early and rapid degassing of rare gases, as water and carbon dioxide are more compatible within both crystals and melts than the noble gases, the degassing of rare gases likely provides an upper bound (and possibly an extreme upper bound) on the rapidity and efficiency of degassing of water and carbon dioxide from the planet. From the accretionary perspective, if a late-accreting noble-gas-carrying planetary component also carries abundant water and carbon (as would be expected, for example, from comets), then the production of the hydrosphere could be intimately tied to rare-gas delivery to the nascent planet.

5.4 Deuterium/Hydrogen Ratios and Earth's Hydrosphere

The D/H ratio is one of the few direct values that provides a constraint on the likely source(s) of Earth's hydrosphere. Indeed, the likely sources of terrestrial water (hydrated meteorites or comets) each have quite distinct ranges of D/H values. **Figure 1** shows a distribution of D/H values for carbonaceous chondrites, comets, and a plausible value for the Earth (Robert, 2001). Clearly, an estimated value for the bulk Earth (Lecuyer *et al.*, 1998) rather closely resembles those of the carbonaceous chondrites; or, phrased another way, it is markedly discrepant from the observed values for comets. The carbonaceous chondrite values are obviously derived from dozens of

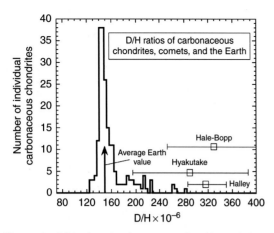

Figure 1 D/H ratios of carbonaceous chondrites relative to an estimated bulk value for the Earth (Lecuyer *et al.*, 1998) and values for the three comets for which D/H ratios have been constrained (Eberhardt *et al.*, 1995; Meier *et al.*, 1998; Bockelee-Morvan *et al.*, 1998). Adapted from Robert F (2001) The origin of water on Earth. *Science* 293: 1056–1058.

measurements; the cometary measurements are, however, derived from only three samples: Halley, Hale-Bopp, and Hyakutake (Eberhardt *et al.*, 1995; Meier *et al.*, 1998; Bockelee-Morvan *et al.*, 1998). Notably, while the three comets are, within error, indistinguishable in terms of their D/H ratio, chondritic meteorites can show significant variations even within a single meteorite, implying that the water(s) from which they formed were not homogenized, and that their variable D/H ratios may reflect the fossilized signature of substantially different sources (Deloule *et al.*, 1998).

The most straightforward and parsimonious interpretation of these data (e.g., Robert, 2001) is that Earth's hydrosphere was largely derived from meteoritic/asteroidal inputs, with comets contributing less than 10% of the planet's water (Dauphas *et al.*, 2000). But, this interpretation is not iron-clad: whether today's population of comets is isotopically identical to that present during planetary accretion is unclear. In particular, a zonation of D/H content is anticipated in the early solar system, with more deuterium-poor material residing closer to the Sun (Drouart *et al.*, 1999). Therefore, if the current (and limited) population of comets represents a population derived from the outer portions of the solar system, then their elevated D/H ratios is simply a consequence of the locality of their origin. If the Earth incorporated significant amounts of a population of comets that formed closer to the Sun, then the D/H ratio would be substantially less (Aikawa and

Herbst, 2001): such comets could be quite rare after 4.5 Gy of solar system evolution. At this level, the distinction between 'chondritic' water and 'cometary' water becomes semantic. If chondritic water formed because of hydrothermal alteration of parent bodies that had incorporated ice balls near the current asteroid belt and 'cometary' water involves incorporation of ice balls formed in this general region into the nascent Earth, then the principal difference between these sources involves the degree to which similarly derived water was processed in small or intermediate-sized rocky bodies: a difficult-to-track problem of protoplanetary hydrology.

5.5 Impacts and Volatile Loss

Impacts into the early Earth likely had a dichotomous effect on Earth's atmosphere. On the one hand, release of atmospheric constituents likely occurred near Earth's surface due to impact-induced devolatilization: an obvious process for ice-rich/cometary material, but requiring moderately high-impact pressures (5–30 GPa) and their resultant temperatures to produce water and carbon dioxide loss from hydrated minerals or carbonates (Lange and Ahrens, 1982; Boslough *et al.*, 1982; Tyburczy *et al.*, 1986, 2001). On the other hand, impact-induced atmospheric blow-off can also occur, resulting in a stripping of atmospheric volatiles from the nascent planet or its precursors.

There are two principal mechanisms that have been proposed for how atmospheric material might be lost during impacts (e.g., Ahrens, 1993): (1) through a blow-off mechanism that is analogous to cratering, or (2) through escape velocities imparted to the atmosphere from being shocked by the impact-induced spallation of the solid Earth (**Figure 2**). The former mechanism operates on the atmosphere close to the impact, while the latter is planetary in scale. The 'atmospheric cratering' model proposed by Vickery and Melosh (1990) involves loss of atmosphere from above a zone defined by a plane lying tangent to the Earth at the point at which the object impacts (**Figure 2(a)**). In this model, lateral momentum associated with the impact-induced vapor cloud is transferred to the surrounding atmosphere, accelerating portions to velocities greater than the planetary escape velocity. Because Earth's atmosphere is quite thin, the absolute volume of atmosphere lost in single-impact events via this mechanism can be small; but some loss is expected even for relatively small impacts during planetary

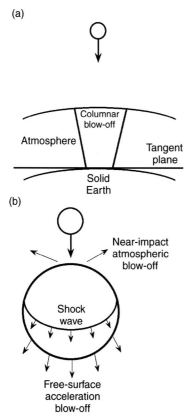

Figure 2 Schematic diagram of two mechanisms for atmospheric blow-off by impacts. (a) Atmospheric loss mechanism for comparatively small impactors; the tangent plane defines the region of atmosphere loss proposed by Vickery and Melosh (1990), while the columnar loss region illustrates the computational results of Newman et al. (1999). (b) The atmospheric loss mechanism from giant impacts, involving propagation of shock waves through the planet, and acceleration of the atmosphere by the free surface of the planet (e.g., Ahrens, 1993).

accretion (of order 1–10 km and more). The extent of atmospheric loss that can occur via this mechanism is, however, controversial: detailed simulations of the ~10 km diameter Cretaceous–Tertiary boundary event impact have indicated that atmospheric loss may only be confined to an almost columnar region (with diameter near 50 km) directly overlying the impact, as illustrated in **Figure 2(a)** (Newman *et al.*, 1999). Thus, the efficiency of this mechanism at removing atmospheric mass hinges critically on the amount and rate of lateral outflow of vaporized material from impacts – if such outflow is small, then many smaller impacts may (depending on the chemistry of the impactors) add volatiles to the accreting planet.

The acceleration of the atmosphere by motion of the free surface of the planet may result in extensive atmospheric loss for impacts by approximately lunar-sized objects (of order 1% Earth mass) at typical impactor velocities of 20 km s^{-1} (**Figure 2(b)**) (Ahrens, 1993; Chen and Ahrens, 1997; Genda and Abe, 2003). Such an impact imparts velocities of ~1–6 km s^{-1} to the antipodal position on Earth's surface (Chen and Ahrens, 1997; Genda and Abe, 2003). The effect of the velocity of the free surface of the solid (or molten) Earth on the atmosphere is dramatically augmented by the reflected shock, or rarefaction wave, that should be generated when the atmospheric shock wave interacts with the top of Earth's atmosphere (Ahrens, 1993). For a solid planetary surface, complete atmospheric loss occurs only when ground velocities exceed the escape velocity (11.2 km s^{-1} for Earth), but percentage and greater losses commence near 2 km s^{-1} (Genda and Abe, 2003). A novel aspect of such simulations is that, while atmospheric loss can be extensive (about 30% for ground velocities of 6 km s^{-1}), it is not complete – and the amount of atmospheric loss for the impactor is of roughly similar magnitude (tens of percent) to that of the main body (Genda and Abe, 2003). Thus, a portion of the impactor's atmosphere could be incorporated into the remaining atmosphere of the main body.

Atmospheric loss due to free-surface velocity effects is profoundly affected by the presence or absence of an ocean at the surface of the planet, with vastly more efficient losses if the planet is water covered (Genda and Abe, 2005). The driving forces for this enhancement are simple: (1) the impedance of water and atmosphere match better than that of rock and atmosphere, and the velocity at the ocean–atmosphere boundary is thus increased relative to that at a rock–atmosphere boundary; and (2) the shock from the solid Earth generates extensive and rapid vaporization of the ocean, which expands and thus provides a large ballistic push to the overlying atmosphere (Genda and Abe, 2005). Thus, there is an important and direct relationship between the timing of large final accretionary events, liquid ocean genesis, and the level of atmospheric loss experienced by the planet. If liquid oceans were present during the final stages of Earth's accretion, then our atmosphere could have been extensively eroded by impact events; conversely, without liquid oceans during final accretion, our early atmosphere could have represented a complicated sum of modestly ablated atmospheres from impacts on, and degassing of, our own planet and those of (perhaps) several large impactors.

5.6   Formation and Evolution of the Atmosphere

The first atmosphere of the planet (if the planet is defined as initiating following the lunar-forming impact event) may have been a thick rock vapor with a temperature in excess of 2300 K (Sleep *et al.*, 2001). The lifetime of this vapor atmosphere was comparatively short – of order 1 ky – but its effects may have been profound. Presumably, volatiles (CO_2, hydrogen, methane, and to some extent, water) would have partitioned between a surface magma ocean and the overlying vapor phase, and presumably entered preferentially into the vapor where they would continue to reside after the condensation of the rock vapor (Sleep *et al.*, 2001). In this instance, the mass of the steam portion of the atmosphere would be dictated by water's solubility within the melt, which results in much of the water being retained within the underlying magma ocean, from which its release occurs as the magma ocean solidifies (Zahnle *et al.*, 1988; Abe and Matsui, 1988). A straightforward extension of this scenario is that the proto-oceans arose when condensation of this early steam atmosphere occurred (Abe and Matsui, 1988; Sleep *et al.*, 2001; Liu, 2004).

It is the other components of this early atmosphere that likely ultimately evolved (added to by accretionary additions and impact-induced subtractions) to form our current atmosphere. That liquid water may have existed at Earth surface between 3.8 and 4.3 Ga (see the following section) provides a constraint on the temperature range of Earth's surface during this era, which when coupled with the decreased luminosity of the early Sun, requires the presence of enhanced levels of greenhouse gases in the atmosphere early in Earth history (e.g., Sagan and Mullen, 1972). The inference that the early Sun was faint (about 75% of its present luminosity at 4 Ga) arises simply from stellar evolution modeling: as the Sun fuses hydrogen to helium, the interior of the star increases in density, which leads to progressive increases in temperature and thus luminosity (e.g., Sagan and Chyba, 1997). The principal suggestions for which greenhouse gases could have been significantly more abundant in Earth's early atmosphere are carbon dioxide, methane, and ammonia (Kasting,

1993; Pavlov *et al.*, 2000; Sagan and Chyba, 1997). Ammonia, the original greenhouse gas suggestion of Sagan and Mullen (1972), readily undergoes photodissociation, and could likely only have persisted in Earth's atmosphere had there both been rapid resupplying of ammonia and shielding from ultraviolet-induced dissociation by a methane-rich atmosphere (Sagan and Chyba, 1997). Given these rather restrictive conditions, methane and CO_2 have become the gases of choice for keeping the early atmosphere sufficiently opaque to infrared radiation that an effective greenhouse could be generated. The appeal of carbon dioxide is that a hot early Earth could have had much of the current crustal reservoir of carbon vaporized into the early atmosphere, producing a possible early CO_2 atmosphere of ~80 bar (e.g., Tajika and Matsui, 1993). But, such an atmosphere is unlikely to have persisted for long, as rapid hydrothermal circulation in the early oceanic crust, resultant carbonate formation, and subduction into the mantle would have reduced carbon dioxide concentrations in the atmosphere to of the order of 100 times the current level (or ~0.036 bar) in of the order of millions of years (Sleep and Zahnle, 2001). Such levels of CO_2, in and of themselves, are insufficient to keep the oceans unfrozen for the solar fluxes inferred to be present prior to ~3.1 Ga (Kasting, 1993).

Geologic evidence derived from carbon isotope ratios in microfossils indicate that atmospheric CO_2 was between 10 and 200 times the present value 1.4 Ga; an amount of CO_2 which could easily produce a sufficient greenhouse to keep temperatures moderate in the face of a Sun that was 90% of its current intensity (Kaufman and Xiao, 2003). The absence of $FeCO_3$-siderite in iron-rich paleosols from between 2.75 and 2.2 Ga yields an upper bound of ~0.04 bar of CO_2 in the atmosphere – a value only marginally sufficient to prevent freezing of the oceans (Rye *et al.*, 1995; Kaufman and Xiao, 2003). But, the possible presence of nahcolite ($NaHCO_3$) in rocks from 3.5–3.2 Gy in age has led to somewhat higher estimates of atmospheric CO_2 contents – perhaps as high as 1000 times the present atmospheric level (Lowe and Tice, 2004). In the case of both the paleosol and evaporite data, the results are crucially dependent on the temperature inferred for their deposition: a notably uncertain quantity (Rye *et al.*, 1995; Lowe and Tice, 2004).

The potential difficulties of maintaining high atmospheric CO_2 pressures coupled with measurements that indicate that CO_2 content may have been insufficient to prevent freezing of the oceans have led to the proposal that elevated methane concentrations could have been a major greenhouse gas in the early Earth (e.g., Pavlov *et al.*, 2000; Catling *et al.*, 2001). The advantages of methane are that it can be generated by either the earliest life forms or abiotically and, as it is a highly efficient greenhouse gas, not very much of it is required: perhaps 100–1000 ppm (today's atmospheric concentration is near 1.7 ppm). Accordingly, a tradeoff between the methane and carbon dioxide concentrations of the ancient atmosphere exists, with each order of magnitude increase in methane concentration requiring between 1 and 2 orders of magnitude less carbon dioxide to generate similar temperatures (Pavlov *et al.*, 2000; Kasting and Catling, 2003).

The two principal components of today's atmosphere, nitrogen and oxygen, appear to have strongly divergent geologic histories. The flux of nitrogen from volcanoes is comparatively small relative to the atmospheric reservoir: the current degassing flux, if constant over geologic history, would only produce ~7% of the atmospheric concentration (Sano *et al.*, 2001). Fluxes may, of course, have been more rapid in the past. But, to contribute large quantities of nitrogen to the atmosphere (and ignoring the small regassing fluxes), the rate of degassing would have had to be dramatically higher in the past. Indeed, calculations of nitrogen fluxes indicate that the nitrogen content of the atmosphere may have changed little (by less than 1%) over the last 600 My – a simple consequence of small and canceling degassing and regassing fluxes (Berner, 2006). Notably, nitrogen may behave compatibly within mantle assemblages, due to the formation of stable nitrides (Javoy, 1997), and the mantle may contain a significant nitrogen content. Characteristic estimates for the average mantle nitrogen content are uncertain, and vary from ~2 to as high as 40 ppm (for reference, the total amount of nitrogen in Earth's atmosphere is equivalent to about 1 ppm sequestered in the mantle) (Cartigny *et al.*, 2001; Marty and Dauphas, 2003a). Mantle nitrogen is known to be depleted in ^{15}N relative to the atmosphere, and the origin of this depletion is controversial (Marty and Dauphas, 2003a, 2003b; Cartigny and Ader, 2003). It is possible that the current nitrogen in the atmosphere may have been largely degassed either during accretion or the hot-atmosphere-magma-ocean phase of early Earth history (Tajika and Matsui, 1993) – as such, it may be the single persistent abundant primordial component of Earth's atmosphere.

In contrast, oxygen has undergone rather dramatic changes in its concentration. These changes were induced by photosynthesis-induced oxygenation of the atmosphere (and ocean and near surface), and possibly by the loss of reduced hydrogen from Earth's atmosphere (Kasting, 1993; Catling *et al.*, 2001). Constraints based on the survival of detrital uraninite indicate that the oxygen partial pressure was less than 10^{-3} bar up to about 2.6 Ga, and calculations of mass-independent fractionation of sulfur isotopes yield partial pressures of less than 2×10^{-6} bar of oxygen prior to 2.3 Ga (Pavlov and Kasting, 2002). The subsequent buildup of oxygen within Earth's atmosphere (and corresponding decline in methane – which in turn may have generated the Huronian glaciations) is a biologic phenomenon, and is reviewed elsewhere (Holland, 1984; Kasting, 1993).

5.7 Ancient Rocks: Signatures of Damp Material in and on the Early Earth

The oxygen isotope compositions of detrital zircons with ages of 4.3–4.4 Gy are the principal probe used to demonstrate that liquid water existed on the early Earth (Mojsis *et al.*, 2001; Wilde *et al.*, 2001; Valley *et al.*, 2002, 2005). The key observations of such studies are that these detrital zircons, which were, as shown by mineral inclusions, formed from a silica-saturated melt, have a moderately heavy (^{18}O-enriched) oxygen-isotopic composition that is comparable to those of more recent zircons. As zircons with ^{18}O-enriched compositions within silicic magmas form from protoliths that have interacted with liquid water or its alteration products, the logical conclusion from these data is that low-temperature metamorphic or hydrothermally altered material was entrained to depth as early as 4.3 Ga. Accordingly, liquid water was probably present in the near surface or at the surface of the planet at this early stage of Earth evolution. The abundance, temperature, and chemistry (salinity, pH, etc.) of this earliest hydrosphere are, however, unconstrained. What can be said with some certainty is that hydrothermal/metamorphic activity involving liquid water occurred early in Earth history, and this material was probably transported to modest depths. This is not to say that liquid water need have persisted continuously through the Hadean: occasional large impacts could have produced transient heating, and perhaps even vaporization, of the ocean sporadically through the first half-billion years of Earth history (e.g., Nisbet and Sleep, 2001). Yet, in between these interludes of planetary trauma, thermal arguments indicate that the Earth's surface was likely to have been at temperatures where liquid water could have readily existed (Sleep *et al.*, 2001).

Beyond the oxygen isotopic evidence, the simple genesis of silica-enriched continental crust provides powerful, albeit indirect, evidence for the injection of water at depth within the planet early in Earth history. The origin of granites and andesites via hydrous melting of a range of source rocks is well documented (e.g., Campbell and Taylor, 1983), and the initiation of continental crust formation may have been within the first 200 My of Earth history (as indicated not only by ancient zircons with silicic inclusions, but also by Hf isotopic constraints on the genesis of the earliest crust (Amelin *et al.*, 1999; Harrison *et al.*, 2005)). Accordingly, an early start for continental crust generation implies that hydrous alteration and subduction of hydrated material to depth was happening – and the critical interplay between water and the solid Earth that gave rise to the planet's bimodal topography had already initiated.

The first-order observation of chemical sediments within the geologic record provides an unequivocal demonstration of the existence of standing water at the surface of the planet. The earliest such sediments are ~3.85 Gy old, and lie close in age to the oldest rocks yet found (as opposed to isolated detrital mineral grains) (Nutman *et al.*, 1997). Accordingly, given the scarcity of the most ancient rocks and their associated preservational difficulties, the near-temporal coincidence of water-generated sediments and the most ancient rocks provides additional support for the idea that oceans existed prior to the start of the rock record.

The water content of portions of Earth's interior may have been notably high during the Archean, as well. Komatiites, igneous rocks that have high magnesium relative to basalts and likely were generated by large degrees of partial melting of the mantle (perhaps in the vicinity of 40% partial melting of the mantle), are among the signature rocks of Archean greenstone belts. Originally, such rocks were viewed as being generated by a substantially hotter mantle, with eruption temperatures near 1600°C (e.g., Green, 1975) – a plausible idea, given that the planet has likely undergone secular cooling through time. Recently, petrographic constraints, including the observation of igneous amphiboles within these materials, have shown that at least some komatiites were associated with remarkably

high water contents – 1–6% within the magma, and 0.5–3% in their source region, coupled with cooler inferred eruption temperatures of 1400°C (Stone *et al.*, 1997; Parman *et al.*, 1997). Thus, at least in some places, the Archean mantle was heavily hydrated: whether this was produced through a subduction-like process or through retention of primordial water within the early Earth is not well constrained. In the latter case, komatiites may have been produced from deeply derived, wet, plume-like upwellings (Kawamoto *et al.*, 1996); in the former case, komatiites would have arisen through subduction of large quantities of hydrated material (Grove and Parman, 2004) that presumably arose from interaction with a surface ocean. Thus, komatiites may also provide a prima facie signature of the presence of an early ocean.

5.8 Constraints on the Volume of the Oceans through Time

Whether the surface hydrosphere has grown, shrunk, or remained almost constant in volume through geologic history is extraordinarily difficult to determine. The ultimate answer is conceptually simple. It is the difference between how much water is subducted into the planet through time, versus how much is degassed from the interior: a balance that is, unfortunately, uncertain even today. Observational constraints on the volume of the hydrosphere are sparse – largely confined to the constancy of freeboard, the mean elevation of continents above sea level – and difficult to uniquely interpret. From the dynamical viewpoint, however, the gross picture of degassing versus regassing is generally straightforward: because of the strong dependence of mantle viscosity on water content (with the upper-mantle dependences assumed to extend throughout the mantle), a single-layered, volatile-rich mantle seems to spit water out rather quickly (McGovern and Schubert, 1989; Franck and Bounama, 1997, 2001; Rupke *et al.*, 2004). The underlying principle is that a dynamic regulation exists between increased mantle water content, lower-mantle viscosity, higher heat flow, faster seafloor spreading (and thus more rapid degassing), which in turn reduces water content. Each of these modeling exercises involves an early, rapid period of mantle convection and thus degassing: scenarios that, if plates were thicker in the early Earth and heat flow less, might not necessarily obtain (Korenaga, 2006). Nevertheless, the end result of such models is typically degassing of ~ 1 hydrosphere from the mantle, with most of the degassing in the first billion years of Earth history, with a significant fraction of the hydrosphere forming in the first few hundred million years. There are possible nuances, however. Scenarios can be constructed that, depending on the rate of regassing, can produce initial large hydrospheres (~ 2 ocean masses) that do regas the interior through time and ultimately yield the present hydrosphere at Earth's surface (Franck and Bounama, 2001). Indeed, it is possible that regassing of the interior could be more efficient today than in Earth's past, when a hotter mantle might have induced dehydration of subducted hydrous phases at shallower depths. While the rate of regassing is uncertain, the consensus of these one-layer dynamic models is that if the total water content of the hydrosphere and mantle was originally all contained within Earth's mantle, early and rapid degassing would occur. Subsequent growth or shrinkage of the hydrosphere through the bulk of Earth history is, however, not well constrained by such models – both types of behavior have been generated in these dynamic models.

The observation that freeboard appears to have been approximately constant is a somewhat loose constraint – the elevations of continents relative to sea level seem to have varied by less than plus or minus a few hundred meters since the Archean. But, sea level is determined by a combination of factors in addition to water volume, including continental crustal volume, the depth of the ocean basins (and thus their mean age and heat flow), and oceanic crustal thickness (Schubert and Reymer, 1985; Galer and Mezger, 1998). Indeed, models of freeboard through time typically simply assume that the volume of water is constant through time – given the wide suite of poorly known input parameters that enter into calculations of ancient freeboard, there is rather little insight that can be garnered from freeboard alone as to whether the oceans have shrunk, expanded (by perhaps as much as a few tens of percent), or simply stayed at constant volume since the Archean.

Kasting and Holm (1992) proposed a novel feedback mechanism for how constant freeboard might be maintained with an ocean of varying volume. They coupled the observation that most fast-spreading ridges lie at pressures that are close to the critical pressure where liquid and vapor become indistinguishable – near 30 MPa for seawater, with the critical temperature being $\sim 400\,°C$. The contention

is that hydrothermal circulation (and thus alteration) is maximized at near-critical conditions. Shallower oceans would produce less alteration in the subridge environment, and the amount of ingassing (via subduction) relative to outgassing would decrease, producing a net growth in the oceans through time. Accordingly, if the critical depth is approached from the shallow side, the oceans might be expected to reach a steady-state depth at which the ingassing rate was equal to the outgassing rate. However, whether a tradeoff between water depth and the amount of oceanic crustal hydration both exists and is capable of producing a robust control on the volume of the oceans remains unclear.

5.9 Degassing and Regassing Rates

The exchange of water between the different regions of the planet (and indeed, with the solar system itself) is dictated by a combination of tectonic processes, magmatic processes, and sequestration of water by metamorphic and weathering processes, as well as contributions from extraterrestrial inputs and loss via photodissociation (**Figure 3**). Unfortunately, many of the fluxes between different water reservoirs of the planet are difficult to characterize: a few, however, can be reasonably accurately estimated. The rate at which water is returned into Earth's interior can be fairly readily estimated from the rate at which sediments and oceanic crust (with an estimated hydration level) enter into subduction zones: such estimates are in general accord at around $1 \times 10^{12}\,\mathrm{kg\,yr^{-1}}$ of water (Peacock, 1990; Bebout, 1995; Rea and Ruff, 1996; Javoy, 1998; Wallmann, 2001). For reference, the mass of the oceans is about 1.5×10^{21} kg, and thus (if current plate rates are simply applied to the past) a total mass of water equal to our present hydrosphere enters into a subduction zone about every 1.5 Gy. The amount of water directly returned to the surface via arc magmatism is less certain, but seems to be about an order of magnitude less (Peacock, 1990; Bebout, 1995). Similarly, the rate of return to the surface of subducted sediment-associated water released by relatively shallow processes (the smectite–illite transition, opal dewatering, and compaction-induced dewatering) is uncertain, but considerable fluid flow does appear to occur both through accretionary prisms and along the decollement between the oceanic crust and the overriding plate (Moore and

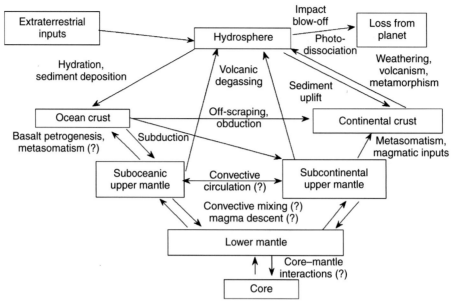

Figure 3 Box diagram of exchange mechanisms (directional arrows) between different reservoirs of water. Adapted from Williams Q and Hemley RJ (2001) Hydrogen in the deep Earth. *Annual Review of Earth and Planetary Science* 29: 365–418.

Vrolijk, 1992; Carson and Screaton, 1998; Kopf *et al.*, 2001). The rate at which the mantle degasses is similarly uncertain, but characteristic estimates of mid-ocean ridge degassing of water yield estimates that are of the order of 1×10^{11} kg yr^{-1}: a full order of magnitude less than the rate of subduction of water, and generally comparable to that of arc magmatism (e.g., Ito *et al.*, 1983). Hot spot related magmas, though wetter, are much smaller in volume, and their net degassing of water is probably about an order of magnitude less than that of the ridge flux (Ito *et al.*, 1983). The underpinning issue produced by these flux estimates is illustrated in **Figure 4**. Two options emerge from the difference in magnitudes between subduction and mantle degassing water fluxes: (1) Earth's oceans are slowly shrinking through time; and/or (2) the ways in which subducted water can be returned to the surface are not well constrained. Certainly, nonmagmatic water inputs to the overlying plate from subduction occur (e.g., Ingebritsen and Manning, 2002), and the long-term magnitude of time-varying phenomena such as dewatering/compaction-driven flow up the interplate decollement or through accretionary prisms are not well known (Moore and Vrolijk, 1992; Carson and Screaton, 1998; Kopf *et al.*, 2001). Ultimately, the question of whether the hydrosphere might be shrinking comes down to how efficiently subducted water is returned to the overlying crust or hydrosphere, versus being

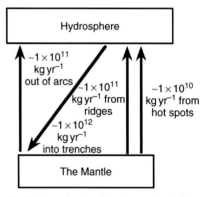

Figure 4 Schematic of order of magnitude fluxes of volcanic degassing from the mantle relative to the flux of water into trenches. See text for data sources.

entrained into the deep mantle. Dixon *et al.* (2002) have proposed that the efficiency of extraction of water at shallow depths from the subducted lithosphere may be greater than 92%, based on water contents of mantle plume components that might be associated with ancient lithosphere. However, as discussed in the section on diffusion below, these results may be severely biased by the high diffusivity of water within silicates (Workman *et al.*, 2006). The bottom line is that there are no simple *a priori* reasons to anticipate that inputs and outputs of water from the mantle should balance, and the raw fluxes of water provide an indication that the mantle may be taking up water with time, but at a rate that is modest (and uncertain) relative to the size of the hydrosphere. If this latter interpretation is correct, then the original Rubey (1951) hypothesis of ocean origination via steady degassing of the solid Earth is falsified for recent Earth history.

5.10   The Residence Mechanisms and Effects of Water on Mantle Properties

The mantle is the large-volume sink or source for water exchange between the hydrosphere and the interior – degassing from the mantle returns or perhaps introduces water into the hydrosphere, while subduction-produced fluxes of water into the mantle have the potential of decreasing the volume of the hydrosphere through time. Accordingly, the manner in which water/hydrogen can be stored in Earth's interior has both been the topic of intense study in the last two decades, and been the subject of a number of reviews (Thompson, 1992; Bell and Rossman, 1992; Ingrin and Skogby, 2000; Abe *et al.*, 2000; Williams and Hemley, 2001). This long-term interest is motivated by both the importance and the complexity of the problem: water/hydrogen can be stored (usually as hydroxyl ions) in a broad suite of crystalline hydrated phases, as defect-type substitutions within nominally anhydrous minerals, as dissolved hydride (H atoms) within iron-rich materials, or

prospectively sequestered within liquids or melts. Indeed, the role of water in modulating the temperature, chemistry, and abundance of silicate melts generated in different tectonic and hot-spot-related environments has also been the topic of a wide suite of studies and reviews (e.g., Poli and Schmidt, 2002; Asimow and Langmuir, 2003; Hirschmann, 2006): the petrologic role of water in generating melting is not covered in detail in this review.

With respect to crystalline hydrated phases, two loose families can be identified. First, in shallow (~50–300 km depth) subduction-related environments, hydrogen is likely stored in a suite of metamorphic phases with notably extensive pressure–temperature stability fields. Second, a sequence of hydrated magnesium silicate phases have been synthesized which may be important for water retention in portions of the mantle spanning from the upper mantle through the transition zone to the lower mantle: the so-called 'alphabet' phases. **Figure 5** provides a general synopsis of which of these metamorphic and synthetic phases, as well as which defect substitutions in nominally anhydrous phases, are likely to be of importance for water storage at various depth ranges within the planet.

The metamorphic phases of likely importance for water transport to depth include lawsonite, the micas phlogopite and phengitic muscovite, and topaz-related phases (each of which are stable to near

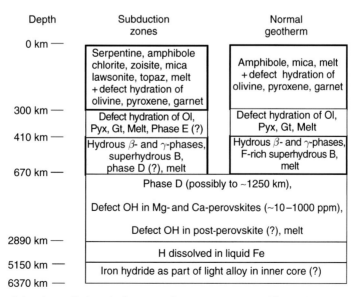

Figure 5   Compilation of the phases likely to be important for water storage at different depth ranges and for different thermal regimes (subduction geotherms relative to the normal mantle geotherm within the planet).

10 GPa and 800–1000°C: Schmidt (1995); Sudo and Tatsumi (1990); Domanik and Holloway (1996); Wunder et al. (1993)) and which are likely stable within basalt, the hydrated mantle wedge overlying subducted slabs, and pelitic sediments, respectively. Phases with more limited stability fields that could nevertheless play important roles in retaining and transporting water across some depth intervals include amphiboles, zoisite, talc, and serpentine (Inoue et al., 1998; Poli and Schmidt, 1998; Yamamoto and Akimoto, 1977; Wunder and Schreyer, 1997): each of these phases have maximum stabilities of 4–5 GPa and 700–1200°C. Both serpentine and amphibole have been proposed to form in the cool region directly above the slab as dewatering material within the slab hydrates the overlying mantle (Tatsumi, 1989; Davies and Stevenson, 1992; Iwamori, 1998): it is the dehydration of this layer near 150 km depth that may control the location of the volcanic front overlying the subduction zone. From the perspective of the importance for water transport to depth in the planet, lawsonite is perhaps the most critical of these phases: with a formula of $CaAl_2Si_2O_7(OH)_2.H_2O$, it not only contains a large quantity of water (11 wt %), but is the principal hydrous phase formed within subducted basalt for a broad range of likely slab geotherms (Poli and Schmidt, 2002).

The notable aspect of these phases is that a subduction-associated environment is required to stabilize them at depth in the planet: in each case, their maximum thermal stability (with the single possible exception of the amphibole K-richterite: Inoue et al. (1998)) lies at lower temperatures than that of 'normal' mantle. Accordingly, their roles are inferred to involve the transport (and, as the subducted material heats, release) of water through the depth range of much of the upper mantle. Additionally, there is both direct and indirect evidence that these phases are present within subducted material: direct, in the sense that these high-pressure metamorphic hydrous phases are observed in xenoliths derived from subduction-related environments (e.g., Usui et al., 2006), and indirect, as derived from seismic characterizations of a thin, low-velocity layer at the top of some subduction zones (Abers, 2000).

The suite of high-pressure hydrous magnesium silicate phases that have, to date, not yet been observed in any natural samples include the so-called phases A, B, D, E, and superhydrous phase B (which was probably identical to the original phase C: the unfortunate naming convention dates back

to Ringwood and Major (1967) – the work that first reported phases A, B, and C). In terms of importance for water recycling into the planet, the most critical of these phases are likely superhydrous B ($Mg_{10}Si_3O_{14}(OH)_4$) and phase D (ideally $MgSi_2H_2O_6$) which have respective stability ranges of between 12 and 24 GPa and ~1300°C in temperature, and 17–~50 GPa and ~2400°C (Shieh et al., 1998, 2000). For such phases, stabilities are for the fully hydrated end-members: within the Earth, partial halogen (particularly F) substitution may stabilize these phases to substantially higher temperature conditions (Gasparik, 1993; Larsen et al., 2003) – such partial fluorination is well known to stabilize micas and amphiboles at lower-pressure conditions. These phases have elevated densities over those of lower-pressure hydrous phases, due to superhydrous B having one-third of its silicon ions in octahedral coordination with respect to oxygen, and phase D having all of its Si being octahedral coordinated (Pacalo and Parise, 1992; Yang et al., 1997). Thus, these phases have structural features in common with phases that occur in the transition zone (such as majoritic garnet, with one-fourth of its Si ions being octahedral) and lower mantle (perovskite-structured phases, with entirely octahedrally coordinated silicon).

The phase assemblages of hydrated peridotite as a function of pressure and temperature are complex (e.g., Schmidt and Poli, 1998; Litasov and Ohtani, 2003; Ohtani et al., 2004). Yet, several zones where hydrous phase decomposition reactions have negative slopes in pressure–temperature space have been identified. These are regions where hydrous phase stabilities in the peridotite-water system might intersect the geotherms of subducting slabs, and thus produce fluids (which might be indistinguishable from water-enriched melts at high pressures: Shen and Keppler (1997)) at depth. Within peridotite, such dewatering reactions are inferred to occur at ~3–5 GPa and temperatures of ~650–750°C due to the dehydration of serpentine; near 13–17 GPa and 900–1150°C associated with the dehydration of phase E; between ~24 and 28 GPa and 1150–1300°C, generated by the dehydration of superhydrous B; and finally between 36 and 42 GPa and 1000–1200°C (Ohtani et al., 2004). The first of these may introduce a 'choke point' at which most subducting slabs will release the vast majority of the water contained in their upper crust and sediments (Iwamori, 2004). Thus, the importance of the peridotite-water system for the deep-water budget of the planet hinges on how much water is released from the serpentinized basalt and hydrated sediments of the downgoing slab into the overlying mantle, and is reincorporated into hydrous phases above the slab. This reincorporation in turn is controlled by the precise geotherm within the mantle overlying the slab: thermal models indicate that any rehydration of the overlying mantle will be confined to a region close to the slab, and at deeper depths (below $\sim$250 km) such rehydration might not even occur (e.g., Iwamori, 1998). Therefore, the importance of hydrous phases within the peridotite-water system (and thus alphabet phases) for water recycling in the deep upper mantle and transition zone remains speculative.

The idea that trace amounts of water can be incorporated into nominally anhydrous minerals is long standing (e.g., Griggs, 1967). The absolute amount of water that is likely incorporated into upper-mantle phases is, from the perspective of the planetary water budget, relatively small, but is of critical importance for the rheology of this region. Estimates of the solubility of water in olivine have been shown to have a strong dependence on the fugacity of water – a parameter that is enhanced by pressure and decreased by higher impurity contents of water-rich fluids (Kohlstedt et al., 1996). The inferred solubility has also depended on analytic technique, with recent estimates yielding maximum solubilities that are substantially elevated relative to previous values (Bell et al., 2003). The key aspect is that the maximum solubility varies from of order $\sim$100 ppm at shallow upper-mantle depths to perhaps as high as a few thousand parts per million at the base of the upper mantle (Kohlstedt et al., 1996; Bell et al., 2003). Pyroxene, on the other hand, appears to have maximum solubilities of near-1000 ppm at a range of upper-mantle conditions (Stalder and Skogby, 2002; Aubaud et al., 2004; Bromiley and Keppler, 2004), with garnet containing somewhat less water than the pyroxenes (Lu and Keppler, 1997). For comparison, the typical amount of water inferred to be present in the source region of normal MORB is $\sim$60–220 ppm, or around 0.01 wt.% (Michael, 1995; Danyushevsky et al., 2000; Saal et al., 2002) – substantially less than the maximum defect solubility of water in an olivine–pyroxene–garnet mantle. Indeed, even the more geochemically enriched EM2 source region is inferred to have a water content of near 400 ppm (Workman et al., 2006), and enriched MORB (E-MORB) of $\sim$575 ($\pm$375) ppm (Sobolev and Chaussidon, 1996). Accordingly, regardless of whether the MORB source region or the EM2 reservoirs represent the bulk of the upper mantle, upper-mantle water is likely to be predominantly sequestered in nominally anhydrous phases (clearly, this does not apply to the hydrated mantle wedges overlying subduction zones).

The situation in the transition zone of the planet is markedly different with respect to water storage within nominally anhydrous phases. In the depth range from 410 to 670 km, wadsleyite and ringwoodite (widely known as β-phase and γ-spinel) predominate. These high-pressure phases of olivine have each been demonstrated to take up relatively large amounts of water as defect substitutions (e.g., Smyth, 1994; Kohlstedt et al., 1996). The solubility of water in these phases is known to decrease strongly with increasing temperature (Williams and Hemley, 2001; Litasov and Ohtani, 2003), and accordingly subduction-associated material (if it retains significant water at these depths) could introduce significant water into the transition zone. This temperature-induced shift in solubility is from near 2 wt.% water-carrying capacity in wadsleyite at 1000°C and 15 GPa to $\sim$0.5% at 1600°C; increasing pressure mildly decreases the carrying capacity as well, to perhaps as low as 0.3% at 1600°C and 18 GPa (Litasov and Ohtani, 2003). In net, both wadsleyite and ringwoodite are each likely to be able to contain around 0.5 wt.% water at normal mantle conditions.

The degree to which phases within the lower mantle can incorporate water is quite controversial, with estimates for how much water can be incorporated into magnesium silicate perovskite, the dominant mineral phase of the lower mantle, varying from almost no defect water solubility in magnesian-perovskites, including those containing aluminum (less than 10 ppm water: Bolfan-Casanova et al. (2000, 2003)) to 60–70 ppm (Meade et al., 1994), to $\sim$0.1 wt.% for aluminum-bearing perovskites (Murakami et al., 2002; Litasov et al., 2003). The likely second-most abundant phase in the lower mantle, (Mg,Fe)O-magnesiowüstite appears to have a small but significant water storage capacity (20–80 ppm) (Bolfan-Casanova et al., 2002; Litasov and Ohtani, 2003). The solubility of water in $CaSiO_3$-perovskite is not well constrained, but appears to be about a factor of 2 higher than that of magnesium silicate-perovskite (Murakami et al., 2002). The pressure and temperature dependences of water solubility in each of these phases, parameters likely to be critical for the storage capacity of water within this volumetrically largest region of the planet, are similarly ill-constrained: given the pressure and temperature dependence of water solubility over rather narrow pressure and temperature ranges for transition zone phases, it is anticipated that dramatic shifts in water solubility are likely to occur as a function of depth within the lower mantle. Moreover, the effect of impurities is likely to enhance water solubility (e.g., Bell and Rossman, 1992). Therefore, systems of natural chemistry are generally expected to have higher water solubilities than those determined from end-member systems.

The differential defect solubility of water within the phases of the transition zone relative to those of the upper mantle and lower mantle has given rise to scenarios designed to describe what might occur when water-enriched transition zone material is convectively advected either upward or downward (Young *et al.*, 1993; Kawamoto *et al.*, 1996; Bercovici and Karato, 2003). If enhanced water contents accompany the high water solubility within transition zone phases, then hot upwellings would release water as they migrate from the transition zone to the upper mantle, and downwellings would correspondingly release water as they move into the lower mantle. This water release has been proposed to produce rheologic softening within the upper mantle, generating secondary, smaller-scale upwellings (Young *et al.*, 1993), or even to generate wide-scale/global melting near the 410 km discontinuity, either in the past (Kawamoto *et al.*, 1996) and/or presently (Bercovici and Karato, 2003). Although melt may be present in isolated regions near 410 km (Revenaugh and Sipkin, 1994; Nolet and Zielhuis, 1994), there is little geophysical evidence to support the existence of a global layer of hydrated, neutrally buoyant melt perched atop the 410 km discontinuity (Bercovici and Karato, 2003). Indeed, the underpinning question for scenarios that hinge on the transition zone operating as a water trap is: does the high water solubility in transition zone phases necessarily indicate that a high water content is present in this zone? It is not at all clear that the answer is affirmative. Modeling of the redistribution of water in the Earth's mantle by both convective and diffusive processes indicates that water is not readily concentrated in the transition zone: advection of water-rich material serves to effectively homogenize the water content between different zones of Earth's mantle (Richard *et al.*, 2002). In effect, for dramatically enhanced water contents to persist within the transition zone, they must (1) be continuously renewed and/or (2) resist convective mixing with the upper and lower mantles. Indeed, even layered convection (with a boundary at 670 km) appears to produce homogenization of water contents between the upper mantle and

the transition zone (Richard *et al.*, 2002). Of course, regions of the transition zone that have been exposed to long-term subduction may be locally water-enriched: however, given descent into the transition zone of (likely fairly dry) upper mantle from above and ascent of (likely dry) lower mantle from below, the transition zone as a whole is unlikely to have a uniformly high level of hydration.

The key point here is that the level of hydration of the deep mantle is highly uncertain. Because of the large mass of this region, even relatively small amounts of stored water become significant for the terrestrial water budget: put another way, each 100 ppm of water that is sequestered in the lower mantle corresponds to about 20% of the water present in our surface hydrosphere. Therefore, if the assumption is made that the lower mantle has a water content that is similar to that of geochemically constrained enriched mantle reservoirs (Sobolev and Chaussidon, 1996; Workman *et al.*, 2006), approximately a hydrosphere of water would be stored in this region. Obviously, larger estimates of lower mantle water contents (e.g., Murakami *et al.*, 2002) would make this region (potentially with the core) the principal water reservoir of the planet.

5.11 Rheology

The role of water in lowering the viscosity of minerals has long been known (e.g., Griggs, 1967), and olivine in particular has had the rheologic effect of defect water examined in particular detail (e.g., Mackwell *et al.*, 1985; Hirth and Kohlstedt, 1996; Mei and Kohlstedt, 2000). Water contents as low as 50 ppm can lower the viscosity of polycrystalline olivine by two orders of magnitude (Hirth and Kohlstedt, 1996). The physical origin for this water-induced weakening probably lies in more rapid rates of dislocation climb in wet samples (Mei and Kohlstedt, 2000): a phenomenon probably related to the high diffusion rates of hydrogen discussed in the next section, and tied to the likely preference for hydrogen to reside in the dislocation core.

The key geodynamic implications of this shift in viscosity lie not only in the role of trace water in reducing the viscosity of the mantle, but also in the effect of the extraction of water by partial melting in dramatically enhancing the viscosity of the residual solids (Hirth and Kohlstedt, 1996). This dehydration-induced strengthening simply arises from the highly incompatible nature of water: when even small amounts of melting are initiated, water is efficiently extracted from solids and enters into the melt. Indeed, the stripping of water from mantle material via partial melting has been proposed to produce a high-viscosity, melt-dried zone beneath ridges that can maintain sufficient stress gradients that melt is driven into a narrow zone (Hirth and Kohlstedt, 1996; Braun et al., 2000). Thus, the effects of hydration (or, more accurately, dehydration) are likely critical in generating the spatially focused magmatic system that generates the preponderance of crust on the planet. Correspondingly, the effect of water may also be important for generating comparatively low-viscosity regions beneath continents (such as beneath the western US: Dixon et al. (2004)), although changes in viscosity produced by water content are difficult to distinguish from those generated by high temperatures.

5.12 Diffusion

Hydrogen is known to be one of the most rapidly diffusing of elements, with diffusion rates within olivine under mantle temperatures of $\sim 10^{-8}\,\mathrm{m^2\,s^{-1}}$, and perhaps approaching the $10^{-7}\,\mathrm{m^2\,s^{-1}}$ range within pyroxenes (Mackwell and Kohlstedt, 1990; Woods et al., 2000). Such rapid diffusion rates have two principal geophysical effects: (1) they can enhance the electrical conductivity of mantle materials, and (2) they can produce significant migration of hydrogen away from initially hydrogen-enriched zones. The precise quantitative effect of hydrogen on the electrical conductivity of olivine is uncertain (Karato, 1990; Yoshino et al., 2006; Wang et al., 2006), including whether hydrogen-induced enhancements in electrical conductivity are highly anisotropic, as is indicated by the diffusion of hydrogen in olivine (Mackwell and Kohlstedt, 1990). Yet, relatively moderate quantities of hydrogen (tens of parts per million: an amount comparable to those inferred to be present in the basalt source region) can elevate electrical conductivities by ~ 2 orders of magnitude relative to dry assemblages: a result in general accord with magnetotelluric sounding results (Karato, 1990; Wang et al., 2006).

With respect to transport of hydrogen over long timescales, the diffusion rates of hydrogen imply that it can diffuse across the ~ 6 km length scale of oceanic crust in a timescale of about 100 My. Therefore, long-lived hydrated, oceanic crustal-sized geochemical heterogeneities should diffusively equilibrate

their hydrogen contents with surrounding material over a several hundred million year timescale (Richard et al., 2002; Workman et al., 2006). The counter-intuitive aspect is that water within the mantle can likely readily diffuse (and, in tandem, be advected as well) away from the material that originally entrained it to depth. Water thus may not be able to be simply tracked with other incompatible elements subducted into the mantle, particularly for lithospheric components that may have resided in the mantle for $O \sim 10^9$ years (Dixon et al., 2002; Workman et al., 2006). Accordingly, the issue of how heterogeneously water (as opposed to other more slowly diffusing incompatible elements) is distributed at depth may depend heavily on how long ago the water was entrained into the mantle – ancient water in the mantle could be, with the tandem effects of diffusion and advection, far more homogeneously distributed than would be inferred from the present-day distributions of the rocks with which the water was originally injected into the mantle.

5.13 Water/Hydrogen in the Core?

The idea that hydrogen might be in the core has been suggested for some time (e.g., Stevenson, 1977), but its credence has been notably enhanced since the recognition that iron hydride exists and routinely occurs as a high-pressure product of reactions of iron with water (Antonov et al., 1980; Fukai, 1984; Badding et al., 1991; Okuchi, 1997). The interaction of water with iron-rich material during accretion and planetary differentiation are likely to proceed via two possible general types of reactions (highly oxidizing rusting reactions that produce trivalent iron are unlikely under the reducing conditions inferred to be present in the early Earth):

$$MgSiO_3 + Fe + H_2O \rightarrow (Mg, Fe)_2SiO_4 + H_2$$

$$3Fe + H_2O \rightarrow 2FeH + FeO$$

The former reaction, involving the production and incorporation of divalent iron in silicates, occurs at low pressure, and eliminates water from the system by creating hydrogen that could react with carbon to form methane, or undergo hydrodynamic escape from the atmosphere. The existence of the latter reaction, which proceeds at pressures above ~ 5 GPa, is the likely means through which hydrogen could be sequestered in Earth's core. In short, either within a magma ocean and/or during the intermediate- to late-stages of planetary accretion, any juxtaposition of iron and water-rich material at depth in the nascent Earth could have generated hydrogen dissolution in iron, which in turn could then have descended into Earth's core. The possible presence of hydrogen in Earth's core produces a complete wild card for both the water budget of the planet and for the possible inferred temperature of Earth's core: if the light alloying component in Earth's core is generated in full or in part through the second reaction above, then literally dozens of hydrospheres could be sequestered within Earth's core. As a result, the Earth could, in net, be a fairly wet planet – but its water budget would be largely sequestered in its innermost layers. Moreover, if hydrogen is abundant in Earth's core, its effect on the melting temperature of iron would be expected to be profound (melting temperature depressions by impurities scale with the number of moles of impurities present – which for hydrogen is far larger than any other possible lighter alloying component if compared on a mass fraction basis (the mole fraction effect of different impurities is independent of nuclear mass for the ideal limit effect on the entropy of mixing)), with possible lowering of the inferred temperature of the core of $\sim$1000 K (Okuchi, 1998; Williams, 1998).

5.14   Conclusions

The preponderance of evidence, from dynamic simulations and petrology, suggests that the Earth's oceans formed early in Earth's history – likely in the planet's first couple hundred million years. Whether they formed predominantly as a consequence of accretion, through impact-induced degassing, or through rapid mantle degassing (or, more likely, through a combination of both) is unclear. There are indications that the interplay between Earth's mantle and hydrosphere, with water emanating from the interior through volcanic degassing and returning to the interior through subduction, may mildly favor regassing of the mantle, and commensurate slow shrinkage of ocean volume. Accordingly, a gradual model of ocean formation through progressive degassing of the mantle appears unlikely. In contrast to the early stabilization of oceans, the principal role of the solid Earth in bulk atmospheric evolution lies in CO_2 degassing – most other atmospheric changes have been generated by the biosphere.

The bulk water content of the planet remains uncertain: the degree to which the core sequestered hydrogen during Earth's differentiation is almost entirely unconstrained, and the level of hydration of (and indeed, even the solubility of water in) the deep mantle is similarly unknown. But, constraints derived from basaltic melts indicate that the water content of at least the upper portion of the mantle (away from subduction zones) is small: of order $\sim$100–500 ppm. Accordingly, when subduction-altered mantle is considered, coupled with a possible mild increase in water content with depth (as is indicated by hot spot magmas generally having elevated water contents relative to MORB), an order-of-magnitude estimate of $\sim$0.1 wt.% water content for the silicate portion of the planet is probably reasonably accurate.

Acknowledgment

The author thanks the US NSF for support through grant EAR-0310342.

References

Abe Y and Matsui T (1988) Evolution of an impact-generated H$_2$O–CO$_2$ atmosphere and formation of a hot proto-ocean on Earth. *Journal of Atmospheric Sciences* 45: 3081–3101.

Abe Y, Ohtani E, Okuchi T, Righter K, and Drake M (2000) Water in the early Earth. In: Canup RM and Righter K (eds.) *Origin of the Earth and Moon*, pp. 413–433. Tucson, AZ: University of Arizona Press.

Abers GA (2000) Hydrated subducted crust at 100–250 km depth. *Earth and Planetary Science Letters* 176: 323–330.

Ahrens TJ (1993) Impact erosion of terrestrial planetary atmospheres. *Annual Review of Earth and Planetary Sciences* 21: 521–555.

Aikawa Y and Herbst E (2001) Two-dimensional distributions and column densities of gaseous molecules in protoplanetary disks. Part II: Deuterated species and UV shielding by ambient clouds. *Astronomy and Astrophysics* 371: 1107–1117.

Allegre CJ, Staudacher T, and Sarda P (1986/87) Rare gas systematics: Formation of the atmosphere, evolution and structure of the Earth's mantle. *Earth and Planetary Science Letters* 81: 127–150.

Amelin Y, Lee DC, Halliday AN, and Pidgeon RT (1999) Nature of the Earth's earliest crust from hafnium isotopes in single detrital zircons. *Nature* 399: 252–255.

Antonov VE, Belash IT, Degtyareva VF, Ponyatovsky EG, and Shiryaev VI (1980) Obtaining iron hydride under high hydrogen pressure. *Doklady Akademii Nauk SSSR* 252: 1384–1387.

Asimow PD and Langmuir CH (2003) The importance of water to ocean ridge melting regimes. *Nature* 421: 815–820.

Aubaud C, Hauri EH, and Hirschmann MM (2004) Hydrogen partition coefficients between nominally anhydrous minerals and basaltic melts. *Geophysical Research Letters* 31 (doi:10.1029/2004GL021341).

Badding JV, Hemley RJ, and Mao HK (1991) High-pressure chemistry of hydrogen in metals: *In situ* study of iron hydride. *Science* 253: 421–424.

Ballentine CJ and Barfod DN (2000) The origin of air-like noble gases in MORB and OIB. *Earth and Planetary Science Letters* 180: 39–48.

Ballentine CJ, van Keken PE, Porcelli D, and Hauri EH (2002) Numerical models, geochemistry and the zero-paradox noble gas mantle. *Philosophical Transactions of the Royal Society A* 360: 2611–2631.

Bebout GE (1995) The impact of subduction-zone metamorphism on mantle–ocean chemical cycling. *Chemical Geology* 126: 191–218.

Bell DR and Rossman GR (1992) Water in Earth's mantle: The role of nominally anhydrous minerals. *Science* 255: 1391–1397.

Bell DR, Rossman GR, Maldener J, Endisch D, and Rauch F (2003) Hydroxide in olivine: A quantitative determination of the absolute amount and calibration of the IR spectrum. *Journal of Geophysical Research* 108 (doi:10.1029/2001JB000679).

Bercovici D and Karato SI (2003) Whole-mantle convection and the transition zone water filter. *Nature* 425: 39–44.

Berner RA (2006) Geological nitrogen cycle and atmospheric N_2 over Phanerozoic time. *Geology* 34: 413–415.

Bockelee-Morvan D, Gautier D, Lis DC, *et al.* (1998) Deuterated water in Comet C/1996 B2 (Hyakutake) and its implications for the origin of comets. *Icarus* 133: 147–162.

Bolfan-Casanova N, Keppler H, and Rubie DC (2000) Water partitioning between nominally anhydrous minerals in the $MgO–SiO_2–H_2O$ system up to 24 GPa: Implications for the distribution of water in the Earth's mantle. *Earth and Planetary Science Letters* 182: 209–221.

Bolfan-Casanova N, Keppler H, and Rubie DC (2003) Water partitioning at 660 km depth and evidence for very low water solubility in magnesium silicate perovskite. *Geophysical Research Letters* 30 (doi:10.1029/2003GL017182).

Bolfan-Casanova N, Mackwell S, Keppler H, McCammon C, and Rubie DC (2002) Pressure dependence of H solubility in magnesiowustite up to 25 GPa: Implications for the storage of water in Earth's lower mantle. *Geophysical Research Letters* 29: 1029–1032.

Boslough MB, Ahrens TJ, Vizgirda J, Becker RH, and Epstein S (1982) Shock-induced devolatilization of calcite. *Earth and Planetary Science Letters* 61: 166–170.

Braun MG, Hirth G, and Parmentier EM (2000) The effects of deep damp melting on mantle flow and melt generation beneath mid-ocean ridges. *Earth and Planetary Science Letters* 176: 339–356.

Bromiley GD and Keppler H (2004) An experimental investigation of hydroxyl solubility in jadeite and Na-rich clinopyroxenes. *Contributions to Mineralogy and Petrology* 147: 189–200.

Brown H (1952) Rare gases and the formation of the Earth's atmosphere. In: Kuiper GP (ed.) *The Atmospheres of the Earth and Planets*, pp. 258–266. Chicago, IL: University of Chicago Press.

Campbell IH and Taylor SR (1983) No water, no granites – No oceans, no continents. *Geophysical Research Letters* 10: 1061–1064.

Carson B and Screaton EJ (1998) Fluid flow in accretionary prisms: Evidence for focused, time-variable discharge. *Reviews of Geophysics* 36: 329–352.

Cartigny P and Ader M (2003) A comment on 'The nitrogen record of crust–mantle interaction and mantle convection from Archean to present' by B. Marty and N. Dauphas. *Earth and Planetary Science Letters* 216: 425–432.

Cartigny P, Harris JW, and Javoy M (2001) Diamond genesis, mantle fractionations and mantle nitrogen content: A study

of $\delta^{13}C–N$ concentrations in diamonds. *Earth and Planetary Science Letters* 185: 85–98.

Catling DC, Zahnle KJ, and McKay CP (2001) Biogenic methane, hydrogen escape, and the irreversible oxidation of early Earth. *Science* 293: 839–843.

Chen GQ and Ahrens TJ (1997) Erosion of terrestrial planet atmosphere by surface motion after a large impact. *Physics of the Earth and Planetary Interiors* 100: 21–26.

Chyba CF (1987) The cometary contribution to the oceans of primitive Earth. *Nature* 330: 632–635.

Class C and Goldstein SL (2005) Evolution of helium isotopes in the Earth's mantle. *Nature* 436: 1095–1096.

Cyr KE, Sears WD, and Lunine JI (1998) Distribution and evolution of water ice in the solar nebula: Implications for solar system body formation. *Icarus* 135: 537–548.

Danyushevsky LV, Eggins SM, Falloon TJ, and Christie DM (2000) H_2O abundance in depleted to moderately enriched mid-ocean ridge magmas. Part I: Incompatible behaviour, implications for mantle storage, and origin of regional variations. *Journal of Petrology* 41: 11329–11364.

Dauphas N (2003) The dual origin of the terrestrial atmosphere. *Icarus* 165: 326–339.

Dauphas N, Robert F, and Marty B (2000) The late asteroidal and cometary bombardment of Earth as recorded in water deuterium to protium ratio. *Icarus* 148: 508–512.

Davies JH and Stevenson DJ (1992) Physical model of source region of subduction zone volcanics. *Journal of Geophysical Research* 97: 2037–2070.

Deloule E, Robert F, and Doukhan JC (1998) Interstellar hydroxyl in meteoritic chondrules: Implications for the origin of water in the inner solar system. *Geochimica et Cosmochimica Acta* 62: 3367–3378.

Dixon JE, Dixon TH, Bell DR, and Malservisi R (2004) Lateral variation in upper mantle viscosity: Role of water. *Earth and Planetary Science Letters* 222: 451–467.

Dixon JE, Leist L, Langmuir C, and Schilling JG (2002) Recycled dehydrated lithosphere observed in plume-influenced mid-ocean-ridge basalt. *Nature* 420: 385–389.

Domanik KJ and Holloway JR (1996) The stability and composition of phengitic muscovite and associated phases from 5.5 to 11 GPa: Implications for deeply subducted sediments. *Geochimica et Cosmochimica Acta* 60: 4133–4150.

Drouart A, Dubrulle B, Gautier D, and Robert F (1999) Structure and transport in the solar nebula from constraints on deuterium enrichment and giant planet formation. *Icarus* 140: 129–155.

Eberhardt P, Reber M, Krankowsky D, and Hodges RR (1995) The D/H and $^{18}O/^{16}O$ ratios in water from Comet P/Halley. *Astronomy and Astrophysics* 302: 301–316.

Fanale FP (1971) A case for catastrophic early degassing of the Earth. *Chemical Geology* 8: 79.

Fegley B, Jr. (2000) Kinetics of gas-grain reactions in the solar nebula. *Space Science Reviews* 92: 177–200.

Franck S and Bounama C (1997) Continental growth and volatile exchange during Earth's evolution. *Physics of the Earth and Planetary Interiors* 100: 189–196.

Franck S and Bounama C (2001) Global water cycle and Earth's thermal evolution. *Journal of Geodynamics* 32: 231–2246.

Fukai Y (1984) The iron-water reaction and the evolution of the Earth. *Nature* 308: 174–175.

Gaidos E, Deschenes B, Dundon L, *et al.* (2005) Beyond the principle of plenitude: A review of terrestrial planet habitability. *Astrobiology* 5: 100–126.

Galer SJG (1991) Interrelationships between continental freeboard, tectonics, and mantle temperature. *Earth and Planetary Science Letters* 105: 214–228.

Galer SJG and Mezger K (1998) Metamorphism, denudation and sea level in the Archean and cooling of the Earth. *Precambrian Research* 92: 389–412.

Gasparik T (1993) The role of volatiles in the transition zone. *Journal of Geophysical Research* 98: 4287–4299.

Genda H and Abe Y (2003) Survival of a proto-atmosphere through the stage of giant impacts: The mechanical aspects. *Icarus* 164: 149–162.

Genda H and Abe Y (2005) Enhanced atmospheric loss on protoplanets at the giant impact phase in the presence of oceans. *Nature* 433: 842–844.

Griggs DT (1967) Hydrolytic weakening of quartz and other silicates. *Geophysical Journal of the Royal Astronomical Society* 14: 19–31.

Green DH (1975) Genesis of Archean peridotitic magmas and constraints on Archean geothermal gradients and tectonics. *Geology* 3: 15–18.

Grove TL and Parman SW (2004) Thermal evolution of the Earth as recorded by komatiites. *Earth and Planetary Science Letters* 219: 173–187.

Harrison TM, Blichert-Toft J, Muller W, Albarede F, Holden P, and Mojsis SJ (2005) Heterogeneous hadean hafnium: Evidence of continental crust at 4.4 to 4.5 Ga. *Science* 310: 1947–1950.

Hayashi C, Nakazawa K, and Mizuno H (1979) Earth's melting due to the blanketing effect of the primordial dense atmosphere. *Earth and Planetary Science Letters* 43: 22–28.

Hirschmann MM (2006) Water, melting and the deep Earth H_2O cycle. *Annual Review of Earth and Planetary Sciences* 34: 629–653.

Hirth G and Kohlstedt DL (1996) Water in the oceanic upper mantle: Implications for rheology, melt extraction and the evolution of the lithosphere. *Earth and Planetary Science Letters* 144: 93–108.

Holland HD (1984) *The Chemical Evolution of the Atmospheres and Oceans*. Princeton, NJ: Princeton University Press.

Holland G and Ballentine CJ (2006) Seawater subduction controls the heavy noble gas composition of the mantle. *Nature* 441: 186–191.

Ikoma M and Genda HG (2006) Constraints on the mass of a habitable planet with water of nebular origin. *Astrophysical Journal* 648: 696–706.

Ingebritsen SE and Manning CE (2002) Diffuse fluid flux through orogenic belts: Implications for the world ocean. *Proceedings of the National Academy of Sciences* 99: 9113–9116.

Ingrin J and Skogby H (2000) Hydrogen in nominally anhydrous upper-mantle minerals: Concentration levels and implications. *European Journal of Mineralogy* 12: 543–570.

Inoue T, Irifune T, Yurimoto H, and Miyashi I (1998) Decomposition of K-amphibole at high pressures and implications for subduction zone volcanism. *Physics of the Earth and Planetary Interiors* 107: 221–231.

Ito E, Harris DM, and Anderson AT, Jr. (1983) Alteration of oceanic crust and geologic cycling of chlorine and water. *Geochimica et Cosmochimica Acta* 47: 1613–1624.

Iwamori H (1998) Transportation of H_2O and melting in subduction zones. *Earth and Planetary Science Letters* 160: 65–80.

Iwamori H (2001) Phase relations of peridotites under H_2O-saturated conditions and ability of subducting plate for transportation of H_2O. *Earth and Planetary Science Letters* 227: 57–71.

Iwamori H (2004) Phase relations of peridotites under H_2O-saturated conditions and ability of subducting plates for transportation of H_2O. *Earth and Planetary Science Letters* 227: 57–71.

Javoy M (1997) The major volatile elements of the Earth: Their origin, behavior, and fate. *Geophysical Research Letters* 24: 177–180.

Javoy M (1998) The birth of the Earth's atmosphere: The behavior and fate of its major elements. *Chemical Geology* 147: 11–25.

Joly J (1914) Denudation. In: Joly J (ed.) *The Birth-Time of the World and Other Scientific Essays*, pp. 30–59. New York: E.P. Dutton.

Jones TD, Libofsky LA, Lewis JS, and Marley MS (1990) The composition and origin of the C, P and D asteroids: Water as a trace of thermal evolution in the outer belt. *Icarus* 88: 172–192.

Kamijo K, Hashizume K, and Matsuda JI (1998) Noble gas constraints on the evolution of the atmosphere–mantle system. *Geochimica et Cosmochimica Acta* 62: 2311–2321.

Karato S (1990) The role of hydrogen in the electrical conductivity of the upper mantle. *Nature* 347: 272–273.

Kasting JF (1993) Earth's early atmosphere. *Science* 259: 920–926.

Kasting JF and Catling D (2003) Evolution of a habitable planet. *Annual Reviews of Astronomy and Astrophysics* 41: 429–463.

Kasting JF and Holm NG (1992) What determines the volume of the oceans? *Earth and Planetary Science Letters* 109: 507–515.

Kaufman AJ and Xiao S (2003) High CO_2 levels in the Proterozoic atmosphere estimated from analyses of individual microfossils. *Nature* 425: 279–282.

Kawamoto T, Hervig RL, and Holloway JR (1996) Experimental evidence for a hydrous transition zone in Earth's early mantle. *Earth and Planetary Science Letters* 142: 587–592.

Kohlstedt DL, Keppler H, and Rubie DC (1996) Solubility of water in the $\alpha-\beta-\gamma$ phases of $(Mg,Fe)_2SiO_4$. *Contributions to Mineralogy and Petrology* 123: 345–357.

Kopf A, Klaeschen D, and Mascle J (2001) Extreme efficiency of mud volcanism in dewatering accretionary prisms. *Earth and Planetary Science Letters* 189: 295–313.

Korenaga J (2006) Archean geodynamics and the thermal evolution of Earth. In: Benn K, Mareschal J-C, and Condie K (eds.) *Archean Geodynamics and Environments*, pp. 7–32. Washington, DC: AGU.

Kramers JD (2003) Volatile element abundance patterns and an early liquid water ocean on Earth. *Precambrian Research* 126: 379–394.

Lange MA and Ahrens TJ (1982) The evolution of an impact-generated atmosphere. *Icarus* 51: 96–120.

Larsen JF, Knittle E, and Williams Q (2003) Constraints on the speciation of hydrogen in Earth's transition zone. *Physics of the Earth and Planetary Interiors* 136: 93–105.

Lecuyer C, Gillet P, and Robert F (1998) The hydrogen isotope composition of seawater and the global water cycle. *Chemical Geology* 145: 249–261.

Lee KKM and Steinle-Neumann G (2006) High-pressure alloying of iron and xenon: 'Missing' xenon in the Earth's core. *Journal of Geophysical Research* 111: (doi:10.1029/2005JB003781).

Litasov K and Ohtani E (2003) Stability of various hydrous phases in CMAS pyrolite-H_2O up to 25 GPa. *Physics and Chemistry of Minerals* 30: 147–156.

Litasov K, Ohtani E, Langenhorst F, Yurimoto H, Kubo T, and Kondo T (2003) Water solubility in Mg-perovskites and water storage capacity in the lower mantle. *Earth and Planetary Science Letters* 211: 189–203.

Liu LG (2004) The inception of the oceans and CO_2-atmosphere in the early history of the Earth. *Earth and Planetary Science Letters* 227: 179–184.

Lowe DR and Tice MM (2004) Geologic evidence for Archean atmospheric and climatic evolution: Fluctuating levels of CO_2, CH_4, and O_2 with an overriding tectonic control. *Geology* 32: 493–496.

Lu R and Keppler H (1997) Water solubility in pyrope to 100 kbar. *Contributions to Mineralology and Petrology* 129: 35–42.

Mackwell SJ and Kohlstedt DL (1990) Diffusion of hydrogen in olivine: Implications for water in the mantle. *Journal of Geophysical Research* 95: 5079–5088.

Mackwell SJ, Kohlstedt DL, and Paterson MS (1985) The role of water in the deformation of olivine single crystals. *Journal of Geophysical Research* 90: 11319–11333.

McGovern PJ and Schubert G (1989) Thermal evolution of the Earth: Effects of volatile exchange between atmosphere and interior. *Earth and Planetary Science Letters* 96: 27–37.

Marty B and Dauphas N (2003a) The nitrogen record of crust–mantle interaction and mantle convection from Archean to present. *Earth and Planetary Science Letters* 206: 397–410.

Marty B and Dauphas N (2003b) 'Nitrogen isotopic compositions of the present mantle and the Archean biosphere': Reply to comment by P. Cartigny and M. Ader. *Earth and Planetary Science Letters* 216: 433–439.

Meade C, Reffner JA, and Ito E (1994) Synchrotron infrared absorbance measurements of hydrogen in MgSiO$_3$ perovskite. *Science* 264: 1558–1560.

Meibom A, Anderson DL, Sleep NH, et al. (2003) Are high ^{3}He/^{4}He ratios in oceanic basalts an indicator of deep-mantle plume components. *Earth and Planetary Science Letters* 208: 197–204.

Meier R, Owen TC, Matthews HE, et al. (1998) A determination of the HDO/H$_2$O ratio in Comet C/1995 O1 (Hale-Bopp). *Science* 279: 842–844.

Michael P (1995) Regionally distinctive sources of depleted MORB: Evidence from trace elements and H$_2$O. *Earth and Planetary Science Letters* 131: 301–320.

Mei S and Kohlstedt DL (2000) Influence of water on plastic deformation of olivine aggregates. 2. Dislocation creep regime. *Journal of Geophysical Research* 105: 21417–21481.

Mojsis SJ, Harrison TM, and Pidgeon RT (2001) Oxygen-isotope evidence from ancient zircons for liquid water at the Earth's surface 4,300 Myr ago. *Nature* 409: 178–181.

Moore JC and Vrolijk P (1992) Fluids in accretionary prisms. *Reviews of Geophysics* 30: 113–135.

Morbidelli A, Chamber J, Lunine JI, et al. (2000) Source regions and timescales for the delivery of water on Earth. *Meteoritics and Planetary Science* 35: 1309–1320.

Moreira M and Sarda P (2000) Noble gas constraints on degassing processes. *Earth and Planetary Science Letters* 176: 375–386.

Morse JW and Mackenzie FT (1998) Hadean ocean carbonate geochemistry. *Aquatic Geochemistry* 4: 301–319.

Murakami M, Hirose K, Yurimoto H, Nakashima S, and Takafuji N (2002) Water in Earth's lower mantle. *Science* 295: 1885–1887.

Newman WI, Symbalisty EM, Ahrens TJ, and Jones EM (1999) Impact erosion of planetary atmospheres: Some surprising results. *Icarus* 138: 224–240.

Nisbet EG and Sleep NH (2001) The habitat and nature of early life. *Nature* 409: 1083–1091.

Nolet G and Zielhuis A (1994) Low S velocities under the Tornquist-Teisseyeare zone: Evidence for water injection into the transition zone by subduction. *Journal of Geophysical Research* 99: 15813–15820.

Nutman AP, Mojzsis SJ, and Friend CRL (1997) Recognition of >3850 Ma water-lain sediments in west Greenland and their significance for the early Archean Earth. *Geochimica et Cosmochimica Acta* 61: 2475–2484.

Ohtani E, Litasov K, Hosoya T, Kubo T, and Kondo T (2004) Water transport into the deep mantle and formation of a hydrous transition zone. *Physics of the Earth and Planetary Interiors* 143–144.

Okuchi T (1997) Hydrogen partitioning into molten iron at high pressure: Implications for Earth's core. *Science* 278: 1781–1784.

Okuchi T (1998) The melting temperature of iron hydride at high pressures and its implications for the temperature of the Earth's core. *Journal of Physics: Condensed Matter* 10: 11595–11598.

Ott U (2002) Noble gases in meteorites – Trapped components. *Reviews in Mineralogy* 47: 71–100.

Pacalo REG and Parise JB (1992) Crystal structure of superhydrous B, a hydrous magnesium silicate synthesized at 1400°C and 20 GPa. *American Mineralogist* 77: 681–684.

Parman SW, Dann JC, Grove TL, and deWit MJ (1997) Emplacement conditions of komatiite magmas from the 3.49 Ga Komati Formation, Barberton Greenstone Belt, South Africa. *Earth and Planetary Science Letters* 150: 303–323.

Pavlov AA and Kasting JF (2002) Mass independent fractionation of sulfur isotopes in Archean sediments: Strong evidence for an anoxic Archean atmosphere. *Astrobiology* 2: 27–41.

Pavlov AA, Kasting JF, Brown LL, Rages KA, and Freedman R (2000) Greenhouse warming by CH$_4$ in the atmosphere of early Earth. *Journal of Geophysical Research* 105: 11981–11990.

Peacock SM (1990) Fluid processes in subduction zones. *Science* 248: 329–337.

Pepin RO (1997) Evolution of Earth's noble gases: Consequences of assuming hydrodynamic loss driven by giant impact. *Icarus* 126: 148–156.

Poli S and Schmidt MW (1998) The high pressure stability of zoisite and phase relations of zoisite-bearing assemblages. *Contributions to Mineralogy and Petrology* 130: 162–175.

Poli S and Schmidt MW (2002) Petrology of subducted slabs. *Annual Review of the Earth and Planetary Sciences* 30: 207–235.

Porcelli D and Wasserburg GJ (1995) Mass transfer of helium, neon, argon and xenon through a steady-state upper mantle. *Geochimica et Cosmochimica Acta* 59: 4921–4937.

Porcelli D, Woollum D, and Cassen P (2001) Deep Earth rare gases: Initial inventories, capture from the solar nebula and losses during moon formation. *Earth and Planetary Science Letters* 193: 237–251.

Raymond SN, Mandell AM, and Sigurdsson S (2006) Exotic Earths: Forming habitable worlds with giant planet migration. *Science* 313: 1413–1416.

Raymond SN, Quinn T, and Lunine JI (2004) Making other Earths: Dynamical simulations of terrestrial planet formation and water delivery. *Icarus* 168: 1–17.

Rea DK and Ruff LJ (1996) Composition and mass flux of sediment entering the world's subduction zones: Implications for global sediment budgets, great earthquakes, and volcanism. *Earth and Planetary Science Letters* 140: 1–12.

Revenaugh J and Sipkin SA (1994) Seismic evidence for silicate melt atop the 410-km discontinuity. *Nature* 369: 474–476.

Richard G, Monnereau M, and Ingrin J (2002) Is the transition zone an empty water reservoir? Inferences from numerical model of mantle dynamics. *Earth and Planetary Science Letters* 205: 37–51.

Ringwood AE and Major A (1967) High-pressure reconnaissance investigations in the system Mg$_2$SiO$_4$–MgO–H$_2$O. *Earth and Planetary Science Letters* 2: 130–133.

Robert F (2001) The origin of water on Earth. *Science* 293: 1056–1058.

Rubey WW (1951) Geologic history of seawater: An attempt to state the problem. *Bulletin of the Geological Society of America* 62: 1111–1148.

Rupke LH, Morgan JP, Hort M, and Connolly JAD (2004) Serpentine and the subduction zone water cycle. *Earth and Planetary Science Letters* 223: 17–34.

Rye R, Kuo PH, and Holland HD (1995) Atmospheric carbon dioxide concentrations before 2.2 billion years ago. *Nature* 378: 603–605.

Saal AE, Hauri EH, Langmuir CH, and Perfit MR (2002) Vapour undersaturation in primitive mid-ocean ridge basalt and the volatile content of Earth's upper mantle. *Nature* 419: 451–455.

Sagan C and Chyba C (1997) The early faint sun paradox: Organic shielding of ultraviolet-labile greenhouse gases. *Science* 276: 1217–1221.

Sagan C and Mullen G (1972) Earth and Mars: Evolution of atmospheres and surface temperatures. *Science* 177: 52–56.

Sano T, Takahata N, Nishio Y, Fischer TP, and Williams SN (2001) Volcanic flux of nitrogen from the Earth. *Chemical Geology* 171: 263–271.

Sarda P (2004) Surface noble gas recycling to the terrestrial mantle. *Earth and Planetary Science Letters* 228: 49–63.

Sarda P, Staudacher T, and Allegre CJ (1985) $^{40}Ar/^{36}Ar$ in MORB glasses: Constraints on atmosphere and mantle evolution. *Earth and Planetary Science Letters* 72: 357–375.

Sasaki S (1990) The primary solar-type atmosphere surrounding the accreting Earth: H_2O-induced high surface temperature. In: Newsom HE and Jones JH (eds.) *Origin of the Earth*, pp. 195–209. Oxford: Oxford University Press.

Schmidt MW (1995) Lawsonite: Upper pressure stability and formation of higher density hydrous phases. *American Minerologist* 80: 1286–1292.

Schmidt MW and Poli S (1998) Experimentally based water budgets for dehydrating slabs and consequences for arc magma generation. *Earth and Planetary Science Letters* 163: 361–379.

Schubert G and Reymer PS (1985) Continental volume and freeboard through geological time. *Nature* 316: 336–339.

Shen AH and Keppler H (1997) Direct observation of complete miscibility in the albite-H_2O system. *Nature* 385: 710–712.

Shieh S, Ming LC, Mao HK, and Hemley RJ (1998) Decomposition of phase D in the lower mantle and the fate of dense hydrous silicates in subducting slabs. *Earth and Planetary Science Letters* 159: 13–23.

Shieh SR, Mao HK, Hemley RJ, and Ming LC (2000) *In situ* X-ray diffraction studies of dense hydrous magnesium silicates at mantle conditions. *Earth and Planetary Science Letters* 177: 69–80.

Sleep NH and Zahnle K (2001) Carbon dioxide cycling and implications for climate on ancient Earth. *Journal of Geophysical Research* 106: 1373–1399.

Sleep NH, Zahnle K, and Neuhoff PS (2001) Initiation of clement surface conditions on the earliest Earth. *Proceedings of the National Academy of Sciences* 98: 3666–3672.

Smyth JR (1994) A crystallographic model for hydrous wadsleyite (β-Mg_2SiO_4): An ocean in the Earth's interior? *Amercan Mineralogist* 79: 1021–1024.

Sobolev AV and Chaussidon M (1996) H_2O concentrations in primary melts from supra-subduction zones and mid-ocean ridge: Implications for H_2O storage and recycling. *Earth and Planetary Science Letters* 137: 45–55.

Stalder R and Skogby H (2002) Hydrogen incorporation in enstatite. *European Journal of Mineralogy* 14: 1139–1144.

Stevenson DJ (1977) Hydrogen in the Earth's core. *Nature* 268: 130–131.

Stone WE, Deloule E, Larson MS, and Lesher CM (1997) Evidence for hydrous high-MgO melts in the Precambrian. *Geology* 25: 143–146.

Sudo A and Tatsumi Y (1990) Phlogopite and K-amphibole in the upper mantle: Implications for magma genesis in subduction zones. *Geophysical Research Letters* 17: 29–32.

Tajika E and Matsui T (1993) Degassing history and carbon cycle of the Earth: From an impact-induced steam atmosphere to the present atmosphere. *Lithos* 30: 267–280.

Tatsumi Y (1989) Migration of fluid phases and genesis of basalt magmas in subduction zones. *Journal of Geophysical Research* 94: 4697–4707.

Thompson AB (1992) Water in the Earth's upper mantle. *Nature* 358: 295–302.

Tolstikhin IN and Marty B (1998) The evolution of terrestrial volatiles: A view from helium, neon, argon and nitrogen isotope modeling. *Chemical Geology* 147: 27–52.

Tyburczy JA, Frisch B, and Ahrens TJ (1986) Shock-induced volatile loss from a carbonaceous chondrite: Implications for planetary accretion. *Earth and Planetary Science Letters* 80: 201–207.

Tyburczy JA, Xu X, Ahrens TJ, and Epstein S (2001) Shock-induced devolatilization and isotopic fractionation of H and C from Murchison meteorite: Some implications for planetary accretion. *Earth and Planetary Science Letters* 192: 23–30.

Usui T, Nakamura E, and Helmstaedt H (2006) Petrology and geochemistry of eclogite xenoliths from the Colorado Plateau: Implications for the evolution of subducted oceanic crust. *Journal of Petrology* 47: 929–964.

Valley JW, et al. (2005) 4.4 billion years of crustal maturation: Oxygen isotope ratios of magmatic zircon. *Contributions to Mineralogy and Petrology* 150: 561–580.

Valley JW, Peck WH, King EM, and Wilde SA (2002) A cool early Earth. *Geology* 30: 351–354.

Vickery AM and Melosh HJ (1990) Atmospheric erosion and impactor retention in large impacts with application to mass extinctions. In: Sharpton VL and Ward PO (eds.) *Geological Society of America Special Paper: Global Catastrophes in Earth History*, pp. 289–300. Boulder, CO: Geological Society of America.

Wallmann K (2001) The geologic water cycle and the evolution of marine $\delta^{18}O$ values. *Geochimica et Cosmochimica Acta* 65: 2469–2485.

Wang D, Mookherjee M, Xu Y, and Karato SI (2006) The effect of water on the electrical conductivity of olivine. *Nature* 443: 977–980.

Wilde SA, Valley JW, Peck WH, and Graham CM (2001) Evidence from detrital zircons for the existence of continental crust and oceans on the Earth 4.4 Gyr ago. *Nature* 409: 175–178.

Williams Q (1998) The temperature contrast across D''. In: Gurnis M, Wysession ME, Knittle E, and Buffett B (eds.) *The Core–Mantle Boundary Region*, pp. 73–81. Washington, DC: American Geophysical Union.

Williams Q and Hemley RJ (2001) Hydrogen in the deep Earth. *Annual Review of Earth and Planetary Science* 29: 365–418.

Woods SC, Mackwell S, and Dyar D (2000) Hydrogen in diopside: Diffusion profiles. *American Mineralogist* 85: 480–487.

Workman RK, Hauri E, Hart SR, Wang J, and Blusztajn J (2006) Volatile and trace elements in basaltic glasses from Samoa: Implications for water distribution in the mantle. *Earth and Planetary Science Letters* 241: 932–951.

Wunder B, Rubie DC, Ross CR, Medenbach O, Seifert F, and Schreyer W (1993) Synthesis, stability and properties of $Al_2SiO_4(OH)_2$: A fully hydrated analogue of topaz. *American Mineralogist* 78: 285–297.

Wunder B and Schreyer W (1997) Antigorite: High pressure stability in the system MgO-SiO_2-H_2O. *Lithos* 41: 213–227.

Yang HX, Presitt CT, and Frost DJ (1997) Crystal structure of the dense hydrous magnesium silicate, phase D. *American Mineralogist* 82: 651–654.

Yamamoto K and Akimoto SI (1977) The system MgO-SiO_2-H_2O at high pressures and high temperatures: Stability field for hydroxyl-chondrodite, hydroxyl-clinohumite and 10 Å-phase. *American Journal of Science* 277: 288–312.

Yoshino T, Matsuzaki T, Yamashita S, and Katsura T (2006) Hydrous olivine unable to account for conductivity anomaly at the top of the asthenosphere. *Nature* 443: 973–976.

Young TE, Green HW, II, Hofmeister AM, and Walker D (1993) Infrared spectroscopic investigation of hydroxyl in β-(Mg,Fe)$_2$SiO$_4$ and coexisting olivine: Implications for mantle evolution and dynamics. *Physics and Chemistry of Minerals* 19: 409–422.

Zahnle K, Kasting JF, and Pollack JB (1988) Evolution of a steam atmosphere during Earth's accretion. *Icarus* 74: 62–97.

6 Plate Tectonics through Time

N. H. Sleep, Stanford University, Stanford, CA, USA

6.1 Introduction

The scientific issues that arose during the advent of the theory of plate tectonics place the modern issue of tectonics on the early Earth in context. The existence of seafloor spreading in the Atlantic and ridge–ridge transform faults were established when the author attended his first international scientific conference at Woods Hole in the summer of 1967. Later, the approximation of rigid plates on a spherical shell provided the global geometry, taking advantage of transform faults. Subduction provided an explanation of deep-focus earthquakes and the sink for the oceanic lithosphere produced by seafloor spreading. It was evident at the Woods Hole meeting that no one there had any idea how seafloor spreading works. This lacuna did not greatly concern the mainly observational scientists in attendance.

The initial hypothesis explained continental drift, relegating continents to plate passengers. Atwater (1970) brought the concept ashore in California demonstrating its great explicative power to continental tectonics. Overall, the plate tectonic revolution shifted attention away from cratons. Yet

the interior of continents had geological records of real events.

In December 1975, the author attended the Penrose Conference on continental interiors, the final stationary bastion under siege. Fixists hijacked the conference, jettisoned the published program, and mounted a determined counterattack. Speaker after speaker droned on that his locality was very complicated, that plate tectonics was of no use in interpreting the geology, that it certainly could not be established just looking at his quadrangle, and thus that the theory is dead wrong.

With regard to tectonics in the Earth's deep past, we are much like land-based geologists in 1975. The biases of geological preservation place us in the same bind that provincial data selection placed the Penrose holdouts during that freezing December in San Diego. We have little intact seafloor older than the ~180 Ma crust along the passive margins of the central Atlantic (Moores, 2002). We cannot observe the lynchpins of modern plate tectonics, magnetic stripes, and ridge–ridge transform faults. We certainly cannot observe earthquake mechanisms, the geoid, and tomography. Essentially, we are stuck with data from continents, the very regions that befuddled earlier geologists.

Yet we can do better with the deep past than Earth science did with the Tertiary in 1960. We can view geology at appropriate scales with the caveat that later tectonics dispersed the Archean record into ~35 blocks (Bleeker, 2003). We have a reasonable understanding of the kinematics of modern Earth, especially the record-producing process of continental breakup eventually followed by continental collision. Geodynamicists picked much of the low-hanging fruit from the orchard opened by plate tectonics over the last 40 years. We know how the modern Earth works well enough that we have hope of exporting concepts in time and to other planets.

Export inflicts some discussion of semantics. Plate tectonics, as originally conceived, was an approximate kinematic theory. It unified continental drift, seafloor spreading, subduction, and transform faulting. Orogeny from continental collision provided an apt explanation of mountain belts. These processes are linked but somewhat disjoint. For example, seafloor spreading does not kinematically require that the plates are rigid, that subduction is one-sided, or even that transform faults exist. Mantle plumes and thick cratonal lithosphere are significant features on the modern Earth that are further afield from plate processes.

With regard to the cardinal kinematic postulate, rigid plates tessellate 85% of the Earth's surface (e.g., Zatman *et al.*, 2005). Diffuse plate boundaries grout the remainder. Diffuse deformation zones divide major oceanic plates into subplates. Typically, the pole of rotation for the two subplates lies near their boundary so that both extension and compression occur. This process may nucleate subduction, but otherwise is lost in the fog of time. Diffuse continental zones, like the Basin and Range, the Lena River region of Siberia, and Tibet occur within hot weak continental crust. Both extension and compression have high preservation potential including sediments, focusing attention on vertical tectonics.

Continental collisions, like India with Tibet, produce strike-slip faults that extend far inland from the point of indention (Molnar and Tapponnier, 1975). The faults disrupt the orogen as well as otherwise stable regions. The record has a high preservation potential that is readily associated with the plate processes.

In all cases, vertical tectonics is more easily recognized than strike-slip tectonics. This is especially true with continent-margin-parallel faults, like the San Andreas Fault and the fault system through the Sumatra volcanic arc. Lateral offset is not obvious because the fault runs along the gross geological strike. Piercing points may be quite distant. Recently uplifted mountain ranges (e.g., in western California) are more evident than the strike-slip component of the tectonics. Historically, this view seemed natural; the US Geological Survey topographic map of Palo Alto perpetuates the archaic term, the San Andreas Rift.

The question of the antiquity of plate processes arose soon after geologists accepted the modern process (Hurley and Rand, 1969). A significant literature exists. The reader is referred to the works of Sleep (1979, 1992) and Sleep and Windley (1982) for discussions of early work. Stevenson (2003) and Korenaga (2006) discuss ancient global dynamics. Bleeker (2003), Sleep (2005), Condie and Benn (2006), and Polat and Kerrich (2006) review the rock record on continents. Moores (2002) reviews possible exposures of ancient oceanic crust. On the other hand, Stern (2005) argues that modern plate processes began in Neoproterozoic time.

The author winnows the subset of geodynamic theory and geological observation that bears on the problem at hand, begins with convection to put the physics in context, and then moves forward in time from the Earth's formation for first evidence of various aspects of plate tectonics. He then considers changes in the more recent past, finally considering the fate of the Earth in the geological future.

This chapter relates to many topics in this *Treatise on Geophysics*. In this volume, Chapters 4, 9, 7, and 1 cover topics that are intimately coupled with this chapter.

6.2   Physical Preliminaries

Plate tectonics is a form of convection where hot material rises at ridges and cold dense slabs sink. Much of geodynamics involves this heat and mass transfer process. This chapter reviews the physics in this section so that it can be referred back when the the Earth's history is discussed.

Since the nature of convection within the Earth changed over time, we loosely divide early Earth history into the period immediately after the Earth's formation, part of the Hadean where a 'magma' ocean filled with mostly crystalline 'mush' covered the planet, and the Archean 2.5–3.8 Ga , where it becomes meaningful to compare theory and observation. After the Archean, we can apply analogies to modern plate processes with more confidence to a reasonable record.

The Earth's mantle is complicated. Plate boundaries fail by faulting. The interior creeps as a very viscous fluid. Partial melting produces buoyant oceanic crust. These processes are complex enough that examination of the rock record is essential. It is useful to see the bases of common physical arguments about the tectonics of the Earth. This chapter begins with a discussion on the heat balance in the mantle.

6.2.1 Global Heat Balance

One can construct a thermal history of the Earth's interior by equating surface heat flow with contributions from transient cooling and radioactive heat generation. Formally, the heat balance for the mantle is

$$4\pi r_E^2 q = \frac{4\pi\left(r_E^3 - r_C^3\right)}{3}\rho H - \frac{4\pi r_E^3}{3}\rho C \frac{\partial T}{\partial t} + 4\pi r_c^2 q_C \quad [1]$$

where q is the heat flow from the mantle, r_E is the Earth's radius, q_C is the heat flow at the core's radius r_C, ρ is density, C is specific heat (essentially at constant pressure), H is radioactive heat generation per mass, T is the interior mantle temperature, and t is time. We make an approximation that the core cools at the same rate as the mantle in example calculations. About one-fifth of the Earth's heat capacity is in the core so its details can be ignored in some qualitative arguments but certainly not those related to plumes.

The specific heat of the mantle is about $1.25\,\mathrm{J\,K^{-1}\,kg^{-1}}$. This implies that a global heat flow of $1\,\mathrm{W\,m^{-2}}$ cools the mantle at $3.2\,\mathrm{K\,My^{-1}}$. Heat flows greater than this quickly cool the mantle and cannot have been sustained through even the Hadean. The present mantle heat flow of $0.070\,\mathrm{W\,m^{-2}}$ would cool the mantle at $225\,\mathrm{K}$ per billion year and the whole Earth at $180\,\mathrm{K}$ per billion year. A major task in geodynamics is to obtain the average surface heat flow as a function of the interior temperature, which leaves radioactivity as the unknown quantity in [1].

This chapter presents the results of simple thermal models of the Earth's history at this stage for use throughout this chapter. The intent is to put potentially geologically observable features and geodynamics in a common context. The models

conserve energy using eqn [1]. The globally averaged heat flow is a multibranched function of the interior temperature. The computed thermal evolution of the Earth is a trajectory as shown in this graph (**Figure 1**). The models generate physically realizable thermal histories, not cartoons. See Appendix 1 for the parameters used to compute these semischematic graphs.

Radioactive decay and the cooling of the Earth's interior over time are now comparable items in the Earth's heat budget. In example calculations, the current heat flow supplied by radioactivity is half the total (i.e., $0.035\,\mathrm{W\,m^{-2}}$). (We use extra digits throughout this chapter to make calculations easier to follow.)

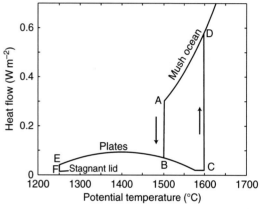

Figure 1 The heat flow from the Earth's mantle is a multibranched function of the potential temperature of the Earth's interior. Thermal histories are paths on this graph. One model has a monotonic thermal history where the heat flow lies along the transition in branch jumps. The other model jumps between branches and has a nonmonotonic thermal history. Both models start at 2000°C at 4.5 Ga. The mantle cools along the mush-ocean path until it reaches point A. The nonmonotonic model jumps to plate tectonics at point B. If the heat flow is greater than the radioactive heat generation, the mantle cools along the plate branch moving toward point E. If the heat flow is lower than the radioactive heat generation, the mantle heats up to point C, where convection jumps to the mush-ocean branch at point D. It then cools to point A. The nonmonotonic model jumps to stagnant-lid convection from point E to F. The mantle then heats up. The monotonic model stays at the jumps with the heat flow equal to radioactive heat generation until radioactive heat generation decreases to heat flow of the cooler mode. Note Thom (1983) popularized this presentation method in his catastrophe theory. The author eschews the word 'catastrophe' in the text as it is pre-empted with a foul stench in the Earth sciences. One could add another axis perpendicular to the plane of the diagram. Continental area would be a good choice (Lenardic et al., 2005). One would then need the net growth (shrinkage) rate of continents as a function of interior temperature, convection mode, and continental area to compute forward models.

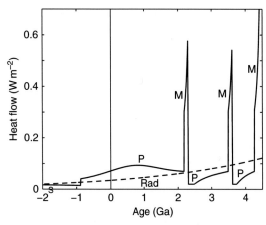

Figure 2 Computed heat flow as a function of age for the nonmonotonic model (dark line) and the radioactive heat generation as heat flow (thin dashed line). The modes of convection are (M) mush ocean, (P) plates, and (S) stagnant lid. The model resolves the expected duration of episodes, but no attempt was made to fine-tune the timing of transitions.

This heat flow is somewhat higher than the bulk silicate Earth model estimate of $0.024\,\mathrm{W\,m^{-2}}$. The heat flow from radioactivity at 4 Ga was a factor of a few more than present (i.e., $\sim 0.13\,\mathrm{W\,m^{-2}}$ in the examples presented here) (**Figure 2**). (Negative ages are in the future.) About one-third of the radioactivity is currently at shallow depths within continental crust where it does not drive mantle convection. Continent-like chemical reservoirs existed by ~ 4.5 Ga (Harrison *et al.*, 2005), but the ancient fraction of radioactivity sequestered within continents in the deep past is unknown. We use the current partition in examples, as the exact amount does not affect the gist of the arguments (**Figure 2**). Antineutrino detectors soon will provide hard data on the absolute abundance of U and Th in the Earth's interior (Araki *et al.*, 2005; Fiorentini *et al.*, 2005; Sleep, 2006). For simplicity, we also ignore the direct effect of continental area on the vigor of convection (Lenardic *et al.*, 2005).

6.2.2 Heat Transfer by Seafloor Spreading

The global heat balance provides necessary attributes for whatever type of global tectonics existed in the Later Hadean and the Archean. Radioactive heat production was small enough that plate tectonics is a viable candidate mechanism. As at present, one needs to consider the main mechanism of heat transfer, the formation, and cooling of oceanic lithosphere. One obtains simple well-known formulas by considering the oceanic lithosphere to be a half-space

(e.g., Turcotte and Schubert, 1982). We present their first-order forms in one place so that we may refer back to them later. The roof of a liquid or mush-filled magma ocean might well cool as a plate above an isothermal half-space.

We use global spreading rate U, the global production rate of new seafloor at ridges, conveniently given in $\mathrm{km^2\,yr^{-1}}$. The time for seafloor spreading to renew the Earth's surface is then $t_E \cong 4\pi r_E^2 / U$. This time represents the age of old seafloor at the time of subduction; we use the subduction age and the renewal time interchangeably in dimensional arguments. The present global spreading rate is $\sim 3\,\mathrm{km^2\,yr^{-1}}$, which would resurface the ocean basins in ~ 100 My and the Earth in 170 My White *et al.* (1992) give a more precise estimate of $3.3\,\mathrm{km^2\,yr^{-1}}$.

The global average heat flow depends on the square root of the renewal time

$$q = \frac{2k\Delta T}{\sqrt{\pi\kappa t_E}} \qquad [2]$$

where ΔT is the temperature contrast between the surface and the hot mantle beneath the lithosphere, k is thermal conductivity, ρ is density, and $\kappa \equiv k/\rho C$ is thermal diffusivity. The current average heat flow in the ocean basins is $\sim 0.10\,\mathrm{W\,m^{-2}}$, comparable to radioactive heat generation in the Archean. Radioactivity alone thus does not mandate a very vigorous form of Archean convection.

Other simple relationships bear on geodynamics. Plate thickness of the oldest lithosphere scales as

$$L = \sqrt{\kappa t_E} \qquad [3]$$

For quick calibration, 1 My lithosphere is ~ 10 km thick. The elevation of buoyant ridges relative to old oceanic crust produces 'ridge push,' a driving force for seafloor spreading. The membrane stress (the difference between horizontal stress and the vertical stress from lithostatic pressure) scales as

$$\tau = \rho g \alpha\, \Delta T (\kappa t_E)^{1/2} \qquad [4]$$

where g is the acceleration of gravity and α is the volume thermal expansion coefficient. Stress is concentrated within the cool brittle upper part of the lithosphere. The integral of this stress over the thickness of the lithosphere is the 'ridge push' force available to drive plate motion. It is

$$F = \tau L = \rho g \alpha \Delta T (\kappa t_E) \qquad [5]$$

The negative buoyancy of downgoing slabs is the other major driver of plate motions.

6.2.3 Parametrized Convection

Equation [1] can be integrated forward or back in time if one has the surface heat flow as a function of the interior temperature (**Figures 1** and **3**) and the concentrations of heat-producing elements (**Figure 2**). From [2], knowing the rate of convection as a function of interior temperature suffices. To do this, we assume that the vigor of convection depends only on the properties of the lithosphere and the uppermost asthenosphere, not those of the deep mantle; that is, the global convective heat flow depends on the temperature at the base of the lithosphere and the fraction of the Earth's surface covered by continental crust (Lenardic *et al.*, 2005). The convective heat flow does not depend on the instantaneous rate of radioactive heat generation.

Fluid dynamists call these assumptions 'boundary layer theory'. It applies in general if the thermal boundary layers comprise a small fraction of the total thickness of the convecting region. The Earth meets this criterion; the lithosphere is much thinner than the mantle.

We briefly review rheology to put the remaining discussion in context. In full tensor form, the strain rate in an isotropic material is

$$\varepsilon'_{ij} = \frac{\tau_{ij}\tau^{n-1}}{2\eta\tau_{\text{ref}}^{n-1}} \qquad [6]$$

where η is the viscosity, τ_{ij} is the deviatoric stress tensor (the stress that produces shear, mathematically

the stress tensor with zero trace), the scalar stress τ is the second invariant of the stress tensor $\sqrt{\tau_{ij}\tau_{ij}}$ (appropriately normalized it is the resolved shear stress in two dimensions), n is the exponent of a power-law rheology, and τ_{ref} is a reference stress that may be chosen for convenience. The scalar form is useful for dimensional calculations:

$$\varepsilon' = \frac{\tau^n}{\eta\tau_{\text{ref}}^{n-1}} \qquad [7]$$

The familiar linear or Newtonian fluid is the case where $n = 1$. A plastic substance fails at a yield stress. This corresponds to the limit $n \to \infty$. We present applications of parametrized convection in the order that their relevancies arose in the Earth's history.

A linearly viscous fluid reasonably represents liquid magma and hence the liquid magma ocean on the early Earth. It may represent convecting mush and the present asthenosphere. The formalism for this rheology illustrates key aspects of the parametrized convection approach before we discuss nonlinear $n < 1$ rheology. The convective heat flow is then

$$q = Ak\,\Delta T \left[\frac{\rho Cg\alpha\Delta T}{\kappa\eta}\right]^{1/3} \qquad [8]$$

The dimensionless multiplicative constant A is of the order of 1; it depends on the upper boundary condition. ΔT is the temperature contrast that actually drives convection and η is the viscosity of the conducting interior (Davaille and Jaupart, 1993a, 1993b, 1994; Solomatov, 1995; Solomatov and Moresi, 2000). These two quantities are the dominant terms in [8]. Temperature contrast is raised to the 4/3 power. Viscosity decreases orders of magnitude with increasing temperature.

The linear rheology assumed in [8] does not apply to plate tectonics because trench and transform boundaries fail in friction rather than viscous flow. One approach is to apply the parametrized formula for a nonlinear fluid:

$$q = A_n k\,\Delta T \left[\frac{\Delta T^n (\rho g\alpha)^n}{\kappa\tau_{\text{ref}}^{n-1}\eta}\right]^{1/(n+2)} \qquad [9]$$

where the full temperature contrast ΔT across the lithosphere drives convection. The dimensionless multiplicative constant A_n depends both on the exponent n and the boundary conditions. As one approaches the limit of a plastic rheology $n \to \infty$, the heat flow becomes independent of the viscosity η and proportional to square of the temperature

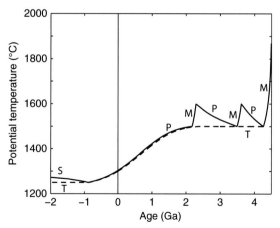

Figure 3 The computed thermal history for the nonmonotonic model (dark line) and the monotonic model (thin dashed line). The modes of convection are mush ocean (M), plates (P), stagnant lid (S), and transition (T) in the monotonic model. The temperature remains constant in the transitions in the monotonic model.

contrast. This inference assumes correctly that the yield stress does not depend greatly on temperature; the 'brittle' strength of plates is concentrated at cool shallow depths (e.g., Zoback and Townend, 2001). Plate-bounding faults, such as the San Andreas Fault, are much weaker than 'normal' rocks failing in friction at hydrostatic pore pressure.

We derive this temperature-squared relationship by letting τ in [4] represent the plastic yield stress τ_Y, where

$$\tau_Y = \rho g \alpha \, \Delta T (\kappa t_E)^{1/2} \qquad [10]$$

Solving [2] and [10] by eliminating the resurfacing time t_E yields the heat flow

$$q = \frac{2 k \rho \alpha g \, \Delta T^2}{\tau_Y \sqrt{\pi}} \qquad [11]$$

As discussed in Section 6.3, the Earth's interior cooled only a few 100 K since a solid lithosphere formed. For example, the heat flow in [11] when the interior was 1800°C versus 1300°C at present was only a factor of $(1800/1300)^2 = 1.9$ greater than present.

Stagnant-lid convection occurs beneath the lithosphere because viscosity is strongly temperature dependent (**Figure 4**). As discussed in Section 6.5, it will become the mode of convection within the Earth in the geological future. It is now the mode of convection within Mars and Venus. The cooler upper part of the lithosphere is essentially rigid on those planets.

Only a thin rheological boundary layer actually partakes in convection (**Figure 4**). The temperature contrast across the layer determines whether underlying convection impinges on the base of the crust. In laboratory and numerical experiments, the viscosity contrast between the top of the rheological boundary layer and the underlying half-space is a factor ~10 for a linear fluid. Simple convenient expressions arise when we represent viscosity in a traditional form,

$$\eta = \eta_{\mathrm{ref}} \exp\left[\frac{T_{\mathrm{ref}} - T}{T_\eta}\right] \qquad [12]$$

where η_{ref} is the reference viscosity at temperature T_{ref} and T_η is the temperature scale for viscosity. The parametrized convection equation [9] in general becomes

$$q = A_n k T_\eta \left[\frac{T_\eta^n (\rho g \alpha)^n}{\kappa \tau_{\mathrm{ref}}^{n-1} \eta}\right]^{1/(n+2)} \qquad [13]$$

where the viscosity is that of the underlying half-space and the reference stress may be defined for convenience. The temperature contrast of the rheological boundary layer is

$$\Delta T_{\mathrm{rheo}} \approx 1.2(n+1) T_\eta \qquad [14]$$

Stagnant-lid convection today supplies modest amounts of heat flow to the base of the oceanic lithosphere. It supplies the bulk of the heat flow through stable continental lithosphere. The asthenosphere and the rheological boundary layer beneath these regions may well behave as a linearly viscous fluid. The temperature scale T_η is not precisely known; 43 K (which provides an order of magnitude over 100 K) is a venerable approximation. It could be as high as 100 K. The temperature contrast of the rheological boundary layer is less than a few hundred kelvin. Stagnant-lid convection does not impinge on the base of the crust within stable parts of the continent.

6.2.4 Changes in the Mode of Convection

The mode of convection in the Earth changed a few times. Changes included mush-filled magma ocean to plate tectonics. (The traditional term 'magma ocean' misleadingly implies a fully molten reservoir. The nonstandard term 'mush ocean' is preferable as it implies a region of mostly crystalline mush.) Convection may have conceivably gone from plate tectonics back to mush ocean. Transition from plate

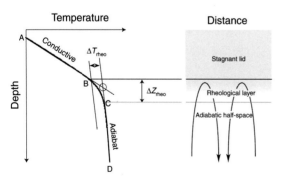

Figure 4 Schematic diagram of stagnant-lid convection at the base of the lithosphere. The geotherm (dark line, left side) has a linear conductive gradient (A to B) within the stagnant lid. It is adiabatic (C to D) below the stagnant lid. The rheological boundary layer is the top of convection (right). The geotherm changes form conductive at B to abiabat at C. The geothermal gradient in the region scales as ~1/2 the conductive gradient. The high viscosity near the top of the rheological layer limits the flow rate.

tectonics to stagnant-lid convection is inevitable in the geological future. The initiation of subduction and the delamination of the base of the lithosphere are local mode changes. ('Delamination' is massive foundering of cold dense mantle lithosphere into the underlying mantle.)

This chapter illustrates a general formalism for representing mode changes involving stagnant lid convection. The physical processes involve the fact that the strain rate within the lithosphere depends nonlinearly on stress. We do not have a quantitative form for transitions involving a mush ocean (See Appendix 2).

To derive [14], Solomatov and Moresi (2000) presumed that stagnant-lid convection self-organizes the rheological temperature contrast ΔT_{rheo} so that flow transports the maximum convective heat flow. We illustrate this property qualitatively before doing mathematics. The convective heat flow depends on the product of the flow rate and the rheological temperature contrast. If the rheological temperature contrast is very small, little heat gets transferred because no temperature variations are available to transfer heat and the buoyancy forces driving convection are quite small. Conversely, there are ample buoyancy forces and temperature variations if the rheological temperature contrast is very large, but the high viscosity of the cool material in the stagnant lid precludes flow. The maximum convective heat flow exists in between where there are both temperature variations and a viscosity only modestly higher than that of the underlying half-space. Laboratory and numerical calculations show that the following formalism provides reasonable predictions of the rheological temperature contrast.

We continue with mathematics, extending the form of the derivation by Sleep (2002) to a more complicated rheology. To keep equations as compact as possible, we assume that linear and nonlinear creep mechanisms act in parallel and use the scalar form [7] to obtain dimensional results. The scalar strain rate is then

$$\varepsilon' = \frac{\tau}{\eta} + a\frac{\tau^n}{\eta\tau_{ref}^{n-1}} \qquad [15]$$

where a is a dimensionless constant. In general, the temperature dependence of the two creep mechanisms may differ and the stress–strain rate relationship may be more complex.

We make the rheological temperature contrast ΔT_{rheo} a free parameter with the intent of outlining

the derivation of [14] by finding its maxima in the convective heat flow (**Figure 5**). The thickness of the rheologically active layer is then (**Figure 4**)

$$\Delta Z_{rheo} = \Delta T_{rheo}\left(\frac{\partial T}{\partial Z}\right)^{-1} \qquad [16]$$

where $\partial T/\partial Z$ is the thermal gradient. The stress from lateral variations of temperature scales with the product of the density contrast within the rheological boundary layer and its thickness. The stress within the rheological layer is dimensionally

$$\tau \approx \rho g\alpha\,\Delta T_{rheo}^2\left[\frac{\partial T}{\partial Z}\right]^{-1} \qquad [17]$$

Note that [4] arises from the analogous product of density contrast with lithosphere thickness.

The velocity in the thermal boundary layer is the integral of the strain rate across the rheologically

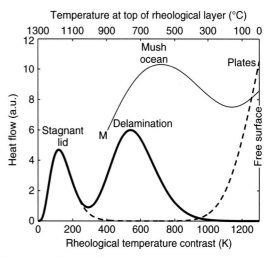

Figure 5 The heat flow in arbitrary units from [19]. The author has adjusted the constant a in each model so that the amplitude of the peaks is comparable for schematic plotting convenience. The second peak of the solid curve corresponds to delamination of the lithosphere, $n = 8$. The dashed curve represents plate tectonics. The computed heat flow is still increasing when the rheological temperature contrast is equal to its maximum possible value, the difference between the asthenosphere and the free surface. This defines a maximum. The author represents a plastic rheology with $n = 24(T\eta = 50\,K)$. The (top) scale for temperature at the top of the rheological boundary layer applies to the present Earth. The mush-ocean curve (thin solid line) is purely schematic and applies when the interior of the Earth was much hotter than today. The author draws the curve with weak dependence of heat flow on rheological temperature contrast. No stagnant-lid branch exists; the temperature contrast cannot be lower than that between upwelling mantle and nearly frozen mush (point M).

active boundary layer thickness ΔZ_{rheo} dimensionally:

$$V = \varepsilon' \Delta T_{\text{rheo}} \left[\frac{\partial T}{\partial Z}\right]^{-1}$$
$$= T_{\text{rheo}} \frac{1}{T'\eta} \left[\left(\frac{\rho g \alpha \Delta T_{\text{rheo}}^2}{T'}\right) + a\left(\frac{(\rho g \alpha \Delta T_{\text{rheo}}^2)^n}{T'^n \tau_{\text{ref}}^{n-1}}\right)\right]$$
[18]

where $T' \equiv \partial T/\partial Z$ compacts notation. The convective heat flow is dimensionally $\rho C V \Delta T_{\text{rheo}}$. Finally, the uppermost part of the rheologically active boundary layer must deform for flow to occur. The effective value of viscosity-resisting flow is that where the temperature is about ΔT_{rheo} below that in the underlying half-space. Applying the viscosity equation [15] yields an equation for the convective heat flow

$$q = \rho C \varepsilon' \Delta T_{\text{rheo}}^2 \left[\frac{\partial T}{\partial Z}\right]^{-1}$$
$$= \Delta T_{\text{rheo}}^2 \frac{\rho C}{T'\eta_0} \left[\left(\frac{\rho g \alpha \Delta T_{\text{rheo}}^2}{T'}\right) + a\left(\frac{(\rho g \alpha \Delta T_{\text{rheo}}^2)^n}{T'^n \tau_{\text{ref}}^{n-1}}\right)\right]$$
$$\times \exp\left[\frac{-\Delta T_{\text{rheo}}}{T_\eta}\right]$$
[19]

where η_0 is the viscosity of the underlying half-space. The expression is at a maximum when $\partial q/\partial T_{\text{rheo}} = 0$. We recover results for power-law rheology by retaining only the a term in [19]. One recovers the $n+1$ dependence in [14]; the factor 1.2 is obtained from experiments. One obtains [13] with the $1/(n+2)$ exponent by setting the convective heat flow in [19] equal to the conductive heat flow $q = k\partial T/\partial Z$ and solving for the temperature gradient.

For application to the Earth, two maxima occur in [19] if the first term dominates at small stresses and small values of the rheological temperature contrast (**Figure 5**). The first maximum represents stagnant-lid convection within a fluid with linear rheology. The second high-stress maximum might represent other situations.

Figure 5 shows a case where the second maximum is within the lithosphere. This represents conditions for delamination of the base of the lithosphere. Ordinarily, the stagnant-lid mode is stable. From [17], its small rheological temperature contrast ΔT_{rheo} implies stresses that are much less than those of the delamination mode. As the rate of nonlinear creep depends on the rheological temperature contrast raised to a high power the upper part of the lithosphere stays rigid.

Delamination requires a large stress perturbation scaling to the stress implied by its ΔT_{rheo} in [17]. In the Earth, this stress could arise from the impingement of a mantle plume, continental collision, low-angle subduction, or continental breakup.

The upper boundary condition on the lithospheric mantle affects delamination. In particular, the hot base of over-thickened continental crust acts as a free surface (Appendix 2). In terms of the heat-flow equation [13], the dimensionless constant A_n is larger for a free surface boundary than a rigid boundary. In terms of [18], the flow velocity is higher if the upper boundary slips freely than if this boundary is rigid. With regard to **Figure 5**, the presence of fluid continental crust affects delamination if the temperature at its base is within the delamination maximum.

Plates are rheological boundary layers extending all the way to the free surface of the Earth. The rheological temperature contrast is thus the full temperature contrast across the lithosphere. On the Earth, existing plate tectonics maintains the high stresses needed to nucleate new plate boundaries as old ones are consumed by continental and ridge–trench collisions. The plate tectonic maximum in **Figure 5** is the intersection of the heat-flow curve with the temperature of the free surface. The author's statements about the transition of stagnant-lid convection to delamination also apply to its transition to plates. Stagnant-lid convection may not produce large stresses in a one-plate planet, making transition to plate tectonics difficult. There is no evidence of it occurring in the geological records of the planets we have observed. However, Solomatov (2004) shows with dimensional and numerical models that this transition may occur under realistic situations.

The transition of mush ocean to plate tectonics is also relevant to the Earth (Appendix 2). **Figure 5** illustrates a time when the Earth's interior was hot enough to maintain a mush ocean. As drawn, maxima correspond to mush-ocean convection where the base of basaltic crust founders and plate motion where the entire crust founders. The rheological temperature contrast is greater than the contrast between solid upwelling mantle and nearly frozen mush. The heat flow is weakly dependent on the rheological temperature contrast. The mush ocean may have behaved as a mixture of these modes.

6.2.5 Pressure-Release Melting at Mid-Oceanic Ridges

Upwelling mantle melts beneath mid-oceanic ridges. The melt ascends and freezes to form the basaltic oceanic crust. Both the basaltic crust and the depleted residual mantle are less dense than the melt-source region from which they differentiated. These density changes are large enough to affect plate dynamics. Lithosphere formed from hot mantle is hard to subduct (e.g., Sleep and Windley, 1982; Davies, 1992; van Thienen *et al.*, 2004).

Following Klein and Langmuir (1987), McKenzie and Bickle (1988), Plank and Langmuir (1992), and Klein (2003), we present a short discussion of how this process affects plate motion. Crudely, the Earth's mantle consists of olivine, orthopyroxene, clinopyroxene, and an alumina-bearing phase. Garnet, a dense mineral, bears the Al_2O_3 at mantle depths below ~100 km. At crustal depths, low-density feldspar contains most of the Al_2O_3. Mid-oceanic ridge basalt that has a near surface density of ~3000 kg m^{-3} is eclogite below 150- km depth at the interior temperature of the mantle with a density of 3600 kg m^{-3} at current plume temperatures (Sobolev *et al.*, 2005). The stability field of eclogite is temperature dependent; subducted basalt begins to form eclogite at ~30 km depth. The density difference of basalt with respect to eclogite is significant. For comparison, the volume thermal contraction coefficient of mantle is 3×10^{-5} K^{-1}; a temperature change of 1000 K produces a density change of only ~100 kg m^{-3}.

Melting occurs as pressure decreases within the mantle upwelling beneath ridges. Eutectic melting illustrates the mass balance at low melt fractions although it is an inadequate representation of melt chemistry. In addition, clinopyroxene and the alumina-bearing phase are exhausted before olivine and orthopyroxene; further temperature increases after ~50% melting result in modest amounts of additional melting.

For additional simplicity we let the melting temperature be a linear function of depth:

$$T_m = T_0 + \beta Z \qquad [20]$$

where T_0 is the melting temperature at the surface and β is the gradient of melting temperature with depth Z. The temperature of the upwelling material in the absence of melting is

$$T_a = T_p + \gamma Z \qquad [21]$$

where T_p is potential temperature and γ is the adiabatic gradient. Melting begins at the depth where the curves intersect:

$$Z_m = \frac{T_p - T_0}{\beta - \gamma} \qquad [22]$$

The temperature above this depth lies on the melting curve. The specific heat for cooling the material from the solid adiabat equals the latent heat for partial melting:

$$\rho C(T_a - T_m) = LM \qquad [23]$$

where L is the latent heat per volume and M is the fraction of partial melt. Solving for the fraction of melt gives

$$M = \frac{\rho C}{L} \left[T_p - T_0 - (\beta - \gamma) Z \right] \qquad [24]$$

The total thickness of melt production is crudely that in a column that reaches the surface

$$\Phi = \int_0^Z M \, dZ = \frac{\rho C (T_p - T_0)^2}{2L(\beta - \gamma)} \qquad [25]$$

The actual mantle is not precisely eutectic. The heat-balance equation becomes

$$\rho C(T_a - T_m(M, Z)) = LM \qquad [26]$$

where the melting temperature is a function of depth and fraction of melt. One solves [26] numerically and integrates the fraction of melt upward (as $M/(1-M)$ to correct geometrically for compaction of the matrix). For a ridge axis, the computed thickness of melt equals the depth to the base of the crust. One truncates the integral at that depth. One truncates at the base of the lithosphere for mid-plate melting.

We calibrate the eutectic model for quick calculations so that it represents melt fractions for the actual melting in the present Earth, $(\beta - \gamma)$ is 3.2 K km^{-1} and $L/\rho C$ is about 840 K. The depth of extensive melting Z_m is 56 km. These parameters give 6 km of average oceanic crust. The temperature change per thickness of crust is

$$\frac{\partial T_p}{\partial \Phi} = \frac{L(\beta - \gamma)}{\rho C (T_p - T_0)} = \frac{L}{\rho C Z_m} \qquad [27]$$

This expression yields the expected value of 15 K km^{-1}.

These considerations imply that oceanic crust was much thicker in the past when the interior of the Earth was hotter (**Figure 6**). Consider a time in the past when the mantle was 200 K hotter than now. The oceanic crust was ~14 km thicker than now or 20 km thick. It everywhere resembled the thick hot-spot crust beneath Iceland. This buoyant crust did not easily subduct until cooling had progressed downward into the mantle lithosphere. Present-day oceanic crust, such as remnants of the Farallon

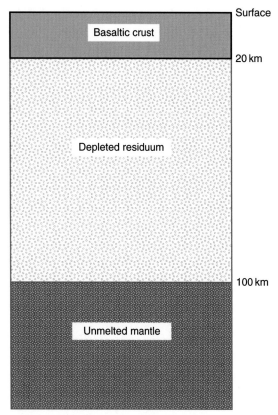

Figure 6 Section through oceanic lithosphere where upwelling mantle starts melting at 100 km depth. This lithosphere is difficult to subduct. Both the crust and depleted residuum are buoyant relative to the underlying mantle. Unlike basaltic mush, the depleted residuum is quite viscous and does not internally convect.

Plate (Atwater, 1970; Wilson *et al.*, 2005), became difficult to subduct at 5–10 My age. As cooling time depends on the square of the depth, 20-km-thick crust would be difficult to subduct at $(20/6)^2 = 11$ times these ages or 55–110 My.

Plate tectonics was even less efficient in subducting thicker crust when the mantle was hotter than in the example. This implies that convection within the partially molten mush carried heat to the surface. We consider this epoch in the next section. Davies discusses an alternative hypothesis for efficient plate tectonic on early Earth. The basaltic component separated from the olivine-rich and accumulated at the base of the mantle. During this epoch, the oceanic crust was much thinner than that in the author's discussion that assumes that the major element composition of the mantle rising at midoceanic ridges did not vary over time.

6.3 Aftermath of the Moon-Forming Impact

The violent birth of the present Earth makes for simple physics (Sleep *et al.*, 2001; *see* Chapters 4 and 1). A Mars-sized body collided with a Venus-sized body at ~4.5 Ga, leaving the Earth–Moon system. The core of the projectile merged with the target's core. Most of new Earth's mantle was hot enough to vaporize at upper-mantle pressures. The vapor radiated heat, quickly forming silicate cloud tops at ~2300 K. The photosphere radiated at this temperature for ~2000 years. During this time, all mantle material circulated through the cloud tops. The Earth lost far more heat during this epoch than during its subsequent history. At the end of this epoch, material arriving at the surface was mostly molten and silicate clouds were only a transient local phenomenon. In the next few 100 years, surface regions cooled so that some solidification occurred around 2000 K. A global solid crust formed at ~1500 K. The water and CO_2 atmosphere then became opaque. The surface heat flow dropped to that of a runaway greenhouse, ~150 W m^{-2}, enough to freeze the mantle and cool it by a few hundred kelvins in a couple of million years. Tidal dissipation from the Moon prolonged this interval by over 3 million years (Zahnle *et al.*, 2007). Thereafter, sunlight and the concentration of greenhouse gases controlled climate with liquid water at the surface. A modest ~200°C surface greenhouse may have persisted until carbonate formation and crustal overturn sequestered the CO_2 in the subsurface.

Asteroids bombarded the early Earth. The precise timing of the impacts is unclear. Current lunar data are compatible with a relatively quiescent period from 4.4 to 4.0 Ga followed by 'late heavy bombardment' between 3.8 and 4.0 Ga (see Ryder (2002), Zahnle and Sleep (2006), Zahnle *et al.* (2007)). Extrapolating from the Moon constrains the size and total number of objects that hit the Earth. Over 100 of them were greater than 100 km in diameter. About 16 were larger than the largest (200 km diameter) object that hit the Moon. A few (0–4) may have been big enough (>300 km diameter) to boil the ocean to the point that only thermophile organisms survived in the subsurface. These impacts on Earth brought brief returns of rock vapor and steam greenhouses.

The effect of impact effect on tectonics, however, was modest. The Moon, which was subjected to the same bombardment, retains its upper crust over much of its surface. Cavosie *et al.* (2004) saw no evidence of impacts in their sample of detrital zircons going back to ∼4.4 Ga. This includes zircons with metamorphic rims, which might show annealed shock features. Conversely, the durations of the rock vapor atmosphere and the hot greenhouse were geologically brief and unlikely to preserve rocks. The existence of 4.4 Ga detrital zircons and ∼4.5 Ga continent-like geochemical reservoirs (Harrison *et al.*, 2005) do not bear on the reality of these epochs.

6.3.1 Early Mush Ocean

The liquid magma ocean ended about the same time that the surface froze (Sleep *et al.*, 2001; Zahnle *et al.*, 2007; *see also* Chapters 4 and 1). Modern fast ridge axes provide analogy. Magma entering the axis freezes quickly (Sinton and Detrick, 1992). The bulk of the 'magma' chamber is mostly crystalline mush. Only a thin (tens of meters) lens of fully molten rock exists at the axis.

These observations show that molten mafic bodies quickly freeze to mush. They provide quantification on the rate that liquid convection within the chamber delivers heat to the surface. The parametrized convection equation [8] for a linear rheology of basaltic melt applies. Current ridge axes provide a minimum estimate for the convective heat flow from the liquid magma ocean. Molten material of mantle composition is orders of magnitude less viscous than modern basalt (Liebske *et al.*, 2005).

Estimating the heat flow through a modern magma lens at the ridge axes driven by latent heat is straightforward. We balance the heat per time per length of ridge with the heat flow per length of ridge axis out of the width of the magma lens. This yields that

$$q = \frac{L Z_c v}{W} \quad [28]$$

where $L = {\sim}1.5 \times 10^9$ J m^{-3} is the latent heat of freezing, W is the width of the magma lens, Z_c is magma chamber thickness, and v is the full spreading rate (Sleep *et al.*, 2001). The fastest spreading centers have the highest heat flow. We retain parameters from the work of Sleep *et al.* (2001). The highest full spreading rate on the modern Earth is 155 mm yr^{-1}, magma chamber thickness is ∼5 km, and the lens width is 1 km. These quantities yield a heat flow of

40 W m^{-2}. The latent heat of the entire mantle could maintain this heat flow only 3 My.

This implies that the fully molten magma ocean was short-lived on the Earth. The fully molten ocean could well have cooled to mush during the runaway greenhouse. At the time the magma ocean froze to mush, it was only a few hundred kelvin hotter than the modern mantle. This fossil heat and radioactivity drove tectonics for the subsequent ∼4.5 billion years.

Like a modern ridge axis, convection in the underlying solid mantle limited heat transfer through the mush ocean. This process has many aspects of stagnant-lid convection (Appendix 2). The viscosity of the solid mantle changes greatly with cooling, controlling the vigor of convection. However, magma from the underlying solid convection sees the floor of the mush chamber as a nearly fluid permeable interface. The multiplicative constant in [9] is thus larger than the multiplicative constant for convection beneath a rigid stagnant lid. The relevant temperature difference driving convection in [9] scales to the difference between the liquidus and the solidus. This quantity decreased modestly as the mush ocean cooled. The crustal lid of the mush ocean foundered and deformed, grossly like plates.

The heat-balance equation [1], the convection equation [9], and the strong dependence of viscosity on temperature [12] imply quick cooling at high temperature and low viscosity followed by slower cooling (**Figure 3**). The Earth transitions from its earliest history where radioactive heat generation is a trivial item in the heat budget to the present status where radioactive heat generation is in crude steady state with surface heat flow.

6.3.2 Steady-State Mush Ocean and Transition to Plate Tectonics

As noted in Section 6.2.1, radioactive heat generation in the Late Hadean was comparable to the modern global average heat flow ∼0.07 W m^{-2}. In the models, the heat flow is ∼0.13 W m^{-2} at 4.0 Ga. Global tectonic rates of $(0.13/0.7)^2 = 3.4$ times present could have vented this heat. This radioactive heat production is low enough to permit a significant hiatus in tectonic vigor at the time of transition between the mush ocean and plates. The radioactive heat generation would increase mantle temperature ∼42 K in 100 My in the absence of heat flow from the core and out of the surface. For example, a total hiatus of 250 My would warm the mantle by only 100 K.

In terms of eqn [1], the heat flow and the interior temperature need not have decreased monotonically with time (Sleep, 2000; Stevenson, 2003). The surface heat flow as a function of temperature may be multi-valued, here with branches for plate tectonics and mush ocean (**Figure 1**). The mush ocean could well have overcooled the mantle (**Figures 2 and 3**). Hot buoyant material rose to the top of the mantle, cooling near the surface of the mush ocean. The cooled material foundered, eventually filling the mantle from bottom up. The mush ocean ceased when the cooled material filled the mantle (point A to point B, **Figure 1**). Sluggish subsequent plate tectonics persisted for several hundred million years until radioactivity heated up the mantle enough to restart the mush ocean (point C to point D). The heat flow and interior temperature looped over points ABCD until the heat flow from plates (point B) was less than radioactive heat generation.

As noted below, the brief mush-ocean episodes could be periods of rapid continent generation because hydrated crust subducts into basaltic mush. The ascending water causes the mush to melt extensively producing a granitic (*sensu latu*) magma.

An alternative possibility is that the mush ocean transitioned monotonically to sluggish plate motions inhibited by thick oceanic crust as discussed in Section 6.2.5 (Korenaga, 2006). Plate rates were slower than present. Mush ocean and plates prevailed in different regions. The mantle temperature remained at the transition (**Figure 3**) and the heat flow remained between the plate and mush-ocean values (**Figure 2**). Eventually, radioactive heat production decreased to where plate tectonics could vent it. Plate tectonics thereafter cooled the Earth.

The transition as drawn is discontinuous. Any strong dependence of global heat flow on interior temperature will trap the mode of convection in transition until radioactive heating wanes enough to be balanced by the less-vigorous mode. In the early days of plate tectonics, it was thought that the dependence of heat flow on interior temperature was so strong that the Earth achieved its present interior temperature soon after it accreted (Tozer, 1970). This concept applies in the monotonic model over much of the early history of the Earth, though not now.

We apply Section 6.2.5 to put the above heat balance inferences into a petrological context. Conditions for maintaining the mush ocean in steady state are relevant to the causes of the mush ocean's eventual demise. The mush ocean consists of a frozen crust with some continental material, melt lens where ascending magma ponds and freezes to mush, a thick layer of mush, and the underlying mantle (**Figure 7**). As already noted, solid-state convection in the mantle limits the rate of heat transfer (Appendix 2).

The base of the mush acts as a permeable interface. Ascending mantle melts. Dikes ascend through the mush and into a magma lens. Some of the magma freezes to mush. Some freezes to form basaltic crust. The crustal volume stays in steady state; cool thick regions of the crust founder into the mush in a process like subduction. The dense base of the mush and foundered crust sink into the mantle, balancing the magmas injected from the mantle.

We focus on the requirement that the supply of melt equals the return of melt to the mantle. Both processes depend on the temperature of the upwelling mantle. If the mantle is hot enough, the thickness of the mush lens self-organizes to a steady state.

First the frozen mush at the base of the ocean needs to be dense enough that it sinks into the mantle. Early in the Earth's history, the mantle was hot enough that upwelling material melted completely. The melt composition hence was mantle composition. Cool material was denser than hot material. Mush sank once frozen.

Later on, the mantle cooled to a temperature where basalt was the most voluminous melt. The cooled basaltic crust has about the same composition as the underlying mush. It can subduct and founder into the mush ocean. However, as already noted, basaltic material and its depleted residuum are less dense than unmelted mantle at shallow depths. However, the dense mineral garnet makes basaltic composition heavier below ~150 km depth. (Sobolev *et al.* (2005) discuss melting of the ecologite component in plumes.) The base of the mush ocean sinks into the mantle if it is deeper than that depth, but not if it is shallower. This feature tends to prevent the base from getting deeper than ~150 km depth.

Second, the balance between recharge from basalt from the upwelling mantle and subduction of slabs into the mantle determines whether the mush chamber thickens or thins. The models assume that the chamber (in the plate mode) thickens once the mantle becomes hotter than 1600°C, the potential temperature where significant melting starts at 150 km depth in [22]. The melt flux needs only to replenish the basaltic material that sinks back into the mantle. It does not have to be hot enough to generate a 150 km thickness of melt in [25].

From this reasoning, the mush chamber becomes unstable to depletion once the mantle temperature

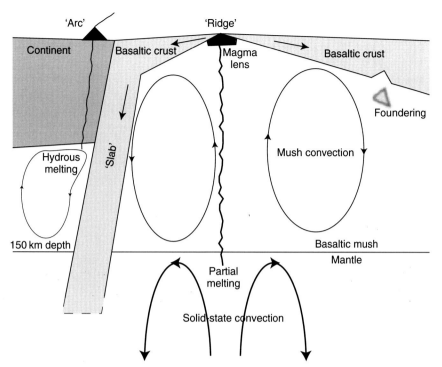

Figure 7 Schematic diagram of mush ocean with vertical exaggeration. Analogs to plate tectonic features are in quotes. Upwelling mantle partially melts. Melt ascends into the magma lens at the ridge axis and freezes, making basaltic crust and basaltic mush. Basaltic crust both subducts as slabs and founders as blocks. Partial melting of hydrated basalt produces arcs and continents. Convection within the mush transfers additional heat from the mantle. Slabs, entrained mush, and foundered blocks return basaltic material to the mantle. Thick lithosphere composed of depleted mantle residuum stabilized some continental regions (not shown).

drops below that needed for recharge. The rounded potential temperature where this happens in the models is 1500°C, where melting in [22] begins at 118 km depth. In the nonmonotonic model (**Figure 1**), convection jumps to plate tectonics. The remaining mush freezes to lithosphere and the foundering regions of crust become slabs. At this time, the ascending mantle generates only 20 km of oceanic crust in the linearized model (**Figure 6**). This is thick enough that plate tectonics cannot vent the heat generated by radioactivity (**Figure 2**). The mantle heats up until enough ascending material melts for the mush ocean to restart (**Figure 3**).

Alternatively, as in the monotonic model, convection becomes sluggish with only local and transient regions of mush ocean, as the mantle approaches the critical temperature to maintain the mush ocean. The convective heat flux balances radioactive heat generation during a long transition period that lasts until after the end of the Archean.

It is obvious that the mush ocean as envisioned provides an efficient mechanism for generating continental crust at arcs. Water from the slab enters hot gabbroic mush, rather than mantle peridotite in standard subduction. The melt is granitic (*sensu latu*). Its detailed geochemistry is beyond the scope of this chapter.

6.3.3 Zircon Evidence

As discussed in the last section, Earth scientists know too little about the material properties of the Earth to make a more quantitative prediction of the vigor of a mush ocean or the timing of its demise. The crust of Mars may provide some analogy with future exploration. Right now the mush ocean is the first epoch of Earth's interior with even a meager geological record. Geochemists have extracted detrital zircon crystals from Archean and Proterozoic quartzites in Australia (e.g., Cavosie *et al.*, 2004; Dunn *et al.*, 2005). The igneous ages of oldest crystals are ~4.4 Ga, a time when the mush ocean might have persisted. The old zircons resemble those from modern granites (*sensu latu*) (Cavosie *et al.*, 2004; and references therein). In particular, some of the granites formed at the expense of sediments derived from meteoric weathering.

Isotopic studies of 4.01–4.37 Ga zircons provide evidence of an older separation of the parent element Lu from the daughter element Hf in the Earth's interior. Within the resolution of a few million years after the moon-forming impact, fractionated continental-type crust with elevated Lu/Hf covered part of the Earth's surface (Harrison *et al.*, 2005). Its depleted low-Lu/Hf complement comprised a significant mantle reservoir.

These data imply some form of crustal overturn analogous to modern subduction. Hydrous mafic rocks either melted directly or dehydrated causing the surrounding hot mush to partially melt. The surface needed to be cool enough for liquid water; a warm greenhouse at ~200°C suffices as hydrous minerals are stable at black-smoker temperatures of 350°C. The hydrated rocks needed to founder. Getting buried by subsequent lava flows or more traditional subduction suffices. They then needed to be heated, no problem given a mush ocean; that is, the typical magmatic rocks associated with subduction are not strong indications of the modern process, especially when they are out of context in detrital zircons.

Cavosie *et al.* (2004) present a constraint on the vigor of crustal recycling processes after 4.4 Ga. Age gaps and clusters exist within their sample suite, just like with a modern orogenic belt (**Figure 8**). If further sampling confirms this finding, terrane-scale regions experienced tens of million-year periods of quiescence. This extends the result of Burke and Kidd (1978) that the base of Archean continental crust did not undergo continuous melting to form granites into deep time. Thermally at ~4.4 Ga, mantle lithosphere existed at least locally beneath continent-like crust. This is consistent with a sluggish mush ocean or with plate tectonics. There is no evidence of any global hiatus in tectonic vigor.

The gaps between events between 4.4 and 3.8 Ga are 50–100 My compared with 500–1000 My for the subsequent history, including modern zircon suites. Taken at face value, crustal recycling rates were ~10 times the present. Sleep and Zahnle (2001), for example, used this rate for modeling Hadean geochemical cycles. The heat flow was ~3 times the present (0.21 W m^{-2}) implying a cooling rate of 280 K per billion year for the radioactive heat generation used above. The lithosphere from [3] was about 40 km thick, enough in analogy to the modern Earth to have plate-like characteristics. This cooling rate is about the limit that could be sustained over 2 billion years until the end of the Archean with the available heat. As already noted, much of the heat probably came out early when the interior was still quite hot.

Alternatively, the gaps and clustering in the ancient zircon suites are analogous with those that exist today on an active continental margin. In this case, the ancient and modern gaps are similar, tens of million years. The crustal sources of the Hadean zircons formed within tectonically active areas. It took time for there to be enough continental crust that stable continental interiors were kinematically

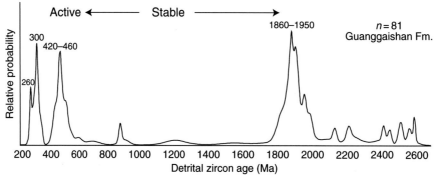

Figure 8 Age distribution of detrital zircons from the Guanggaishan formation of Ladianian (Triassic) age in eastern China illustrates problems in obtaining the age of Hadean tectonics from ancient populations. The sample of 81 ages resolves four peaks. The source region was active before the deposition age of ~228 Ma and stable for a billion-year interval before then. It is not known whether Hadean source regions represent the relatively stable or the relatively active regions at that time. It is inappropriate to infer global events from this Chinese locality and quite premature to infer global events from a few zircon populations from the early Earth. Data from Weislogel AL, Graham SA, Chang EZ, Wooden JL, Gehrels GE, and Yang H (2006) Detrital zircon provenance of the Late Triassic Songpan - Ganzi complex: Sedimentary record of collision of the North and South China blocks. *Geology* 34(2): 97–100.

possible. In any case, tectonics after 4.4 Ga were sluggish enough that some Hadean continental crust survived to at least the end of the Archean.

6.4 Dawn of Plate Tectonics

It is not clear when modern plate tectonics started. We see none of the Earth's original crust and when we do get zircon samples at ~4.4 Ga, the evidence is compatible with plate-like processes. We see no evidence for any period where the surface of the Earth was a one-plate planet like Mars. The viable candidates are a mush ocean with significant regions of continental crust and lithosphere and plate tectonics with solid lithosphere above solid mantle. As noted already, the transition between these two modes may have well been protracted, regional, and gradual. Various aspects of modern tectonics may have begun at different times.

The widespread occurrence of vertical tectonics in the Archean does not preclude plate processes. The author's discussion follows comments by Sleep (2005) and Condie and Benn (2006). As already noted, vertical tectonics occur over broad swaths of the present Earth. Second, the Earth's interior was obviously hotter in the Archean than at present. Volcanic sequences tended to be thicker and the granitic crust was more easily mobilized by mafic and ultramafic intrusions. It is likely that some continental areas, like the Minto province in Canada, formed from remobilized oceanic plateaus (Bédard et al., 2003). Alternatively, the Minto basaltic lithosphere may be a remnant of a mush-ocean plate.

Van Kranendonk et al. (2004) discuss the eastern Pilbara craton. A prolonged period of crustal formation occurred from 3.47 to 3.24 Ga. Notably supracrustal rocks on the domes are not extensively metamorphosed. This indicates that the surrounding greenstone belts subsided and that the tops of the domes were never deeply buried. It is incompatible with assemblage as a series of alpine-style thrust sheets. An analogous situation exists in the 2.7 Ga Belingwe Belt in Zimbabwe. Sediments on the flanks of the belt are weakly metamorphosed (Grassineau et al., 2005). This is evidence that the center of the belt subsided in situ without deeply burying its flanks. It is evidence against this greenstone belt being a 6-km-thick allochthonous oceanic plateau as proposed by Hofmann and Kusky (2004).

Careful work indicates that horizontal tectonics and vertical tectonics coexisted in the Archean.

Hickman (2004) shows that vertical tectonics prevailed in the eastern Pilbara craton before 2.9 Ga, while horizontal tectonics are evident in the western Pilbara craton. Lin (2005) demonstrated the formation of a greenstone belt as the keel of a strike-slip deformation zone in the Superior Province.

We continue by reviewing evidence for the earliest documentation of various plate tectonic features. To do this, we need to consider both preservation bias and mechanics. Overall, we see plate-like features as soon as we have a record that could reasonably provide them.

6.4.1 Seafloor Spreading and Oceanic Crust

Almost all oceanic crust eventually subducts into the mantle. The tidbit that survives provides a biased sample. Parautochthonous ophiolites from back-arc, forearc, and breakup margins start adjacent to continental crust. These rocks may well show continental chemical affinities (Hofmann and Kusky, 2004). Thin oceanic crust formed during cool slow breakups is most likely to stay intact. Conversely, we are not likely to find intact 'normal' Archean oceanic crust if it was in fact as thick as modern oceanic plateau crust (Moores, 2002; Condie and Benn, 2006).

To bring up another uncertainty, mafic rocks formed away from ridge axes mimic ophiolites in tectonically disrupted exposures. The rift zones of major edifices like Hawaii constitute a serious problem. The rifts spread at a few centimeters per year. In a submarine section, pillow basalts overlie dikes and then gabbro (e.g., Okubo et al., 1997). Subaerial pahoehoe may even be difficult to distinguish from submarine pillows at highly deformed and metamorphosed outcrops. Cover tectonics on the flanks of edifices resemble foreland thrusts when viewed locally (e.g., Morgan et al., 2003).

There are certainly abundant pillow basalts and other mafic and ultramafic rocks in the Archean. Moores (2002) reviews candidates for oceanic crust. The oldest putative ophiolite is the 2.5 Ga Dongwanzi complex in China (Kusky et al., 2001). Too little is known to appraise this claim. Fumes et al. (2007) pillow basalts, sheet dikes, and gabbros from the 3.8 Ga Issua belt in southwestern Greenland. As with many modern ophiolites, these rocks show arc affinity. This exposure suffices to show that sea-spreading and ridge-axis hydrothermal circulation occurred on the early Earth. However, these rocks provide little constraint on the thickness of ancient oceanic crust as the base of the ophiolite is not exposed.

Blocks of apparent oceanic crust in the Pilbara block are better studied and are older >3.2 Ga candidates (Hickman, 2004; Van Kranendonk *et al.*, 2004). Upper-crustal rocks at Marble Bar are well preserved, providing an early 3.46 Ga example of carbonate formation by deep marine hydrothermal fluids (Nakamura and Kato, 2004). Kitijima *et al.* (2001) studied similar nearby 3.46 Ga oceanic basalts at North Pole, Australia. Using boiling relationships, they obtained that the water depth was comparable to modern ridge axes with a best estimate of 1600 m.

Abyssal peridotites distinguish conventional oceanic lithosphere from lithosphere formed in a mush ocean. This rock type occurs as soon as there is geological record at 3.8 Ga in Greenland (Friend *et al.*, 2002). However, younger examples are not described until the end of the Archean, for example, as part of a putative ophiolite in China (Kusky *et al.*, 2001). This dearth might indicate that conventional oceanic lithosphere was uncommon in the Archean. It might indicate the geometrical difficulties of tectonically exhuming the base of thick oceanic crust. There are cumulate ultramafics in the Archean (e.g., Friend *et al.*, 2002; Furnes *et al.*, 2007) so poor preservation potential once emplaced does not seem likely. There is cause for concern about reporting biases, including lack of recognition, lack of study, and publication in ways not retrievable by online search.

6.4.2 Transform Faults

The strike-slip motion on transform faults obviously dissipates energy but provides no gravitational driving force (Sleep, 1992). Mathematically, strike slip is toroidal motion around a vertical axis. Such motion cannot occur unless rheology is strongly laterally heterogeneous, that is, rigid plates and weak plate boundaries. Polat and Kerrich (2006) summarize known Archean strike-slip faults. We have already noted the difficulty of recognizing strike-slip motion in the presence of associated vertical tectonics (Chen *et al.*, 2001; Lin, 2005).

Continental margin transform faults preserve a record of oceanic tectonics on their outboard sides (Sleep, 1992). Their length provides a minimum limit of plate size. The duration of sense of slip constrains the ratio of plate size and plate velocity. One could get the relative plate velocity directly with piercing points or with the demise of volcanism as a Mendocino-type triple junction passed. The gross endeavor may be feasible. Atwater (1970) applied these tectonic concepts to explain Tertiary California geology. There are limits to resolution. Coastal California is quite complicated when examined in detail (Wilson *et al.*, 2005).

The Inyoka Fault in southern Africa is the oldest 3.2 Ga well-preserved example. It cuts syntectonic basins of the Moodies group (Huebeck and Lowe, 1994; de Ronde and de Wit, 1994). Like the San Andreas Fault, motion is parallel to the gross strike, piercing points are not recognized in spite of detailed mapping (de Ronde and de Wit, 1994; de Ronde and Kamo, 2000). Motion is associated with detachment faulting (Huebeck and Lowe, 1994; Kisters *et al.*, 2003).

6.4.3 Subduction and Continental Lithosphere

We have already noted that igneous rocks analogous to arc magmas are ubiquitous back to the time ~4.5 Ga of the first hint of a geological record (Harrison *et al.*, 2005; Polat and Kerrich, 2006). We have also noted that foundered crust within a mush ocean is likely to be analogous to the crust in modern slabs. Thrusting and rifting are likely to resemble the modern analog at shallow depths of exhumation as brittle failure is involved. On the other hand, Stern (2005) contends that the lack of high-pressure and ultrahigh-pressure metamorphism on the ancient Earth indicates that modern subduction did not occur.

Subduction, however, merely requires that surface rocks penetrate to great depth, while ultrahigh metamorphism requires that these rocks return to the surface in a recognizable form. The higher temperature in the Archean mantle clearly heated subducted material. Any subducted material that made it back to the surface was likely to be hotter than modern returning material (Condie and Benn, 2006).

We journey downward into deep cratonal lithosphere in search of a better rock record. Lithospheric mantle beneath cratons is less dense than normal mantle. This along with its higher viscosity keeps continental lithosphere from being entrained into the deeper mantle by stagnant-lid convection (Doin *et al.*, 1997; Sleep, 2003). Its great yield strength guards it during continental collisions (Lenardic *et al.*, 2003). The oldest in-place diamond pipe is 2.7 Ga in the southern Superior Province of Canada (O'Neill and Wyman, 2006). Detrital diamonds in the

2.5-Ga Witwatersrand sequence also indicate that thick stable Archean lithosphere had already cooled into the diamond stability field by that time (Nisbet, 1991, p. 61).

The Archean subsurface crystallization age of xenoliths in diamond pipes and diamond inclusions is well known. For present purposes, sulfides in ~2.9-Ga diamond inclusions from South Africa studied by Farquhar *et al.* (2002) are relevant. The sulfides have mass-independent sulfur fractionation. This indicates that this sulfur was exposed to UV radiation in an anoxic (<2 ppm O_2) atmosphere (Pavlov and Kasting, 2002). The sediments and altered volcanic rocks containing this sulfur were subducted into the ~200 km-deep lithosphere. This material remained sequestered until tapped by a kimberlite.

In addition, some deep lithospheric samples provide evidence that their depths of formation were shallower than their depths of sequestration (Jacob *et al.*, 2003; Menzies *et al.*, 2003; Saltzer *et al.*, 2001; Schulze *et al.*, 2003a, 2003b; Canil, 2004). This feature and the chemistry of the samples indicate entrainment to depth in the wedge above the slab. Conversely, they do not require highly elevated temperatures that would generate komatiites and their residuum at ~200 km depth.

6.5 The Rate of Plate Tectonics over Time

The record after 3 Ga becomes increasing interpretable in terms of the rates of plate motions. One would like to deduce the change in the temperature of the Earth's interior over time along with the variation in the evolution of plate motions. In the absence of magnetic lineations, methods grab at straws.

One approach utilizes the longevity of geological features. This has already been discussed in Section 6.3.3 with the clusters and gaps in detrital zircon population ages. Application of this method becomes simpler after 3 Ga in that one can more easily ascribe a tectonic process to an age. The length of the Wilson cycle between passive margin breakup, to arc collision, and final continental collision is attractive. The interval between supercontinent assemblies is a more macroscopic aspect. Mention has already been made to the duration of strike-slip events on active margins. The formation-to-accretion duration of seamounts and the ridge-to-trench lifetime of ophiolites provide direct information on oceanic plates (Moores, 2002).

Korenaga (2006) briefly reviewed this evidence. Additional data since Sleep (1992) discussed tectonic rates have not provided much better constraints. The main difficulty is that plate velocities and the durations of margins, ocean crust, and seamounts are highly variable on the modern Earth. Other than that, rates have not systematically varied by a factor of 2 from current rates over the last 3 Ga; we do not attempt to decipher a trend. We note that Korenaga (2006) contends that rates have actually increased since 3 Ga.

Model-dependent estimates of ancient plate rates come from dynamic considerations. These estimates are coupled with estimates of the interior temperature of the mantle, which we discussed first. There is general agreement that the mantle was hotter at 3 Ga than now. The amount of change is not well resolved. A major difficulty is that the modern source temperature of mantle-derived volcanic rocks is variable, with plume-related material hotter. Taking this into account Abbott *et al.* (1994) and Galer and Mezger (1998) give their preferred temperature decrease of 150 K. A second difficulty is that hydrous magma melts at a lower temperature than anhydrous magma. Grove and Parman (2004) contend that komatiites are hydrous magmas and the mantle was only 100 K hotter. Arndt (2003) notes that this is unlikely with komatiites in general. There has been progress in distinguishing arc from mid-oceanic lavas in the Archean (Polat and Kerrich, 2006). This along with recognition of seamount lavas should help with obtaining a representative source region temperature in the Archean.

Archean cratonal xenoliths provide additional evidence of the temperature in the Archean mantle. Canil (2004) showed that trace elements in his samples are compatible with formation within a subduction zone, but not with formation from a hotter magma at their present lithospheric depth.

The freeboard of the continents provides both a kinematic constraint and a dynamic constraint. Basically, many Archean terranes have never been deeply eroded or deeply buried by sediments once they were stabilized (e.g., Galer and Mezger, 1998). Thus, the elevation of these regions above sea level has not changed much. If the amount of seawater has been relatively constant, the available room in ocean basins determines sea level. Thick Archean crust is more buoyant than modern crust and should cause a sea-level rise. Sea level rises as global continental area increases. Vigorous plate tectonics implies that the oceanic basin is filled with young shallow crust. This causes a sea-level rise. Overall, the data are compatible with plate rates comparable to present since 3 Ga.

Mechanics provide a more reliable version of this constraint that does not involve ocean volume (England and Bickle, 1984). The initial thickness of Archean continental crust is similar to that of young crust. Continental crust is buoyant and would spread laterally like oil over water if it were fluid. A compressional orogeny must overcome this force to form the continent. The available force scales to the ridge push force in [5] and hence to the age of seafloor at subduction t_E. The lack of change in compressional orogenies since 3 Ga indicates that rate of plate overturn has not changed much.

Paleomagnetic data provide constraints on mainly continental plate velocities. For example, Blake *et al.* (2004) obtained precise radiometric ages on strata in the eastern Pilbara craton for which paleolatitudes had been determined. Samples at 2721 ± 4 Ma and 2718 ± 3 Ma differ by 27° or 3000 km. Taken at face value, the drift rate was 1000 mm yr^{-1}. The drift rate was at least 100 mm yr^{-1} within the error of the data. The faster rate is ~10 times modern rates. Sleep and Windley (1982) hypothesized brief episodes of rapid Archean plate motion when old thick lithosphere subducted into the underlying fluid mantle. Available plate size (here at least 3000 km) ended the rapid drift episodes often by continental collisions. Transient mush oceans would have a similar effect. Such periods of rapid drift (if real) need to be included in the global average plate motion.

6.6   Death of Plate Tectonics

We have already shown that the rock vapor atmosphere and the mush ocean ceased when the available energy flux failed to maintain them. Plate tectonics will meet the same fate as the mantle cools. The process, called ridge-lock, involves melt generation at ridge axes. If the interior temperature is lower by 50 K, much less melt is generated, see equation [25]. Solid peridotite rather than a fluid mush chamber and sheet dikes must deform for spreading to occur.

In the model (**Figures 1–3**), the mantle becomes too cool for plate tectonics in the future at −0.9 Ga. The heat flow drops to the stagnant-lid branch (**Figure 1**, point E to point F). The mantle slowly heats up but does not become hot enough to restart plates. At −2 Ga, radioactivity heating is in equilibrium with heat flow. The temperature peaks (**Figure 3**). The mantle cools thereafter without

returning to plate tectonics. Alternatively, the transition is gradual. The mantle stays at the transition temperature until radioactive heat generation and heat-flow balance, again at −2 Ga (**Figure 3**).

Two harbingers of doom for plate tectonics exist on the present Earth. First, full spreading rates are less than 20 mm yr^{-1} along the Arctic spreading center. This rate is slow enough that material cools as it ascends, violating the adiabatic ascent assumption in deriving [25]. Much of the crust is serpentinite formed from hydrated peridotite (Dick *et al.*, 2003). Gabbros freeze at depth. The flow then carries them to the surface. The radiometric ages of such rocks can be millions of years older than the nominal age of the seafloor.

Second, the Australian–Antarctic spreading center produces a 'discordance zone' with thin ocean crust and rugged topography. The crustal thickness and the unusually large depth at the ridge axis indicate that the upwelling mantle is cooler than that at normal ridges. Ritzwoller *et al.* (2003) present tomography showing an excess potential temperature of the upwelling mantle of 100 K relative to normal ridge axes. They compile petrological excess temperatures of 60–150 K. Herzberg *et al.* (2007) review the global variation of magma source temperatures.

Both the Arctic ridge and the Australian–Antarctic discordance are modest segments of boundaries between major plates. Forces from the rest of these plates could well supply the additional driving force for spreading. When the Earth's interior has cooled, both slow spreading and cool upwelling will tend to lock the entire length of all ridge boundaries. Neither Mars nor Venus provides a qualitative analogy at the present level of ignorance.

A second more subtle effect is already occurring (Sleep, 2005). Stagnant-lid convection provides heat flow to platform regions where the deep lithosphere is not chemically buoyant. The lithospheric thickness in these regions from [13] has increased as the mantle cooled. It is now similar to that beneath cratons where chemical buoyancy defines the base of the lithosphere. Further cooling of the mantle will bring the rheological boundary layer in [14] below the zone of chemical buoyancy everywhere on the planet. The stagnant-lid versus chemical-lid distinction between platforms and cratons then will cease to exist.

6.7   Biological Implications

The minimal requirements for the survival of life, energy sources and habitable temperatures, are linked. Nonphotosynthetic life couples a Gibbs energy-producing reaction, like methanogenesis, with an energy consuming reaction, like ADP⇒ATP. High temperatures speed reaction rates, making energy-producing kinetically inhibited reactions less available. They tend to cause decomposition of the organic matter within all life forms, both squandering the energy that the organism has gone to great ends to convert into a useful form and trashing its genetic information.

We follow the discussion of Sleep *et al.* (2001) to put the establishment of habitable conditions into a tectonic context. The Earth cooled to a solar greenhouse within several million years after its formation. At first, over 100 bars of CO_2 maintained temperatures above 200°C. Carbonates were stable within a thin rind at the surface. The volume of the rind, however, was too small to take up a sufficient quantity. At some stage (4.44 Ga in the model), global heat flow waned to $1 \, W \, m^{-2}$. The Earth resembled oceanic crust younger than 1 My, the age t_E in [2] of subduction at that time. The crust was a significant CO_2 sink. Carbonate formation now occurs within the uppermost several 100 m of young oceanic crust. Mass balance indicates that crustal overturn was necessary to sequester the CO_2. Available divalent cations in the uppermost crust could take up only ~1/6 of the global CO_2 inventory. Degassing of foundered crust returned an unknown amount of CO_2 to the surface. With a global resurfacing time of 1 My, 6 My was needed to completely consume the dense CO_2 atmosphere in the absence of any degassing or return flux.

Once the primordial CO_2 was sequestered in the subsurface, ejecta from impacts was an effective CO_2 sink. The climate was freezing until life produced methane, a potent greenhouse gas (Sleep and Zahnle, 2001). Life could well have arrived within Mars rocks. Mars was habitable up to 100 My before the Earth. Martian origin is more testable than Earth origin. The Mars record is intact, just hard to retrieve.

After the demise of the thick CO_2 atmosphere, the Earth was habitable except for brief (10^4 years) periods after asteroid impacts (Sleep and Zahnle, 1998; Zahnle and Sleep, 2006; Zahnle *et al.*, 2007). A few impacts large enough to vaporize much of the ocean are likely. They left only thermophile survivors below 1 km deep in the subsurface. Terrestrial life could well root in these survivors.

The main safety requirement is that multiple areas with thermal gradients less than $100 \, K \, km^{-1}$

existed on the planet. For a hard-rock conductivity of $2.4 \, W \, m^{-1} \, K^{-1}$, this requires a heat flow less than $0.24 \, W \, m^{-2}$. This is the heat flow within older oceanic crust when the average heat flow is $0.48 \, W \, m^{-2}$, occurring at 4.37 Ga in the model. Continent-like regions, however, date from ~4.5 Ga. These low heat-flow regions could well have provided earlier refugia.

Life placed less onerous conditions on tectonics over the next 4 billion years. It mainly demanded that tectonics recycled volatiles and nutrients so that we did not end up like Venus or Mars. Life assisted tectonics through weathering (Rosing *et al.*, 2006). Biology partitioned C and S between the crust and mantle.

Terrestrial organic evolution is a mixed blessing for inferring what happened on other planets. One cannot extrapolate cosmically from the period where we have a good geological record. Life has evolved to be fit for the conditions it actually experienced during the Earth's history. For example, the Earth has the right amount of water so that tectonics and erosion both tend to generate surfaces near sea level (Sleep, 2005). Abundant shallow seas made the transition to land easy. Overall, evolution for fitness to current conditions produces the illusions of design and providence. Any intelligent organism, as a survivor, will find that its personal family tree and its species have had a harrowing history, giving the illusion of good fortune to the point of miracle.

Past performance does not imply future results. Tectonic vigor wanes. Our Sun waxes. Unless a high-tech society intervenes, our greenhouse will runaway because water vapor is a potent greenhouse gas (e.g., Kasting and Catling, 2003). An increase in global sea-surface temperature produces more water vapor and hence a hotter climate. Above a threshold the feedback is unstable. Our life-giving ocean will be our curse.

Tectonics may extricate us from disaster by jettisoning the ocean in the mantle. The mass balance is reasonable. Right now, basalt covers about 90% of the ocean's surface and serpentinite covers the other 10%. Basalt hydrates at ridge axes to form greenstone. This is a modest water sink. An equivalent thickness of 1–2 km is hydrated to 3% water by mass or 10% by volume (Staudigel, 2003; Rüpke *et al.*, 2004). The basalt flux by slabs is equivalent to 100–200 m of ocean depth globally every 170 My. Serpentinite is ~40% water by volume (Rüpke *et al.*, 2004). Its 10% of the seafloor carries an equivalent of 40–80 m every 170 My. In total, it takes 1.5–3.0

billion years to subduct the ocean equivalent thickness (2500 m) of water. However, most of this water returns to the surface in arc volcanoes (Wallace, 2005).

A prolonged period of sluggish seafloor spreading occurs in the models (**Figures 1–3**). The Earth may transition to rather than jump to the stagnant-lid mode of convection. The cool mantle during this epoch implies that much of the seafloor will be serpentinite and that any basaltic crust is highly fractured. The mantle will be too cool for most arcs to be an effective water-return mechanism. The spreading rate at −0.9 Ga in the model is still 83% of the current rate at the transition. A fully serpentinized ocean would subduct the available water in 0.65–1.3 Ga at that rate.

The Earth will resemble the fictional planet *Dune* once the oceans have been subducted. There will still be some groundwater. There will be open water and ice only at the poles. This cold trap will set the global dew point to a low temperature. The meager amount of water vapor will be an ineffective greenhouse gas. Our descendants may well help these processes by pumping water into serpentinite and harvesting H_2 gas from the effluent.

On the other hand, plate tectonics may end suddenly and soon. In that case, erosion will bevel the continents. Some volcanoes and dry land may persist, but not the bounty associated with active tectonics. We will not need a set to remake *Waterworld*.

6.8   Conclusions and Musings

Geologists' view of the ancient Earth and their repertoire of methods have changed beyond recognition since 1963. Geochemistry and geochrononology were mostly aids to mapping rather than powerful quantitative techniques directed at physical processes. The idea that high-temperature metamorphism was exclusively Pre-Cambrian flourished. There was a general hesitancy for considering geology beyond a local scale. Physics did not seem relevant. Geosynclines produced orogenies when and where they were wont to do so.

Geodynamic models of the early Earth were generic in the 1970s. They did not utilize available information including the presence of high-temperature komatiite lavas in the Archean. 'Don't think' signs remained posted around the Archean 'basement complex' until the late 1970s. For example, a senior faculty member criticized the author for pointing out

transposed dikes on a departmental field trip to the Superior Province. These features, when of venerable Archean age, were to be called by the uninformative generic term 'schieren'. Brian Windley visited Northwestern right after this trip. He had much better, more deformed, and far older examples in Greenland that he traced into intact dikes. We began to think about the Archean in a quantitative physical manner.

The high-hanging fruits of years of careful fieldwork with geochronology and geochemistry are now in the basket. Geologists often recognize the basic features of plate processes when they look in the Archean. However, Archean tectonics did differ from the modern process. The continental crust frequently acted as a fluid. Dense mafic rocks foundered into the lower crust. Broad areas like the Minto Block in Canada and the eastern Pilbara craton were dominated by vertical tectonics.

There is enough information to rationally discuss the evolution of tectonic rates. Tiny zircon crystals show that continent-like material formed and survived soon after the moon-forming impact. The statistics of their ages indicate that the events punctuated the local evolution of the crust, just as they do now. The rate of crustal overturn was probably higher than present by as much as a factor of 10. The heat flow may have been as much as a factor of $\sqrt{10} = \sim3$ higher.

By 3 Ga, the interior was less than 200 K hotter than at present. There is more record and it is easier to talk about rates. The answer may not be satisfying. The global rate of plate tectonics was comparable to the present rate. It remained so thereafter.

Chemically buoyant continental lithospheric mantle kept some overlying continental crust out of harm's way. Its basal region sequestered a record of the conditions at the time of its formation that ascends to the surface as xenoliths. Geochemists are now examining this treasure trove. Their work establishes the antiquity of subduction. It is yet to provide tight constraint on the variation of mantle temperature over time.

The simple quantitative models in this chapter provide heat balance constraints and a guide to possibilities. The geological record is a better master. The transition from mush ocean to plate tectonics is complicated. Before the transition, partly molten material at the base of the mush ocean facilitated convection in the underlying mantle and foundering of cooled mafic crust. After the transition, the need for the base of the oceanic crust to freeze hindered subduction. The Earth could have lingered over a billion years at this transition. A semantic discussion about what constitutes plate tectonics is unproductive.

Further quantification from geodynamics requires obtaining the globally averaged heat flow by much better than a factor of 2. We need to know whether global heat flow varies continuously and monotonically; that is, does interior temperature and the heat flow–temperature relationship have branches and sharp transitions? These tasks are not simple. Archean tectonics are more complicated than modern plates or even an early vigorous mush ocean. There is meager preservation of the oceanic record. Continental tectonics often involved fluid-like deformation in the crust. Geodynamicists do not have quantitative understanding of analogous vertical tectonic regions like the Basin and Range. The feedback between global heat flow and continental area has only recently been studied numerically (Lenardic *et al.*, 2005).

Current dynamic models do not even do a good job of representing the spontaneous formation of modern plate boundaries. A damage rheology where faults form and heal is a minimum requirement (Tackley, 2000a, 2000b). In this case, the instantaneous rheology implies an increase in stress for an increase in strain rate. The steady-state rheology is strain-rate strengthening below a critical strain rate and 'yield' stress. It is strain rate weakening above that stress and strain rate. This represents the observation that plate boundaries are weaker than typical rocks failing in friction (Zoback and Townend, 2001). The numerical models will need to nicely represent 'normal' plate boundaries, hot-spot ridges like Iceland, mobile continental crust, arc–continent collisions, oceanic subplate boundaries, and oceanic plateaus before they can be exported to the ancient Earth.

We generalize from the observation that plate-like behavior occurred on the Earth for much of its history to infer that plate tectonics is a viable candidate mechanism on other worlds. Pressure-release melting clearly occurs over an extensive depth range in the low gravity of Mars. The base of the mush ocean would stabilize at ~150 km/0.4 = 375 km depth. The 'room' problem for basaltic crust to subduct into the mush is less severe than on the Earth. However, quantitative geodynamics is not ready for extraterrestrial export to the early active history of Mars and Venus. Rather with exploration,

geodynamics is likely to import calibration from the excellent geological record of Mars.

The Cretaceous–Tertiary asteroid impact linked geology, astronomy, and biology. The conditions following an ocean-boiling impact are even more foreign to the peaceful environments studied by traditional paleontology. Geology now provides a shopping list of feasible conditions for the early Earth. Molecular biology provides eyewitness accounts in the genomes of extant organisms. It has the potential to resurrect information lost from our rock record.

Acknowledgments

Amy Weislogel pointed out and provided her detrital zircon data. This work was supported by NSF grant EAR-0406658. This work was performed as part of collaboration with the NASA Astrobiology Institute Virtual Planetary Laboratory Lead Team.

Appendix 1: Thermal Models

Figures 1–3 represent generalizable features of the Earth's thermal evolution. The calculations represent realizable cases. We made no effort to tweak the model parameters other than adjusting the monotonic model so that the current potential temperature is 1300°C and the current heat flow is 0.07 W m^{-2}. (Petrologists do much better with the relative variations of temperature than they do with the absolute temperature (Putirka, 2005; Herzberg, 2007)). The author uses 1300°C to maintain consistency with his earlier papers.) Heat flow is in W m^{-2} and age B is billion years in the following equations. The radioactive heat production is

$$q_{rad} = 0.035 \cdot 2^{B/2.5} \qquad [29]$$

The rate of temperature change per billion years is 180 K per billion year at a net heat flow of 0.07 W m^{-2}. Mush-ocean heat flow is

$$q = 0.3\exp\left[\frac{T - 1500°C}{150\,K}\right] \qquad [30]$$

This corresponds to $T_\eta = 50$ K. Plate tectonic heat flow is the maximum of

$$q = 0.0934 - 0.0234\left[\frac{T - 1400°C}{100\,K}\right]^2 \qquad [31]$$

and 0.02 W m^{-2}. This provides a weak temperature dependence with a maximum at 1400°C. The extra digits tweaked the model to the assumed current conditions. The stagnant-lid heat flow is

$$q = 0.02\exp\left[\frac{T - 1300°C}{150\,K}\right] \qquad [32]$$

This also corresponds to $T_\eta = 50\,K$. The branch jumps on cooling are 1500°C for mush ocean to plates and 1250°C for plates to stagnant lid. The monotonic model remained at the transition temperature until radioactive heat generation was lower than the heat flow of the cooler mode. The nonmonotonic model heated up in the plate model and jumped to mush ocean at 1600°C. No jump back to plates occurred in this model, as the interior did not heat up much.

Appendix 2: Convection beneath Free-Slip Lid

As envisioned, solid-state convection within the underlying mantle was the process that limited heat flow through the mush ocean. Magma entering the mush ocean froze. The temperature at the base of the mush ocean was thus less than that of the entering magma and of the ambient mantle. The difference between this temperature and the temperature of the ambient mantle drove mantle convection. The difference was probably a few hundred kelvin. This appendix shows that the heat flow was insensitive to this rheological temperature contrast ΔT_M. The author defines the dimensionless variable $A \equiv \Delta T_M / T_\eta$ to compact notation.

Here the numerically simple situation of convection beneath a conductive layer with a free-slip bottom as a proxy for a complicated mush chamber is used (**Figure 9**). The numerical method is discusssed at the end of this appendix. From [8], the heat flow should scale to $A^{4/3}$ when the temperature contrast is smaller than 1. The stagnant-lid heat flow in [8] provides a natural scaling. This normalized heat flow may be greater than 1 as the upper part of the rheological boundary layer flows efficiently. The eyeball curve in the form of the linear rheology part of [18] is

$$\left[\frac{q_{CL}}{q_{SL}}\right]^{3/4} = \left[\frac{A}{3}\right]\exp[(4.5 - A)/3.8] \qquad [33]$$

The heat flow reaches a maximum at a normalized rheological temperature contrast of ~3.5. The

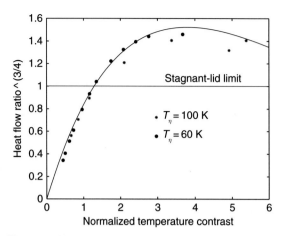

Figure 9   Numerical results for convection beneath a free-slip conducting lid in normalized heat flow and normal rheological temperature contrast A. The author varied the rheological temperature contrast to show that the scaling relationship applies.

transition to ordinary stagnant-lid convection occurs at higher values. Solomatov and Moresi (2000) give the transition at a normalized contrast of 8 for convection beneath a free-slip isothermal boundary. The curve [33] extrapolates to 1 at 8.5, compatible with this constraint.

The heat flow is slightly greater than the stagnant-lid limit over a broad range of rheological temperature contrast. One thus does not have to know the details of the gabbroic mush to compute the stagnant-lid heat flow. One does have to add the contribution of subduction of basaltic slabs into the mantle. Both the mush-ocean parametrization (29) and the stagnant-lid parametrization (31) have the form of [8] with a ratio of 3.95, rather than ~1.5 in [33]; that is, we have implicitly included a significant slab contribution in the parametrized model.

The author uses two-dimensional (2-D) thermal models to find the scaling relationship [33] using the stream-function code by Sleep (2002, 2003, 2005). We impose a conducting lid from 197.5 km depth to the surface where it was 0°C. The models have a 5 km grid. We define temperature at integers times 5 km points in depth and horizontal coordinate. We define the stream function at integer times 5 km plus 2.5 km points. This boundary condition naturally applies at stream function points. We apply free-slip boundary conditions where the vertical velocity was zero and the horizontal shear traction zero to represent a lubricated base of the conducting lid.

The domain of the calculation is 900 km wide. We apply free-slip mechanical boundary conditions at each end and no horizontal heat flow as a thermal boundary condition. We applied a permeable bottom boundary condition at 500 km depth to represent a large underlying adiabatic region. Fluid enters the domain at the potential temperature of 1300°C. There is constant pressure at the boundary so that it does no work on the domain of the model. There is also no horizontal velocity at the boundary.

We did not vary well-constrained parameters in our generic models. These include the thermal expansion coefficient $\alpha = 3 \times 10^{-5}\,\text{K}^{-1}$, volume specific heat $\rho C = 4 \times 10^{6}\,\text{J m}^{-3}\,\text{K}^{-1}$, thermal conductivity $k = 3\,\text{W m}^{-1}\,\text{K}^{-1}$, the potential temperature of the mantle adiabat $T_L = 1300°C$, the density $\rho = 3400$ kg m^{-3}, and the acceleration of gravity $9.8\,\text{m s}^{-2}$.

This situation is also relevant to upper boundary layers in a stratified magma chamber and convection within D'' if it is chemically denser and more viscous than the overlying mantle.

References

Abbott DL, Burgess L, Longhi J, and Smith WHF (1994) An empirical thermal history of the Earth's upper mantle. *Journal of Geophysical Research* 99: 13835–13850.

Araki T, Enomoto S, Furuno K, *et al.* (2005) Experimental investigation of geologically produced antineutrinos with KamLAND. *Nature* 436: 499–503.

Arndt N (2003) Komatiites, kimberlites, and boninites. *Journal of Geophysical Research* 108: (doi:10.1029/2002JB002157).

Atwater T (1970) Implications of plate tectonics for Cenozoic tectonic evolution of Western North America. *Geological Society of America Bulletin* 81: 3513–3536.

Bédard JH, Brouillette P, Madore L, and Berclaz A (2003) Archaean cratonization and deformation in the northern Superior Province, Canada: An evaluation of plate tectonic versus vertical tectonic models. *Precambrian Research* 127: 61–87.

Blake TS, Buick R, Brown SJA, and Barley ME (2004) Geochronology of a Late Archaean flood basalt province in the Pilbara Craton, Australia: Constraints on basin evolution, volcanic and sedimentary accumulation, and continental drift rates. *Precambrian Research* 133: 143–173.

Bleeker W (2003) The late Archean record: A puzzle in c. 35 pieces. *Lithos* 71: 99–134.

Burke K and Kidd WSF (1978) Were Archean continental geothermal gradients much steeper than those of today? *Nature* 272: 240–241.

Canil D (2004) Mildly incompatible elements in peridotites and the origins of mantle lithosphere. *Lithos* 77: 375–393.

Cavosie AJ, Wilde SA, Liu D, Weiblen PW, and Valley JW (2004) Internal zoning and U–Th–Pb chemistry of Jack Hills detrital zircons: A mineral record of early Archean to Mesoproterozoic (4348–1576 Ma) magmatism. *Precambrian Research* 135: 251–279.

Chen SF, Witt WK, and Liu SF (2001) Transpression and restraining jogs in the northeastern Yilgarn craton, western Australia. *Precambrian Research* 106: 309–328.

Condie KC and Benn K (2006) Archean geodynamics: Similar to or different from modern geodynamics. In: Benn K, Mareschal J-C, and Condie KC (eds.) *Geophysical Monograph Series 164: Archean Geodynamics and Environments*, pp. 47–59. Washington, DC: American Geophysical Union.

Davaille A and Jaupart C (1993a) Thermal convection in lava lakes. *Geophysical Research Letters* 20: 1827–1830.

Davaille A and Jaupart C (1993b) Transient high-Rayleigh-number thermal convection with large viscosity variations. *Journal of Fluid Mechanics* 253: 141–166.

Davaille A and Jaupart C (1994) The onset of thermal convection in fluids with temperature-dependent viscosity: Application to the oceanic mantle. *Journal of Geophysical Research* 99: 19853–19866.

Davies GF (1992) On the emergence of plate tectonics. *Geology* 20: 963–966.

de Ronde CEJ and de Wit MJ (1994) Tectonic history of the Barberton greenstone belt, South Africa: 490 million years of Archean crustal evolution. *Tectonics* 13: 983–1005.

de Ronde CEJ and Kamo SL (2000) An Archean arc–arc collision event: A short-lived (Ca 3 Myr) episode, Weltevreden area, Barbarton greenstone belt, South Africa. *Journal of African Earth Sciences* 30: 219–248.

Dick HJB, Lin J, and Schouten H (2003) An ultraslow-spreading class of ocean ridge. *Nature* 426: 405–412.

Doin M-P, Fleitout L, and Christensen U (1997) Mantle convection and stability of depleted and undepleted continental lithosphere. *Journal of Geophysical Research* 102: 2771–2787.

Dunn SJ, Nemchin AA, Cawood PA, and Pidgeon RT (2005) Provenance record of the Jack Hills metasedimentary belt: Source of the Earth's oldest zircons. *Precambrian Research* 138: 235–254.

England P and Bickle M (1984) Continental thermal and tectonic regimes during the Archean. *Journal of Geology* 92: 353–367.

Farquhar J, Wing BA, McKeegan KD, Harris JW, Cartigny P, and Thiemens MH (2002) Mass-independent sulfur of inclusions in diamond and sulfur recycling on the early Earth. *Science* 298: 2369–2372.

Fiorentini G, Lissia M, Mantovani F, and Vannucci R (2005) Geo-neutrinos: A new probe of Earth's interior. *Earth and Planetary Science Letters* 238: 235–247.

Friend CRL, Bennett VC, and Nutman AP (2002) Abyssal peridotites >3800 Ma from southern West Greenland: Field relationships, petrography, geochronology, whole-rock and mineral chemistry of dunite and harzburgite inclusions in the Itsaq Gneiss Complex. *Contributions to Mineralogy and Petrology* 143: 71–92.

Furnes H, de Wit M, Staudigel H, Rosing M, and Muehlenbachs K (2007) A vestige of Earth's oldest ophidite. *Science* 315: 1704–1707.

Galer SJG and Mezger K (1998) Metamorphism denudation and sea level in the Archean and cooling of the Earth. *Precambrian Research* 92: 387–412.

Grassineau NV, Abell PWU, Fowler CMR, and Nisbet EG (2005) Distinguishing biological from hydrothermal signatures via sulphur and carbon isotopes. In: Archaean mineralizations at 3.8 and 2.7 Ga. In: McDonald I, Boyce AL, Butler IB, Herrington RJ, and Polya DA (eds.) *Geological Society London, Special Publication 248: Mineral Deposits and Earth Evolution*, pp. 195–212. Bath, UK: Geological Society Publishing House.

Grove TL and Parman SW (2004) Thermal evolution of the Earth as recorded by komatiites. *Earth and Planetary Science Letters* 219: 173–187.

Harrison TM, Blichert-Toft J, Muller W, Albarede F, Holden P, and Mojzsis SJ (2005) Heterogeneous hadean hafnium: Evidence of continental crust at 4.4 to 4.5 Ga. *Science* 310: 1947–1950.

Herzberg C, Asimow PD, Arndt N, *et al.* (2007) Temperatures in ambient mantle and plumes: Constraints from basalts, picrites and komatiites. *Geochemistry Geophysics Geosystems* 8: Q02006.

Hickman AH (2004) Two contrasting granite-greenstone terranes in the Pilbara Craton, Australia: Evidence for vertical and horizontal tectonic regimes prior to 2900. *Precambrian Research* 131: 153–172.

Hofmann A and Kusky TM (2004) The Belingwe Greenstone belt: Ensialic or oceanic? In: Kusky TM (ed.) *Developments in Precambrian Geology, 13: Precambrian Ophiolites and Related Rocks*, pp. 487–538. Amsterdam: Elsevier.

Huebeck C and Lowe DR (1994) Late syndepositional deformation and detachment tectonics in the Barberton Greenstone Belt, South Africa. *Tectonics* 13: 1514–1536.

Hurley PM and Rand JR (1969) Pre-drift continental nuclei. *Science* 164: 1229–1242.

Jacob DE, Schmickler B, and Schulze DJ (2003) Trace element geochemistry of coesite-bearing eclogites from the Roberts Victor kimberlite, Kaapvaal craton. *Lithos* 71: 337–351.

Kasting JF and Catling D (2003) Evolution of a habitable planet. *Annual Review of Astronomy and Astrophysics* 41: 429–463.

Kisters AFM, Stevens G, Dziggel A, and Armstrong RA (2003) Extensional detachment faulting and core-complex formation in the southern Barberton granite-greenstone terrain, South Africa: Evidence for a 3.2 Ga orogenic collapse. *Precambrian Research* 127: 355–378.

Kitijima K, Maruyama S, Utsunomiya S, and Liou JG (2001) Seafloor hydrothermal alteration at an Archaean mid-ocean ridge. *Journal of Metamorphic Geology* 19: 581–597.

Klein EM (2003) Geochemistry of the igneous oceanic crust. In: Holland HD and Turekian KK (eds.) *Treatise on Geochemistry: The Crust,* vol. 3 (Rudnick RL), ch. 3.13, pp. 433–463. Amsterdam: Elsevier.

Klein EM and Langmuir CH (1987) Global correlations of ocean ridge basalt chemistry with axial depth and crustal thickness. *Journal of Geophysical Research* 92: 8089–8115.

Korenaga J (2006) Archean geodynamics and the thermal evolution of the Earth. In: Benn K, Mareschal J-C, and Condie KC (eds.) *Geophysical Monograph Series 164: Archean Geodynamics and Environments*, pp. 7–32. Washington, DC: American Geophysical Union.

Kusky TM, Li JH, and Tucker RD (2001) The archean dongwanzi ophiolite complex, North China craton: 2.505-billion-year-old oceanic crust and mantle. *Science* 292: 1142–1145.

Lenardic A, Moresi LN, Jellinek AM, and Manga M (2005) Continental insulation, mantle cooling, and the surface area of oceans and continents. *Earth and Planetary Science Letters* 234: 317–333.

Lenardic A, Moresi L-N, and Mühlhaus H (2003) Longevity and stability of cratonic lithosphere: Insights from numerical simulations of coupled mantle convection and continental tectonics. *Journal of Geophysical Research* 108: (doi:10.1029/2002JB001859).

Liebske C, Schmickler B, Terasaki H, *et al.* (2005) Viscosity of peridotite liquid up to 13 GPa: Implications for magma ocean viscosities. *Earth and Planetary Science Letters* 240: 589–604.

Lin SF (2005) Synchronous vertical and horizontal tectonism in the Neoarchean: Kinematic evidence from a synclinal keel in the northwestern Superior craton, Canada. *Precambrian Research* 139: 181–194.

McKenzie D and Bickle MJ (1988) The volume and composition of melt generated by extension of the lithosphere. *Journal of Petrology* 29: 625–679.

Menzies AH, Carlson RW, Shirey SB, and Gurney JJ (2003) Re–Os systematics of diamond-bearing eclogites from the Newlands kimberlite. *Lithos* 71: 323–336.

Molnar P and Tapponnier P (1975) Cenozoic tectonics of Asia: Effects of a continental collision. *Science* 189: 419–425.

Moores EM (2002) Pre-1 Ga (pre Rodinian) ophiolites: Their tectonic and environmental implications. *Geological Society of America Bulletin* 114: 80–95.

Morgan JK, Moore GF, and Clague DA (2003) Slope failure and volcanic spreading along the submarine south flank of Kilauea volcano, Hawaii. *Journal of Geophysical Research* 108: 2415 (doi:10.1029/2003JB002411).

Nakamura K and Kato Y (2004) Carbonatization of oceanic crust by the seafloor hydrothermal activity and its significance as a CO_2 sink in the Early Archean. *Geochimica et Cosmochimica Acta* 68: 4595–4618.

Nisbet EG (1991) *Living Earth: A Short History of Life and Its Home*, 237 pp. London: Harper Collins Academic.

Okubo PG, Benz HM, and Chouet BA (1997) Imaging the crustal magma sources beneath Mauna Loa and Kilauea volcanoes, Hawaii. *Geology* 25: 867–870.

O'Neill C and Wyman DA (2006) Geodynamic modeling of Late Archean subduction: Pressure-temperature constraints from greenstone belt diamond deposits. In: Benn K, Mareschal J-C, and Condie KC (eds.) *Geophysical Monograph Series 164: Archean Geodynamics and Environments*, pp. 177–188. Washington, DC: American Geophysical Union.

Pavlov A and Kasting JF (2002) Mass-independent fractionation of sulfur isotopes in Archean sediments: Strong evidence for an anoxic Archean atmosphere. *Astrobiology* 2: 27–41.

Plank T and Langmuir CH (1992) Effects of the melting regime on the composition of the oceanic crust. *Journal of Geophysical Research* 97: 19749–19770.

Polat A and Kerrich R (2006) Reading the geochemical fingerprints of Archean hot subduction volcanic rocks: Evidence for accretion and crustal recycling in a mobile tectonic regime. In: Benn K, Mareschal J-C, and Condie KC (eds.) *Geophysical Monograph Series 164: Archean Geodynamics and Environments*, pp. 189–213. Washington, DC: American Geophysical Union.

Putirka KD (2005) Mantle potential temperatures at Hawaii, Iceland, and the mid-ocean ridge system, as inferred from olivine phenocrysts: Evidence for thermally driven mantle plumes. *Geochemistry Geophysics Geosystems* 6: Q05L08.

Ritzwoller MH, Shapiro NM, and Leahy GM (2003) A resolved mantle anomaly as the cause of the Australian–Antarctic Discordance. *Journal of Geophysical Research* 108: B12, 2559.

Rosing MT, Bird DK, Sleep NH, Glassley W, and Albarede F (2006) The rise of continents – An essay on the geologic consequences of photosynthesis. *Palaeogeography Palaeoclimatology Palaeoecology* 232(2-4): 99–113.

Rüpke LH, Phipps Morgan J, Hort M, and Connolly JAD (2004) Serpentine and the subduction zone water cycle. *Earth and Planetary Science Letters* 223: 17–34.

Ryder G (2002) Mass flux in the ancient Earth–Moon system and benign implications for the origin of life on Earth. *Journal of Geophysical Research* 107(E4): 5022.

Saltzer RL, Chatterjee N, and Grove TL (2001) The spatial distribution of garnets and pyroxenes in mantle peridoties: Pressure-temperaturer history of peridotites from the Kaapvaal craton. *Journal of Petrology* 42: 2215–2229.

Schulze DJ, Harte B, Valley JW, Brenan JM, and Channer DM de R (2003a) Extreme crustal oxygen isotope signatures preserved in coesite in diamond. *Nature* 423: 68–70.

Schulze DJ, Valley JW, Spicuzza MJ, and Channer DM de R (2003b) The oxygen isotope composition of eclogitic and peridotitic garnet xenoliths from the La Ceniza kimberlite, Guaniamo, Venezuela. *International Geology Review* 45: 968–975.

Sinton JM and Detrick RS (1992) Midocean ridge magma chambers. *Journal of Geophysical Research* 97: 197–216.

Sleep NH (1979) Thermal history and degassing of the Earth: Some simple calculations. *Journal of Geology* 87: 671–686.

Sleep NH (1992) Archean plate-tectonics – what can be learned from continental geology. *Canadian Journal of Earth Sciences* 29: 2066–2071.

Sleep NH (2000) Evolution of the mode of convection within terrestrial planets. *Journal of Geophysical Research* 105: 17563–17578.

Sleep NH (2002) Local lithospheric relief associated with fracture zones and ponded plume material. *Geochemistry Geophysics Geosystems* 3: 8506 (doi:2001GC000376).

Sleep NH (2003) Survival of Archean cratonal lithosphere. *Journal of Geophysical Research* 108 (doi:10.1029/2001JB000169).

Sleep NH (2005) Evolution of continental lithosphere. *Annual Review of Earth and Planetary Science* 33: 369–393.

Sleep NH (2006) Strategy for applying neutrino geophysics to the earth sciences including planetary habitability. *Earth Moon and Planet* 99: 343–358.

Sleep NH and Windley BF (1982) Archean plate tectonics: Constraints and inferences. *Journal of Geology* 90: 363–379.

Sleep NH and Zahnle K (1998) Refugia from asteroid impacts on early Mars and the early Earth. *Journal of Geophysical Research* 103: 28529–28544.

Sleep NH and Zahnle K (2001) Carbon dioxide cycling and implications for climate on ancient Earth. *Journal of Geophysical Research* 106: 1373–1399.

Sleep NH, Zahnle K, and Neuhoff PS (2001) Initiation of clement surface conditions on the early Earth. *PNAS* 98: 3666–3672.

Sobolev AV, Hofmann AW, Sobolev SV, and Nikogosian IK (2005) An olivine-free mantle source of Hawaiian shield basalts. *Nature* 434: 590–597.

Solomatov VS (1995) Scaling of temperature- and stress-dependent viscosity convection. *Physics of Fluids* 7: 266–274.

Solomatov VS (2004) Initiation of subduction by small-scale convection. *Journal of Geophysical Research* 109: B01412.

Solomatov VS and Moresi L-N (2000) Scaling of time-dependent stagnant lid convection: Application to small-scale convection on Earth and other terrestrial planets. *Journal of Geophysical Research* 105: 21795–21817.

Staudigel H (2003) Hydrothermal alteration processes in the oceanic crust. In: Holland HD and Turekian KK (eds.) *Treatise on Geochemistry: The Crust*, vol. 3 (Rudnick RL), pp. 511–535. Amsterdam: Elsevier.

Stern RJ (2005) Evidence from ophiolites, blueschists, and ultrahigh-pressure metamorphic terranes that the modern episode of subduction tectonics began in Neoproterozoic time. *Geology* 33: 557–560.

Stevenson DJ (2003) Styles of mantle convection and their influence on planetary evolution. *Comptes Rendus Geoscience* 335: 99–111.

Tackley PJ (2000a) Self-consistent generation of tectonic plates in time-dependent, three-dimensional mantle convection simulations. Part 1: Pseudoplastic yielding. *Geochemistry Geophysics Geosystems* 1 (doi:10.1029/2000GC000036).

Tackley PJ (2000b) Self-consistent generation of tectonic plates in time-dependent, three-dimensional mantle convection simulations. Part 2: Strain weakening and asthenosphere. *Geochemistry Geophysics Geosystems* 1 (doi:10.1029/2000GC000043).

Thom R (1983) *Parabolas et Catastrophes* 193 pp. Paris Flammarion.

Tozer DC (1970) Factors determining the temperature evolution of thermally convecting Earth models. *Physics of the Earth and Planetary Interiors* 2: 393–398.

Turcotte DL and Schubert G (1982) *Geodynamics: Applications of Continuum Physics to Geological Problems*, 450 pp. New York: John Wiley.

Van Kranendonk MJ, Collins WJ, Hickman A, and Pawley MJ (2004) Critical tests of vertical vs. horizontal tectonic models for the Archaean East Pilbara Granite-Greenstone Terrane, Pilbara Craton, western Australia. *Precambrian Research* 131: 173–211.

van Thienen P, Vlaar NJ, and van den Berg AP (2004) Plate tectonics on the terrestrial planets. *Physics of the Earth and Planetary Interiors* 142: 61–74.

Wallace PJ (2005) Volatiles in subduction zone magmas: Concentrations and fluxes based on melt inclusion and volcanic gas data. *Journal of Volcanology and Geothermal Research* 140: 217–240.

Weislogel AL, Graham SA, Chang EZ, Wooden JL, Gehrels GE, and Yang H (2006) Detrital - zircon provenance of the Late Triassic Songpan - Ganzi complex: Sedimentary record of collision of the North and South China blocks. *Geology* 34(2): 97–100.

White RS, McKenzie D, and O'Nions RK (1992) Oceanic crustal thickness from seismic measurements and rare Earth element inversions. *Journal of Geophysical Research* 97: 19683–19715.

Wilson DS, McCrory PA, and Stanley RG (2005) Implications of volcanism in coastal California for the Neogene deformation history of Western North America. *Tectonics* 24(3): TC3008.

Zahnle K, Arndt N, Cockell C, *et al.* (2007) Emergence of a habitable planet. *Space Science Review* (in press).

Zahnle K and Sleep NH (2006) Impacts and the early evolution of life. In: Thomas PJ, Hicks RD, Chyba CF, and McKay CP (eds.) *Comets and the Origin and Evolution of Life,* 2nd edn., pp. 207–251. Berlin: Springer.

Zatman S, Gordon RG, and Mutnuri K (2005) Dynamics of diffuse oceanic plate boundaries: Insensitivity to rheology. *Geophysical Journal International* 162: 239–248.

Zoback MD and Townend J (2001) Implications of hydrostatic pore pressures and high crustal strength for the deformation of intraplate lithosphere. *Tectonophysics* 336: 19–30.

7 Mechanisms of Continental Crust Growth

M. Stein, Geological Survey of Israel, Jerusalem, Israel

Z. Ben-Avraham, Tel Aviv University, Ramat Aviv, Israel

7.1 Introduction

The continental crust comprises the outermost 20–80 km (average thickness is about 36 km) of the solid surface of the Earth covering $\sim$41% of the Earth surface area. Most of this area ($\sim$71%) is currently elevated above sea level while the rest is defined by the topography of the continental shelves. The continental crust density ranges from 2.7 to 2.9 g cm^{-3} and increases with depth. The vertical extent of the continental crust is defined by the compressional seismic wave velocity that jumps from $\sim$7 to >7.6–8 km s^{-1} across the Mohorovicic (or 'Moho') discontinuity. The Moho discontinuity discovered by the Croatian seismologist Andrija Mohorovicic is the boundary between the crust and the mantle, which reflects the different densities of crust and the mantle, normally occurring at an 'average' depth of $\sim$35 km beneath the continental surfaces and about 8 km beneath the ocean basins. The original recognition and definition of this boundary reflected seismic properties. However, in some regions the

Moho is not always well defined and appears to be transitional in its physical–chemical characteristics (cf. Griffin and O'Reilly, 1987).

The oceanic crust covering ~59% of the Earth's surface is significantly thinner than the continental crust and is characterized by its distinct chemical composition and younger age (the oldest oceanic crust is Jurassic in age). Nevertheless, it appears that growth of continental crust involves accretion of thick magmatic sequences that were formed in the intraoceanic environment (e.g., oceanic plateaus), thus the geodynamic and geochemical evolution of oceanic and continental crusts are related.

The temperature gradient is typically less beneath continents than beneath oceanic basins reflecting the thicker immobile regions underlying the continents. These regions are termed the lithospheric mantle, and they typically form a thick colder 'root' beneath ancient shields and a relatively thinner root beneath younger shields. Yet, in intracontinent rifting regions the asthenospheric mantle is rising and interacts with the lithospheric mantle and continental crust and replaces it with new mantle-derived material. Moreover, it appears that many of the current ocean basins (and oceanic crust) commenced their history within the continental environment calling for some important relationship between stability and magmatic-thermal history of the continental lithosphere and the plate-tectonic cycle. The relationship between asthenospheric and lithospheric mantle, as well as the questions concerning the growth, maturation, and fate of continental lithosphere and crust throughout geological time are fundamental problems in Earth sciences that will be reviewed in this chapter.

The continental crust is of great antiquity and contains the record of most of the geological (physical and chemical) evolution of the Earth. A major question in Earth sciences is when, in what modes, and at what rates did continental crust form and evolve. It is broadly accepted that most of the present-day continental masses are composed of rocks that were produced at the end of the Archean and the beginning of the Proterozoic, within the time period lasting from ~3.2 to ~2.0 Ga. The oldest surviving rocks were found in northwestern Canada and date from about four Ga, 400–500 My after the formation of the Earth (Taylor and McLennan, 1995). Yet, although the continental crust covers about 40% of the terrestrial surface, rocks older than 2.5 Ga account for less than ~14% of the exposed continental area (Windley, 1995). Based on Nd model ages the 'average' age of the continental crust was calculated as ~2.2 ± 0.1 Ga, which indicates that 50–60% of the continental crust was formed before 2.6 Ga (cf. Nelson and DePaolo, 1985; DePaolo et al., 1991; McCulloch and Bennett, 1993; Taylor and McLennan, 1995). These chronological determinations are critical to the long-standing debate on the history and mechanism of crust formation, where models of very early creation and subsequent recycling of continental crust material contrast with continuous/episodic crustal growth (the 'Armstrong–Moorbath debate' elaborated below).

Much of the present-day continental formation appears to occur at convergent plate margins along the subduction zones; however, several studies indicate that large amounts of continent were created in the geological past during short 'superevents' at rates that are difficult to explain by 'typical' subduction activity (Section 7.3.3). This would imply that large portions of the continents were formed by other means other than magmatism at convergent margins, probably processes related to hot spot or mantle plume magmatism (e.g., oceanic plateau formation, accretion of juvenile mantle terranes to the continents, and basaltic underplating). How these accreted terranes were transformed into continental crust and retain geophysical properties of crustal material represents another central question that has been open to debate (the 'Ontong Java paradox' outlined below). Crust formation is related to heat generation in the Earth, which has clearly declined since early Archean time through the Proterozoic and Phanerozoic. What was the reflection of this change on the modes and mechanism of crust production, for example, intracrustal or lithospheric melting versus subduction related magmatism and arc accretion?

In the chapter we summarize some of the main results that emerge from the extensive efforts that were devoted over the past decades to determine the composition of the upper and lower and bulk continental crust (this subject and the relevant literature were thoroughly reviewed by Taylor and McLennan, 1995 and recently by Rudnick and Gao, 2003). Then, we go through the Armstrong–Moorbath debate on early versus episodic crustal growth and the Reymer and Schubert dilemma concerning arc-plume modes of crustal growth. We describe mantle overturn and accretion models (cf. Stein and Hofmann, 1994; Ben-Avraham et al., 1981) that invoke episodic activity of large mantle plumes and accretion of juvenile mantle material to the continents, and alternative arc-tectonics models (Patchett and Chase, 2002).

Finally, we describe the production of juvenile continental crust in major segments of the continents (termed major orogenies): the Archean Baltic Shield and Superior province, the Proterozoic Birimian orogen, and the late Proterozoic Arabian–Nubian Shield (ANS).

7.2 Structure and Chemical Composition of the Continental Crust

7.2.1 General

Based on seismic wave velocities, the continental crust can be vertically divided into two parts: the upper crust between the surface and the Conrad discontinuity, which is poorly defined at 10–20 km from the surface and the lower crust, which principally extends from the Conrad to the Moho discontinuity.

The problem of defining the location of the intracontinental crustal boundaries and obtaining an estimate of the chemical composition of the bulk continental crust reflects the complex history of the crust in different regions of the Earth. This evaluation requires knowledge of the types of the rock comprising the upper and lower crust in various crustal domains. However, only a small portion of the continental crust is exposed (e.g., due to tectonic exhumation of mainly upper-crust segments and deep erosion) and accessible for direct research. The composition and physical properties of the lower crust are mainly evaluated from geophysical data combined with information derived from xenoliths and outcrops of uplifted granulite terranes (cf. Rudnick and Fountain, 1995).

Nevertheless, many researchers devoted extensive efforts over the years to estimate the bulk chemical composition of the crust. This was done by determining the distribution and chemical composition of the dominant rock types that are exposed at the surface (e.g., the pioneering studies of Clarke and co-workers; (Clarke, 1889; Clarke and Washington, 1924) and recent works by Gao *et al.* (1998) (see the comprehensive summaries of this topic by Condie (1993) and Rudnick and Gao (2003)). Another approach was to use sediments as natural samplers of large areas of the upper continental crust. Goldschmidt (1933, 1958) introduced this approach in his pioneering studies on glacial sediments from the Baltic Shield.

Later, other researchers applied similar ideas and methodologies using other samplers including fine-grained sediments such as desert dust or deep-sea terrigenous sediments. Important examples are the works of Taylor and McLennan (1985); Plank and Langmuir (1998); Gallet *et al.* (1998); Barth *et al.* (2000); and Hattori *et al.* (2003). The results of these and other related studies were used to estimate the chemical composition of the upper crust (e.g., Taylor and McLennan, 1995; Rudnick and Gao, 2003).

The estimates on the bulk composition of continental crust require, however, knowledge on the composition of the deeper parts of the crust. This information is provided mainly by xenoliths of mafic granulites that are transported to the surface by intraplate alkali basalts (mostly in rift-related environments) and tectonically uplifted blocks that consist of granulite facies metamorphic terranes. The post-Archean metamorphic granulite terranes appear to represent upper-crustal segments or transitional zones between the upper and lower crust (e.g., depressed in Himalayan-type collision zones (Bohlen and Mezger, 1989; Mezger, 1992). Xenoliths of mafic granulites are considered as representatives of the lower crust. The xenoliths were interpreted as cumulates from basaltic magmas that were rising to lower-crust levels and were later subjected to the granulite facies metamorphism (cf. Kay and Kay, 1981; Rudnick, 1992). Since the xenolith data are rather sporadic, Rudnick and Fountain (1995) combined it with seismic velocities to get an estimate of lower-crust bulk composition. Nevertheless, as noted above, the xenoliths of mafic granulite were mostly transported to the surface by rift-related alkali basalts and they may be associated only with the particular tectonic–magmatic environment that also supports alkali basalt production. Rifting environments are associated with the rise of asthenospheric (mid-ocean ridge basalt (MORB)-like) mantle into the spreading lithosphere (see the example of the opening of the Red Sea below), and thus the production of the source lithologies of the mafic garnulite-xenoliths may be associated with additions of new material to an older continental lithosphere. This scenario is supported by younger Re–Os isotope ages of lower crustal xenoliths compared to the above lying lithosphere (Rudnick, 1992).

Rudnick and Gao (2003) combined the estimates of upper, middle, and lower crust compositions to produce a bulk continental crust composition. Trace-element patterns (normalized to primitive mantle values) of lower, upper, and bulk crust are illustrated in **Figure 1**. The order of appearance of the trace

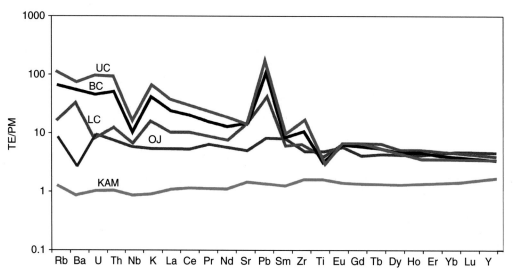

Figure 1 Primitive mantle normalized trace-element patterns of upper, lower, and bulk crust estimates (UC, LC, BC, respectively), the Cretaceous Ontong Java plateau basalt (OJ) and Archean Kambalda komatiite (KAM). Note the overall enrichment of the UC and BC crustal samples in the incompatible trace elements, the distinctive positive Pb and negative Nb anomalies in the crustal samples and the flat patterns of the plateau and almost primitive in KAM. A significant differentiation is required to transform OJ or KAM-type oceanic crust to the upper continental crust. The fractionation of trace elements such as Nb and Pb could reflect processes in the subduction environment. Nevertheless, if accreted plateaus similar to OJ basalts comprise an important part of the continental crust they are either not well represented by the bulk continental crust composition estimates or that they underwent (before or during accretion) internal magmatic differentiation (which is reflected by their continental-like seismic velocities, see **Figure 6** below). Data sources: Rudnick and Gao (2003) and Mahoney *et al.* (1993a).

elements on the abscissa reflects the tendency of the incompatible trace elements to move preferentially from the source to the melt during magma generation. This behavior can be applied to the processes that are involved in the production of continental crust from mantle sources (Hofmann, 1988). Noticeable features in this figure are the strong positive Pb anomalies and the negative Nb anomalies in the lower, upper, and bulk continental crust curves. These anomalies are probably related to mantle–crust differentiation processes that can fractionate Pb and Nb from the rare earth elements (henceforth REE), Th, and U (cf. Hofmann *et al.*, 1986). Such fractionation could occur in the subduction environment ('the subduction factory') along with the production of 'arc-type' calc-alkaline magmas (cf. Stein *et al.*, 1997). The Mesozoic Ontong Java plateau basalts and the Archean Kambalda komateiite display trace-element patterns that are distinctly different from those of the 'lower, upper, and bulk' continental crust. These types of magmas show almost 'flat' primitive trace-element patterns (although the Ontong Java sample displays small negative Ba anomaly that seems to oppose the strong positive Ba anomaly of the 'average lower crust'). If the Kambalda and Ontong Java-type magmas represent

important constituents of continental crust, they are not well represented by the estimated 'bulk crust' patterns.

7.2.2 Composition of the Continental Crust: Open Questions

Below we outlined several topics and open questions that emerged from the extensive efforts to evaluate the chemical composition of the continental crust:

7.2.2.1 Problems with the simple 'andesite model'

The overall geochemical similarity of continental crust to arc-type magmas (e.g., the estimated average ∼60% SiO_2 andesitic composition of bulk continental crust, as well as similar values of indicative trace-element ratios such as Ce/Pb and Nb/Th in arc – crust magmas) led many workers to suggest that the arc-subduction environment where andesitic magmas are produced is the main locus of juvenile crust production (the andesite model of continental crust formation; e.g., Taylor (1967)), and that accretion of arcs at convergent margins is a major means of continental crust growth. Transport and differentiation of melts and fluids that eventually control the composition of continental crust were associated with the subduction environment where mantle-wedge melts and fluids with trace elements move upward, while interacting and mixing with the bulk peridotitic assemblages (e.g., Kelemen, 1995). However, most arcs are too mafic to build continental crust by simple accretion (cf. Anderson, 1982; Nye and Reid, 1986; Pearcy et al., 1990). Moreover, Taylor and McLennan (1985) pointed out several problems with a simplistic application of the andesite model, such as the failure of this model to explain the abundances of Mg, Ni, and Cr and the Th/U ratio in upper crustal rocks. Rudnick (1995) suggested that the lower Sr/Nd and La/Nb ratios in continental crust compared to arc magmas are compatible with involvement of intraplate magmatism in continental crust production. The andesitic magmatism and in a broader sense the intermediate 'grandioritic' bulk composition of the upper crust can be considered as a product of the differentiation processes that are involved in mantle–continental crust transformation. Taylor and McLennan (1985) suggested that only ∼20–25% of the crust reflects the currently operating 'andesite growth mode', whereas the dominant mechanism that produced the Archean crust was the derivation of the bimodal basic-felsic magma suites from mantle sources (e.g., formation of Archean tonalitic–trodhjemitic–granodioritic (TTG) magmas). It is argued below that the production of thick oceanic crust and lithosphere by mantle plume activity and their accretion to the continents was an essential part of continental crust production throughout the past 2.7–2.9 Ga of the Earth's history.

7.2.2.2 Plumes and arcs

Contribution of juvenile material from 'plume' or 'arc' sources to the continents can be assessed by using trace-element ratios of Ce/Pb and Nb/Th that are distinctly different in arcs and plume-related mantle domains such as OIB, and oceanic plateaus (Hofmann et al., 1986; Stein and Hofmann, 1994; Kerr, 2003). Meta-tholeiites and greenstone sequences from major ancient orogens (e.g., the Superior, Birimian, and ANS), as well as plateau basalts such as the Ontong Java tholeiites display flat (chrondite-like) REE patterns (**Figure 2**) and where data exist show plume-type Nb/Th ratios (**Figures 1** and **3**). It appears that these juvenile crustal units are not well represented by the major sediment samplers. The upper-crust composition is largely dictated by the subduction and post-accretionary processes such as crustal melting (that can lead to granite production), lithospheric

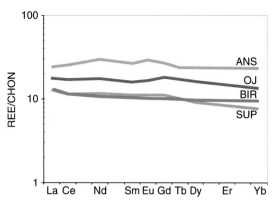

Figure 2 Chondrite normalized REE pattern in oceanic basalt from the Ontong Java plateau (OJ) and several meta-tholeiites that were interpreted as plateaus basalts (from the Arabian-Nubian Shield (ANS); Superior Province (SUP); and Birimian orogen (BIR). All these basalts erupted rapidly and formed thick magmatic sequences marking the early stage in a major orogenic cycle and production within oceanic basin. Data sources: Mahoney et al. (1993a), Abouchami et al. (1990), Reischmann et al. (1983), and Arth and Hanson, (1975).

delamination, and asthenospheric rising. These processes involved the preferential recycling of major elements (Mg, Ca) to the mantle at subduction zones and the removal of other major elements such as Si, Al, Na, K to the continental crust (cf. Albarede, 1998). The subduction factory can be envisaged as a big chromatographic column that fractionates the incompatible trace elements Nb, Pb, Rb from REE and Th, U (Stein et al., 1997). Overall, these processes modified the original geochemical signature of the juvenile mantle material and eventually shifted the upper continental crust to its average andesitic composition (Rudnick, 1995; Stein and Goldstein, 1996; Albarede, 1998).

7.2.2.3 The Eu dilemma

The application of the REE to the study of the composition and evolution of continental crust is going back to the works of Goldschmidt (1933) and the subject was thoroughly developed and discussed by Taylor and McLennan (1985). Most of the fine-grain sedimentary rocks that are used to estimate bulk upper-crust compositions typically show a negative Eu anomaly (e.g., loess sample in **Figure 4**), pointing to significant intracrustal melting processes that require the existence of a residual reservoir with an REE pattern showing a positive Eu anomaly (Taylor and McLennan, 1995). However, such a reservoir remains largely unknown,

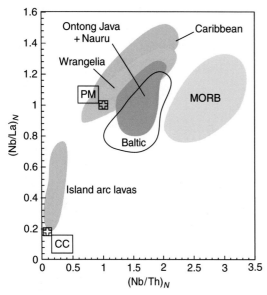

Figure 3   The location of basalts from several plateau basalt provinces on an (Nb/La)$_N$ versus (Nb/Th)$_N$ diagram (after Puchtel *et al.*, 1998). The plateau basalts define fields that are distinctly different from island arc magmas and MORB and lie between primitive mantle compositions (PM) and Nb enriched that are similar to OIB and some rift-related alkali basalts (see **Figure 4**). Hofmann *et al.* (1986) attributed the high Nb/U and Nb/Th ratios to production of a 'residual' mantle after extraction of the low Nb/Th continental crust. Stein *et al.* (1997) proposed that Nb could be fractionated from REE, U, and Th due to production of an amphibole 'front' in the mantle wedge, which they viewed as a giant chromatographic column. The Nb-rich zone is recycled back to the mantle or is frozen at the lower part of the lithospheric mantle after cessation of subduction. Does the convergence of several plateau basalts to primitive mantle ratios indicate that the plume represents mantle with primitive Nb–Th–La ratios? In any event, Archean plateau magmas such as those of the Baltic Shield and the Phanerozoic magmas overlap to some extent in the diagram but are clearly not identical.

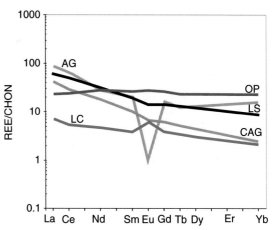

Figure 4   REE (chondrite normalized) pattern of magmas representing main stages in the growth of a 'typical' orogenic cycle: tholeiitic magmas related to oceanic plateaus (OP), calc-alkaline granites (CAG), and alkaline granites (AG), as well as xenoliths of mafic granulites that are interpreted as lower crustal samples (LC) and Late Pleistocene loess material that sample the exposed-eroded upper crust in the Sahara desert (LS). The pattern of the loess material is identical to the estimated bulk upper-crust estimate (not shown). Note the appearance of a pronounced negative Eu anomaly related to plagioclase fractionation in the alkali granites and the almost smooth pattern of the calc-alkaline granite. The production of the alkaline granites in the Arabian–Nubian Shield was attributed to significant fractionation of mafic (Mushkin *et al.*, 2003). The calc-alkaline granites could be the product of mid-crust (amphibolite-type) melting. Both would produce residual assemblages. The topic of production of the vast calc-alkaline granitic batholiths is not well understood. The production of granites requires water ('no water, no granites'), but the source of heat for the melting and the fate of the residual crust are not well known and the whole subject requires further study.

because the xenoliths of mafic granulite that are considered as lower-crust samples, and indeed show positive Eu anomalies are not interpreted as partial melting residues (cf. Rudnick and Taylor, 1987; Rudnick, 1992). It should be noted, however, that among the upper-crust magmatic rock suites, it is mainly the alkaline granites and related volcanics that display distinctive negative Eu anomalies, while most of the calc-alkaline granites from juvenile crustal terranes display REE patterns with small or no Eu anomalies (e.g., the calc-alkaline batholiths of the Birimian and the ANS orogens (**Figure 4**). The alkaline granites and related rhyolitic volcanics typically characterize the closing stages of large orogenic

cycles and their petrogenesis reflects fractionation of mafic magmas in shallow crustal levels (e.g., Mushkin *et al.*, 2003 and see elaboration of this topic below). These suites comprise, however, a rather small fraction of the total upper-crust magmatic inventory that is probably irrelevant to the total crustal budget.

7.2.2.4   Crustal foundering

The search after 'residual reservoir' is associated with models of crustal foundering. This topic was recently discussed by Plank (2005) in light of the high Th/La ratio in the continental crust (e.g., **Figure 1**). She argued that the high Th/La ratio is unlikely to reflect fractionation processes within the subduction environment, nor could it be related to processes of intraplate magmatism, or to Archean magmatism

(Condie (1993) showed that Archean continents were rather characterized by low Th/La). Alternatively, the Th/La continental crust ratio can be related to production of lower crustal cumulates (with low Th/La ratios), and foundering of such cumulates and partial melting restites (both represent high-density heavy crust) into the mantle. This process would create over geological time bulk continental crust with high Th/La ratio.

7.2.2.5   Archean TTG magmatic suites

Archean magmatic suites were characterized by the bimodal mafic-felsic compositions with scarce appearance of andesitic rocks (Taylor and McLennan, 1985). Thus, the simple andesite model cannot be applied to the production of the major inventory of continental crust which is Archean in age. The felsic members of these suites show fractionated heavy rare earth element (HREE) patterns with no Eu anomaly indicating melting in the garnet stability zone and no evidence for intracrustal melting involving plagioclase fractionation. The felsic precursors of these sediments were interpreted to reflect partial melting products of basaltic or TTG magma sources. The TTG magmas for themselves were interpreted as fractionation products of intermediate parental magmas that were derived from enriched and metasomized mantle peridotite (Shirey and Hanson, 1984) or melting products of the subducting slab. For example, high-Al TTG plutons that are characterized by high-La/Yb ratios, like those exposed in the Superior province greenstone-granitoid terranes were interpreted as melting of young, hot subducting oceanic slabs (Drummond *et al.*, 1996). In this relation, the production of modern magnesian andesites, Nb-rich basalts, and adakites could be attributed to subduction of hot oceanic slabs (e.g., central Andes; Stern and Kilian (1996): Papua New Guinea; Haschke and Ben-Avraham (2005)). The processes involved with the TTG magma production required an environment with high heat flow (supporting production of Mg-rich basalts and komatiites) and probably rapid stirring of the mantle. Taylor and McLennan (1985) envisage the formation of the Archean continental nuclei as a process involving eruption piling and sinking of tholeiitic basalts followed by partial melting and formation of the felsic magmas. To a limited extent this scenario resembles the descriptions of fast production of thick oceanic crust in relation to mantle plume activity and the subsequent processing of the juvenile-enriched crust by possible subduction and

intraplate processes possible since late Archean time. The major tectonic-magmatic element in this description that apparently was not significant during the Archean is the subduction activity and the production of andesitic-calc-alkaline magmas.

7.2.2.6   Post-Archean granites

While the felsic members of the Archean bimodel suites show highly fractionated HREE, the post-Archean granitoids typically show flat or slightly fractionated HREE patterns, indicating a shallower depth of source melting above the garnet stability zone. The depth of melting and the different sources thus represent a fundamental difference between Archean and post-Archean crust formation (Taylor and McLennan, 1985). It should be stressed that TTG magmas are petrogenetically different from the post-Archean granitoids. For example, if the subducting plate melts in the eclogite stability zone, there is a possibility to produce TTG magmas and thus granitic rocks without water. The post-Archean granites are melting and differentiation products of water-bearing mafic sources at shallower depth of the crust. This melting involves fractionation of amphibole instead of clinopyroxene and orthopyroxene. Since the amphibole is low in SiO_2 the residual melt is more silicic driving the differentiation from basalts (the 'normal': melt produced in subduction zones) to silicic melts all the way to granite. The apex of this discussion is the requirement of water carrying subduction processes in the formation of the post-Archean granitic magmas.

Campbell and Taylor (1983) emphasized the 'essential role of water in the formation of granites' and, in turn, continents and introduced to the literature the important concept of 'no water, no granites, no oceans, no continents'. They argued that the Earth, which is the only inner planet with abundant water, is the only planet with granite and continents, whereas the Moon and the other inner planets have little or no water and no granites or continents. This concept could be extended to: 'no water, no plate tectonics, no continents', since with no water plate tectonics is unlikely to work. Thus, there is a fundamental linkage between formation of granitic (granodioritic) continental crust and plate-tectonic mechanism.

7.2.3   The Lithospheric Mantle and Continent Stability

Beneath the Moho and the continental crust lies the lithospheric mantle whose thickness reaches 300–400 km beneath old continents and is relatively thin (up to 100–120 km) beneath young continents (e.g., beneath the Canadian and ANS, respectively). Subcontinental lithospheric xenoliths show large variations and heterogeneities in the trace-element abundances that over time can lead to large variations in the radiogenic isotope systems such as Pb, Nd, Sr (cf. McKenzie and O'Nions, 1983; Hawkesworth *et al.*, 1993). The magmatic and thermal history of the lithospheric mantle is related to that of the above-lying continental crust. Based on Pb, Nd, and Sr iotopes in ophiolites, galenas, and basalts, Stein and Goldstein (1996) suggested that the lithospheric mantle beneath the Arabian continent was generated during the events of the late Proterozoic crust formation. The lithospheric mantle can be regarded as a 'frozen' mantle wedge, which previously accommodated the production of calc-alkaline magmas and fractionation–transportation processes of trace elements en route to shallower levels of the crust (the subject is elaborated below). After cessation of subduction, the water-bearing lithospheric mantle could have become a source for alkali magmas that rose to the shallower levels of the continental crust (e.g., alkali granites) or erupted at the surface (cf. Mushkin *et al.*, 2003; Weinstein *et al.*, 2006). Thus, the lithospheric mantle is not only a potential source of information on the 'subduction factory' processes, but can be regarded as an important source of magmas that constitute a part of the continental crust inventory. Yet, the source of heat and conditions of lithospheric melting is not obvious. One possibility is that asthenospheric mantle rises into the rifting continent, supplying heat for melting. The production of magmas may have eventually caused the depletion in the lithospheric mantle of incompatible trace elements and basaltic components, leading to its exhaustion.

7.3   Models of Crust Formation and Continental Growth

7.3.1   General

Models that describe the mechanism of continental crust growth and the evolution of continental masses can be divided into two major groups: (1) models of early crustal growth and subsequent recycling and (2) models of continuous or episodic crustal growth.

The most prominent researcher that pushed the 'early crust growth/recycling' model was R.L. Armstrong. He proposed that soon after the early mantle–crust differentiation and formation of

continental crust, the Earth reached a steady state with accretion and destruction of continental masses occurring at approximately equal rate with essentially constant volumes of ocean and crust through geological time (Armstrong, 1981a, 1991). The key process in this model is the recycling of the continental masses back into the mantle. Armstrong based his model on the principles of the constancy of continental freeboard and the uniformity of thickness of stable continental crust with age. These two critical parameters imply negligible continental crustal growth since 2.9 Gy BP. Defending his model, Armstrong addressed some of the arguments that were put against the steady state model, mainly those that relied on the behavior of the radiogenic isotopes (e.g., Nd, Sr, and Pb; Armstrong (1981b)).

Among the pioneering researchers that developed the ideas of continuous/episodic crust growth were W.W. Rubey (who advocated the underplating theory as well), A.E. Engel, P.M. Hurley, and S.R. Taylor. The prominent researcher of this school of thinking was S. Moorbath who pointed out the difficulty of subducting large amounts of continental crust due to its buoyancy and argued that the chronological evidence derived from continental crustal rocks suggests creation of juvenile continental crust throughout Earth's history. Moorbath also developed the concept of continental accretion as a major building process of continental crust (Moorbath, 1975, 1978).

The steady-state-recycling theory of crustal growth requires evidence for substantial sediment recycling. Sediments are clearly recycled in the subduction environment. Sediment involvement in subduction was demonstrated by various geochemical tracers of arc magmas: for example, the radiogenic isotope ratios of Nd, Pb, Sr, and Hf (cf. Patchett *et al.*, 1984), [10]Be isotopes (cf. Tera *et al.*, 1986), and trace-element fluxes (Plank and Langmuir, 1998). Plank (2005) introduced the Th/La ratio as a tracer of sediment recycling in subduction zones. Th/La in the subducting sediments reflects the mixing between terrigenous material with high 'upper crustal' Th/La ratios (0.3–0.4) and metalliferous sediments and volcaniclastic sediments with low Th/La ratios (<0.1). White and Patchett (1984) argued that a significant but small (at least 1–2%) contribution of older continental material could be potentially recycled into new continents. Nevertheless, based on mass-age distribution and Nd isotope data of sediments McLennan (1988) estimated the mass of sediment available for subduction to be $<1.6 \times 10^{15}\,\mathrm{g\,yr^{-1}}$, and argued that, that is insufficient to support a steady-state crustal mass according to the Armstrong model (see also Taylor and McLennan (1995)).

Hofmann *et al.* (1986) suggested that the Nb/U ratio in mafic–ultramafic magmas can be used to deduce the history of continental crust growth because the Nb/U ratio changed from the primitive mantle ratio of ~30 to a higher ratio ~47 in the 'residual' mantle due to extraction of crust with Nb/U of ~10. Some of the Archean komatiites (e.g., samples from Kambalda and Kostomoksha provinces) show high Nb/U ratios ~47 (Sylvester *et al.*, 1997; Puchtel *et al.*, 1998) that may support the Armstrong steady-state model. Yet, other komatiites and early Proterozoic tholeiites show Nb/U ratios that are similar to the primitive value (**Figure 1**). Puchtel *et al.*, (1998) commented on this topic that: "the Nb/U data alone cannot unequivocally resolve this important controversy."

Although the continents appear to have undergone episodic changes throughout all of geologic time, the available geochronology of crust formation shows time periods when large areas of continental crust were formed at a fast rate, alternating with apparently more quiescent periods of low crust-formation rates (marked in **Figure 5**). It may be that these quiescent periods merely represent missing continental material because the crust was destroyed and recycled into the mantle (according to the Armstrong model) but actual evidence for this process is indirect at best.

7.3.2 Timing and Rates of Crustal Growth

Crust formation ages were determined by direct dating of major crustal terranes (numerous works applying the Rb–Sr, Sm–Nd dating methods and U–Pb and zircon ages) and by using the concept of Nd crustal model ages that mark the time of major chemical fractionation and the Sm/Nd ratio change accompanying the extraction of melts from the mantle and their incorporation into the continental crust (e.g., McCulloch and Wasserburg, 1978; Nelson and DePaolo, 1985; DePaolo, 1981; DePaolo *et al.*, 1991). The Sm/Nd crustal formation ages suggest that 35–60% of the currently exposed crustal masses were produced in the Archean (while actual exposed Archean crust accounts only for 14%; Goodwin

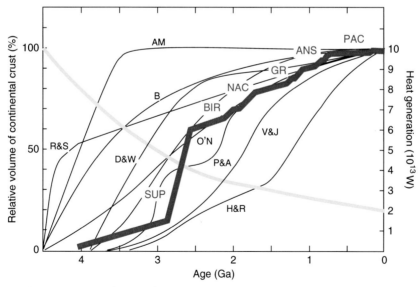

Figure 5 Models of continental crust (cumulative) growth and heat generation (yellow curve) throughout Earth's history (after Reymer and Schubert, 1984; Taylor and McLennan, 1985, 1995; Condie, 1990; Patchett, 1996). Crustal curves: R&S = Reymer and Schubert (1984); AM = Armstrong (1981a,1981b); B = Brown (1979); D&W = Dewey and Windley (1981); O'N = O'Nions *et al.* (1979); P&A = Patchett and Arndt (1986); V&J = Veizer and Jansen (1979); H&R = Hurley and Rand (1969). The thick red curve is after Taylor and McLennan (1995), with marks of the major orogenic episodes that were related by Stein and Hofmann (1994) to major upwelling events and plume head uprise in the mantle (the MOMO model illustrated in **Figure 9**). SUP = Superior province; BIR = Birimian orogeny; NAC = North Atlantic Continent; GR = Grenville orogeny; ANS = Arabian–Nubian Shield; PAC = Pacifica 'super-plume'.

(1991)). **Figure 5** shows that the most dramatic shift in the generation of continental crust happened at the end of the Archean, 2.5–2.7 Gya. The late Archean, early Proterozoic pulse of crustal growth is represented in all continents (Condie, 1993). Very little continental crust was preserved before 3 Ga. The oldest preserved terrestrial rocks are the 3.96 Ga Acasta Gneiss in the Northwest Territories of Canada (Bowring *et al.*, 1990) and detrital zircons from the Yilgarn Block, Western Australia, were dated to a much older age of 4.4 Ga (Wilde *et al.*, 2001).

7.3.3 The Reymer and Schubert Dilemma

Reymer and Schubert (1984, 1986) evaluated the rate of growth of major continental crust segments (e.g., the Superior province in North America, and the ANS) and showed that they significantly exceed the rate of continental addition that prevailed along the subduction margins during Phanerozoic time ($\sim 1 \, \text{km}^3 \, \text{yr}^{-1}$). The Phanerozoic rate was based on estimates of crustal additions along Mesozoic–Cenozoic arcs, hot spots, and some other additional sources (e.g., underplating). They calculated a total addition rate (mainly along the arcs) of 1.65 and total subtraction rate 0.59 km³ yr⁻¹, yielding a net growth rate of 1.06 km³ yr⁻¹. In addition, Reymer and Schubert calculated a growth rate by an independent model based on the constancy of freeboard relative to mantle with declining radiogenic heat production.

The term freeboard was derived from civil engineering, describing the additional height above a normal operating water level and the top of the water-holding structure. The geological–geophysical use of this concept encompasses the complicated relation between continental crustal thickness, volume, mantle temperature, and Earth's heat budget. Wise (1974) tied the freeboard concept to crustal growth rates, by suggesting that approximately constant continental crustal volumes and areas pertained since the Archaean–Proterozoic boundary, thus maintaining approximately the same elevation or freeboard of the continents above mean sea level. The freeboard calculation of Reymer and Schubert yielded a Proterozoic–Phanerozoic growth rate of 0.9 km³ yr⁻¹, very similar to the above-calculated value. The Phanerozoic growth rate is 3 times less than the average Archean growth rate. Assuming formation between 900 and 600 Ma during the Pan-African orogeny, Reymer and Schubert showed that

if the ANS was generated by convergent margin magmatism it required an addition rate of 310 km³-km of convergence⁻¹-Ma⁻¹, as compared to a global total of $\sim$40 km³-km⁻¹-Ma⁻¹ (**Figure 6**). Thus, they speculated that an additional large-scale mechanism, probably plume-related magmatism, played an important role in the formation of the ANS, as well as in other major orogenic events.

At the very least, the results of Reymer and Schubert indicate that the local crust-formation rates of the major continental provinces are as high as or higher than the present-day rate averaged over the entire globe. As a consequence of their analysis, Reymer and Schubert proposed that juvenile additions to continental crust require additional mechanism besides the arc magmatism and this mechanism could involve activity of hot spots or mantle plumes. The geodynamic conditions during these intervals of juvenile crust creation may resemble to some extent the Archean conditions.

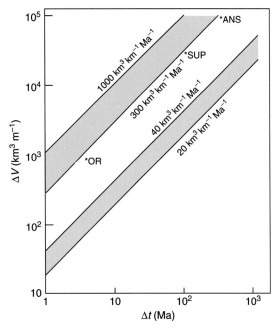

Figure 6 Arc-crust (volume) addition curves (after Reymer and Schubert, 1984) showing the large difference between Phanerozoic arc activity (20–40 km³ km⁻¹ Ma⁻¹) and the growth of some major orogens (e.g., SUP = Superior; ANS = Arabian–Nubian Shield $\sim$300 km³ km⁻¹ Ma⁻¹; OR = Oregon coastal range). The figures demonstrate the Reymer and Schubert dilemma that points to the necessity to invoke additional sources and mechanism to support the production of continental crust during the episodes of major orogens production.

Galer (1991) performed another evaluation of the interrelationships between continental freeboard, tectonics, and mantle temperature. He concluded that before ~3.8 Ga the potential temperature exceeded 1600°C in the shallow mantle and therefore little or no hypsometric or tectonic distinction existed between 'continental' and 'oceanic' regimes. Thus, the pre-3.8 Ga period was subjected to tectonic and sedimentary processes that were distinctly different to those characterized by the Earth surface since mid-Archean time. An important component in his evaluation is the assumption that the early Archean oceanic crust was significantly thicker than typical present crust. The Reymer and Schubert (1984, 1986) assessment indicates that mantle plumes could play an important role in post-Archean crustal history of the Earth. Below we discuss the possible role of thick oceanic crust (e.g., the oceanic plateaus) in continental crustal growth. We can ask the question whether post-Archean episodes of crustal growth have some important components that 'mimic' to some extent the early Archean conditions.

7.3.4 Continental Growth and Heat Production

The heat production in the Earth, which is the main driving force for mantle convection and upwelling processes has declined by a factor of 5 since the Early Archean time (see yellow curve in **Figure 5**). This decrease reflects the decay of the long-living radiogenic isotopes. **Figure 5** indicates that continental crust accumulation does not reflect the Earth thermal history very well, particularly in its earliest Archean history since only in the late Archean does a rough correspondence appear between the heat generation and crustal growth (Patchett, 1996). The heat engine of the mantle should have supported extensive mantle movements in the Archean such as plume head upwelling. Yet, the low cumulative crustal growth in the Archean may indicate that the system was as efficient at destroying continental crust as making it (Taylor and McLennan, 1995).

The intensive pulses of continental crust formation in the late Archean and early Proterozoic probably took place in a mantle that convected less vigorously than at earlier times. One possibility is that the tectonic regime became more similar to modern plate tectonics, where water played an essential role. By late Archean and early Proterozoic times we might enter the regime of 'water, plate tectonics, continents'.

7.4 Oceanic Plateaus and Accreted Terranes

7.4.1 General

With the advances made in geophysical techniques over the last 40 years, a different and new picture of the Earth began to emerge. For the first time, it was possible to image and sample the planet's crust and even layers deeper down. During the 1970s it started to become clear that the ocean floor is not homogenous, but contains areas of significantly thicker crust than the norm. The discovery of one such area, the Ontong Java plateau with a crustal thickens of over 30 km, gave rise to the term 'oceanic plateau' (Kroenke, 1974). Today, about 100 such anomalous regions ranging in size from 1000 km to a few kilometers have been documented (**Figure 7**).

The formation of these plateaus is still not clearly understood. They were considered to represent remnants of extinct arcs, abandoned spreading ridges, detached and submerged continental fragments, anomalous volcanic piles, mantle plume head traces, and uplifted oceanic crust. However, both isotope dating and modeling support the theory that they formed rapidly, often in less than 2–3 My (Richards *et al.*, 1989), with the majority forming at or near mid-ocean ridges (Kerr, 2003). These areas are conducive to large decompression melting (Eldholm and Coffin, 2000), providing a source of magma needed to form these features.

At least from a geophysical point of view, values of crustal seismic velocities, V_p, in the upper 5–15 km (where V_p ranges from 6.0 to 6.3 km s^{-1}, insert in **Figure 7**), together with the anomalous thickness, relief, and gravity data, would indicate a continental structure for some of these plateaus (Nur and Ben-Avraham, 1977, 1978; Ben-Avraham *et al.*, 1981; Carlson *et al.*, 1980). Other plateaus are from clearly oceanic origin, probably the result of continuous extrusion of basalts, probably from an active hot spot or mantle plume head.

Nur and Ben-Avraham (1977) and Ben-Avraham *et al.* (1981) suggested that the plateaus with the 'crustal seismic velocities' (e.g., Ontong Java plateau) represent continental fragments rifted away from a parent continent that they named Pacifica. Stein and Hofmann (1994) and Abbott and Mooney (1995) on the other hand suggested that the plateaus were first extracted from the mantle within the oceanic basins above mantle plumes and were later accreted to the continents. In this context, it is interesting to note the

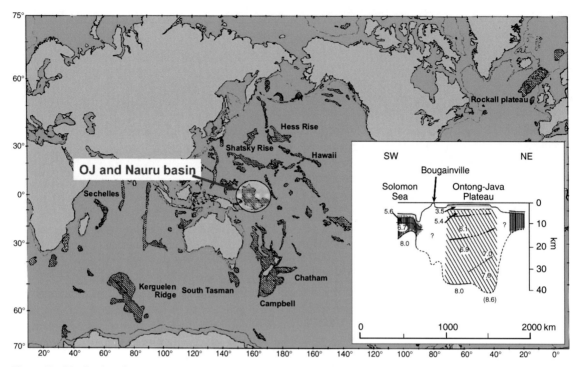

Figure 7   Distribution of oceanic plateaus and location of the Ontong Java plateau and Nauru Basin in the Pacific (after Ben-Avraham *et al.*, 1981). The Mesozoic Pacific was the locus of enhanced magmatism that was related to plume activity ("The last pulse of the Earth." Larson (1991)). It also appears that shortly after the production of the thick oceanic plateaus subduction zones were developed on their margins. The Nauru Basin comprised oceanic crust that erupted in an ocean-spreading environment but shows geochemical similarities to the adjacent Ontong Java plateau. Thus, it appears that both regions derived their magmas from similar enriched sources. This configuration supports the existence of an enriched upper mantle beneath the Pacific. Stein and Hofmann (1994) argued that the Nd–Sr isotope composition of this mantle is similar to the PREMA isotope composition (e.g., εNd $\sim +6$ and $^{87}Sr/^{86}Sr \sim 0.7035$ as defined by Worner *et al.* (1986) for various volcanic rocks around the world). It appears that the Pacific-PREMA composition characterized the entire upper mantle during the time of the production of the Ontong Java and Nauru Basin basalts and thus could mark the overturn or plume uprise events according to the MOMO model. Anderson (1994) proposed an alternative interpretation to the production of the Mesozoic Pacific plateau that requires no plume activity. He suggests that the upper mantle of the Mesozoic Pacific and the Indian Ocean is hot because these regions have not been cooled by subduction for more than 200 My. Anderson argues that the timing of formation and location of the plateaus are related to the organization of the tectonic plates. This would require that the PREMA-type enriched isotope compositions characterized the upper mantle rather than MORB, but magmatism in evolved continental rift environments (e.g., Ross Sea, Antractica; and Red Sea, Arabia) clearly show transition from enriched isotope composition to typical depleted MORB (e.g., Stein and Hofmann (1992); Rocholl *et al.* (1995) and see Figure 11 and 'Red Sea model' in **Figure 12** below). (b) Seismic velocities across the Ontong Java plateau (after Ben-Avraham *et al.*, 1981). Surprisingly, the velocities suggest continental crust composition and structure. This observation illustrates a kind of 'paradox' since it is expected that Ontong Java will show seismic structure of a thick oceanic crust. Does this mean that crustal formation processes operate in the mid-oceanic environment? We regard the 'Ontong Java paradox' as a major open question that should be addressed by future studies.

co-appearance of the basaltic sequences of the Ontong Java plateau and those of the adjacent Nauru Basin in the Pacific Ocean (cf. Mahoney *et al.*, 1993a, 1993b). The tectonic setting of the Nauru Basin is clearly of a mid-oceanic spreading center type, with the typical appearance of magnetic anomalies and eruption of tholeiitic basalts. The important point is that the Ontong Java and Nauru Basin basalts display similar geochemical characteristics such as Nd and Sr isotope

ratios that are enriched relative to the depleted MORB-type mantle (e.g., the basalts from the East Pacific Rise). Thus, it appears that the basalts erupted in the Nauru Basin and the Ontong Java plateau represent a similar uppermost mantle that is enriched relative to the 'normal' depleted asthenospheric mantle. This enriched asthenosphere could be related to a large plume head (or a mantle upwelling) event that brought enriched material to the shallower mantle.

Recently, Richardson *et al.* (2000) produced a three-dimensional tomographic model of the seismic structure beneath the Ontong Java plateau that indicated the existence of a low-velocity mantle 'root' reaching the depth of ~300 km. This was interpreted as a remnant of the Cretaceous Ontong Java plume that was attached to the plateau.

Ishikawa *et al.* (2004) address the question of this seismically anomalous low-velocity root beneath the Ontong Java Plateau and its lithospheric mantle composition and structure by studying suite mantle xenoliths (peridotites and pyroxenites) from Malaita, Solomon Islands. The shallower mantle (Moho to 95 km) is composed of variably metasomatized peridotite with subordinate pyroxenite derived from metacumulates, while the deeper mantle (95–120 km) is represented by pyroxenite and variably depleted peridotites. The shallower and deeper zones are separated by a garnet-poor zone (90–100 km), which is dominated by refractory spinel harzburgites. Ishikawa *et al.* attributed this depth-related variation to different degree of melting for a basalt–peridotite hybrid source at different level of arrival depth within a single adiabatically ascending mantle plume: the lack of pyroxenites at shallower depths was related to extraction of hybrid melt from completely molten basalt through the partially molten ambient peridotite, which caused the voluminous eruption of the Ontong Java plateau basalts. The authors concluded that the lithosphere forms a genetically unrelated two-layered structure, comprising shallower oceanic lithosphere and deeper impinged plume material, which involved a recycled basaltic component, now present as a pyroxenitic heterogeneity. The possible relevancy of this model to the shallower seismic structure of the Ontong Java plateau (**Figure 7**) could be that evolved magmas were produced within the shallower layers.

Regardless of the mechanisms responsible for the formation of oceanic plateaus, they are assumed by many researchers to have an important role in the formation of new continental crust. Because they are part of moving plates and are embedded in them, they are destined to arrive at subducting plate boundaries and to accumulate, at least in some cases, onto the continental margin as accreted allochthonous terranes.

The potential of oceanic crust to be accreted to the continents rather than being subducted depends on its magmatic and thermal history and is reflected in its buoyancy. Under conditions of 'normal seafloor spreading operation' the oceanic lithosphere begins its history as a buoyant hot plate that cools with time and becomes susceptible to subduction. The buoyancy of the oceanic lithosphere depends on its age and density distribution (Oxburgh and Parmentier, 1977). The density distribution reflects the composition and the thickness of the crustal and mantle layers of the lithosphere, which in turn are controlled by the mantle temperature at the place of crust formation. A hotter mantle will produce a thicker oceanic crust with more depleted lithospheric mantle, which can become less dense than more fertile mantle and thus, more buoyant (McKenzie and Bickle, 1988; Oxburgh and Parmentier, 1977; Langmuir *et al.*, 1992). This is probably the case in the production of oceanic plateaus by rising plume heads. The plumes are ~200–300°C hotter than an ambient mantle and can produce 15–40 km thick oceanic plateaus (compared to ~7 km thickness of normal oceanic crust). Therefore, some oceanic plateaus are too buoyant to subduct and they can either obduct onto the continent (e.g., Caribbean and Wrangelia; Kerr *et al.* (1997); Lassiter *et al.* (1995)) or when arriving to the subducting magins start subduction in the opposite direction (e.g., Ontong Java plateau; Neal *et al.* (1997)). The Archean mantle whose temperature was a few hundred degrees higher than the post-Archean mantle could produce thicker oceanic crust of 20–25 km (Sleep and Windley, 1982) and the plume-related plateaus could reach an average thickness of ~30 km (e.g., Puchtel *et al.*, 1997). Some of these thick oceanic plateaus might be too warm and too buoyant when they reach the subduction margins and remain susceptible to subduction. Cloos (1993) and Abbott and Mooney (1995) argued that oceanic plates with crust thickness over ~25 km are not subductable regardless of their age.

Nevertheless, several oceanic plateaus are currently being consumed, either by collision or by subduction. Among these are the Nazca and Juan Fernandez Ridges off South America and the Louisville and Marcus-Necker Rises in the western Pacific (Pilger, 1978; Cross and Pilger, 1978; Nur and Ben-Avraham, 1981). These plateaus, which are presently being subducted, can clearly be associated with gaps in seismicity of the downgoing slab (e.g., Nur and Ben-Avraham, 1982). However, even more pronounced than these gaps are gaps in volcanism, associated with these plateaus. Close spatial association was found between zones of collision or subduction of plateaus and gaps in volcanism in the Pacific (McGeary *et al.*, 1985). Thus, these currently active areas of continental accretion are clearly identifiable.

Large portions of the Pacific margins, especially the northeast margin, are made of accreted or allochthonous terranes (**Figure 8(a)**). The best-understood accreted terranes are those in the northern cordillera of western North America, particularly in southern Alaska and British Columbia (Coney *et al.*, 1980; Monger, 1993). Paleomagnetic evidence indicate that many of the North Pacific accreted terranes in Alaska and northeast Asia migrated several thousand kilometers over periods of tens of millions of years prior to their accretion to the margins. The accreted terranes comprising the Canadian–Alaskan cordillera are predominantly juvenile in composition (e.g., Samson and Patchett (1991) estimated that ~50% by mass of the Canadian segment of the cordillera was juvenile crustal material). This evaluation is based on the initial $^{87}Sr/^{86}Sr$ and ε_{Nd} values of lithological assemblages from both outboard and inboard terrains (e.g., Alexander, Cache

Creek, and Slide Mountain), as well as arc assemblages like Quesnel (Armstrong (1988); Armstrong and Ward (1993); Samson *et al.* (1989, 1990, 1991); and other references listed by Patchett and Samson (2003) in their recent review on this topic).

Future allochthonous terranes may be found in the oceans, in the various plateaus, which are present on the ocean floor (Ben-Avraham *et al.*, 1981). The plateaus are modern accreted terranes in migration, moving with the oceanic plates in which they are embedded and fated eventually to be accreted to continents adjacent to subduction zones. The present distribution of oceanic plateaus in the Pacific Ocean and the relative motion between the Pacific Plates and the Eurasian Plate suggest that the next large episode of continental growth will take place in the northwest Pacific margin, from China to Siberia. Similar situations may have occurred in the past in other parts of the world.

Thus, it may be possible to identify ancient accreted terranes in the geological record, based on

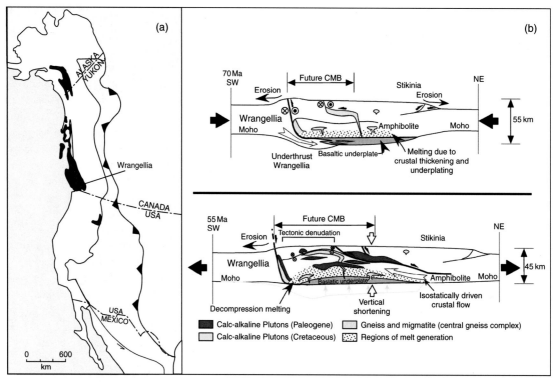

Figure 8 (a) Accretion plateaus along the western margins of North America and Alaska. The location of the Wrangallia plateau is marked in brown (after Ben-Avraham *et al.*, 1981). (b) Model of plateau accretion and transformation of the oceanic crust to new continental crust and lithosphere (after Hollister and Andronicos, 2006). The figure shows processes of accretion, shortening, and melting. Note the stratigraphic organization of the Cretaceous and the Paleogene calc-alkaline granites and location of zones of melting due to crustal thickening and basaltic underplating. Amphibole appears above the Moho and its production could control Nb/Th(REE) fractionation (Stein *et al.*, 1997).

chemical and geological evidence. The oldest preserved oceanic plateau sequences have been dated to 3.5 Ga in the Kaapvaal Shield, South Africa (Smith and Erlank, 1982; De Wit *et al.*, 1987) and the Pilbara craton in Australia (Green *et al.*, 2000). Greenstone belts of the Canadian Superior province, ranging in age from 3.0 to 2.7 Ga also contain lava groups that have been interpreted to be remnants of accreted oceanic plateaus (see recent review by Kerr (2003)). The evidence for an oceanic plateau origin is based on the occurrence of pillow basalts and komatiites without terrestrial sedimentary intercalations or sheeted like swarms, possessing the characteristics of Cretaceous oceanic plateaus. Below, we elaborate on the examples of crustal growth in the ANS and the Baltic Shield where evidence was found for formation of oceanic plateau and accretion of juvenile terrains to the existing older continents.

7.4.2 The Transformation of Oceanic Plateau to Continental Crust

The accretion of oceanic plateaus with a clear oceanic affinity to continental margins is of fundamental importance in the mechanism of crustal growth. The crustal structure of these plateaus, which was originally composed of one layer made of basalt, is modified with time and a typical structure of upper continental crust is developed. The most dramatic example of this process can be seen in British Columbia, northwest Canada. Here many accreted (or 'exotic') terranes were added to the continent during lithospheric plate convergence from the Archean to the present. The tectonic evolution of northwestern Canada involved a series of accretion events, alternating with periods of continental extension. Thus, it is tempting to suggest that the continent has 'grown' westward by the surface area of the accreted terranes (Cook and Erdmer, 2005). Recently, a very large-scale geophysical and geological transect, Slave-Northern Cordillera Lithospheric Evolution (SNORCLE), was carried out in northwestern Canada (Cook and Erdmer, 2005). The results show quite clearly that several of the large accreted terranes, which have traveled long distances embedded within oceanic plates before colliding with the continent, now have a continental crustal structure. For example, Wrangelia and Stikinia terranes, which were accreted during the Phanerozoic, and Hottah terrane, which was accreted during the Proterozoic, have continental crustal

structure with an upper crustal layer having granitic Vp velocities (Clowes *et al.*, 2005).

In a recent paper Hollister and Andronicos (2006) have proposed that crustal growth in the Coast Mountains, along the leading edge of the Canadian Cordillera, was the result of processes associated with horizontal flow of material during transpression and subsequent transtension, and the vertical accretion of mantle-derived melts. Their model has two distinct tectonic phases. The first occurred during a period of transpression when the continental crust was thickened to about 55 km, and mafic lower crust of the oceanic plateau Wrangellia was pushed under the thickened crust where basalt from the mantle heated it to temperatures hot enough for melting. The basalt and the melts mixed and mingled and rose into the arc along transpressional shear zones, forming calc-alkaline plutons. The second phase occurred as the arc collapsed when a change of relative plate motions resulted in transtension that accompanied intrusion of voluminous calc-alkaline plutons and the exhumation of the core of the Coast Mountains Batholith (CMB) (**Figure 8(b)**).

The calculation made by Schubert and Sandwell (1989) suggests that accretion of all oceanic plateaus to the continents on a timescale of 100 My, would result in a high rate of continental growth of $3.7 \, km^3 \, yr^{-1}$. This rate is much higher than continental growth rate based only on accretion of island arcs, which is $1.1 \, km^3 \, yr^{-1}$ (Reymer and Schubert, 1986), thus providing an explanation for the above-mentioned crustal growth rate dilemma.

7.4.3 Mantle Overturn and Crust Formation Episodes

In an attempt to link between various characteristics of mantle geochemistry to the dynamics and the apparent episodic mode of enhanced crustal growth (the Reymer and Schubert dilemma), Stein and Hofmann (1994) proposed that the geological history of the mantle-crust system has alternated between two modes of mantle convection and dynamic evolution, one approximating a two-layer convective style when 'normal mode' of plate tectonics prevails (Wilsonian tectonics) and the other characterized by significant exchange between the lower and upper mantle, when large plume heads form oceanic plateaus. The exchange periods apparently occurred over short time intervals – a few tens of million years (as estimated from the chronology of magmatism that is involved in the production of the oceanic plateaus), whereas the Wilsonian periods extend over several hundreds of millions of years (e.g., ~800 My in the case of the ANS, see below). The idea is that during mantle overturn episodes (termed by Stein and Hofmann as MOMO episodes) substantial amount of lower mantle material arrives to the Earth's surface (mainly via plume activity), replenishes the upper mantle in trace elements, and forms a new basaltic crust which eventually contributes juvenile material to the continental crust (via both subduction and accretion processes, **Figure 9**). Shortly after mantle upwelling and the production of thick oceanic crust, new subduction zones can be developed along the oceanic plateau margins as happened in the Mesozoic Pacific province. This mechanism views the formation of new subduction zones as a consequence of the production of thick oceanic crust and it's cooling.

The final product of the plume-thick oceanic-crust subduction system is a mass of continental crust that can be generated over a relatively short period of geological time. The sequence of events involving formation of an anomalously thick, oceanic lithosphere, which is modified by subduction at its margin explains both the formation of major continental segments during discrete episodes of time and their calc-alkaline affinities.

In an alternative view of the Reymer and Schubert dilemma, Patchett and Chase (2002) and Patchett and Samson (2003) suggested that the evidence for enhanced crust production during discrete episodes can reflect the tectonics of accretion processes alone, arguing for the role of transform faulting, which serves to pile up various terrains of juvenile crust in one restricted region (e.g., the Canadian–Alaskan cordillera, where slices of juvenile accreted crustal material were piled up following northward along-margin transport associated with transform faulting; see **Figure 8**). The transport and accretion of juvenile mantle terranes is a central part of the plume-accretion model of crustal growth (**Figure 9**). Yet, the additional important component in this model that is not required by the arc-accretion models is the important role of oceanic plateaus as the nucleus of the juvenile continental crust growth and possibly the locus of newly developed subduction zones. Moreover, the crustal terranes that were formed during the episodes of 'major orogeny' are not limited to geographically restricted regions but rather are distributed globally. The best example is that of the late Proterzoic Pan-African orogeny, when juvenile crust terranes were produced in Arabia, Nubia (the ANS discussed below), as well as in North Africa, New Zealand, and Patagonia. All these geographical-crustal segments were parts of the early Phanerozoic Gondwana. Considering a reconstruction of this old continent (**Figure 10**) it appears that continental segments within the Gondawana that can be

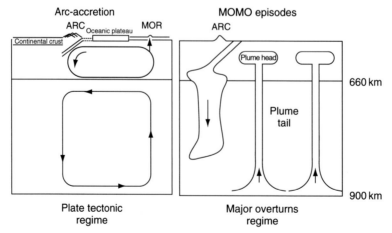

Figure 9 Mantle overturn and crust addition model (modified after the original MOMO model by Stein and Hofmann (1994)). The original MOMO model assumes that the mantle–crust system evolves through two modes of convection and dynamic evolution, one (MOMO overturn episodes) is characterized by significant exchange between lower and upper mantle and crust and the other by two-layer convective style when the typical plate tectonic Wilsonian cycle operates. The question of layered vs whole mantle convection has been continuously discussed in the literature and it is beyond the scope of this review. Here we present a modified representation of the model where we stress the accretion of the oceanic plateau to the existing continent and development of subduction zones on the margins of the plateau (similar to the scenario shown in **Figure 8**). Most of the existing subduction zones in the world are associated with plateaus though a few subduction zones are not clearly associated with plateaus.

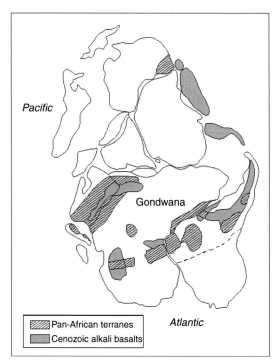

Figure 10 Schematic representation of regions that expose terranes of Late Proterozoic Pan-African juvenile magmatic sequences and Cenozoic alkali basalts over a reconstructed map of Gondwana. The juvenile magmatic sequences are characterized by Nd and Sr isotope compositions (e.g., εNd $= +5 \pm 1$ and $^{87}Sr/^{86}Sr =$ 0.7028 ± 2; Stein and Hofmann (1994); Stein (2003)). The Pb isotopic compositions of the Gondwana basalts are marked in **Figure 11**. These basalts show the plume-lithosphere connection) that were derived from the juvenile Pan-African lithospheric mantle. The areas that are best documented are those of the Arabian–Nubian Shield on both sides of the current Red Sea (closed in the reconstruction).The figure suggests that during the Late Proterozoic Pan-African orogeny juvenile terranes were accreted to an existing continental nucleus in a similar manner to the growth that was described for the North America continent. The distribution of the juvenile Pan-Africa terranes may mark the loci of the subduction zones that were developed at that time along the margins of the Pan-African oceanic plateaus.

identified as juvenile Pan-African crustal or lithospheric mantle terranes (see Stein (2003) and caption of **Figure 10**) form a 'belt' along the possible margins of a mid-or-late Proterozoic continent. All of these segments are characterized by juvenile Pan African Nd and Sr isotope compositions that in turn are consistent with plume (MOMO) type mantle (see MOMO evolution line in Stein and Hofmann (1994)).

7.5 Rapid Growth of Major Continental Segments

In this section we describe the magmatic histories of several prominent continental crust segments (termed as major orogens). We first elaborate on the evolution of continental crust comprising the ANS, which is one of the best-exposed crustal terranes (due to the Cenozoic uplift of the Red Sea shoulders). Then, we describe in chronological order the other major orogens. Overall, we show the general similarity in the evolutionary pattern of all these orogenic episodes.

7.5.1 The ANS Orogeny ($\sim$0.9–0.6 Ga)

The excellent preserved exposures of the Late Proterozoic crystalline basement of the ANS and the sequences of Phanerozoic alkali basalts provide an opportunity to monitor the magmatic history of the ANS over the past $\sim$900 My. The low initial $^{87}Sr/^{86}Sr$ ratios of the Late Proterozoic magmas of Sinai, Jordan, northern Saudi Arabia, and the Eastern Desert of Egypt preclude significant involvement of an old pre-Pan-African felsic continental precursor in this part of the northern ANS, indicating that basement magmas were derived from either a pre-existing mafic crust or directly from the mantle. The absence of older continental crust is further indicated by mantle-type oxygen isotope ratios (e.g., $\delta^{18}O = +5.5$; Bielski (1982)).

The Late Proterozoic magmatic history of the ANS can be divided into four evolutionary phases.

7.5.1.1 Phase I
Production of thick sequences of tholeiitic during a very short interval of time ($\sim$900 $\pm$ 30 My). The (meta) tholeiites are characterized by chondritic REE patterns and 'mantle-type' Th/Nb ratios and were interpreted as oceanic basalts similar to the plateaus basalts (Stein and Goldstein (1996); Stein (2003) and references therein). It is important to emphasize that along with the production of the plateau magmas, thinner sequences of meta-tholeiites were interpreted as 'normal' mid-oceanic tholeiites erupting along spreading ridges (the Old Meta Volcanics in Egypt, described by Stern (1981)). The two ANS meta-tholeiite suites resemble the magmatic–tectonic association of the above-described Pacific Ontong Java plateau–Nauru Basin basalts.

7.5.1.2 Phase II

Upon reaching a continental margin, the thick oceanic plateau lithosphere resisted subduction and plate convergence occurred on its margins, forming the ANS calc-alkaline magmas. The subduction activity lasted for ~200 My (between ~870 and 650 Ma), similar to the duration of the activity of Phanerozoic arc magmatism in Japan. The arcs and oceanic plateaus were accreted to the old Gondwanaland producing a thick and mainly mafic crust and lithospheric mantle, which served later as sources for the calc-alkaline granitic batholiths, alkali granites, and Phanerozoic alkali basalts.

The enriched plume-type mantle of the early ANS was exhausted rapidly by the subduction processes and the Gabel Gerf ophiolitic magmas (~800–750 Ma), which are assumed to represent the arc environment, show isotope mixtures between the enriched ('plume lithosphere mantle') and depleted ('MORB-type mantle') components (see evolution lines in **Figure 11**).

7.5.1.3 Phase III

Rapid production (~640−600 Ma) of vast amounts of calc-alkaline granitic and dioritic magmas. They form batholiths that cover large areas, mainly in the northern ANS and their mechanism of production is not entirely clear. Melting of a metamorphosed mafic (e.g., amphibolitic) crust probably led to the production of the calc-alkaline granites.

However, the production of the calc-alkaline granitic magmas certainly required the supply of 'heat and water', a problem that is not clearly understood. One possibility is a relation between asthenospheric rise associated with delamination of a thick lower crust and lithosphere that were produced in the earlier evolutionary stage of the island arc magmatism.

7.5.1.4 Phase IV

The orogenic cycle terminated by production of alkali granites and related rhyolites (~600−530 Ma; Bielski (1982); Mushkin *et al.* (2003)). The parental magmas of the alkaline magmas were probably derived from the lithospheric mantle and underwent extensive differentiation at shallow crustal levels (Mushkin *et al.*, 2003).

The transition from calc-alkaline magmatism to alkaline magmatism associated with the lithospheric mantle melting after 600 Ma reflects the ending of plate convergence. In the ANS this transition is

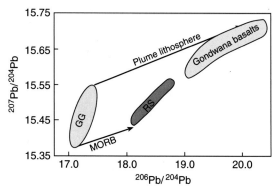

Figure 11 The distribution of Late Proterozoic oceanic basalts (the Gabel Gerf (GG) ophiolites, Red Sea basalts (RS), and Gondwana basalts in the Pb–Pb diagram (after Stein and Goldstein, 1996; Stein, 2003). Stein and Goldstein proposed that the GG ophiolites lie in the diagram on a mixing trend between the asthenopsheric MORB mantle and asthenopsheric plume-type mantle (in similar tectonic-magmatic settings to the configuration of the Nauru Basin and East Pacific Rise in the Pacific Ocean). The Gondwana basalts (location marked in **Figure 10**) lie on an evolution trend together with the Late Proterozoic enriched ophiolites and other magmas (e.g., galenas and K-feldspars not shown here, see Stein and Goldstein, (1996)) and suggest that the lithospheric mantle beneath the Arabian continent was generated during the Late Proterozoic crust-formation events. It can be regarded as a 'frozen' mantle wedge, which previously accommodated the production of calc-alkaline magmas and fractionation–transportation processes of trace elements en route to shallower levels of the crust (the subject is elaborated below). The Red Sea basalts lie on a mixing trend between the lithospheric derived basalts and the MORB-type asthenosphere. Stein and Hofmann (1992) noted that enriched-type Red Sea basalts erupt on the southern and northern edges of the Red Sea Basin while depleted MORB erupt only in the central trough where 'real' spreading occurs. They regarded this configuration as a strong evidence for the existence of the plume-type mantle above the asthenospheric MORB. The latter is rising into the rifting lithosphere and continental crust. This also provides strong evidence for the MORB-type composition of the upper mantle. The Gondwana basalts show a mixing relation with the Red Sea asthenosphere suggesting that MORB is rising into the rifting lithosphere (e.g., in the Ross Sea rift, Ahaggar, and the Dead Sea rift). This upwelling has important implications for the heating of the lithosphere and the lower crust, as well as for production of basaltic melts from the lithosphere and provide evidence for asthenospheric underplating (Stein, 2006).

associated with a major uplift (Black and Liegeois, 1993). Similar rapid transitions in the stress field associated with continental collisions were described in other localities such as the Variscan province in Europe, the Basin and Range province in the USA, and the Tibetan plateau (Klemperer *et al.*, 1986; Rey,

1993; Costa and Rey, 1995). In the Himalayas, major uplift and the appearance of alkaline volcanism has been related to the thinning and melting of the lithosphere after the ending of plate convergence.

7.5.2 The Sumozero–Kenozero and Kostomuksha Orogeny (~2.8–2.9 Ga)

The magmatic history of the ANS appears to represent a complete MOMO cycle illustrating a primary means of continental crust formation and cratonization throughout Earth history. The model may be particularly relevant in cases of rapid continental growth as demanded by the Reymer and Schubert dilemma. Evidence for the existence of magmatic

provinces that resemble oceanic plateaus was found in other older continental crust provinces (**Figure 12**). The 2.9 Ga Sumozero–Kenozero greenstone belt in the SE Baltic Shield comprises a several kilometer-thick lower unit that is made of submarine mafic–ultramafic units (with plume-type Nb/U ratios of 43 ± 6) that were interpreted as representing oceanic plateau magmas and the upper unit consisting of arc-type magmas (Puchtel *et al.*, 1999). The authors suggested that the over-thickened plume-related oceanic crust and the overlying arc-type magmas were later accreted to the exisiting TTG-type older crust of the Vodla block.

The ~2.8 ka Kostomuksha greenstone belt in the NW Baltic Shield consists of komatiitic–basaltic

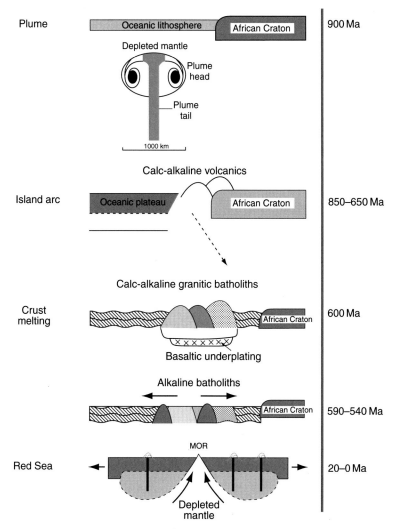

Figure 12 Schematic model illustrating the growth and fate of the continental crust and lithospheric mantle in the Arabian–Nubian Shield. Similar (but clearly not identical) structure of magmatic histories can be ascribed to other major crustal segments (e.g., Canadian Shield, Birimian orogen, the Mesozoic Pacific).

submarine lavas and volcano-clastics whose formation was attributed by Puchtel *et al.* (1997) to plume activity and the production of a thick oceanic crust–oceanic plateau. The plateau reached the Baltic continental margin but was too buoyant to subduct and became a new part of the continent. Using the example of the Cretaceous Caribbean plateau (Kerr *et al.*, 1997), Puchtel and colleagues suggested that only the upper part of the thick komatiitic–basaltic crust of the proposed Baltic oceanic crust was imbricated and obducted on the ancient continent, whereas the deeper zones were subducted to the mantle. The subduction of this lower crust gave rise to subduction-related volcanism that erupted within the accreted sequences. This mechanism distinguished the thick plateau type oceanic crust from the thinner crust that is represented in the ophiolite sections. Moreover, it provides a plausible explanation for the sequence of events that occurred within a 'cycle' of plateau formation–accretion–delamination–subduction–calc-alkaline arc-type magmatism.

7.5.3   The Superior Province (~2.7 Ga)

The ~2.7 Ga Superior province, the largest Archean craton on Earth comprises several continental and oceanic mega-terranes that were formed and accreted rapidly (the ages of the various superprovinces lie in the range of 3.1–2.67 Ga and acceretionary tectonics occurred between 2.74 and 2.65 Ga (cf. Percival and Williams, 1989). The geological history and magmatic stratigraphy of the Superior province was divided by Arth and Hanson (1975) into four stages: (1) production of tholeiitic basalts by shallow melting of mantle peridotites, followed by fractional crystallization that produces basaltic-andesites; (2) formation of trondhjemite, tonalite, and dacites by melting of eclogites in mantle depth; (3) intrusion of quartz monzonites; and (4) intrusion of post-kinematic syenites.

The general pattern of this history is strikingly similar to the four growth phases and magmatic evolution of the above-described late Proterozoic ANS or the Proterozoic Birimian orogeny. In the framework of the MOMO model this means: production of thick sequences of tholeiitic magmas in the plume-related intraplate (oceanic) environment, followed by development of subduction zones and their related magmatism, followed by intracrustal and intralithospheric melting that recycled the material formed in the early two phases. The main

difference between these orogenic suites lies in the appearance of the Archean TTG assemblage.

Polat and Kerrich (2001a, 2001b) and Polat *et al.* (1998) described the formation of the 2.7 Ga Wawa and Abitibi greenstone subprovinces of the Superior province as an example of episodic continental growth, by lateral spread of subduction-accretion complexes. They recognized two principal volcanic associations: (1) tholeiitic basalt-komatiite; and (2) tholeiitic to calc-alkaline bimodal basalt-rhyolite. Tholeiitic basalts of the former association are characterized by near-flat REE patterns and Th/U, Nb/Th, Nb/La, and La/Sm-pm ratios that span the primitive mantle values. Transitional to alkaline basalts and Al-depleted komatiites have fractionated REE patterns and OIB-like trace-element signatures. Polat and his co-authors interpreted these assemblages as representing a oceanic plateau derived from a heterogeneous mantle plume. The bimodal association has fractionated REE patterns and negative Nb, Ta, P, and Ti anomalies typical of arc magmas. Tonalite plutons derived from partial melting of subducted oceanic slabs intrude the subduction-accretion complex as the magmatic arc axis migrated toward the trench. The eruption of voluminous ocean plateau and island arc volcanic sequences and the following deposition of kilometer-thick siliciclastic trench turbidites, and intrusion of the supracrustal units by syn- to post-kinematic TTG plutons occurred in the relatively short time period of 2.75–2.65 Ga (a classic representation of the Reymer and Schubert super-rapid crustal formation events). The above-mentioned events were accompanied by contemporaneous intense poly-phase deformation, regional greenschist to amphibolite facies metamorphism, and terrain accretion along a north-northwest-dipping subduction zone. The description of the history of the Superior province is consistent with the plume upwelling – terrane accretion model where the growth of the new continental crust involved formation and accretion of oceanic plateaus, production of magmatic arcs, and accretion of older continental fragments.

7.5.4   The Birimian Orogeny (~2.2 Ga)

Another well-studied case of episodic crustal formation is that of the Birimian orogeny that lasted between ~2.2 and ~2.1 Ga (Abouchami *et al.*, 1990; Boher *et al.*, 1992). The early stage in the growth of the Birimian juvenile crust was marked by eruption of thick sequence of basalts in oceanic basin that were interpreted as oceanic plateau (formed in an environment remote from any African-Archean crust). Shortly after, calc-alkaline magmas were produced, probably in island arcs that were developed on the margins of the oceanic plateau, which then collided with the Man Archean Craton. Overall, the magmatic-geodynamic history of the Birimian orogeny strikingly resembles that of the ANS and to some extent that of the Superior province.

7.6 Summary

The continents come from the mantle. Large submarine oceanic plateaus whose production was associated with mantle plumes may have formed the basis for the growth of continental crust, at least since the late Archean time. The chronological evidence suggests that plateau formation was rapid and episodic and that it was followed by enhanced activity of arcs and calc-alkaline magmatism. Thick oceanic plateaus are buoyant and resist subduction. Yet, some plateaus were subducted, reflecting probably internal differentiation within the plateaus and the geodynamic conditions (e.g., thermal state of the mantle). Subduction of segments of plateaus can in turn enhance arc magmatism, because its altered parts can increase the water flux at depth of magma generation. It should be also emphasized that oceanic plateaus may well be an important component of current arc magmatism, as some of the currently active arcs are located close to anomalously thick oceanic crust (e.g., the central Aleutians and Lesser Antilles).

Along with the production of new crust, the lithospheric mantle is evolved . It probably represents a frozen mantle wedge where production of magmas and migration of fluids and incompatible trace elements occurs, thus both the continental crust and the lithospheric mantle preserved the complicated history of continent growth. Besides the plateau and arc accretion and calc-alkaline magmatism, delamination of lithospheric mantle and lower crust and underplating by rising asthenospheric magmas, as well as lithospheric melting are all important processes that lead to the magmatic and thermal maturation of the continents. The fate of the continent relates eventually to the geodynamics and thermal conditions in the mantle.

Many of the abovementioned topics require further study. In particular, we emphasize the importance of understanding the origin of the 'crustal type' Vp velocities that indicate a continental structure for some of the oceanic plateaus. Are these velocities associated with significant magmatic differentiation before accretion of the plateau to the existing continents? The mechanism of production of the calc-alkaline granitic batholiths is not as well established as is the composition of 'mid-crust'. The processes of asthenospheric underplating, crustal foundering, and crustal recycling and their relation to the plate tectonic cycles and mantle dynamics through geological time require more attention.

Finally, among the numerous works and thorough reviews on continental crust composition and evolution we consider and 'rephrase' the Taylor and Campbell statement: no water, no oceans, no plate tectonics, no granites, no continents, as a major characteristic of the environment of production of the Earth continental crust.

References

Abbott D and Mooney W (1995) The structural and geochemical evolution of the continental-crust – Support for the oceanic plateau model of continental growth. *Reviews of Geophysics* 33: 231–242.

Abouchami W, Boher M, Michard A, and Albarede F (1990) A major 2.1 Ga event of mafic magmatism in west Africa – An early stage of crustal accretion. *Journal of Geophysical Research-Solid Earth and Planets* 95: 17605–17629.

Albarède F (1998) The growth of continental crust. *Tectonophysics* 296: 1–14.

Anderson AT (1982) Parental melts in subduction zones: Implications for crustal evolution. *Journal of Geophysical Research-Solid Earth and Planets* 87: 7047–7070.

Anderson DL (1994) Superplumes or supercontinents? *Geology* 22: 39–42.

Armstrong RL (1968) A model for the evolution of strontium and lead isotopes in a dynamic Earth. *Reviews of Geophysics* 6: 175–199.

Armstrong RL (1981a) Radiogenic Isotopes – The case for crustal recycling on a near-steady-state no-continental-growth Earth. *Philosophical Transactions of the Royal Society of London Series A-Mathematical Physical and Engineering Sciences* 301: 443–472.

Armstrong RL (1981b) Comment on 'Crustal growth and mantle evolution: Inferences from models of element transport and Nd and Sr isotopes'. *Geochimica et Cosmochimica Acta* 45: 1251.

Armstrong RL (1988) Mesozoic and early Cenozoic magmatic evolution of the Canadian Cordillera. *Geological Society of America Special Paper* 218: 55–91.

Armstrong RL (1991) The persistent myth of crustal growth. *Australian Journal of Earth Sciences* 38: 613–640.

Armstrong RL and Ward PL (1993) Late Triassic to earliest Eocene magmatism in the North American Cordillera: Implications for the Western Interior Basin. In: Caldwell WGE and Kauffman EG (eds.) *Geololgical Association Canada Special Paper 39: Evolution of the Western Interior Basin*, pp. 49–72. Boulder, CO: Geological Society of America.

Arth JG and Hanson GN (1975) Geochemistry and origin of Early Precambrian crust of Northeastern Minnesota. *Geochimica et Cosmochimica Acta* 39: 325–362.

Barth MG, McDonough WF, and Rudnick RL (2000) Tracking the budget of Nb and Ta in the continental crust. *Chemical Geology* 165: 197–213.

Ben-Avraham Z, Nur A, Jones D, and Cox A (1981) Continental accretion and orogeny: From oceanic plateaus to allochthonous terranes. *Science* 213: 47–54.

Bennett VC and Depaolo DJ (1987) Proterozoic crustal history of the Western United-States as determined by neodymium isotopic mapping. *Geological Society of America Bulletin* 99: 674–685.

Bielski M (1982) *Stages in the Evolution of the Arabian–Nubian Massif in Sinai*. PhD Thesis, 155pp, The Hebrew University of Jerusalem.

Black R and Liegeois J (1993) Cratons, mobile belts, alkaline rocks and continental lithospheric mantle: The Pan-African testimony. *Journal of the Geological Society* 150: 89–98.

Boher M, Abouchami W, Michard A, Albarede F, and Arndt NT (1992) Crustal growth in west Africa at 2.1 Ga. *Journal of Geophysical Research-Solid Earth* 97: 345–369.

Bohlen SR and Mezger K (1989) Origin of granulite terranes and the formation of the lowermost continental-crust. *Science* 244: 326–329.

Bowring SA and Housh T (1995) The Earth's early evolution. *Science* 269: 1535–1540.

Bowring SA, Housh TB, and Isachsen CE (1990) The Acasta Gneisses: Remnants of Earth earliest crust. In: Newsom HE and Jones JH (eds.) *Origin of the Earth*, pp. 319–343. New York: Oxford University Press.

Brown GC (1979) The changing pattern of batholith emplacement during Earth history. In: Atherton MP and Tarney U (eds.) *Origin of Granite Batholiths: Geochemical Evidence*, pp. 106–115. Orpington: Shiva.

Campbell IH and Taylor SR (1983) No water, no granites – No oceans, no continents. *Geophysical Research Letters* 10: 1061–1064.

Carlson RL, Christensen NI, and Moore RP (1980) Anomalous crustal structures on ocean basins: Continental fragments and oceanic plateaus. *Earth and Planetary Science Letters* 51: 171–180.

Clarke FW (1889) The relative abundance of the chemical elements. *Philosophical Society of Washington Bulletin* XI: 131–142.

Clarke FW and Washington HS (1924) The composition of the Earth's crust. *USGS Professional Paper* 127: 117.

Cloos M (1993) Lithospheric buoyancy and collisional orogenesis – Subduction of oceanic plateaus, continental margins, island arcs, spreading ridges, and seamounts. *Geological Society of America Bulletin* 105: 715–737.

Clowes RMC, Hammer P T, Fernández-Viejo G, and Welford JK (2005) Lithospheric structure in northwestern Canada from Lithoprobe seismic refraction and related studies: A synthesis. *Canadian Journal of Earth Sciences* 42: 1277–1293.

Collerson KD and Kamber BS (1999) Evolution of the continents and the atmosphere inferred from Th-U-Nb systematics of the depleted mantle. *Science* 283: 1519–1522.

Condie KC (1990) Growth and accretion of continental crust: Inferences based on Laurentia. *Chemical Geology* 83: 83–194.

Condie KC (1993) Chemical-composition and evolution of the upper continental-crust – Contrasting results from surface samples and shales. *Chemical Geology* 104: 1–37.

Condie KC (1997) *Plate Tectonics and Crustal Evolution*. Oxford, UK: Butterworth-Heinemann.

Condie KC (1998) Episodic continental growth and supercontinents: A mantle avalanche connection? *Earth and Planetary Science Letters* 163: 97–108.

Condie KC (2000) Episodic continental growth models: Afterthoughts and extensions. *Tectonophysics* 322: 153–162.

Condie KC (2002) Continental growth during a 1.9-Ga superplume event. *Journal of Geodynamics* 34: 249–264.

Coney PJ, Jones DL, and Monger JWH (1980) Cordilleran suspect terranes. *Nature* 288: 329–333.

Cook FA and Erdmer P (2005) An 1800 km cross section of the lithosphere through the northwestern North American plate: Lessons from 4.0 billion years of Earth's history. *Canadian Journal of Earth Sciences* 42: 1295–1311.

Costa S and Rey P (1995) Lower crustal rejuvenation and growth during post-thickening collapse – Insights from a crustal cross-section through a variscan metamorphic core complex. *Geology* 23: 905–908.

Cross TA and Pilger RH (1978) Constraints on absolute motion and plate interaction inferred from cenozoic igneous activity in Western United States. *American Journal of Science* 278: 865–902.

De Wit MJ, Hart RA, and Hart RJ (1987) The Jamestown ophiolite complex Barberton mountain belt: A section through 3.5 Ga oceanic crust. *Journal of Africa Earth Sciences* 6: 681–730.

DePaolo DJ (1981) Neodymium isotopes in the Colorado Front Range and crust–mantle evolution in the Proterozoic. *Nature* 291: 193–196.

DePaolo DJ, Linn AM, and Schubert G (1991) The continental crustal age distribution – Methods of determining mantle separation ages from Sm-Nd isotopic data and application to the Southwestern United States. *Journal of Geophysical Research-Solid Earth and Planets* 96: 2071–2088.

Dewey JF and Windley BF (1981) Growth and differentiation of the continental-crust. *Philosophical Transactions of the Royal Society of London: Series A-Mathematical Physical and Engineering Sciences* 301: 189–206.

Drummond MS, Defant MJ, and Kepezhinskas PK (1996) The petrogenesis of slab derived tronhejmite – tonalite – dacite/adakite magmas. *Transactions of the Royal Society of Edinborough.*

Eldholm O and Coffin MF (2000) Large igneous provinces and plate tectonics. *AGU Monograph* 121: 309–326.

Galer SJG (1991) Interrelationships between continental freeboard, tectonics and mantle temperature. *Earth and Planetary Science Letters* 105: 214–228.

Gallet S, Jahn BM, Lanoe BV, Dia A, and Rossello E (1998) Loess geochemistry and its implications for particle origin and composition of the upper continental crust. *Earth and Planetary Science Letters* 156: 157–172.

Gao S, Luo TC, Zhang BR, *et al.* (1998) Chemical composition of the continental crust as revealed by studies in East China. *Geochimica et Cosmochimica Acta* 62: 1959–1975.

Goldschmidt VM (1933) Grundlagen der quantitativen Geochemie. *Fortschr. Mienral. Kirst. Petrogr* 17: 112.

Goldschmidt VM (1958) *Geochemistry*. Oxford: Oxford University Press.

Goodwin AM (1991) *Precambrian Geology*. London: Academic Press.

Green M (2001) Early Archaean crustal evolution: Evidence from ~3.5 billion year old greenstone successions in the Pilgangoora Belt, Pilbara Craton, Australia, The University of Sydney.

Green MG, Sylvester PJ, and Buick R (2000) Growth and recycling of early Archaean continental crust: Geochemical evidence from the Coonterunah and Warrawoona Groups, Pilbara Craton, Australia. *Tectonophysics* 322: 69–88.

Griffin WL and O'reilly SY (1987) Is the Continental Moho the crust–mantle boundary? *Geology* 15: 241–244.

Haschke M and Ben-Avraham Z (2005) Adakites along oceanic transforms? *Geophysical Research Letters* 32: L15302 (10: 1029/2005GLO23468).

Hattori Y, Suzuki K, Honda M, and Shimizu H (2003) Re-Os isotope systematics of the Taklimakan Desert sands, moraines and river sediments around the Taklimakan Desert, and of Tibetan soils. *Geochimica et Cosmochimica Acta* 67: 1195–1205.

Hawkesworth CJ, Gallagher K, Hergt JM, and Mcdermott F (1993) Mantle and slab contributions in arc magmas. *Annual Review of Earth and Planetary Sciences* 21: 175–204.

Hofmann AW (1988) Chemical differentiation of the Earth: The relationship between mantle, continental crust, and oceanic crust. *Earth and Planetary Science Letters* 90: 297–314.

Hofmann AW, Jochum KP, Seufert M, and White WM (1986) Nb and Pb in oceanic basalts: New constraints on mantle evolution. *Earth and Planetary Science Letters* 79: 33–45.

Hollister LS and Andronicos CL (2006) Formation of new continental crust in western British Columbia during transpression and transtension. *Earth and Planetary Science Letters* 249: 29–38.

Hurley PM and Rand JR (1969) Pre-drift continental nuclei. *Science* 164: 1229–1242.

Ishikawa A, Shigenori M, and Komiya T (2004) layered lithospheric mantle beneath the Ontong Java plateau: Implications from xenoliths in Alonite, Malaita, Solomon Islands. *Journal of Petrology* 47: 2011–2044.

Kay RW and Kay SM (1981) The nature of the lower continental-crust – Inferences from geophysics, surface geology, and crustal xenoliths. *Reviews of Geophysics* 19: 271–297.

Kay RW and Mahlburgkay S (1991) Creation and destruction of lower continental-crust. *Geologische Rundschau* 80: 259–278.

Kelemen PB (1995) Genesis of high Mg-number Andesites and the continental-crust. *Contributions to Mineralogy and Petrology* 120: 1–19.

Kerr AC (2003) Oceanic plateaus. In: Rudnick R (ed.) *Treatise on Geochemistry, vol. 10: The Crust*, pp. 537–565. Amsterdam: Elsevier.

Kerr AC, Marriner GF, Arndt NT, *et al.* (1996) The petrogenesis of Gorgona komatiites, picrites and basalts: New field, petrographic and geochemical constraints. *Lithos* 37: 245–260.

Kerr AC, Tarney J, Marriner GF, Klaver GT, Saunders AD, and Thirlwall MF (1996) The geochemistry and petrogenesis of the late-Cretaceous picrites and basalts of Curacao, Netherlands Antilles: A remnant of an oceanic plateau. *Contributions to Mineralogy and Petrology* 124: 29–43.

Kerr AC, Tarney J, Marriner GF, Nivia A, and Saunders AD (1997) The Caribbean–Colombian Cretaceous igneous province: The internal anatomy of an oceanic plateau. In: John JM and Millard FC (eds.) *Geophysical Union Monograph: Large Igneous Provinces; Continental, Oceanic and Planetary Flood Volcanism*, pp. 45–93. Washington, DC: American Geophysical Union.

Kessel R, Stein M, and Navon O (1998) Petrogenesis of late Neoproterozoic dikes in the northern Arabian–Nubian shield – Implications for the origin of A-type granites. *Precambrian Research* 92: 195–213.

Kimura G and Ludden J (1995) Peeling oceanic-crust in subduction zones. *Geology* 23: 217–220.

Klemperer SL, Hauge TA, Hauser EC, Oliver JE, and Potter CJ (1986) The Moho in the northern Basin and Range province along the COCORP 40°N seismic reflection transect. *Geological Society America Bulletin* 97: 603–618.

Kroenke LW (1974) Origin of continents through development and Coalescence of oceanic flood basalt plateaus. *Transactions-American Geophysical Union* 55: 443–443.

Kroner A, Compston W, and Williams IS (1989) Growth of Early Archean crust in the Ancient Gneiss Complex of Swaziland as revealed by Single Zircon Dating. *Tectonophysics* 161: 271–298.

Langmuir CH, Klein EM, and Plank T (1992) Petrological constraints on mid-ocean ridge basalts. In: Morgan JP and Blackman DK (eds.) *Geophysical Monograph: Mantle Flow and Melt Generation at Mid-Ocean Ridges*, pp. 182–280. Washington, DC: American Geophysical Union.

Larson RL (1991) Latest pulse of Earth – Evidence for a midcretaceous superplume. *Geology* 19: 547–550.

Lassiter JC, Depaolo DJ, and Mahoney JJ (1995) Geochemistry of the Wrangellia Flood-Basalt Province – Implications for the role of continental and oceanic lithosphere in flood-basalt genesis. *Journal of Petrology* 36: 983–1009.

Mahoney JJ, Storey M, Duncan RA, Spencer KJ, and Pringle M (1993a) Geochemistry and age of the Ontong Java Plateau. In: Pringle MS, Sager WW, Sliter WV, and Stein S (eds.) *American Geophysical Union Monograph: The Mesozoic Pacific: Geology, Tectonics, and Volcanism*, pp. 233–261. Washington, DC: American Geophysical Union.

Mahoney JJ, Storey M, Duncan RA, Spencer KJ, and Pringle M (1993b) Geochemistry and geochronology of Leg 130 basement lavas; nature and origin of the Ontong Java Plateau. *Proceedings of the Ocean Drilling Program, Scientific Results* 130: 3–22.

McCulloch MT and Bennett VC (1993) Evolution of the early Earth – Constraints from Nd-143-Nd-142 isotopic systematics. *Lithos* 30: 237–255.

McCulloch MT and Bennett VC (1994) Progressive growth of the Earths continental-crust and depleted mantle – Geochemical constraints. *Geochimica et Cosmochimica Acta* 58: 4717–4738.

McCulloch MT and Wasserburg GJ (1978) Sm-Nd and Rb-Sr chronology of continental crust formation. *Science* 200: 1003–1011.

McGeary S, Nur A, and Benavraham Z (1985) Spatial gaps in arc volcanism – The effect of collision or subduction of oceanic plateaus. *Tectonophysics* 119: 195–221.

McKenzie D and Bickle MJ (1988) The volume and composition of melt generated by extension of the lithosphere. *Journal of Petrology* 29: 625–679.

McKenzie D and O'Nions RK (1983) Mantle reservoirs and ocean island basalts. *Nature* 301: 229–231.

McLennan SM (1988) Recycling of the continental-crust. *Pure and Applied Geophysics* 128: 683–724.

Mezger K (1992) Temporal evolution of regional granulite terranes: Implications for the evolution of the lowermost crust. In: Fountain DM, Arculus R, and Kay RV (eds.) *Continental Lower Crust*, pp. 447–478. New York: Elsevier Science.

Monger JWH (1993) Canadian cordilleran tectonics: From geosynclines to crustal collage. *Canadian Journal Earth Science* 30: 209–231.

Moorbath S (1975) Evolution of Precambrian crust from strontium isotopic evidence. *Nature* 254: 395–398.

Moorbath S (1978) Age and isotope evidence for evolution of continental crust. *Philosophical Transactions of the Royal Society of London Series A-Mathematical Physical and Engineering Sciences* 288: 401–112.

Moorbath S, Gale NH, Pankhurs RJ, Mcgregor VR, and Onions RK (1972) Further rubidium–strontium age determinations on very Early Precambrian rocks of Godthaab District, west Greenland. *Nature-Physical Science* 240: 78–82.

Moorbath S, Whitehouse MJ, and Kamber BS (1997) Extreme Nd-isotope heterogeneity in the early Archaean – Fact or fiction? Case histories from northern Canada and west Greenland. *Chemical Geology* 135: 213–231.

Mushkin A, Navon O, Halicz L, Hartmann G, and Stein M (2003) The petrogenesis of A-type magmas from the Amram Massif, southern Israel. *Journal of Petrology* 44: 815–832.

Neal CR, Mahoney JJ, Kroenke LW, Duncan RA, and Petterson MG (1997) The Ontong Java Plateau. In: Mahoney JJ and Coffin M (eds.) *American Geophysical Union Monograph : On Large Igneous Provinces*, pp. 183–216. Washington, DC: American Geophysical Union.

Nelson BK and DePaolo DJ (1984) 1,700-Myr Greenstone volcanic successions in Southwestern North-America and isotopic evolution of proterozoic mantle. *Nature* 312: 143–146.

Nelson BK and DePaolo DJ (1985) Rapid production of continental-crust 1.7 to 1.9B.Y. Ago – Nd isotopic evidence from the basement of the North-American mid-continent. *Geological Society of America Bulletin* 96: 746–754.

Nur A and Ben-Avraham Z (1977) Lost Pacifica continent. *Nature* 270: 41–43.

Nur A and Ben-Avraham Z (1978) Speculations on mountain building and the lost Pacifica continent. *Journal of Physucs of the Earth Supplement* 26: S21–S37.

Nur A and Ben-Avraham Z (1981) Volcanic gaps and the consumption of aseismic ridges in South America. *Memoirs of the Geological Society of America* 154: 729–740.

Nur A and Ben-Avraham Z (1982) Displaced terranes and mountain building. In: Hsu KJ (ed.) *Mountain Building Processes*, pp. 73–874. London: Academic Press.

Nye CJ and Reid MR (1986) Geochemistry of primary and least fractionated lavas from Okmok Volcano, central Aleutians – Implications for arc magmagenesis. *Journal of Geophysical Research-Solid Earth and Planets* 91: 271–287.

O'Nions RK, Evensen NM, and Hamilton PJ (1979) Geochemical modeling of mantle differentiation and crustal growth. *Journal of Geophysical Research* 84: 6091–610.

Oxburgh ER and Parmentier EM (1977) Compositional density stratification in the oceanic lithosphere-causes and consequences. *Journal of the Geological Society of London* 133: 343–355.

Patchett PJ (1996) Crustal growth – Scum of the Earth after all. *Nature* 382: 758–759.

Patchett JP and Arndt NT (1986) Nd isotopes and tectonics of 1.9-1.7 Ga crustal genesis. *Earth and Planetary Science Letters* 78: 329–338.

Patchett PJ and Bridgwater D (1984) Origin of continental-crust of 1.9-1.7 Ga age defined by Nd isotopes in the Ketilidian Terrain of south Greenland. *Contributions to Mineralogy and Petrology* 87: 311–318.

Patchett PJ and Chase CG (2002) Role of transform continental margins in major crustal growth episodes. *Geology* 30: 39–4.

Patchett PJ and Samson SD (2003) *Treatise on Geochemistry: Ages and Growth of the Continental Crust from Radiogenic Isotopes*. Amsterdam: Elsevier.

Patchett PJ, White WM, Feldmann H, Kielinczuk S, and Hofmann AW (1984) Hafnium rare-Earth element fractionation in the sedimentary system and crustal recycling into the Earth's mantle. *Earth and Planetary Science Letters* 69: 365–78.

Pearcy LG, Debari SM, and Sleep NH (1990) Mass balance calculations for two sections of island arc crust and implications for the formation of continents. *Earth and Planetary Science Letters* 96: 427–442.

Percival JA and Williams HR (1989) Late Archean Quetico accretionary complex, Superior Province. *Canada Geology* 17: 23–25.

Pilger RH (1978) Method for finite plate reconstructions, with applications to Pacific-Nazca plate evolution. *Geophysical Research Letters* 5: 469–472.

Plank T (2005) Constraints from thorium/lanthanum on sediment recycling at subduction zones and the evolution of the continents. *Journal of Petrology* 46: 921–944.

Plank T and Langmuir CH (1998) The chemical composition of subducting sediment and its consequences for the crust and mantle. *Chemical Geology* 145: 325–394.

Polat A and Kerrich R (2001a) Geodynamic processes, continental growth, and mantle evolution recorded in late Archean greenstone belts of the southern Superior Province, Canada. *Precambrian Research* 112: 5–25.

Polat A and Kerrich R (2001b) Magnesian andesites, Nb-enriched basalt-andesites, and adakites from late-Archean 2.7 Ga Wawa greenstone belts, Superior Province, Canada: Implications for late Archean subduction zone petrogenetic processes. *Contributions to Mineralogy and Petrology* 141: 36–52.

Polat A, Kerrich R, and Wyman DA (1998) The late Archean Schreiber-Hemlo and White River Dayohessarah greenstone belts, Superior Province: Collages of oceanic plateaus, oceanic arcs, and subduction-accretion complexes. *Tectonophysics* 289: 295–326.

Puchtel IS, Arndt NT, Hofmann AW, et al. (1998) Petrology of mafic lavas within the Onega plateau, central Karelia: Evidence for 2.0 Ga plume-related continental crustal growth in the Baltic Shield. *Contributions to Mineralogy and Petrology* 130: 134–153.

Puchtel IS, Haase KM, Hofmann AW, et al. (1997) Petrology and geochemistry of crustally contaminated komatiitic basalts from the Vetreny Belt, southeastern Baltic Shield: Evidence for an early Proterozoic mantle plume beneath rifted Archean continental lithosphere. *Geochimica et Cosmochimica Acta* 61: 1205–1222.

Puchtel IS, Hofmann AW, Amelin YV, Garbe-Schonberg CD, Samsonov AV, and Schipansky AA (1999) Combined mantle plume-island arc model for the formation of the 2.9 Ga Sumozero-Kenozero greenstone belt, SE Baltic Shield: Isotope and trace element constraints. *Geochimica et Cosmochimica Acta* 63: 3579–3595.

Puchtel IS, Hofmann AW, Mezger K, Jochum KP, Shchipansky AA, and Samsonov AV (1998) Oceanic plateau model for continental crustal growth in the archaean, a case study from the Kostomuksha greenstone belt, NW Baltic Shield. *Earth and Planetary Science Letters* 155: 57–74.

Reischmann T, Kroner A, and Basahel A (1983) Petrography, geochemistry and tectonic setting of metavolcanic sequences from the Al Lith area, southwestern Arabian Shield. *Bulletin of Faculty of Earth Sciences, King Abdulaziz University* 6: 365–378.

Rey P (1993) Seismic and tectonometamorphic characters of the lower continental-crust in Phanerozoic areas – A consequence of post-thickening extension. *Tectonics* 12: 580–590.

Reymer A and Schubert G (1984) Phanerozoic addition rates to the continental-crust and crustal growth. *Tectonics* 3: 63–77.

Reymer A and Schubert G (1986) Rapid growth of some major segemtns of continental crust. *Geology* 14: 299–302.

Richards MA, Duncan RA, and Courtillot VE (1989) Flood basalts and hot-spot tracks – Plume heads and tails. *Science* 246: 103–107.

Richardson WP, Emile A, and Van der Lee S (2000) Rayleigh – wave tomography of the Ontong – Java Plateau. *Physics of the Earth and Planetary Interiors* 118: 29–51.

Rocholl A, Stein M, Molzahn M, Hart SR, and Worner G (1995) Geochemical evolution of rift magmas by progressive tapping of a stratified mantle source beneath the Ross Sea Rift, Northern Victoria Land, Antarctica. *Earth and Planetary Science Letters* 131(3): 207–224.

Rudnick RL (1992) Xenoliths—samples of the lower continental crust. In: Fountain DM, Arculus R, and Kay RW (eds.) *Continental Lower Crust*, pp. 269–316. Amsterdam: Elsevier.

Rudnick RL (1995) Making continental-crust. *Nature* 378: 571–578.

Rudnick RL and Taylor R (1987) The composition and petrogenesis of the lower continental crust: A xenolith study. *Journal of Geopysical Research* 92: 13981–14005.

Rudnick RL and Fountain DM (1995) Nature and composition of the continental crust: A lower crustal prespective. *Reviews of Geophysics* 33: 267–.310.

Rudnick RL and Gao S (2003) *Treatise on Geochemistry: Composition of the Continental Crust*. Amsterdam: Elsevier.

Samson SD and Patchett PJ (1991) The Canadian Cordillera as a modern analogue of Proterozoic crustal growth. *Australian Journal of Earth Sciences* 38: 595–611.

Samson SD, McClelland WC, Patchett PJ, Gehrels GE, and Anderson RG (1989) Evidence from neodymium isotopes for mantle contributions to Phanerozoic crustal genesis in the Canadian Cordillera. *Nature* 337: 705–709.

Samson SD, Patchett PJ, Gehrels GE, and Anderson RG (1990) Nd and Sr isotopic characterization of the Wrangellia terrane and implications for crustal growth of the Canadian cordillera. *Journal of Geology* 98: 749–762.

Samson SD, Patchett PJ, McClelland WC, and Gehrels GE (1991) Nd isotopic characterization of metamorphic rocks in the coast mountains, Alaskan and Canadian cordillera: Ancient crust bounded by juvenile terranes. *Tectonics* 10: 770–780.

Schubert G and Reymer APS (1985) Continental volume and freeboard through geological time. *Nature* 316: 336–339.

Schubert G and Sandwell D (1989) Crustal volumes of the continents and of oceanic and continental submarine plateaus. *Earth and Planetary Science Letters* 92: 234–246.

Shirey SB and Hanson GN (1984) Mantle-derived Archean Monozodiorites and Trachyandesites. *Nature* 310: 222–224.

Sleep NH and Windley BF (1982) Archean plate tectonics constraints and inferences. *Journal of Geology* 90: 363–379.

Smith HS and Erlank AJ (1982) Geochemistry and petrogenesis of komatiites from the Barberton greenstone belt. In: Arndt NT and Nisbet EG (eds.) *Komatiites*, pp. 347–398. London: Allen and Unwin.

Stein M (2003) Tracing the plume material in the Arabian–Nubian shield. *Precambrian Research* 123: 223–234.

Stein M (2006) The rise of the asthenopsheric mantle into the Arabian lithosphere and its cosnsequences: Heating , melting uplifting. IAVCEI meeting, Guangzhou.

Stein M and Hofmann AW (1992) Fossil plume head beneath the Arabian lithosphere? *Earth and Planetary Science Letters* 114: 193–209.

Stein M and Hofmann AW (1994) Mantle plumes and episodic crustal growth. *Nature* 372: 63–68.

Stein M and Goldstein SL (1996) From plume head to continental lithosphere in the Arabian–Nubian shield. *Nature* 382: 773–778.

Stein M, Navon O, and Kessel R (1997) Chromatographic metasomatism of the Arabian–Nubian lithosphere. *Earth and Planetary Science Letters* 152: 75–91.

Stern CR and Kilian R (1996) Role of the subducted slab, mantle wedge and continental crust in the generation of adakites from the Andean Austral Volcanic Zone. *Contributions to Mineralogy and Petrology* 123: 263–281.

Stern RA, Syme EC, and Lucas SB (1995) Geochemistry of 1.9 Ga morb-like and oib-like basalts from the Amisk Collage, Flin-Flon Belt, Canada – Evidence for an intraoceanic origin. *Geochimica et Cosmochimica Acta* 59: 3131–3154.

Stern RJ (1981) Petrogenesis and tectonic setting of late Precambrian ensimatic volcanic rocks, central eastern desert of Egypt. *Precambrian Research* 16: 195–230.

Sylvester PJ (2000) Continent formation, growth and recycling – Preface. *Tectonophysics* 322: Vii–Viii.

Sylvester PJ, Campbell IH, and Bowyer DA (1997) Niobium/uranium evidence for early formation of the continental crust. *Science* 275: 521–523.

Taylor SR (1967) The origin and growth of continents. *Tectonophysics* 4: 17–34.

Taylor SR and McLennan SM (1985) *The Continental Crust: Its Composition and Its Evolution*. Oxford: Blackwell.

Taylor SR and McLennan SM (1995) The geochemical evolution of the continental-crust. *Reviews of Geophysics* 33: 241–265.

Taylor SR and McLennan SM (1996) The evolution of continental crust. *Scientific American* 274: 76–81.

Tera F, Brown L, Morris J, Selwayn Sacks I, Klein J, and Middleton R (1986) Sediment incorporation in island-arc magmas: Inferences from 10Be. *Geochimica et Cosmochimica Acta* 50: 535–550.

Veizer J and Jansen SI (1979) Basement and sedimentary recycling and continental evolution. *Journal of Geology* 87: 341–370.

Weinstein Y, Navon O, Alther R, and Stein M (2006) The role of fluids and of pyroxenitic veins in the generation of alkali-basaltic suites, northern Arabian plate. *Journal of Petrology* 47: 1017–1050.

Wilde SA, Valley JW, Peck WH, and Graham CM (2001) Evidence from detrital zircons for the existence of continental crust and oceans on the Earth 4.4 Gyr ago. *Nature* 409: 175–178.

Windley BF (1995) *The Evolving Continents*. Chichester: Wiley.

White WM and Patchett PJ (1984) Hf–Nd–Sr isotopes and incompatible-element abundances in island arcs; implications for magma origins and crust–mantle evolution. *Earth Planetary Science Letters* 67: 167–185.

Wise (1974) Continental margins, freeboard and volumes of continents and oceans through time. In: Burke CA and Drake CL (eds.) *The Geology of Continental Margins*, pp. 45–58. Berlin: Springer.

Wörner G, Zindler A, Staudigel H, and Schmincke HU (1986) Sr, Nd, and Pb isotope geochemistry of Tertiary and Quarternary alkaline volcanics from West Germany. *Earth Planetary Science Letters* 79: 107–119.

8 Thermal Evolution of the Mantle

G. F. Davies, Australian National University, Canberra, ACT, Australia

8.1 Introduction

The thermal evolution of Earth's interior has featured fundamentally in the geological sciences, and it continues to do so. This is because Earth's internal heat fuels the driving mechanism of tectonics, and hence of all geological processes except those driven at the surface by the heat of the Sun. The thermal evolution also featured famously in a nineteenth-century debate about the age of Earth.

8.1.1 Cooling and the Age of Earth

According to Hallam (1989), from the time when the origin of Earth was considered by scientists of the European Enlightenment it was conceived as having started hot, notably in the nebular theories of Kant and Laplace. This led to conjectures and very rough estimates of how long it may have taken to cool down, and thus to estimates of the age of Earth. However, it was not until Fourier had formulated his 'law' of conduction and the science of thermodynamics was established that a quantitatively reliable, though physically inappropriate, estimate of Earth's age emerged. This was done by William Thomson, better known as Lord Kelvin, who played a key role in the development of thermodynamics, notably as the formulator of the 'second law'.

Kelvin actually calculated an age for the Sun first, using estimates of the gravitational energy released by the Sun's accretion from a nebular cloud, the heat content of the present Sun, and the rate of radiation of heat, concluding that the Sun was likely to be younger than 100 Ma and most unlikely to be older than 500 Ma (Kelvin, 1862). The following year Kelvin calculated an age of Earth using different physics (Kelvin, 1863). He reasoned that the known geothermal gradient of about $20°C\,km^{-1}$ near Earth's surface would have declined from an essentially infinite gradient at the time of Earth's formation. Making use of Fourier's law to solve the resulting thermal diffusion problem (Davies, 1999) he obtained an age between 20 and 400 Ma with his most probable value being 98 Ma. There ensued several decades of heated debate between and among geologists and physicists, with Kelvin's last estimate (Kelvin, 1899) being only 24 Ma.

A number of Kelvin's assumptions were challenged during the course of this long controversy, the most telling challenge in retrospect coming from The Reverend Osmond Fisher (Fisher, 1881) who pointed out that if Earth's interior were 'plastic' (meaning in some degree fluid), then a much greater reservoir of heat would be tapped and the resulting age could be much greater. Fisher's point was that conduction alone would have cooled only the outer few hundred kilometers of Earth, whereas a circulating fluid could bring heat from much deeper as well, so it would take longer for the larger amount of heat to be removed at the currently observed rate. Fisher's point did little to settle the nineteenth-century controversy, but it returned in a slightly different guise a century later.

The controversy with Kelvin was finally settled through the discovery of radioactivity, which provided a previously unknown source of heat within Earth, so heat could be replenished as it is lost through the surface, and the flow of heat thus maintained for much longer than Kelvin's estimates. The actual determination of Earth's age relied on exploiting a different aspect of radioactivity, the production of daughter isotopes. This task was pursued most notably by Holmes from early in the twentieth century (Holmes, 1911, 1913) but not completed to general satisfaction until meteorites were dated by the lead–lead method and Earth was shown to fit within the meteorite trend (Patterson, 1956). The age of Earth is still determined only indirectly.

8.1.2   A Static Conductive Mantle?

The discovery of the heat generated by radioactivity provided Earth with an enduring source of heat, but it created another problem by breaking the consistency in Kelvin's argument between age and the removal of heat by conduction. If heat escapes from Earth's interior by conduction, then Kelvin's argument would still seem to require Earth to be less than 100 My old, otherwise the surface heat flow would be much less than is observed. Put another way, if Earth is 4.5 Gy old, roughly 100 times Kelvin's estimate, then the geothermal gradient ought to be 10 times less than the 15–20°C km^{-1} commonly observed in deep mines and boreholes. This is because in the relevant thermal diffusion process the heat flow (and the temperature gradient) declines in proportion to the square root of time (Davies, 1999).

Until the 1960s, it was most generally held that Earth's mantle is solid and unyielding (Jeffreys, 1976). This conclusion was based on several lines of evidence, the most obvious being that the mantle transmits seismic shear waves, and shear waves cannot propagate

through a liquid because a liquid does not have any shear strength. A second argument is that Earth's response to tidal forces implies quite high rigidity. Finally, Earth has, on average, a slightly greater equatorial bulge than is accounted for by its rotation, and this was intepreted to imply that the mantle has a viscosity (a fluid's resistance to flow) sufficiently high to preclude significant internal motion. Against this, the concept of a deformable 'asthenosphere' extending down for several hundred kilometers was widely held among geologists as necessary to accommodate the considerable movements they inferred to have occurred in the crust (Barrell, 1914).

The initial presumption used to reconcile the observed heat flow with a static mantle was that the radioactivity must be confined to within about 20 km of the surface, and observations in continental areas support this (Jeffreys, 1976). The implication is that most of the heat emerging from continental crust is generated there and can readily be conducted to the surface. The amount of heat coming from deeper was presumed or implied to be very small. However, this encounters the problem that heat flow from ocean basins is greater than from continents, but the thin, less radioactive oceanic crust can account for only a small fraction of the total (Sclater et al., 1980).

A later presumption was that thermal conductivity must increase substantially with depth, so that a sufficient heat flow could be sustained for a much greater time. This possibility was bolstered by the proposal that radiative transfer of heat could be important in the mantle (Clark, 1957). The idea is that if mantle minerals are sufficiently transparent, then the blackbody radiation of the hot materials could transmit significant distances. Since the transmission would be like a random walk it constitutes a diffusion process and so can be described by an enhanced effective conductivity. However, the presence of iron in minerals tends to reduce the transparency of the minerals at the relevant wavelengths (Shankland et al., 1979). Thus, it was not at all obvious that radiative transfer could resolve the issue. There have been recent claims for a significant effect (Hofmeister, 2005), though if they are significant it would probably only be for the deepest mantle.

8.2   The Convecting Mantle

The emergence of plate tectonics in the 1960s directly implied that at least the upper part of the mantle is mobile. In 1969 Goldreich and Toomre

(1969) pointed out that the excess bulge at the equator varied about as much with longitude as with latitude, so the old explanation that it was a fossil from past faster rotation could not be correct. In their reinterpretation, the Earth adjusts its rotation to bring its largest bulges to the equator, and this implies an 'upper' bound on viscosity that is quite compatible with a mobile mantle. With later interpretations of postglacial rebound constraints and positive geoids over subduction zones, the current picture is that the upper mantle has a viscosity of $(1-3) \times 10^{20}$ Pa s while the lower mantle is perhaps 30 times more viscous than this (Hager, 1984; Mitrovica, 1996; Mitrovica and Forte, 1997).

A robust argument in favor of mantle convection was put by Tozer, starting in the 1960s at a time when the possibility was still hotly disputed (Tozer, 1965, 1972). He noted that the viscosity of mantle material is strongly dependent on temperature, as we will be considering shortly, which means its resistance to convection decreases rapidly as its temperature increases. He argued that either the mantle started hot and would therefore be soft enough to convect, or radioactivity would heat it until it reached a temperature at which it would become soft and mobile, and convect. This is essentially our present understanding, as we will shortly see.

The idea of some kind of mantle convection quickly became widely accepted, though for a time there was some reluctance to admit the lower mantle into the convection mileu (Isacks et al., 1968; McKenzie et al., 1974). Eventually, the lower mantle was also conceived as mobile (McKenzie and Weiss, 1975; Davies, 1977; O'Connell, 1977) and it was then realized that convection would be efficient enough at transporting heat that the surface heat flux would fairly closely approximate the radioactive heat generation in the interior. However, it was difficult to quantify the heat transported by convection because the required numerical models challenged the computers available at that time and because the temperature dependence of mantle viscosity creates technical challenges. Nevertheless, there is a simple and fairly general relationship that relates the heat transported by a convecting fluid to the temperature difference across the fluid layer (e.g., Turcotte and Oxburgh, 1967; Rossby, 1969; McKenzie et al., 1974). McKenzie and Weiss (1975) and Davies (1979) used this relationship to estimate temperature changes in the mantle on the assumption that the heat loss at Earth's surface reflects the heat generation in the interior.

Subsequently several groups realized, more-or-less independently, that the relationship could be used to calculate the thermal evolution of the Earth without assuming that heat loss is equal to heat production. Sharpe and Peltier (1978, 1979) were the first, although they implemented it in terms of an enhanced thermal diffusion, which does not accurately represent the internal temperature profile. Others soon used it to characterize the total heat transport in terms of a representative internal temperature (Cassen et al., 1979; Schubert, 1979; Schubert et al., 1979a, 1979b, 1980; Stevenson and Turner, 1979; Davies, 1980; Stacey, 1980). The treatments in these papers were closely equivalent, although differing in minor detail. Systematic presentations are given by Davies (1999), Schubert et al. (2001), and Turcotte and Schubert (2001). The presumption in the simple theory is that the essential relationship between temperature difference and heat flow is not sensitive to the local details, like geometry, whether the flow is turbulent or laminar, and so on. Experience in other fields had already shown that it can be a good first approximation (Rossby, 1969). The influence of some factors will be illustrated after a reference case is presented.

The relationship between heat flow and internal temperature is expressed most generally in terms of the Nusselt number, Nu, and the Rayleigh number, Ra:

$$Nu = a(Ra/Ra_c)^p \quad [1]$$

where a and p are constants and Ra_c is the critical Rayleigh number, that is, the Rayleigh number at which convection just begins. Typically $a \sim 1$, $p \sim 1/3$ and $Ra_c \sim 1000$. Here

$$Nu = qD/K\Delta T \quad [2]$$

and

$$Ra = \frac{g\rho\alpha\Delta T D^3}{\kappa\mu} \quad [3]$$

where q is the surface heat flux, D is the depth of the fluid layer, K is thermal conductivity, ΔT is the temperature difference across the fluid layer, g is gravity, ρ is density, α is thermal expansion, $\kappa = K/\rho C_P$ is thermal diffusivity, μ is viscosity, and C_P is specific heat at constant pressure. Equation [1] can be rearranged to give

$$q = q_r\left(\frac{\Delta T}{\Delta T_r}\right)^{1+p}\left(\frac{\mu}{\mu_r}\right)^{-p} \quad [4]$$

where subscript r refers to a reference state, which could be the present Earth, for example. This gives the heat flux in terms of the temperature difference driving the convection.

The viscosity is kept explicit in eqn [4] because it is a strong function of temperature in the mantle, and this has an important effect. The viscosity can be written as

$$\mu = \mu_r \exp\left[T_A\left(\frac{1}{T} - \frac{1}{T_r}\right)\right] \qquad [5]$$

where

$$T_A = (E^* + PV^*)/R_G = H^*/R_G \qquad [6]$$

and E^*, V^*, and H^* are the activation energy, volume, and enthalpy, respectively, P is pressure, R_G is the gas constant and μ_r is the viscosity at a reference temperature T_r. T_A could then be called an activation temperature. With $E^* = 400\,\text{kJ}\,\text{mol}^{-1}$ and $R_G = 8.31\,\text{J}\,\text{mol}^{-1}\,\text{K}^{-1}$, $T_A = 48\,100\,\text{K}$, which is the source of the strong dependence of μ on T.

To compute the thermal evolution of the mantle we need the energy equation for the mantle, considered as a covecting layer. With the situation depicted in **Figure 1**, the energy equation can be written as

$$\frac{dT_u}{dt} = \frac{M_m H_m + Q_c - Q_m}{\chi_m M_m C_m} \qquad [7]$$

where T_u is the upper-mantle temperature, H_m is the rate of heat generation per unit mass of the mantle,

M_m and C_m are the mass and specific heat of the mantle, respectively, and $\chi_m = T_m/T_u$, where T_m is the mean temperature of the mantle. Q_c and Q_m are the heat flows out of the core and the mantle, respectively. For the moment we will take Q_c to be zero. Q_m is just given as

$$Q_m = 4\pi R_e^2 q \qquad [8]$$

where R_e is the radius of Earth and q is given by eqn [4]. In this context it is appropriate to take $\Delta T = (T_u - T_s)$, where T_s is the temperature at Earth's surface (**Figure 1**).

The heat generation is taken to be due to radioactive heating and is given by

$$H_m = \frac{U_e}{Ur}\sum_{i=1}^{4} b_i \exp[\lambda_i(t_E - t)] \qquad [9]$$

where U_e is the equivalent uranium concentration that would yield the observed heat loss, the index i refers to the isotopes ^{238}U, ^{235}U, ^{232}Th, and ^{40}K, b_i is the heat production per unit mass of uranium, λ_i is the decay constant, and t_E is the age of the Earth. Ur is the Urey ratio, defined as the ratio of present heat loss to present heat generation in the Earth, that is,

$$Ur = Q_m/H_m M_m \qquad [10]$$

This is the reciprocal of the way Ur is sometimes defined. Values for the required parameters are given in **Table 1**.

8.2.1 A Reference Thermal History

A solution for eqn [7] that reasonably satisfies observational constraints is shown in **Figure 2** (heavy curves). Parameters used generally in calculations here are given in **Table 2**, while observational constraints are given in **Table 3**. The level of radioactive heating has been adjusted, through the Urey ratio, to yield a present heat loss of about 36 TW (36×10^{12} W), which is the portion of Earth's total heat loss (41 TW) emerging from the mantle, the balance being generated in the continental crust and lost directly to the surface (Davies, 1999).

The initial temperature is taken rather arbitrarily to be 1800°C. The peak temperature during the formation of Earth is difficult to estimate, because it was presumably determined by a competition between heat deposition by large impacts and heat removal by the 'gardening' effect of further impacts, by conduction near the surface, by mantle convection, and

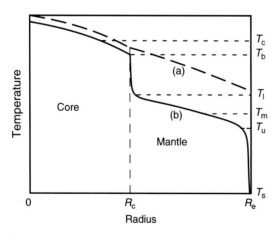

Figure 1 Sketch of temperature profiles through Earth (geotherms). (a) Initial state. (b) Present state. T_c, mean core temperature; T_b, core–mantle boundary temperature; T_l, lower-mantle temperature; T_m, mean mantle temperature; T_u, upper-mantle temperature; T_s, surface temperature.

Table 1 Parameters of heat-producing isotopes

Isotope (i)	Half life (Ga)	Decay const.[a] (λ_i) Ga^{-1}	Power[a] (μW) (kg Element)$^{-1}$	Element/ U[b] (g/g)	Power (hi) (μW (kg U)$^{-1}$)
^{238}U	4.468	0.155	94.35	1	94.35
^{235}U	0.7038	0.985	4.05	1	4.05
^{232}Th	14.01	0.049	26.6	3.8	101.1
^{40}K	1.250	0.554	0.0035	1.3×104	45.5
				Total	245

[a]Stacey (1992).
[b]Galer et al. (1989).

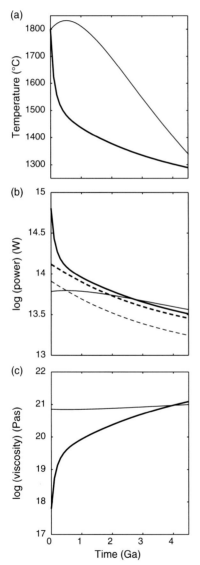

Figure 2 A typical thermal evolution of the mantle (heavy curves) and a thermal evolution in which the viscosity is almost independent of temperature (light curves). (a) Temperature. (b) Heat generation (dashed) and heat loss (solid). (c) Viscosity. The parameters of the calculations are adjusted to fit presently observed conditions given in **Table 3**.

by rapid cooling of surface melt, including the possibility of a transient magma ocean (Davies, 1990). Heat transport by magma is generally so much more efficient than the other mechanisms that it would presumably have prevented the temperature from rising too high, but all of the processes are complex and poorly constrained enough to make only rough estimates. Anyway the point here is just to illustrate the kind of behavior during later thermal evolution and the reasons for it.

Figure 2 illustrates typical features of this kind of solution – there is an early transient phase of rapid cooling, lasting about 0.5 Ga, and thereafter the heat loss tracks the heat generation, which slowly declines due to radiaoactive decay. As the heat generation declines, the mantle adjusts by slowly cooling so that its heat loss also declines. However, the slow cooling releases some internal heat that must also be removed, so the heat loss is larger than the heat generation. Another way to look at this is that the cooling occurs because the heat loss is a little larger than the heat generation, as described by eqn [7]. If the heat generation were constant, the heat loss would asymptotically approach the heat generation, and the ultimate state would be a steady state in which heat loss balanced heat generation.

The character of this solution is determined by the strong temperature dependence of the mantle viscosity, which changes by three orders of magnitude during the evolution (**Figure 2(c)**). At the beginning the viscosity is much lower than at present, and this reduces the resistance to mantle convection (see eqn [4]) which can therefore remove heat very rapidly (**Figure 2(b)**). This high heat loss causes the temperature to drop rapidly (**Figure 2(a)**), but then the viscosity rises rapidly (**Figure 2(c)**) and reduces the heat loss (**Figure 2(b)**). This early transient stage continues until the heat loss approaches the heat generation, at which point the initially large

Table 2   Quantities used in calculations

Symbol	Quantity	Value
M_c	Mass of the core	2.10^{24} kg
M_m	Mass of the mantle	4.10^{24} kg
R_e	Radius of the Earth	6371 km
R_c	Radius of the core	3485 km
C_c	Specific heat of the core	$1000\,\mathrm{J\,kg^{-1}\,K^{-1}}$
C_m	Specific heat of the mantle	$1000\,\mathrm{J\,kg^{-1}\,K^{-1}}$
ρ_m	Density of the mantle	$3500\,\mathrm{kg\,m^{-3}}$
α_m	Thermal expansion of the mantle	$2\times10^{-5}\,\mathrm{K^{-1}}$
κ_m	Thermal diffusivity of the mantle	$0.86\times10^{-6}\,\mathrm{m^2\,s^{-1}}$
η_u	Viscosity of the mantle	10^{21} Pa s
E^*	Activation energy of the mantle	$400\,\mathrm{kJ\,mol^{-1}}$
V^*	Activation volume	$2.5\times10^{-6}\,\mathrm{m^3\,mol^{-1}}$
T_s	Surface temperature	$0°C$
T_u	Present upper-mantle temperature	$1300°C$
n_m	Heat flow – temperature exponent (mantle)	1/3
U_e	Uranium abundance equivalent to present mantle heat loss	$38\,\mathrm{ng\,g^{-1}}$

Table 3   Empirical constraints on the present thermal regime

Quantity	Value	Reference
Upper-mantle temperature	1300°C	McKenzie and Bickle (1988)
Mantle heat loss	36 TW	Davies (1999)
Plume heat transport	3.5–7 TW	Davies (1999), Bunge (2005)

imbalance between them becomes small and therefore the rate of decline of temperature slows (**Figure 2(a)**).

The behavior is rather different if the viscosity is only a weak function of temperature, as is also illustrated in **Figure 2** (light curves) for a case in which the activation energy is reduced from 400 to $20\,\mathrm{kJ\,mol^{-1}}$. In this case the initial heat loss is only a little greater than the present heat loss (61 vs 37 TW), because only the temperature term in eqn [4] is significantly different, the small variation in viscosity hardly affecting the heat loss. In fact, the heat loss in this example starts out lower than the heat generation, which means the mantle actually heats up initially. Only after the heat generation decays to

lower values does the mantle temperature start to decrease. Thereafter the temperature declines, but never as fast as in the early transient phase of the black curve in **Figure 2(a)**. In fact, the transient phase is still in progress in the constant viscosity case. This is also reflected in the large difference between the heat loss and the heat generation (**Figure 2(b)**) – the Urey ratio in this case is 2.0.

The difference between these two solutions arises because in the former case a relatively small temperature change, about 200°C, is sufficient to reduce the heat loss as the rate of heat generation declines from about 130 to 29 TW, a decline by a factor of 4.5. The change in heat loss comes mainly through the large increase in viscosity (**Figure 2(c)** and eqns [4] and [5]) accompanying this modest change of temperature. In the latter case, the change in heat loss must be accomplished solely through ΔT, which must decrease by $(4.5)^{3/4}$, or a factor of 3. Thus, the mantle would be required to cool by 1200°C (from 1800 to 600°C) to reduce the heat loss to its presently observed value. This requires the removal of much more internal heat, which is why heat production is only half of the heat loss rate in this case, the other half coming from internal heat.

Thus, the strong temperature dependence of mantle viscosity results in the early transient cooling phase being relatively brief, about 500 Ma. Thereafter, the thermal regime tracks the slow decay of radiogenic heat. Incidentally, it is not very useful in light of this to think of contributions from primordial heat versus radiogenic heat, since there is no clear way to separate them in the present Earth. It is more useful to separate the early transient phase (which 'can' usefully be thought of as removing 'excess' primordial heat) from the later phase regulated by radioactivity. It is also useful to separate current radiogenic heat production and the heat released by the decreasing internal heat of the mantle, as its temperature declines.

8.3 Internal Radioactivity and the Present Cooling Rate

The approach taken above of adjusting the Urey ratio amounts to inferring the internal heat production rate that would account for the present rate of heat loss, assuming the present theory is accurate. The early conclusion from models like those in **Figure 1** was that the Urey ratio is significantly greater than 1, meaning the heat loss rate is greater

than the heat generation rate. For example, Schubert
et al. (1980) found $Ur = 1.2–1.5$, while Davies found
Ur could be as high as 2 (Davies, 1980), though
subsequent work yielded values in the range 1.25–
1.35 (Davies, 1993).

Another early conclusion was that these results
are marginally consistent with Earth having a complement of uranium and thorium like that of
chondritic meteorites, with a potassium complement
such that $K/U = (1–2) \times 10^4$. The latter ratio is lower
than for chondrites, but consistent with the ratio
found in surface rocks. The depletion of potassium
relative to chondrites fits a trend of stronger depletion for more volatile elements (McDonough and
Sun, 1995; Palme and O'Neill, 2004).

Subsequently, it has emerged that there seems to
be a significant discrepancy. The U content of the
primitive mantle (precrust extraction) is estimated to
be 22 ng g^{-1} (Palme and O'Neill, 2004). With a current heat production from U, Th, and K of 245
µW kg^{-1} of U (**Table 1**), this implies a heat production of 5.4 pW kg^{-1} of mantle and a total heat
production of 22 TW. On the other hand, the
observed mantle heat loss is about 36 TW, which
implies $Ur = 1.6$, higher than the estimates from thermal evolution modelling. $Ur = 1.3$ would require the
total heat production to be about 28 TW.

The discrepancy is actually worse than that,
because about half of Earth's complement of heat-
producing elements is estimated to be in the continental crust (Rudnick and Fountain, 1995), which
implies that the average mantle U content is only
about 10 ng g^{-1}. The actual U content of the source
of mid-ocean ridge basalts is even smaller, less than
5 ng g^{-1} (Jochum et al., 1983), which implies that
there may be some deeper region of the mantle
that has higher concentrations. If we take only the
inferred mean mantle U content of 10 ng g^{-1} then
the implied heat production is only about 10 TW
and $Ur = 3.6$.

Such a large imbalance would imply that the
mantle is cooling rapidly. **Table 4** summarizes several examples of heat input, Urey ratio, and the
resulting cooling rates. Chondritic mantle heating
implies that the mantle is cooling by about
110°C Ga^{-1}, whereas if only half of the chondritic
heat sources are in the mantle, the cooling rate would
be over 200°C Ga^{-1}. Thermal evolution models that
match the present heat flow yield cooling rates of 50–
70°C Ga^{-1}, although the example given in **Figure 2**,
which falls a little short of the observed heat loss rate,
has a cooling rate of only 30°C Ga^{-1}. The case in

Table 4   Implied cooling rates of the mantle

Case	Thermal	Chondr.	.5 Chond.	Fig. 2	Fig. 2
Q_m(TW)	36	36	36	32.4	36.9
H_m(TW)	28	22	10	28.6	17.9
Q_{net}(TW)	8	14	26	3.8	19.0
Ur	1.25	1.6	3.6	1.13	2.1
dT/dt (°C Ga^{-1})	−63	−110	−205	−30	−150

Figure 2 with little variation in viscosity has a present cooling rate of 150°C Ga^{-1}.

It is difficult to see how the larger cooling rates
just quoted can be consistent with the geological
record, since they imply that the late Archean mantle
would have been 300–600°C hotter than at present.
As pointed out by Campbell and Griffiths (1992),
although there are Archean komatiites with source
potential temperatures of 1800–1900°C, the komatiites comprise only a small fraction of Archean mafic
rocks, the great majority of which are basalts with
source potential temperatures no more than 100–
200°C above the present mantle temperature. The
komatiites are interpreted to come from the core of
plumes or analogous high-temperature mantle
upwellings, and would therefore not be representative of mean mantle temperatures. The lower cooling
rates of 30–60°C Ga^{-1} in **Table 4** would be consistent with this interpretation of the dominant basaltic
Archean rocks.

8.4 Variations on Standard Cooling Models

A number of variations on this kind of thermal history model have been explored. These have the same
general character as the reference model in **Figure 2**,
but the details differ.

8.4.1 Effect of Volatiles on Mantle Rheology

The viscosity of mantle materials is affected by water
content, as well as by temperature and pressure
(Hirth and Kohlstedt, 1996). Reducing the water
content from typical values of a few hundred micro-
grams per gram to zero can increase the viscosity by
about an order of magnitude. The effect of this on
thermal evolution was explored by McGovern and

Schubert (1989) and Schubert *et al.* (2001). Typical results are illustrated in **Figure 3**, in which 'degassing' and 'regassing' models are compared with a model in which the viscosity is not affected by water (Schubert *et al.*, 2001). Their parametrization of the degassing and regassing processes was rather simplified, but it suffices to illustrate the effect here. The main effect of the removal by degassing of up to 1.5 ocean masses from the mantle is to increase the mantle temperature by about 100°C. Returning water to the mantle ('regassing') has the reverse effect.

This result can be simply understood as the mantle self-regulating to compensate for the stiffening of its material. If all the water were to be removed suddenly, the mantle viscosity would increase and convection would slow. This would reduce the rate of heat loss, so the mantle would begin to heat up (or

to cool more slowly) through its internal radioactivity. As it heated the viscosity would drop, and when the viscosity reached the value it had before the removal of the water the convection would have returned to its previous vigor and the previous rate of heat loss would be restored. The authors confirm that the heat flow and the viscosity are similar in the 'dry' and 'wet' models, the only significant difference being in their temperature.

8.4.2   Two-Layer Mantle Convection

The hypothesis that the mantle convected in two layers separated at the transition zone at 660 km depth has fallen from favor since seismic tomography yielded images of subducted lithosphere penetrating deep into the lower mantle (Grand *et al.*, 1997). However, the possibility of layering deeper in the mantle is still debated (Kellogg *et al.*, 1999). Although the example given here is of the earlier form of two-layer convection, it adequately illustrates the effect of layering.

Spohn and Schubert (1982) and Schubert *et al.* (2001) formulated the viscosity in terms of the so-called homologous temperature, which is the ratio of temperature to melting temperature (or more accurately for the mantle, the solidus temperature, at which melting begins). They also assumed that the lower-mantle material is 60 times more viscous than the upper mantle at the same temperature and pressure, due to it being in high-pressure phases. The resulting temperature structure is shown in **Figure 4**(a), with a whole-mantle convection model for comparison. There is a steep increase in temperature by nearly 500°C near 660 km depth. This increase actually comprises two thermal boundary layers on opposite sides of the interface – a lower thermal boundary layer of the upper mantle and an upper thermal boundary layer of the lower mantle. There is also a small thermal boundary layer at the base of the lower mantle and a strong one at the top of the upper mantle, for a total of four thermal boundary layers.

In spite of the higher intrinsic viscosity of the lower mantle in this model, the viscosity ends up being almost the same as in the upper mantle because of the higher temperature in the lower mantle. As in the previous example, the mantle has self-regulated by adjusting its temperature until the viscosity is such as to allow the required heat transport to occur. (In this particular model, the rate of heat generation in the lower mantle had to be kept quite low so the

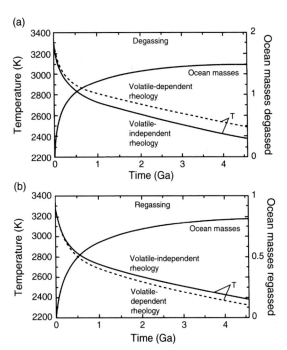

Figure 3   The effect of volatile-dependent viscosity on mantle cooling history (dashed curves). Volatile-independent case is included for comparison (solid curves). The main volatile component is assumed to be water, which reduces the viscosity of mantle materials. (a) Degassing: 1.5 ocean masses of water are progressively removed. Mantle temperature increases to compensate for the stiffening resulting from degassing. (b) Regassing: about 0.8 ocean masses of water are progressively added. Mantle temperature decreases as regassing reduces viscosity. From Schubert G, Turcotte DL, and Olson P (2001) *Mantle Convection in the Earth and Planets.* Cambridge: Cambridge University Press.

(a)

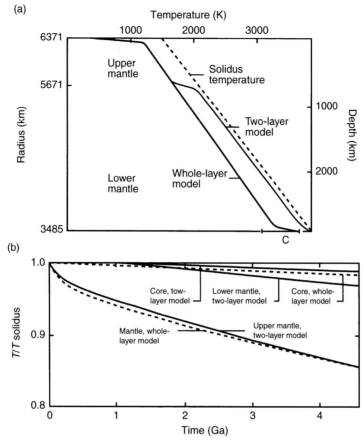

Figure 4 The effect of mantle convection operating in two separate layers. (a) Temperature profiles. (b) Temperature histories. A single-layer ('whole-layer') model is included for comparison. The lower-mantle layer is about 500°C hotter and has a similar viscosity to the upper mantle, in spite of being intrinsically more viscous. See text for details. From Schubert G, Turcotte DL, and Olson P (2001) *Mantle Convection in the Earth and Planets*. Cambridge: Cambridge University Press.

lower mantle did not melt. However, the mantle solidus is not very well determined at very high pressures so the real mantle might be more tolerant of higher heating rates.)

The evolution of homologous temperature is shown in **Figure 4**(b). The upper-mantle cooling is similar to the reference example given earlier – there is a brief, rapid transient cooling followed by a slow decline as radioactivity declines. The lower mantle cools much less. This is because it can only begin to cool as the upper-mantle temperature falls, and because the temperature drop across the top thermal boundary layer of the lower mantle is relatively small, so it drives weaker convection and so transports less heat. (The authors also included a separate core layer in both models. An example in which the core is treated explicitly is given below.)

8.4.3 Static Lithosphere Convection, and Venus

Earth's lithosphere turns out to be unusual within the solar system, in that it is mobile in spite of being cold and strong. Its mobility comes about because it is also, loosely speaking, brittle. Evidently internal stresses are enough to fracture it into pieces, and the pieces comprise the tectonic plates. Other solid planets and satellites in the solar system seem to have strong but unfractured lithospheres, with the possible exception of Venus. These are sometimes called one-plate planets. The big difference, from the point of view of mantle convection, is that most of the lithospere does not partake in the convection. It is expected that only the lower part of the lithosphere will be warm enough to be mobile and hence 'drip' away and drive convection. Since the lithosphere is also the location of most of the (negative) buoyancy that might drive convection, it means that less buoyancy is available to drive convection than is available in Earth's mantle, in which the whole lithosphere founders into the mantle.

The effect of a static lithosphere can be illustrated by using a higher surface temperature, T_s, in calculating the temperature difference ΔT across the top thermal boundary layer in eqn [4]. In effect T_s is taken to define the boundary between the static and mobile parts of the lithosphere. An example with $T_s = 900°C$ is shown in **Figure 5**. The main effect is to raise the temperature of the mantle by nearly 200°C. As in previous examples, we see that the mantle self-regulates by adjusting its internal viscosity, through its temperature, until the convection is vigorous enough to transport the required amount of heat. The viscosity is reduced by about a factor of 25 (**Figure 5(c)**). This type of model might also apply to Venus, whose surface temperature really is high, about 475°C.

The more difficult question in considering a static lithosphere is in estimating how much of the lower lithosphere will take part in convection, especially in different circumstances or as a mantle slowly cools. This has been considered in a series of studies, and various formulations have been proposed. One approach is to use the temperature dependence of viscosity to define a length scale over which the viscosity varies by a characteristic amount, say a factor of 10. Several studies have found scaling laws using a mobile layer thickness defined in terms of this length scale (Davaille and Jaupart, 1993; Solomatov, 1995; Ratcliffe *et al.*, 1997; Grasset and Parmentier, 1998) and using the viscosity of the actively convecting fluid below the lithosphere. On the other hand, Manga *et al.* (2001) used the viscosity at the mean temperature across the lithosphere and concluded this more usefully characterized their results.

8.5   Coupled Core–Mantle Evolution and the Geodynamo

Stevenson and Schubert (1983) and Stacey and Loper (1984) first treated the core as a separate layer with its own thermal history, and the latter authors made the fundamental point that plumes or other upwellings from the base of the mantle are driven by heat emerging from the core. Subsequently, others have further explored such models (Davies, 1993, 2007b; Schubert *et al.*, 2001; Nimmo, 2007). **Figure 6** shows an example, taken from Davies (2007b). The evolution of mantle

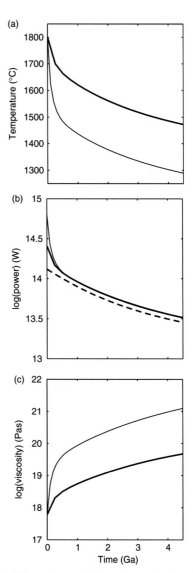

Figure 5 Effect of a static lithosphere, simulated through a surface temperature of 900°C (heavy curves). A mobile lithosphere case is included for comparison (light curves). (a) Temperature. (b) Rate of heat loss. Dashed curve, heat generation. (c) Viscosity. The high surface temperature effectively defines the boundary between the lower, warmer part of the lithosphere that is soft enough to partake in convection and the upper, cooler part that is too strong to move. Only the lower, mobile part of the lithosphere is available to drive convection, and the mantle compensates for the reduced driving buoyancy by running hotter, by nearly 200°C, so the viscous resistance is reduced in proportion. The heat loss (panel (b)) is the same as for the mobile-lithosphere model. This type of model could also apply to Venus, whose surface temparture is high (475°C).

temperature and surface heat flow is very similar to that of the reference model (**Figure 1**). The model is adjusted so the present heat flow from the core,

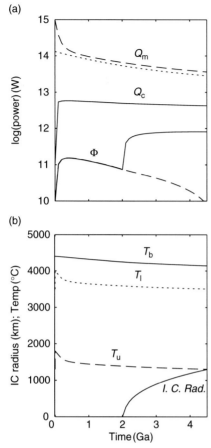

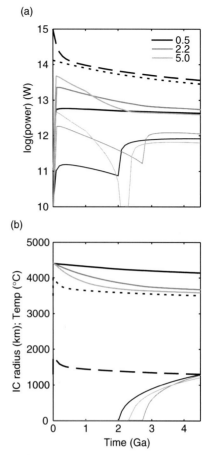

Figure 6 Coupled thermal history of the mantle and core, including dissipation in the core and growth of the inner core. (a) Heat flow. Q_m, mantle heat loss; Q_c, core heat loss; Φ, dissipation in the core with (solid) and without (dashed) the effects of the inner core; dotted curve, heat generation. (b) Temperature histories and inner core radius. Symbols are defined in **Figure 1**. From Davies GF (2007b) Mantle regulation of core cooling: A geodynamo without core radioactivity? *Physics of the Earth and Planetary Interiors* 160: 215–29.

Figure 7 Core thermal histories with different efficiency factors governing removal of heat by the mantle. Curve identifications are the same as in **Figure 6**. From Davies GF (2007b) Mantle regulation of core cooling: A geodynamo without core radioactivity? *Physics of the Earth and Planetary Interiors* 160: 215–229.

4.3 TW, is within the range inferred from hot spot swells (**Table 3**), on the assumption that hot spots are caused by plumes rising from the base of the mantle (Davies, 1988; Sleep, 1990). In this model the core heat flow is remarkably steady through Earth history, with a maximum of only 5.8 TW. This low and steady heat flow causes only moderate core cooling, by 260°C. The temperature difference between the mantle and the core also remains fairly steady.

It turns out that there are other possible thermal histories that yield similar present heat flows. This is illustrated in **Figure 7**, which compares three cases in which the efficiency with which mantle plumes transport heat away from the core is varied. There

is some uncertainty in a numerical factor controlling this efficiency, with which the cases are labeled (Davies, 2007b). With higher plume efficiency, the early core heat flow is higher (**Figure 7**(**a**)) and the core cools more (**Figure 7**(**b**)). The mantle cooling history is not greatly affected and only one case is shown, for clarity. Rapid core cooling reduces the temperature difference between the core and the mantle and, since this also determines the plume heat flow, the core heat loss declines (**Figure 7**(**a**)) and ends up close to the first case. The value of the present core heat loss actually passes through a maximum in this kind of model as the efficiency of plume transport is varied (final heat losses are, respectively, 4.3, 5.3, 3.8 TW). To remove this ambiguity of past core heat loss we will need either better characterization of the efficiency of plumes (or other upwelling), or perhaps constraints from paleomagnetism, or both.

The thermal history of the core is closely tied to the generation of the Earth's magnetic field, which is believed to occur by the dynamo action of convection in the core (e.g., Labrosse and Macouin, 2003; Roberts *et al.*, 2003). A puzzle has arisen recently concerning how to reconcile the calculated core cooling rate with the energy required to maintain the dynamo and also with the inferred age of the inner core. The inner core is involved because compositional convection arising from the solidification of the inner core couples to the dynamo more efficiently, which is therefore readily maintained. Before the inner core began to solidify fairly strong thermal convection might have been required to maintain the dynamo, but this would imply faster core cooling and, eventually, rapid growth of the inner core. Unless the inner core began to grow relatively recently (only about 1 Ga), it would have grown bigger than is observed. This only intensifies the difficulty of driving the dynamo before the inner core started to form, and yet the paleomagnetic evidence is that a magnetic field of comparable strength to the present field has existed at least since 2.5 Ga, and possibly since 3.5 Ga. To reconcile these difficulties, it has been proposed that the core contains a radioactive heat source, in the form of dissolved potassium (Buffett, 2002; Nimmo *et al.*, 2004). None of the models in **Figures 6** and **7** include core radioactivity.

Included in **Figures 6(a)** and **7(a)** is the rate of energy dissipation in the core, Φ, which is the maximum energy available to the dynamo. Also included (**Figures 6(b)** and **7(b)**) is the radius of the inner core, with the initiation temperature adjusted to yield the present inner core radius of 1220 km. There is some uncertainty in the energy required to maintain the dynamo, with estimates ranging from 0.1 to 2 TW (Buffett, 2002; Roberts *et al.*, 2003). Two of the models have dissipation around 1 TW for much of Earth history, so they would be viable even with a fairly high dynamo requirement. The low core heat model of **Figure 6** would only be viable before the inner core formed if a low-energy requirement applies. This argument is explored in more detail elsewhere (Nimmo *et al.*, 2004; Davies, 2007b; Nimmo, 2007). The main point here is to illustrate how the core thermal history is regulated by mantle dynamics and therefore quite sensitive to the details of mantle behavior.

8.6 Alternative Models of Thermal History

The models considered so far are characterized by assuming mantle convection to be basically similar to convection in more familiar fluids and by their smoothly decreasing temperature and heat flow. On the other hand, Earth's tectonic history appears to have been quite episodic, since the distribution of ages in the continental crust is strongly peaked, as is illustrated in **Figure 8**. Since plate tectonics is a manifestation of mantle convection, it is not obvious that such a smoothly varying history could give rise to such a peaked age distribution. Therefore, mechanisms that might produce an episodic thermal evolution have been considered. We have also seen that estimates of Earth's internal radioactivity based on conventional thermal evolution models are hard to reconcile with estimates from cosmochemistry. This has given rise to one quite novel kind of model, which is considered below.

8.6.1 Episodic Histories

It was suggested fairly soon after plate tectonics became accepted that pressure-induced phase transformations in the mantle transition zone (400–660 km depth) could affect convection by either hindering or enhancing the rising and sinking of buoyant and negatively buoyant fluid (Schubert *et al.*, 1975). Machatel and Weber (1991) demonstrated that, depending on the thermodynamic parameters involved, phase

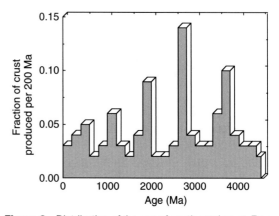

Figure 8 Distribution of the age of continental crust. From McCulloch MT and Bennett VC (1994) Progressive growth of the Earth's continental crust and depleted mantle: Geochemical constraints. *Geochimica et Cosmochimica Acta* 58: 4717–4738.

transformations could so hinder convection that it separates into two layers. This was confirmed in more detail in subsequent studies (e.g., Honda *et al.*, 1993; Tackley *et al.*, 1993; Nakakuki *et al.*, 1994; Solheim and Peltier, 1994b, 1994a; Weinstein, 1995). However, studies that incorporated strong temperature dependence of viscosity, thus allowing plates and plumes to be incorporate into the models, showed less propensity for layering (Nakakuki *et al.*, 1994; Davies, 1995a). Also it seems that the thermodynamic parameters of the phase transformations are only marginally adequate to cause layering, at least in the present mantle (Bina and Helffrich, 1994; Tackley *et al.*, 2005).

However, it is also plausible that conditions in the past were more conducive to layering. This led Honda (1995) to develop a parametrized convection model in which the mantle is initially layered but later undergoes a transition to whole-mantle convection. Davies (1995b) developed a more elaborate model, featuring multiple episodes of layering followed by breakdown into whole-mantle convection. One such history is illustrated in **Figure 9**. There is considerable uncertainty in several of the parameters entering such models, so this should be taken as no more than a conjectural illustration of the kind of episodes that might have been induced by phase transformations.

8.6.2 Strong Compositional Lithosphere

Motivated in part by the discrepancy in inferred radioactive heat sources, Korenaga (2006) has devised a substantially different kind of thermal evolution model. He assumes that because of enhanced melting in a hotter mantle, the thickness of the lithosphere, defined as the strong outer region of Earth, would be controlled not by the thermal boundary layer but by the zone of strong depletion due to melting. He also assumes that the bending of plates prior to subduction is a major source of resistance to plate tectonics, and that the radius of curvature of the bending is proportional to plate thickness. The result of this combination of assumptions is that plates go slower in a hotter mantle, and that in turn means heat loss is inhibited at high mantle temperatures.

The result is illustrated in **Figure 10**. The model assumes a Urey ratio such that $\gamma = 1/Ur$ is between 0.15 and 0.3 (his definition of Urey ratio is the inverse of the one used here). The thermal history was integrated backward from the present, on the grounds that the present is better determined. The model has

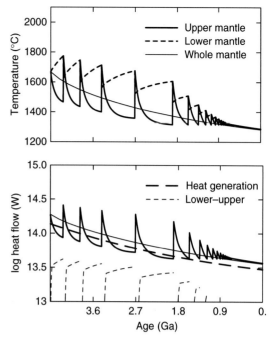

Figure 9 An episodic thermal history of the mantle, with episodes induced by parametrized effects of phase transformations. An unlayered ('whole mantle') case is included for comparison. In the lower panel the heat flow out of the mantle (heavy curve) and between the lower and upper layers (light dashed curves) are included. During the early evolution mantle convection stratifies into two layers, but eventually the temperature difference between the layers becomes large enough to cause a breakthrough and thermal homogenization, before layering is re-established. This example is conjectural and illustrative only (see text). From Davies GF (1995b) Punctuated tectonic evolution of the earth. *Earth and Planetary Science Letters* 136: 363–379.

a number of novel features. The temperature would have been quite high in the Archean, perhaps even reaching a maximum, then declining at an increasing rate toward the present (in this respect the model is reminiscent of the viscosity-invariant model of **Figure 2**). The recent cooling rate is over $100°C\,Ga^{-1}$. On the other hand, the heat loss would have been little different in the Archean and would have peaked only about 0.5 Ga. Correspondingly, plate velocities would also have peaked recently and been lower in the Archean, while the inverse Urey ratio would have declined toward the present. A 'conventional' thermal history with such a low inverse Urey ratio would have had to be extremely hot in the late Archean and unrealistically hot in the early Archean (**Figure 10(a)**).

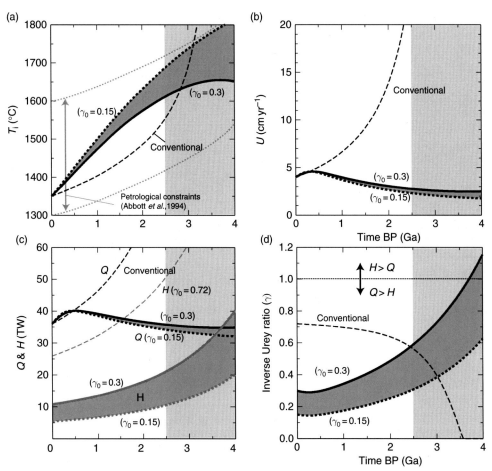

Figure 10 Mantle thermal evolution under the assumptions used by Korenaga (2006) and assuming low inverse Urey ratio $\gamma = 1/Ur = 0.15$ and 0.3 – see text. 'Conventional' cases are included for comparison (short-dashed). (a) Mantle temperature. Estimated petrological constraints from Abbott (Abbott *et al.*, 1994) are included. (b) Plate velocity. (c) Heat flow, Q, and heat generation, H. (d) Inverse Urey ratio. Gray shading on right indicates region of high uncertainty. From Korenaga J (2006) *The Archean Earth*. Washington, DC: American Geophysical Union.

Although there are aspects of this model that are not very well established, it does illustrate the possibility of quite different controls on thermal history than have usually been assumed, and it has some significant merits. For example, it reconciles the thermal history with a high Urey number, and it might help to explain why continental lithosphere surviving from the Archean is so thick. Nevertheless, its results are sensitive to some of its assumptions and parameter values. For example, assuming the bending radius of subducting lithosphere is proportional to thickness is plausible at first sight, but nonlinear rheological effects could cause the plate to bend more sharply with less dissipation, and this could considerably change the results. In addition, Korenaga argues that the high mantle temperatures

in his model are compatible with the range of temperatures inferred by Abbott *et al.* (1994). On the other hand, as noted earlier, Campbell argues that komatiites yielding the higher inferred temperatures are unusual and that most of the Archean mantle was producing basalt, and therefore within 100°C or so of the present temperature (Campbell and Griffiths, 1992). This model needs to be further debated.

8.7 Implications for Tectonic Evolution

The two main expressions of mantle convection in the present Earth are plate tectonics and volcanic hot spots, many of which are plausibly due to mantle plumes (Davies, 2005). If mantle convection changes then its tectonic expression may also change. The main difference between the early mantle and the present mantle, according to the models presented here, is that the early mantle was hotter. The question is then whether the higher temperature would change the tectonic expression of mantle convection.

Plate tectonics can be traced back with reasonable confidence from geological evidence to about 2 Ga (Hoffman and Bowring, 1984), but whether plates operated earlier than that is debated (Kröner, 1981; Harrison et al., 2005; Stern, 2005). The antiquity of plumes is also a matter requiring careful interpretation of the geological record (Campbell and Griffiths, 1992).

Korenaga's theory (Korenaga, 2006) implies that tectonics were different in the Archean, at least in terms of what regulates plate velocities. Plates could have operated in the Archean, though their dynamics would have been significantly different than at present.

The tectonic regime is determined by how the mantle gets rid of its heat (currently through plate tectonics) and how the mantle removes heat emerging from the core (currently plumes, although some debate that; Anderson (2004); Foulger (2005)). Any proposed tectonic mechanism must meet two fundamental conditions. First, there must be forces sufficient to drive it. Second, it must remove heat at a sufficient rate to cool the mantle to its present temperature (or, more generally, it must be explained how the mantle reached its present temperature). An understanding of mantle dynamics is thus an indispensable part of resolving the question of past tectonic regimes. The subject is far from being able to predict with any confidence which modes would apply. Nevertheless, it has progressed to the point of being able to examine some key questions.

8.7.1   Viability of Plate Tectonics

Davies (1992) proposed that plate tectonics might not have been viable when the mantle was hotter because greater melting under spreading centers would produce thicker oceanic crust, the buoyancy of which would hinder plate subduction. Even at present, plates initially have a net positive buoyancy because the oceanic crust is less dense than the mantle. They do not become negatively buoyant until conductive cooling has thickened them to the point where their negative thermal buoyancy overcomes the crust's positive compositional buoyancy. At present this occurs when they are about 15 My old. Since at present the average age of plates at subduction is

about 100 Ma, this does not interfere significantly with the plates' ability to cool the mantle. However, as recently as 1.6 Ga the plates would, on average, have become negatively buoyant only as they arrived at a subduction zone. This is not only because the crust would have been thicker but also because plates would have been going faster.

Prior to this 'cross-over' time, plates would still be positively buoyant at the time they would need to subduct if the mantle was to be cooled. Plate tectonics might still have operated, but the plates would need to age for longer before they became negatively buoyant and thus able to subduct. This means plate tectonics would have been less efficient at cooling the mantle. The hotter the mantle, the less efficient plate tectonics would be. This leads to the paradox that the early hot mantle would not have been able to cool to its present temperature, in fact it might have gone into thermal runaway. The implication is that some other mode of tectonics would have been required. Another way to say this is that the top thermal boundary layer of the mantle, which drives the cooling mode of mantle convection, would have to have operated in a different dynamical mode.

Davies (1992) noted that one way out of this paradox might be provided by the fact that basaltic oceanic crust transforms to eclogite at about 60 km depth. Since eclogite is denser than the average mantle, it would then hasten plate subduction rather than hindering it. Once plate tectonics was operating it might be able to continue to operate because the deeper plate would pull the surface plate down even if its surface buoyancy were still slightly positive. However, there would remain the barrier of initially getting low-density basaltic material to sufficient depth (and temperature) to transform, and the hotter the mantle the bigger this barrier would be. This question needs to be addressed with quantitative modeling, though this is not straightforward because the sources of plate resistance are not understood in detail. A related situation in the modern Earth, the subduction of oceanic plateaus, has been examined numerically by van Hunen et al. (2002), and this approach can be extended to the Archean context.

In the meantime another possible way out of the paradox has emerged. Numerical models of convection in the early, hot, low-viscosity mantle have shown that subducted oceanic crust, in its denser eclogite form, would tend to settle out of the upper mantle, leaving it depleted of basaltic-composition components (Davies, 2006) This would reduce the meltable portion of the upper mantle and therefore reduce the thickness of the oceanic crust. Initial modeling indicated that the oceanic crust might be only a few kilometers thick for a mantle temperature of 1550°C, rather than about 30 km if the upper mantle were as fertile as at present (Davies, 2006). More thorough subsequent testing suggests that the crustal thickness might be 6–10 km in these conditions (Davies, 2007a). This would still be enough to make plate tectonics more viable than the earlier argument suggested. Whether the reduction is enough for mantle to plate tectonics to cool the mantle is not yet clear.

8.7.2 Alternatives to Plate Tectonics

If plates were too buoyant to subduct, what would happen? One possibility is that the mantle part of the lithosphere founders and leaves the crust at the surface. Two ways in which this might happen are illustrated in **Figure 11** (Davies, 1992). In both cases the buoyant crust is sheared off as the mantle part of the lithosphere founders. The difference between the cases is that in (1) the thermal boundary layer is still strong enough to behave like a plate, whereas in (2) it is assumed to be deformable. Since asymmetric subduction, as in (1), results from the

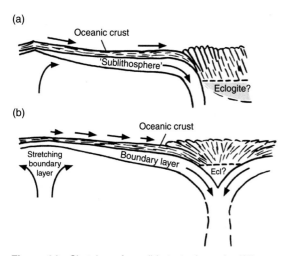

Figure 11 Sketches of possible tectonic modes if the buoyancy of oceanic crust prevents it from subducting with the rest of the lithosphere. (a) Lithosphere still thick enough to behave as a plate, giving rise to asymmetric subduction. (b) Deformable lithosphere, giving rise to symmetric foundering. From Davies GF (1992) On the emergence of plate tectonics. *Geology* 20: 963–966.

formation of a fault zone, this may not happen in (2) and the foundering may be symmetric. It is possible in either case that the accumulating basaltic crust becomes thick enough for its lower reaches to transform to eclogite, in which case the whole basaltic body might become unstable and founder, progressively transforming to eclogite as it does so. This might result in tectonic activity being strongly episodic.

8.7.3 History of Plumes

The cooling model of **Figure 6** yields a relatively constant rate of heat loss from the core. Since plumes are the expected form of upwelling in the mantle, due to the strong temperature dependence of viscosity (Davies, 2005), this implies a fairly steady level of plume activity through Earth history. On the other hand, higher early core heat losses and greater early plume activity are also possible (**Figure 7**).

Campbell *et al.* (1989) and Campbell and Griffiths (1992) have argued that Archean greenstone belts can be interpreted as the melting products of plume heads, so they would be analogs of Phanerozoic flood basalts. Assuming that the highest-temperature magmas are the best available sample of the plume, Campbell notes that there are two fundamental changes around the end of the Archean. First, the highest-temperature magmas in the Archean are komatiites with inferred source potential temperatures of 1800–1900°C, whereas the hottest post-Archean magmas are picrites with source potential tempatures of 1400–1600°C (**Figure 12**). Second, the Archean komatiites are depleted in incompatible elements, or neutral, whereas the later picrites are enriched.

The higher source temperature implied by komatiites has been widely remarked upon (e.g., Abbott *et al.*, 1994), and some have taken it to indicate a much hotter mantle in the Archean. However, as noted earlier, komatiites comprise only a small fraction of Archean mafic rocks, the vast majority of which are basalts, so a more straightforward interpretation seems to be that most of the mantle was only 100–200°C hotter in the Archean, and the komatiites come from plumes. This implies that plumes were significantly hotter in the Archean than subsequently, and suggests in turn that the models with more rapid early core cooling (**Figure 7**) might be favored. On the other hand, Campbell and Griffiths note that the change from high to low

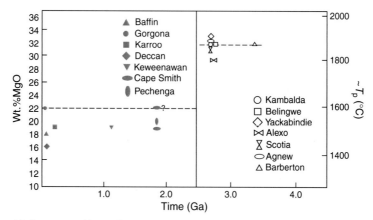

Figure 12 Maximum MgO content of komatiites and picrites versus age ('time'). The approximate mantle potential temperature is shown on the right. From Campbell IH and Griffiths RW (1992) The changing nature of mantle hotspots through time: Implications for the chemical evolution of the mantle. *Journal of Geology* 92: 497–523.

temperature seemed to occur relatively rapidly, between 2.7 and 1.9 Ga (**Figure 12**), suggesting a more discontinuous cause.

Noting the change in trace-element chemistry, from depleted to enriched, Campbell and Griffiths (1992) suggest that there was a change in material accumulating at the base of the mantle, in the D″ zone. Their scenario is sketched in **Figure 13**. The post-Archean regime is like the present, with subducted oceanic crust presumed to settle toward the base and accumulating as a trace-element enriched layer (Hofmann and White, 1982; Hofmann, 1997). On the other hand, they suppose that during the pre-Archean or Hadean (4.5–4 Ga), the mafic crust was too buoyant to founder, and only the underlying mantle part of the thermal boundary layer foundered. Since this material would have been depleted by the extraction of mafic crust and was also cool, it would sink to form a depleted layer at the base of the mantle. The post-Archean change in chemistry is then attributed to the replacement of the early depleted D″ layer by enriched subducted mafic crust. Their proposed cause of the change in temperature is less clear-cut in this scenario. It may be that in the earlier phase the D″ layer covered only part of the core, leaving hot core directly in contact with mantle elsewhere and thus generating very hot plumes. Alternatively, they speculate that a difference in chemistry may lead the later material to become positively buoyant at lower temperatures. In any case the change in character of inferred mantle plumes is attributed ultimately to a change in surface tectonics, which change the nature of the material sinking to the base of the mantle.

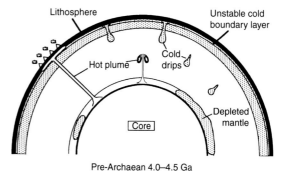

Figure 13 Sketch of proposed styles of mantle convection for pre- and post-Archean. The stippled regions are depleted of incompatible elements by melting. It is supposed that in the pre-Archean the mafic crust renders the upper lithosphere buoyant. The lower lithosphere is depleted, and cools and 'drips' away, accumulating at the base of the mantle. The post-Archean regime features modern-style plates and plumes, with subducted oceanic crust accumulating at the base of the mantle and displacing the previously accumulated depleted material. From Campbell IH and Griffiths RW (1992) The changing nature of mantle hotspots through time: Implications for the chemical evolution of the mantle. *Journal of Geology* 92: 497–523.

8.8 Conclusion

Our understanding of Earth's thermal history seems to have been given a firm foundation with the discovery of radioactivity and the realization that the mantle is mobile enough to transport heat by convection. We now have a quantitative and relatively simply theory of cooling since Earth's presumed hot formation by accretion of planetesimals (Safronov, 1978; Wetherill, 1985). However, there remains an important discrepancy between the heating inferred from this theory and the content of radioactive heat sources inferred from cosmochemistry. Whether this reflects an incomplete understanding of the mechanism of heat removal or of Earth's relationship with meteorites remains to be seen.

The strong temperature dependence of the viscosity of the mantle imparts a distinctive character to Earth's cooling history, as it does also to the mantle's two inferred modes of convection, involving plates and plumes (Davies, 1999). This factor allows the mantle to adjust its rate of convection, and therefore of heat transport, by relatively small adjustments in temperature. The result, in 'conventional' thermal history calculations, is a transient phase of rapid cooling, lasting perhaps 500 Ma, followed by a long phase in which the mantle heat loss follows the slow decline of heat sources due to radioactive decay.

The mantle's properties and dynamics also control the cooling of the core, whose heat must be transported away by the mantle. With present uncertainties about the efficiency of this process, there is an ambiguity in core histories, which could involve either relatively constant heat loss or steadily declining heat loss. This ambiguity affects our understanding of the history of mantle plumes and of the energetics involved with maintaining the geodynamo, apparently for at least 3.5 Ga.

One important question remains unresolved, namely whether plate tectonics could have operated in a significantly hotter mantle. There is a mechanism that might have hindered the plates, and there are suggestions of how other effects might have countered this mechanism, but more investigation is required.

The geological record seems to be plausibly, if so far only approximately, consistent with the broad picture presented here. Archean mafic rocks suggest a dominant mantle source perhaps 100–200°C hotter than the present mantle, with relatively uncommon komatiites and picrites suggesting plumes with temperatures 500–600°C hotter than the present mantle during the Archean.

The possibility of substantial deviations from this general picture must still be entertained, and some examples have been mentioned here.

References

Abbott D, Burgess L, Longhi J, and Smith WHF (1994) An empirical thermal history of the Earth's upper mantle. *Journal of Geophysical Research* 99: 13835–13850.

Anderson DL (2004) Simple scaling relations in geodynamics: The role of pressure in mantle convection and plume formation. *Chinese Science Bulletin* 49: 2017–2021.

Barrell J (1914) The strength of the Earth's crust. *Journal of Geology* 22: 655–683.

Bina CR and Helffrich G (1994) Phase transition Clapeyron slopes and transition zone seismic discontinuity topography. *Journal of Geophysical Research* 99: 15853–15860.

Buffett BA (2002) Estimates of heat flow in the deep mantle based on the power requirements for the geodynamo. *Geophysical Research Letters* 29: 1566.

Bunge H-P (2005) Low plume excess temperature and high core heat flux inferred from non-adiabatic geotherms in internally heated mantle circulation models. *Physics of the Earth and Planetary Interiors* 153: 3–10.

Campbell IH and Griffiths RW (1992) The changing nature of mantle hotspots through time: Implications for the chemical evolution of the mantle. *Journal of Geology* 92: 497–523.

Campbell IH, Griffiths R, and Hill RI (1989) Melting in an Archaean mantle plume: Heads it's basalts, tails it's komatiites. *Nature* 339: 697–699.

Cassen P, Reynolds RT, Graziani F, Summers A, McNellis J, and Blalock L (1979) Convection and lunar thermal history. *Physics of the Earth and Planetary Interiors* 19: 183–196.

Clark SP, Jr. (1957) Radiative transfer in the Earth's mantle. *Transactions of the American Geophysical Union* 38: 931–938.

Davaille A and Jaupart C (1993) Transient high-Rayleigh-number thermal convection with large viscosity variations. *Journal of Fluid Mechanics* 253: 141–166.

Davies GF (1977) Whole mantle convection and plate tectonics. *Geophysical Journal of the Royal Astronomical Society* 49: 459–486.

Davies GF (1979) Thickness and thermal history of continental crust and root zones. *Earth and Planetary Science Letters* 44: 231–238.

Davies GF (1980) Thermal histories of convective Earth models and constraints on radiogenic heat production in the Earth. *Journal of Geophysical Research* 85: 2517–2530.

Davies GF (1988) Ocean bathymetry and mantle convection, 1. Large-scale flow and hotspots. *Journal of Geophysical Research* 93: 10467–10480.

Davies GF (1990) Heat and mass transport in the early Earth. In: Newsome HE and Jones JH (eds.) *Origin of the Earth*, pp. 175–194. New York: Oxford University Press.

Davies GF (1992) On the emergence of plate tectonics. *Geology* 20: 963–966.

Davies GF (1993) Cooling the core and mantle by plume and plate flows. *Geophysical Journal International* 115: 132–146.

Davies GF (1995a) Penetration of plates and plumes through the mantle transition zone. *Earth and Planetary Science Letters* 133: 507–516.

Davies GF (1995b) Punctuated tectonic evolution of the Earth. *Earth and Planetary Science Letters* 136: 363–379.

Davies GF (1999) *Dynamic Earth: Plates, Plumes and Mantle Convection*. Cambridge: Cambridge University Press.

Davies GF (2005) A case for mantle plumes. *Chinese Science Bulletin* 50: 1541–1554.

Davies GF (2006) Gravitational depletion of the early Earth's upper mantle and the viability of early plate tectonics. *Earth and Planetary Science Letters* 243: 376–382.

Davies GF (2007a) Controls on density stratification in the early Earth. *Geochemistry, Geophysics, Geosystems* (in press).

Davies GF (2007b) Mantle regulation of core cooling: A geodynamo without core radioactivity? *Physics of the Earth and Planetary Interiors* 160: 215–229.

Fisher O (1881) *Physics of the Earth's Crust*. London: Murray.

Foulger GR (2005) Mantle plumes: Why the current skepticism? *Chinese Science Bulletin* 50: 1555–1560.

Galer SJG, Goldstein SL, and O'Nions RK (1989) Limits on chemical and convective isolation in the Earth's interior. *Chemical Geology* 75: 257–290.

Goldreich P and Toomre A (1969) Some remarks on polar wandering. *Journal of Geophysical Research* 74: 2555–2567.

Grand S, van der Hilst RD, and Widiyantoro S (1997) Global seismic tomography: A snapshot of convection in the Earth. *Geological Society of America Today* 7: 1–7.

Grasset O and Parmentier EM (1998) Thermal convection in a volumetrically heated, infinite Prandtl number fluid with strongly temperature-dependent viscosity: Implications for planetary thermal evolution. *Journal of Geophysical Research* 103: 18171–18,181.

Hager BH (1984) Subducted slabs and the geoid: Constraints on mantle rheology and flow. *Journal of Geophysical Research* 89: 6003–6015.

Hallam A (1989) *Great Geological Controversies*. Oxford: Oxford University Press.

Harrison TM, Blichert-Toft J, Müller W, Albarède F, Holden P, and Mojzsis SJ (2005) Heterogeneous Hadean hafnium: Evidence of continental crust at 4.4 to 4.5 Ga. *Science* 310: 1947–1950.

Hirth G and Kohlstedt DL (1996) Water in the oceanic upper mantle: Implications for rheology, melt extraction and the evolution of the lithosphere. *Earth and Planetary Science Letters* 144: 93–108.

Hofmann AW (1997) Mantle chemistry: The message from oceanic volcanism. *Nature* 385: 219–229.

Hoffman PF and Bowring SA (1984) Short-lived 1.9 Ga continental margin and its destruction, Wopmay orogen, northwest Canada. *Geology* 12: 68–72.

Hofmann AW and White WM (1982) Mantle plumes from ancient oceanic crust. *Earth and Planetary Science Letters* 57: 421–436.

Hofmeister AM (2005) Dependence of diffusive radiative transfer on grain size, temperature, and Fe-content: Implications. *Journal of Geodynamics* 40: 51–72.

Holmes A (1911) *Proceedings of the Royal Society of London A* 85: 248.

Holmes A (1913) *The Age of the Earth*. London: Harper & Brothers.

Honda S (1995) A simple paramaterized model of Earth's thermal history with the transition from layered to whole mantle convection. *Earth and Planetary Science Letters* 131: 357–369.

Honda S, Yuen DA, Balachandar S, and Reuteler D (1993) Three-dimensional instabilities of mantle convection with multiple phase transitions. *Science* 259: 1308–1311.

Isacks B, Oliver J, and Sykes LR (1968) Seismology and the new global tectonics. *Journal of Geophysical Research* 73: 5855–5899.

Jeffreys H (1976) *The Earth, Its Origin, History and Physical Constitution*. Cambridge: Cambridge University Press.

Jochum KP, Hofmann AW, Ito E, Seufert HM, and White WM (1983) K, U and Th in mid-ocean ridge basalt glasses and heat production, K/U and K/Rb in the mantle. *Nature* 306: 431–436.

Kellogg LH, Hager BH, and van der Hilst RD (1999) Compositional stratification in the deep mantle. *Science* 283: 1881–1884.

Kelvin L (1862) On the age of the sun's heat. *Macmillans Magazine* 5: 288.

Kelvin L (1863) On the secular cooling of the Earth. *Philosophical Magazine (Series 4)* 25: 1.

Kelvin L (1899) The age of the Earth as an abode fitted for life. *Journal of Victoria Institute* 31: 11.

Korenaga J (2006) *The Archean Earth*. Washington, DC: American Geophysical Union.

Kröner A (ed.) (1981) *Precambrian Plate Tectonics*, pp. 59–60. Amsterdam: Elsevier.

Labrosse S and Macouin M (2003) The inner core and the geodynamo. *Comptes Rendus Geoscience* 335: 37–50.

Machatel P and Weber P (1991) Intermittent layered convection in a model mantle with an endothermic phase change at 670 km. *Nature* 350: 55–57.

Manga M, Weeraratne D, and Morris SJS (2001) Boundary layer thickness and instabilities in Benard convection of a liquid with temperature-dependent viscosity. *Physics of the Fluids* 13: 802–805.

McCulloch MT and Bennett VC (1994) Progressive growth of the Earth's continental crust and depleted mantle: Geochemical constraints. *Geochimica et Cosmochimica Acta* 58: 4717–4738.

McDonough WF and Sun S-s (1995) The composition of the Earth. *Chemical Geology* 120: 223–253.

McGovern PJ and Schubert G (1989) Thermal evolution of the Earth: Effects of volatile exchange between atmosphere and interior. *Earth and Planetary Science Letters* 96: 27–37.

McKenzie DP and Bickle MJ (1988) The volume and composition of melt generated by extension of the lithosphere. *Journal of Petrology* 29: 625–679.

McKenzie DP, Roberts JM, and Weiss NO (1974) Convection in the Earth's mantle: Towards a numerical solution. *Journal of Fluid Mechanics* 62: 465–538.

McKenzie DP and Weiss N (1975) Speculations on the thermal and tectonic history of the Earth. *Geophysical Journal of the Royal Astronomical Society* 42: 131–174.

Mitrovica JX (1996) Haskell (1935) revisited. *Journal of Geophysical Research* 101: 555–569.

Mitrovica JX and Forte AM (1997) Radial profile of mantle viscosity: Results from the joint inversion of convection and postglacial rebound observables. *Journal of Geophysical Research* 102: 2751–2769.

Nakakuki T, Sato H, and Fujimoto H (1994) Interaction of the upwelling plume with the phase and chemical boundary at the 670 km discontinuity: Effects of temperature-dependent viscosity. *Earth and Planetary Science Letters* 121: 369–384.

Nimmo F (2007) Thermal and compositional evolution of the core this volume.

Nimmo F, Price GD, Brodholt J, and Gubbins D (2004) The influence of potassium on core and geodynamo evolution. *Geophysical Journal International* 156: 363–376.

O'Connell RJ (1977) On the scale of mantle convection. *Tectonophysics* 38: 119–136.

Palme H and O'Neill HSC (2004) Cosmochemical estimates of mantle composition. In: Carlson RW, Holland HD, and Turekian KK (eds.) *Treatise on Geochemistry*, pp. 1–38. Oxford: Elsevier.

Patterson C (1956) Age of meteorites and of the Earth. *Geochimica et Cosmochimica Acta* 10: 230–237.

Ratcliffe JT, Tackley PJ, Schubert G, and Zebib A (1997) Transitions in thermal convection with strongly variable viscosity. *Physics of the Earth and Planetary Interiors* 102: 201–212.

Roberts PH, Jones CA, and Calderwood AR (2003) *Earth's Core and Lower Mantle*. London: Taylor and Francis.

Rossby HT (1969) A study of Benard convection with and without rotation. *Journal of Fluid Mechanics* 36: 309–336.

Rudnick RL and Fountain DM (1995) Nature and composition of the continental crust: A lower crustal perspective. *Reviews of Geophysics* 33: 267–309.

Safronov VS (1978) The heating of the Earth during its formation. *Icarus* 33: 8–12.

Schubert G (1979) Subsolidus convection in the mantles of terrestrial planets. *Annual Review of Earth and Planetary Sciences* 7: 289–342.

Schubert G, Cassen P, and Young RE (1979a) Core cooling by subsolidus mantle convection. *Physics of the Earth and Planetary Interiors* 20: 194–208.

Schubert G, Cassen P, and Young RE (1979b) Subsolidus convective cooling histories of terrestrial planets. *Icarus* 38: 192–211.

Schubert G, Stevenson DJ, and Cassen P (1980) Whole planet cooling and the radiogenic heat source contents of the Earth and moon. *Journal of Geophysical Research* 85: 2531–2538.

Schubert G, Turcotte DL, and Olson P (2001) *Mantle Convection in the Earth and Planets*. Cambridge: Cambridge University Press.

Schubert G, Yuen DA, and Turcotte DL (1975) Role of phase transitions in a dynamic mantle. *Geophysical Journal of the Royal Astronomical Society* 42: 705–735.

Sclater JG, Jaupart C, and Galson D (1980) The heat flow through the oceanic and continental crust and the heat loss of the Earth. *Reviews of Geophysics* 18: 269–312.

Shankland TJ, Nitsan U, and Duba AG (1979) Optical absorption and radiative heat transport in olivine at high temperature. *Journal of Geophysical Research* 84: 1603–1610.

Sharpe HN and Peltier WR (1978) Paramaterized mantle convection and the Earth's thermal history. *Geophysical Research Letters* 5: 737–740.

Sharpe HN and Peltier WR (1979) A thermal history model for the Earth with parameterised convection. *Geophysical Journal of the Royal Astronomical Society* 59: 171–203.

Sleep NH (1990) Hotspots and mantle plumes: Some phenomenology. *Journal of Geophysical Research* 95: 6715–6736.

Solheim LP and Peltier WR (1994a) Avalanche effects in phase transition modulated thermal convection: A model of Earth's mantle. *Journal of Geophysical Research* 99: 6997–7018.

Solheim LP and Peltier WR (1994b) Correction to Avalanche effects in phase transition modulated thermal convection: A model of Earth's mantle. *Journal of Geophysical Research* 99: 18203.

Solomatov VS (1995) Scaling of temperature- and stress-dependent viscosity convection. *Physics of the Fluids* 7: 266–274.

Spohn T and Schubert G (1982) Modes of mantle convection and the removal of heat from the Earth's interior. *Journal of Geophysical Research* 87: 4682–4696.

Stacey FD (1980) The cooling earth: A reappraisal. *Physics of the Earth and Planetary Interiors* 22: 89–96.

Stacey FD (1992) *Physics of the Earth*. Brisbane: Brookfield Press.

Stacey FD and Loper DE (1984) Thermal histories of the core and mantle. *Physics of the Earth and Planetary Interiors* 36: 99–115.

Stern RJ (2005) Evidence from ophiolites, blueschists, and ultrahigh-pressure metamorphic terranes that the modern episode of subduction tectonics began in Neoproterozoic time. *Geology* 33: 557–560.

Stevenson DJ and Schubert G (1983) Magnetism and thermal evolution of the terrestrial planets. *Icarus* 54: 466–489.

Stevenson DJ and Turner JS (1979) Fluid models of mantle convection. In: McElhinny M (ed.) *The Earth, Its Origin, Evolution and Structure*, pp. 227–263. New York: Academic Press.

Tackley PJ, Stevenson DJ, Glatzmaier GA, and Schubert G (1993) Effects of an endothermic phase transition at 670 km depth in a spherical model of convection in the Earth's mantle. *Nature* 361: 699–704.

Tackley PJ, Xie S, Nakagawa T, and Hernlund JW (2005) Numerical and laboratory studies of mantle convection: Philosophy, accomplishments, and thermochemical structure and evolution. In: van der Hilst RD, Bass JD, Matas J, and Trampert J (eds.) *Earth's Deep Mantle: Structure, Compositon, and Evolution*, pp. 83–99. Washington, DC: American Geophysical Union.

Tozer DC (1965) Heat transfer and convection currents. *Philosophical Transactions of the Royal Society of London, Series A* 258: 252–271.

Tozer DC (1972) The present thermal state of the terrestrial planets. *Physics of the Earth and Planetary Interiors* 6: 182–197.

Turcotte DL and Oxburgh ER (1967) Finite amplitude convection cells and continental drift. *Journal of Fluid Mechanics* 28: 29–42.

Turcotte DL and Schubert G (2001) *Geodynamics: Applications of Continuum Physics to Geological Problems*. Cambridge: Cambridge University Press.

van Hunen J, van den Berg AP, and Vlaar NJ (2002) On the role of subducting oceanic plateaus in the development of shallow flat subduction. *Tectonophysics* 352: 317–333.

Weinstein S (1995) The effects of a deep mantle endothermic phase change on the structure of thermal convection in silicate planets. *Journal of Geophysical Research* 100: 11719–11728.

Wetherill GW (1985) Occurrence of giant impacts during the growth of the terrestrial planets. *Science* 228: 877–879.

9 Thermal and Compositional Evolution of the Core

F. Nimmo, University of California, Santa Cruz, CA, USA

9.1 Introduction

The evolution of the Earth's core is important for three main reasons. First, the formation of the core was one of the central events in the ancient, but geologically rapid, period over which the Earth accreted, and generated observational constraints on this poorly understood epoch. Second, the initial conditions, both thermal and compositional, established during this period have largely controlled the subsequent evolution of the core, and may have also significantly affected the mantle. Finally, the evolution of the Earth's core resulted in the generation of a long-lived global magnetic field, which did not occur for the superficially similar cases of Mars or Venus. The objective of this chapter is to describe our current understanding of the evolution of the core, from shortly after its formation to the present day.

The first section of this chapter will summarize the present-day state of the core, since it is this state which is the end product of the core's evolution. The bulk of the chapter will then examine how the core evolved from its initial thermal and compositional state. Most of the arguments will be based on physics rather than chemistry, as compositional constraints on the core's long-term evolution are rare and often controversial.

The material covered in this chapter follows on directly from the chapter by Chapter 3, in which the origin and formation of the core are discussed. Much of the discussion of the core's energy and entropy budgets is derived from a more thorough treatment. Other aspects of the core's behavior are described in chapters in this treatise. The companion *Treatise on Geochemistry* contains useful articles on planetary accretion (Chambers, 2003) and various aspects of core composition (Righter and Drake, 2003; Li and Fei, 2003; McDonough, 2003).

9.2   Present-Day State of the Core

Prior to investigating the earliest history and evolution of the core, it is important to briefly describe its present-day features. More detail can be found in the chapters referred to above; here the focus is on those parameters which are most important when considering the thermal and compositional evolution of the core. In particular, the uncertainties associated with these parameters will be assessed; doing so is important when assessing the likely range of thermal evolution outcomes (Section 9.3.2). The values and uncertainties adopted are discussed below; they are based on those used in previous investigations by Buffett *et al.* (1996), Roberts *et al.* (2003), Labrosse (2003), and Nimmo *et al.* (2004).

9.2.1   Density and Pressure

The radially averaged density structure of the core may be derived directly from seismological observations. The density of the core increases monotonically with depth, due to the increasing pressure. However, there is also a sharp density discontinuity at the inner-core boundary (ICB), which arises because of two effects. First, solid core material is inherently denser than liquid core material at the same pressure and temperature (P, T) conditions. Second, the outer core contains more of one or more light elements than the inner core (e.g., Poirier, 1994; McDonough, 2003), and would therefore be less dense even if there were no phase change. This compositional density contrast $\Delta\rho_c$ has a dominant role in driving compositional convection in the core; unfortunately, its magnitude is uncertain by a factor ≈ 2.

The total density contrast across the ICB is somewhat uncertain. A recent normal mode study (Masters and Gubbins, 2003) gives a total density contrast of 640–1000 kg m^{-3}, or 5.3–8.3%, which agrees rather well with the result of 600–900 kg m^{-3} obtained using body waves (Cao and Romanowicz, 2004), but is somewhat higher than the value obtained by Koper and Dombrovskaya (2005). The density

contrast between pure solid and liquid Fe at the ICB is estimated at 1.8% (Alfe *et al.*, 1999). These results imply a compositional density contrast of 3.5–6.5%, or $\Delta\rho_c = 400$–800 kg m^{-3}, and may in turn be used to estimate the difference in light element(s) concentrations between inner and outer core, which helps to sustain the dynamo (see Section 9.3.2.2).

For the theoretical models described later (Section 9.3), it is important to have a simple description of the density variation within the Earth. One such description is given by Labrosse *et al.* (2001), where the variation of density ρ with radial distance r from the centre of the Earth is given by

$$\rho(r) = \rho_{cen}\exp\left(-r^2/L^2\right) \qquad [1]$$

where ρ_{cen} is the density at the center of the Earth and L is a length scale given by

$$L = \sqrt{\frac{3K_0\left(\ln\left(\rho_{cen}/\rho_0\right)+1\right)}{2\pi G\rho_0\rho_{cen}}} \qquad [2]$$

Here K_0 and ρ_0 are the compressibility and density at zero pressure, respectively, G is the universal gravitational constant and $L = 7272$ km using the parameters given. Although this expression neglects the density jump at the ICB, the error introduced is negligible compared to other uncertainties.

The corresponding pressure is given by

$$P(r) = P_c + \frac{4\pi G\rho_{cen}^2}{3}\left[\left(\frac{3r^2}{10}-\frac{L^2}{5}\right)\exp\left(-r^2/L^2\right)\right]_r^R \qquad [3]$$

where P_c is the pressure at the CMB and R is the core radius.

9.2.2   Thermodynamic Properties

From the point of view of the thermal evolution of the core, the most important parameters are those which determine the temperature structure and heat flux within the core, in particular the thermal conductivity k and expansivity α (see Section 9.2.4).

The thermal conductivity of iron at core conditions is obtained by using shock-wave experiments and converting the measured electrical conductivity to thermal conductivity using the Wiedemann–Franz relationship (Stacey and Anderson, 2001). The canonical value for k at the CMB of 46 W m^{-1} K^{-1} (Stacey and Anderson, 2001) was based on shock measurements by Matassov (1977). More recent shock experiments by Bi *et al.* (2002) suggest a conductivity closer to $30 \, \mathrm{W \, m^{-1} \, K^{-1}}$. Here a value of $40 \pm 20 \, \mathrm{W \, m^{-1} \, K^{-1}}$ as spanning the likely uncertainties is assumed.

The thermal expansivity within the core may be obtained from seismology if the Gruneisen parameter is known. Recent results suggest that this parameter remains constant at roughly 1.5 throughout the core (e.g., Anderson, 1998; Alfe *et al.*, 2002b). Because the seismic parameter increases with depth, α increases by a factor of 1.5–2 from the center of the Earth to the CMB (Labrosse, 2003; Roberts *et al.*, 2003), but little accuracy is sacrificed if a constant mean value of α is adopted. Following the latter two authors, in this chapter a range $(0.8–1.9) \times 10^{-5} \, \mathrm{K^{-1}}$ is adopted. A list of estimated values for important parameters is given in **Table 2**.

9.2.3 Composition

The composition of the core is important because it potentially provides constraints on its origin and mode of formation. Unfortunately, as will be seen below, the constraints provided are currently rather weak, as few elements have well-known core abundances.

9.2.3.1 Light elements

It is clear from seismology and experiments that the outer core is 6–10% less dense than pure liquid iron would be under the estimated P, T conditions (e.g., Alfe *et al.*, 2002a). While the core almost certainly contains a few weight percent nickel (e.g., McDonough, 2003), this metal has an almost identical density to iron and is thus not the source of the density deficit (e.g., Li and Fei, 2003). The inner core also appears to be less dense than a pure iron composition would suggest (Jephcoat and Olson, 1987), though here the difference is smaller. Both the outer and inner core must therefore contain some fraction of light elements, of which the most common suspects are sulfur, silicon, oxygen, carbon, and hydrogen (see Poirier (1994), Hillgren *et al.* (2000), and Li and Fei (2003) for reviews). For any particular element, or mixture of them, the inferred density deficit may be used to infer the molar fraction of the light element(s) present. Because the density deficit is larger in the outer core, it is thought that light elements are being expelled during crystallization of the inner core. This expulsion is of great importance, because it generates compositional convection which helps to drive the geodynamo (see Section 9.3.2.2).

Apart from their role in driving the dynamo, these light elements are important for two other reasons. First, they probably reduce the melting temperature of the core by several hundred degrees kelvin (see Section 9.2.4). Second, if the actual light elements in the core could be reliably identified, they would provide a strong constraint on the conditions under which the core formed.

Table 1 gives several examples of model core compositions. All these models are derived by comparing estimates of the bulk silicate Earth elemental abundances (inferred from upper-mantle nodules and crustal samples), with estimates of the initial solar nebular composition (based mainly on chondritic meteorite samples). Although there is some agreement on the abundances of Fe, Ni, and Co, the relative abundances of the light elements (Si, S, O, C) vary widely. The abundance of H in the core cannot be modeled in this way because of its extreme volatility; in practice, it will be determined by the P, T conditions and amount of H in the Earth's mantle prior to and during differentiation (see, e.g., Abe *et al.* (2000) and Okuchi (1997)).

These cosmochemical models do not take into account the ease with which different elements partition into iron under the relevant conditions. Available experiments suggest that O and S can both enter the core under oxidizing conditions, while Si requires reducing conditions and is mutually incompatible with O (Kilburn and Wood, 1997; Hillgren *et al.*, 2000; Li and Fei, 2003; Malavergne *et al.*, 2004). An Fe–O–S liquid with 10.5 ± 3.5 wt.% S and 1.5 ± 1.5 wt.% O is also compatible with seismological observations (Helffrich and Kaneshima, 2004), although cosmochemical models do not favor such large amounts of sulfur (**Table 1**). An Fe–O–S core thus suggests core formation conditions which were relatively oxidizing (hence ruling out, for instance, the presence of substantial amounts of H in

Table 1 Model core compositions

	MA	WD	A+	McD-1	McD-2
Fe (wt.%)	84.5	80.3	79.4	85.5	88.3
Ni	5.6	5.5	4.9	5.2	5.4
Si	–	14.0	7.4	6.0	–
S	9.0	–	2.3	1.9	1.9
O	–	–	4.1	–	3.0
C	–	–	–	0.2	0.2
Co	0.26	0.27	0.25	–	–

MA = Morgan and Anders 1980; WD = Wanke and Dreibus 1988; A+ = Allegre *et al.* 1995a; McD-1 and McD-2 refer to two different models given in McDonough (2003).

the core), and also relatively high temperature. These inferred conditions are roughly consistent with estimates based on mantle siderophile element abundances (*see* Chapter 3).

An additional constraint is that the outer core appears to contain more of the light element(s) than the inner core (Section 9.2.3.1). This implies that one or more of the light elements must partition strongly into liquid iron during freezing, which is potentially diagnostic behavior. For instance, Alfe *et al.* (2002a) used molecular dynamics simulations to find that oxygen, due to its small atomic radius, tends to be expelled during freezing. Conversely, S and Si have atomic radii similar to that of iron at core pressures, and thus substitute freely for iron in the solid inner core. These results thus support the case for O being one of the light elements, in agreement with the Fe–O–S core hypothesized by Helffrich and Kaneshima (2004). Unfortunately, similar models have not yet been carried out for either H or C, which might also behave in a similar manner to O. Furthermore, the results concerning S, Si, and O need additional confirmation, preferably by experiments. Nonetheless, the implications for core formation and composition are potentially important.

For completeness, it is noted that some other gases, such as nitrogen (Adler and Williams, 2005) and xenon (Lee and Steinle-Neumann 2006), may also partition into the core. However, other studies have found negligible partitioning (Matsuda *et al.*, 1993; Ostanin *et al.*, 2006). More to the point, since neither the initial abundance of such gases, nor their current concentrations in the core, are currently known, they do not in general provide any constraints on core evolution.

9.2.3.2 Radioactive isotopes

Although the bulk of the core consists of Fe, Ni and a few weight percent light elements, some trace elements are also of importance. First, the radioactive isotopes of K, Th, and U can potentially have a significant effect on both the age of the inner core and the maintenance of the dynamo (e.g., Labrosse *et al.*, 2001; Buffett, 2002; Nimmo *et al.*, 2004). Unfortunately, there is as yet little agreement on whether or not such elements are really present in the core. Longer discussions on this issue may be found in McDonough (2003) and Roberts *et al.* (2003); only a brief summary is given here, while the role of radioactive elements in core evolution is discussed in Section 9.3.2.

There is little evidence, either from cosmochemistry or partitioning experiments, to expect either U or Th to partition into the core. On the other hand, the Earth's mantle is clearly depleted in K relative to chondrites (e.g., Lassiter 2004). However, since K is a volatile element, it is unclear whether this depletion is due to sequestration of K in the core, or simple loss of K from the Earth as a whole early in its history. Experimental investigations (Gessmann and Wood, 2002; Murthy *et al.*, 2003) show that partitioning of K into core materials is possible, but also depends in a complex fashion on other factors such as the amount of sulfur present. The removal of K to the core would also likely involve the removal of other elements with similar affinities for iron, but it is not yet clear what constraints the observed abundances of these other elements place on the amount of K in the core. It currently appears that up to a few hundred ppm K in the core is permitted, but not required, by both the experiments and the geochemical observations. The detection of antineutrinos produced by radioactive decay in the Earth's interior (Araki *et al.*, 2005) may help to ultimately resolve this question.

A second set of potentially very important isotopes are those of osmium, because they may constrain the onset of inner-core formation (e.g., Walker *et al.*, 1995). The arguments for and against this somewhat controversial hypothesis are discussed in Section 9.3.3.1. The reason the arguments are important is that the onset of inner-core formation is currently very poorly constrained by theoretical models (Section 9.3.2.5); thus, the addition of an observational constraint would significantly improve our understanding of the core's evolution.

A final important isotopic system is ^{182}Hf–^{182}W. This system permits the age of core formation to be deduced (e.g., Harper and Jacobsen, 1996; Kleine *et al.*, 2002; Nimmo and Agnor, 2006) and is discussed in some detail in Chapter 3. Other isotopic systems can potentially be used in similar ways (see Allegre *et al.* (1995b)). However, the Pd-Ag system is experimentally very challenging (Carlson and Hauri, 2001), and the U–Pb system suffers from the potential loss of lead due to its high volatility (e.g., Halliday, 2004).

9.2.4 Temperature Structure

Both the temperature structure within the core, and the shape of the melting curve, play an important role in determining the thermal evolution of the core. As long as the core is convecting, its mean temperature profile will be that of an adiabat, except at the very thin top and bottom boundary layers. Since the temperature at the ICB must equal the melting

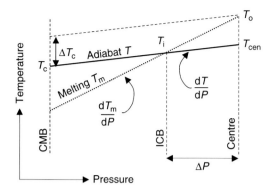

Figure 1 Definition sketch for temperature structure within the core. Note that in reality neither temperature profile is linear.

temperature of the core at that pressure (**Figure 1**), the temperature elsewhere in the core may be extrapolated from the ICB conditions by using the appropriate adiabat. Thus, determining the melting behavior of core material is crucial to establishing the temperature structure of the core.

The adiabatic temperature T within the core is given by (Labrosse et al., 2001)

$$T(r) = T_{cen} \exp(-r^2/D^2) \qquad [4]$$

where T_{cen} is the temperature at the center of the Earth and D is a length scale given by

$$D = \sqrt{3C_p/2\pi\alpha\rho_{cen}G} \qquad [5]$$

Here C_p is the specific heat capacity, α is the thermal expansivity (Section 9.2.2) and $D = 5969$ km using the parameters given in **Table 2**, model 2.

The melting behavior of pure iron is difficult to establish: experiments at the P, T conditions required

(e.g., Brown and McQueen, 1986; Yoo et al., 1993; Boehler, 1993) are challenging, and computational (first-principles) methods (e.g., Laio et al., 2000; Belonoshko et al., 2000; Alfe et al., 2002b) are time-consuming and hard to verify. Furthermore, the presence of the light element(s) likely reduces the melting temperature from that of pure iron, but by an uncertain amount. These issues are discussed in detail elsewhere in this volume, and only a summary is provided here.

Based on first-principles calculations, Alfe et al. (2003) predict a temperature at the ICB T_i of 5650 ± 600 K, taking into account the reduction in temperature due to the light element(s). The gradient in melting temperature is roughly 8.5 K GPa^{-1} at the ICB. These results are broadly consistent with the low-pressure diamond anvil cell results of Shen et al. (1998) and Ma et al. (2004), though not those of Boehler (1993). Similarly, the results agree with the higher-pressure shock-wave results of Brown and McQueen (1986) and Nguyen and Holmes (2004), though not those of Yoo et al. (1993). The numerical results of Belonoshko et al. (2000) and Laio et al. (2000) also give similar answers once corrections due to the different molecular dynamics techniques used have been applied. Further discussions of the differing results and the reliability of different theoretical approaches may be found in Alfe et al. (2004) and Bukowinski and Akber-Knutson (2005).

Given the ICB temperature and the relevant thermodynamic quantities, the temperature at the core side of the CMB, T_c, may be deduced and is approximately 4000 K. This value is actually of only secondary importance as far as the thermal evolution of the core is concerned; of much greater interest are

Table 2 Parameter values adopted

	Model						Model				
	1	**2**	**3**	**Units**	**Eq.**		**1**	**2**	**3**	**Units**	**Eq.**
k	20	40	60	W m^{-1} K^{-1}	2.2	$\Delta\rho_c$	600	400	200	kg m^{-3}	2.1
ΔT_c	150	100	50	K	(7)	α	0.8	1.35	1.9	$\times 10^{-5}$ K^{-1}	(5)
D	7754	5969	5031	km	(5)	T_{cen}	4893	5619	6454	K	(4)
T_i	4773	5389	6085	K	2.4	T_o	5076	5760	6535	K	2.4
T_{m0}	1460	1322	1165	K	(6)	dT_m/dP	10.1	12.4	15.0	K GPa^{-1}	(6)
Q_k	1.4	4.8	10.0	TW	(12)	$\tilde{Q}_T$	2.9	3.2	3.8	$\times 10^{27}$ J	(11)

Eq. is the equation or section in which each parameter is defined.
Models 1 and 3 are end-member cases using parameter values designed to generate ancient and recent inner cores, respectively; model 2 is a best guess at the real parameter values. Variables below the horizontal line have values derived from the initial parameter choices (k, α, $\Delta\rho_c$, ΔT_c). Other parameters not specified here are assumed constant in all three models and are generally the same as those adopted in 00128. In particular, $T_c = 4000$ K, latent heat $L_H = 750$ kJ kg^{-1}, $C_p = 840$ J kg^{-1} K^{-1}, $\rho_{cen} = 12\,500$ kg m^{-3}.

the relative slopes of the adiabat and the melting curve (see below).

Although more complicated approaches may be adopted (e.g., Buffett *et al.*, 1996; Labrosse *et al.*, 2001; Roberts *et al.*, 2003), a reasonable approximation is that the core melting temperature is simply a linear function of pressure. Thus, the core melting temperature T_m may be written as

$$T_m(P) = T_{m0} + \frac{dT_m}{dP}P \qquad [6]$$

where dT_m/dP and T_{m0} are constants and T_{m0} incorporates the reduction in melting temperature due to the light element(s). Note that this linearization is only intended to work over the core pressure range, and that T_{m0} thus does not represent the actual zero-pressure melting temperature.

When considering the growth history of the inner core, the crucial parameter is the difference in gradients between the melting curve and the adiabat. One way of expressing this quantity is to define ΔT_c, the change in the CMB temperature since the onset of inner-core solidification. As shown in **Figure 1**, ΔT_c may be defined as follows:

$$\Delta T_c = \frac{\Delta P}{f_{ad}}\left(\frac{dT_m}{dP} - \frac{dT}{dP}\right)$$
$$= 22\,K\left(\frac{dT_m/dP - dT/dP}{1\,K\,GPa^{-1}}\right) \qquad [7]$$

where ΔP is the pressure difference between the present ICB and the center of the Earth, T and T_m are the adiabatic and melting temperatures, respectively, f_{ad} is a factor converting the temperature at the CMB to that at the ICB and the curves are assumed linear over the relevant pressure range. The numerical values are obtained from model 2 in **Table 2**. Because the adiabatic and melting gradients are both uncertain and of similar sizes, the uncertainty in ΔT_c tends to be amplified. Values for ΔT_c from four recent studies (Buffett *et al.*, 1996; Roberts *et al.*, 2003; Labrosse, 2003; Nimmo *et al.*, 2004) range from 31–146 K, with smaller values implying younger inner cores. Here I assume a range of 50–150 K as representative of the likely uncertainties. By choosing values for T_c, ΔT_c and the adiabatic gradient, the ICB temperature T_i and dT_m/dP are then specified (see **Table 2**). Note that the melting gradients obtained exceed the value given by Alfe *et al.* (2003); a reduction in this gradient would result in a smaller ΔT_c (eqn [7]) and thus a younger inner core. Hence, the range of values for ΔT_c chosen here is conservative.

9.2.5   The CMB Region

The CMB region is relevant to core evolution for two reasons. First, it is the behavior of this region, and in particular its temperature structure, which controls the rate at which heat is extracted from the core. As a consequence, the thermal evolution of the core is intimately tied to that of the mantle. This interdependence between the core and mantle is one of the reasons that theoretical investigations of core thermal evolution are subject to such large uncertainties.

Second, the CMB region is interesting from a compositional point of view, since it is the point at which regions containing elements with very different chemical potentials meet. The extent to which the resulting reactions have influenced the behavior of the deepest mantle (or the outermost core) is unclear. However, because these reactions provide indications of core evolution, the question of whether core or CMB material can plausibly be entrained to the surface is of considerable interest (Section 9.3.3.2).

Whether or not a dynamo can be sustained ultimately depends on the CMB heat flow, that is, the rate at which heat is extracted from the core (Section 9.3.2.2). The CMB heat flow, in turn, is determined by the ability of the mantle to remove heat. Importantly, independent estimates on this cooling rate exist, based on our understanding of mantle behavior.

One approach to estimating the heat flow across the base of the mantle relies on the conduction of heat across the bottom boundary layer. As discussed in Section 9.2.4, the temperature at the bottom of this layer (the core) arises from extrapolating the temperature at the ICB outwards along an adiabat, and is about 4000 K. The temperature at the top of the layer is obtained from extrapolating the mantle potential temperature inwards along an adiabat, and is about 2700 K (Boehler, 2000). Williams (1998) concluded that the likely temperature contrast across the CMB is 1000–2000 K. Based on seismological observations, the thickness of D'', which might represent a thermal boundary layer, is 100–200 km. For likely lower-mantle thermal conductivities, the resulting conductive heat flow is probably in the range 9 ± 3 TW (Buffett, 2003). Unfortunately, as discussed below, the CMB region is complicated enough that this simple estimate may not be robust.

A second method of estimating CMB heat flow is to add up the near-surface contributions from inferred convective plumes (Davies, 1988; Sleep, 1990). These early estimates gave heat flows a factor of 2–4 smaller than the simple conductive argument. However, it is now becoming clear that these estimates are probably too low, both because the temperature contrast between plumes and the background mantle varies with depth (Bunge, 2005; Zhong, 2006; Mittelstaedt and Tackley, 2006) and because not all plumes may reach the surface (Labrosse, 2002). Two recent theoretical studies of convection including, respectively, compositional layering and the postperovskite phase transition result in CMB heat flows of ~13 TW (Zhong, 2006) and 7–17 TW (Hernlund et al., 2005).

One additional complication is that D'' may contain enhanced concentrations of radioactive elements (e.g., Coltice and Ricard, 1999; Tolstikhin and Hofmann, 2005; Boyet and Carlson, 2005). In this case, the heat flow out of the core will be less than the heat flows inferred from the models of Zhong (2006) and Hernlund et al. (2005). Bearing this caveat in mind, these model results are roughly consistent with the simple conductive heat flow estimate, and suggest that a range of 10 ± 4 TW is likely to encompass the real present-day CMB heat flow. This range of heat flows suggests a current core cooling rate dT_c/dt of 65–150 K Gy^{-1}, using the parameters for model 2.

The compositional nature of the CMB region may also have an effect on the evolution of the core. The CMB region may be at least partially molten, an inference supported by the presence of a (laterally discontinuous) ultralow velocity zone (e.g., Garnero et al., 1998). The presence of such a melt layer, which is probably denser than the surrounding solid material (Knittle, 1998; Akins et al., 2004), is likely to affect heat transfer from the core to the mantle. Such a layer is also likely to have been more extensive in the past, when core temperatures were higher (see Section 9.3.2.6 and Chapter 3). Another possibility is the presence of high-density, compositionally distinct material, probably subducted oceanic crust. Again, this material, especially if enriched in radioactive materials (Buffett, 2002), is likely to have affected long-term core evolution (Nakagawa and Tackley, 2004a). Finally, the CMB region may include a phase transition to a postperovskite structure (e.g., Murakami et al., 2004), which will also affect the CMB heat flux (Nakagawa and Tackley, 2004b; Hernlund et al., 2005). The manner in which the CMB may have evolved with time in response to

the evolution of the core is discussed further in Section 9.3.3.2.

9.2.6 Dynamo Behavior Over Time

One might expect that the behavior of the Earth's magnetic field over time would provide information on the evolution of the dynamo and core. However, despite much work on this subject (see reviews by Jacobs (1998) and Valet (2003)), the information is limited to the following: (1) a reversing, predominantly dipolar field has existed, at least intermittently, for at least the last 3.5 Gy; (2) the amplitude of the field does not appear to have changed in a systematic fashion over time.

There are several reasons why there are so few constraints. First, the magnetic field that we can measure at the surface is different in both frequency content and amplitude from the field within the core. In particular, ohmic heating is dominated by small-scale magnetic fields which are not observable at the surface (see below). Second, the number of observations on paleomagnetic fields decline dramatically prior to ≈150 My BP because of the almost complete absence of unsubducted oceanic crust. Third, there is little theoretical understanding of how changes in core behavior relate to changes in the observed magnetic field.

The first two problems are unlikely to be resolved in the forseeable future. However, there has been some progress with the third, thanks to increasingly realistic simulations of the geodynamo (see reviews by Busse (2000), Glatzmaier (2002), Kono and Roberts (2002), and a recent paper by Olson and Christensen (2006)). In particular, a study by Roberts and Glatzmaier (2001) found that increasing the inner-core size tended to result in a less axisymmetric field and (surprisingly) greater time variability. Thus, at least in theory, observed changes in the time variability of the magnetic field with time could be used to place constraints on the evolution of the Earth's core. An observed variation in the amplitude with time (e.g., Labrosse and Macouin, 2003), however, is less likely to be useful: Roberts and Glatzmaier (2001) found that models with inner cores 0.25 and 2 times the radii of the current inner core both produced similar mean field amplitudes, and a similar result was found by Bloxham (2000). Furthermore, it is not clear that changes in global variables, such as core cooling rate or inner core size, will have a larger effect on the field behavior than local factors such as the heat flux boundary condition (e.g., Christensen and Olson, 2003).

In spite of the difficulties in extracting detailed information on core evolution from the paleomagnetic record, an important result is that the geodynamo appears to have persisted, without long-term interruptions, for at least 3.5 Gy (McElhinny and Senanayake, 1980). The pattern of magnetic reversals for the Proterozoic is well known, but not well understood. For instance, although reversals occur roughly every 0.25 My on average (Lowrie and Kent, 2004), there were no reversals at all in the period 125–85 Ma, for reasons which are obscure but may well have to do with the behavior of the mantle over that interval (e.g., Glatzmaier et al., 1999). The earliest documented apparent paleomagnetic reversal is at 3.2 Gy BP (Layer et al., 1996). Although the amplitude of the field has varied with time (Selkin and Tauxe, 2000; Prevot et al., 1990), the maximum field intensity appears never to have exceeded the present-day value by more than a factor of 5 (Valet, 2003; Dunlop and Yu, 2004).

In summary, the fact that a reversing dynamo has apparently persisted for >3.5 Gy can be used to constrain the evolution of the core over time (see Section 9.3.2 below). Unfortunately, other observations which might potentially provide additional constraints, such as the evolution of the field intensity, are either poorly sampled or difficult to relate to the global energy budget, or both.

9.2.6.1 Ohmic dissipation

As discussed below, the power dissipated in the core by ohmic heating is a critical parameter to determining whether a dynamo can operate: a more dissipative dynamo requires more rapid core cooling and a higher CMB heat flux. Unfortunately, this heating rate is currently very poorly constrained. The heating is likely to occur at length scales which are sufficiently small that they can neither be observed at the surface, nor resolved in numerical models (Roberts et al., 2003). Moreover, the toroidal field, which is undetectable at the surface, may dominate the heating.

The ohmic dissipation Q_Φ may be converted to an entropy production rate E_Φ using $E_\Phi = Q_\Phi/T_D$ (Roberts et al., 2003), where the characteristic temperature T_D is unknown but intermediate between T_i and T_c and is here assumed to be 5000 K. The entropy production rate is simply a convenient way of assessing the potential for generating a dynamo, and is discussed in more detail in Section 9.3.2.2 . One approach to estimating the required rate is to extrapolate from numerical dynamo simulations.

Roberts et al. (2003) used the results of the Glatzmaier and Roberts (1996) simulation to infer that 1–2 TW are required to power the dynamo, equivalent to an entropy production rate of 200–400 MW K^{-1}. The dynamo model of Kuang and Bloxham (1997) gives an entropy production rate of 40 MW K^{-1}. Christensen and Tilgner (2004) gave a range of 0.2–0.5 TW, based on numerical and laboratory experiments, equivalent to 40–100 MW K^{-1}, and Buffett (2002) suggested 0.1–0.5 TW, equivalent to 20–100 MW K^{-1}. Labrosse (2003) argues for a range 350–700 MW K^{-1}, and Gubbins et al. (2003) favor 500–800 MW K^{-1}. We shall regard the required ohmic dissipation rate as currently unknown, but think it likely that entropy production rates in excess of 50 MW K^{-1} are sufficient to guarantee a geodynamo.

9.2.7 Summary

The present-day temperature structure and composition of the core establish boundary conditions which constrain both the core's initial mode of formation, and subsequent evolution. In particular, the size of the inner core and the persistence of the geodynamo for at least 3.5 Gy place constraints on the CMB heat flux. Light elements in the Earth's core not only help to power the dynamo, but also constrain the conditions under which the Earth formed. Radioisotopes are a potential additional source of power, and also provide the ability to date core formation and (potentially) constrain the age of the inner core.

The next section will examine how the core evolved from the initial conditions established by the accretion process to its inferred present-day state.

9.3 Evolution of the Core

9.3.1 Formation and Initial State

The initial thermal and chemical conditions of the Earth's core were determined by the manner in which the Earth accreted, a relatively geologically rapid process. This period of the core's history is discussed extensively in Chapter 3 and only a brief summary is given here.

Theoretical arguments and geochemical observations suggest that the Earth accumulated the bulk of its mass through a few, large impacts within about 50 My of solar system formation, and that each of these impacts generated a global, if transient, magma ocean. Although the impacting bodies were undoubtedly differentiated, pre-existing chemical signals appear to have been overprinted by the impact process. Siderophile element concentrations are consistent with a magma ocean extending to mid-mantle depths (500–1000 km, 2000–3000 K). The impactor cores likely underwent emulsification as they traversed the magma ocean, resulting in near-complete chemical re-equilibration. This re-equilibration ceased as the metal pooled at the base of the magma ocean; subsequent transport of the resulting large-scale iron masses to the pre-existing core was rapid and resulted in increased core temperatures. Following the Moon-forming impact, the initial core temperature was probably at least 5500 K, suggesting extensive melting in the lowermost mantle.

After this initial period of large and geologically rapid transfers of mass and energy, the subsequent thermal and compositional evolution of the core – the focus of this section – was much less dramatic. Unfortunately, there are few observational constraints on the details of this longer-term evolution. As discussed below, present-day observations (in particular, the size of the inner core and estimates of the CMB heat flux) provide some constraints. The fact that a geodynamo has apparently operated for at least 3.5 Gy provides a lower bound on the rate at which the core must have cooled. However, it is important to note that the long-lived field does not necessarily require a similarly ancient inner core. Isotopic signals, however, may provide a constraint on the inner core age, though this is highly controversial (Section 9.3.3.1).

The first half of this section will investigate the thermal evolution of the core. In particular, it will focus on three questions: how much has the core cooled over time?; when did the inner core start to grow?; and how was the dynamo maintained? Because of the paucity of observational constraints, this section will focus on theoretical approaches, and in particular on the uncertainties introduced by uncertainties in the relevant parameters. The subsequent section will focus on the compositional evolution of the core, in particular the chemical effects of inner-core formation, and possible reactions taking place at the CMB.

9.3.2 Thermal Evolution

The thermal evolution of the Earth's core has been the subject of considerable interest over the last decade. As outlined in Section 9.2, experimental uncertainties have led different groups to adopt

different values for parameters of interest, such as the thermal conductivity. Accordingly, the calculations carried out in this section will make use of three different sets of parameters (**Table 2**): one end-member designed to maximize the likelihood of an ancient inner core (model 1); one using the best-guess parameter values (model 2); and one using values designed to minimize the inner-core age (model 3). In this way, the uncertainties involved in the theoretical calculations will be made clear, while conclusions which are robust under all three models are likely to prove durable.

9.3.2.1 Core cooling

In one sense, the thermal evolution of the core is relatively simple. Heat is extracted out of the core at the CMB, at a rate which depends primarily on processes within the mantle. As a result, in the absence of an internal heat source, the core cools with time. At some point, the core adiabatic crosses the melting curve, and inner-core solidification begins (**Figure 1**).

The instantaneous energy balance within the core may be written (e.g., Buffett et al., 1996; Roberts et al., 2003; Gubbins et al., 2003) as

$$Q_{cmb} = Q_s + Q_g + Q_L + Q_R = \tilde{Q}_T \frac{dT_c}{dt} + Q_R \quad [8]$$

Here Q_{cmb} is the heat flow across the CMB, the core contributions Q_s, Q_g, Q_L, and Q_R are, respectively, from secular cooling, gravitational energy release, latent heat release, and radioactive decay, and the outer core is assumed to be adiabatic and homogeneous. Note that the assumption that the outer core is well-mixed and convecting throughout may not be the case if a stable conductive layer (e.g., Labrosse et al., 1997) or a compositionally buoyant layer (e.g., Braginsky, 2006) develop at the top of the core.

The first three terms in eqn [8] are all proportional to the core cooling rate dT_c/dt, where T_c is the core temperature at the CMB and $\tilde{Q}_T$ is a measure of the total energy released per unit change in core temperature. Both Q_g and Q_L depend on the inner-core size, and are zero in the absence of an inner core. This equation allows the evolution of the core temperature to be calculated if the CMB heat flux through time is known.

This energy balance has several important consequences. First, when inner-core formation begins, the same CMB heat flux results in a reduced core cooling rate, because of the extra energy terms (Q_g, Q_L). Second, the result of radioactive heating is likewise to reduce the core cooling rate for the same CMB heat flux.

Radiogenic elements can have a strong effect on the core cooling rate, and thus the age of the inner core (e.g., Labrosse *et al.*, 2001). Rewriting eqn [8] we obtain a core cooling rate of

$$\frac{\mathrm{d}T_\mathrm{c}}{\mathrm{d}t} = \frac{Q_\mathrm{cmb} - Q_\mathrm{R}}{\tilde{Q}_\mathrm{T}} \qquad [9]$$

It is clear that the effect of the Q_R term is to reduce the rate of core cooling, and hence prolong the life of the inner core. This is an issue we return to below.

A major disadvantage with eqn [8] is that it does not include an ohmic dissipation term, because the transformation of kinetic energy to magnetic energy to heat occurs without changing the global energy balance (see Gubbins *et al.* (2003)). This equation is therefore not useful in determining the evolution of the geodynamo.

9.3.2.2 *Maintaining the geodynamo*

How the geodynamo is maintained has been a question of considerable interest since the initial work of Bullard (1950), Verhoogen (1961), and Braginsky (1963). The entropy balance approach described below was developed in the 1970s (Backus, 1975; Hewitt *et al.*, 1975; Gubbins, 1977; Loper, 1978; Gubbins *et al.*, 1979; Hage and Muller, 1979) and has been re-invigorated in the last few years (Braginsky and Roberts, 1995; Buffett *et al.*, 1996; Buffett, 2002; Lister, 2003; Labrosse, 2003; Roberts *et al.*, 2003; Gubbins *et al.*, 2003, 2004).

Ultimately, the geodynamo is maintained by the work done on the field by convective motions. This convection is driven partly by the extraction of heat into the overlying mantle, and partly by the fact that the resulting inner-core growth releases light elements into the base of the outer core. Thus, both thermal and compositional convection are important, with the relative contributions depending on the different parameter values adopted, in particular the size of the inner core.

Just as eqn [8] describes the energy balance in the core, an equivalent equation can be derived for the entropy balance (e.g., Roberts *et al.*, 2003; Labrosse, 2003; Lister, 2003; Gubbins *et al.*, 2003, 2004). The latter equation does include ohmic dissipation (dissipation is nonreversible and is thus a source of entropy). The entropy may be thought of as the power divided by a characteristic temperature and multiplied by a thermodynamic efficiency factor.

Different mechanisms (e.g., thermal and compositional convection) have different efficiency factors (e.g., Buffett *et al.*, 1996; Lister, 2003). Unfortunately, it is not currently understood how to relate the entropy production rate to global magnetic field characteristics, such as reversal frequency (Section 9.2.6).

The entropy rate available to drive the dynamo may be written as (e.g., Labrosse, 2003; Gubbins *et al.*, 2004)

$$\begin{aligned}\Delta E &= E_\mathrm{S} + E_\mathrm{L} + E_\mathrm{g} + E_\mathrm{H} + E_\mathrm{R} - E_\mathrm{k}\\ &= \tilde{E}_\mathrm{T}\frac{\mathrm{d}T_\mathrm{c}}{\mathrm{d}t} + E_\mathrm{R} - E_\mathrm{k}\end{aligned} \qquad [10]$$

where $E_\mathrm{s}, E_\mathrm{L}, E_\mathrm{g}$, and E_H are the contributions due to cooling, latent heat and gravitational energy release and heat of reaction, respectively, E_R depends on the presence of radioactive elements in the core, and E_k depends on the adiabatic heat flux at the CMB. The first four terms are all proportional to the core cooling rate $\mathrm{d}T_\mathrm{c}/\mathrm{d}t$, and $\tilde{E}_\mathrm{T}$ is simply a convenient way of lumping these terms together. This equation illustrates two important points. First, as expected, a higher cooling rate or a higher rate of radioactive heat production increases the entropy rate available to drive a dynamo. Second, a larger adiabatic contribution (e.g., higher thermal conductivity) reduces the available entropy.

By combining eqns [8] and [10], an expression may be obtained which gives the core heat flow required to sustain a dynamo characterized by a particular entropy production rate E_Φ:

$$Q_\mathrm{cmb} = Q_\mathrm{R}\left(1 - \frac{T_\mathrm{T}}{T_\mathrm{R}}\right) + T_\mathrm{T}(E_\Phi + E_\mathrm{k}) \qquad [11]$$

where T_R is the effective temperature such that $T_\mathrm{R} = Q_\mathrm{R}/E_\mathrm{R}$ and likewise $T_\mathrm{T} = \tilde{Q}_\mathrm{T}/\tilde{E}_\mathrm{T}$. This equation encapsulates the basic physics of the dynamo problem.

Equation [11] shows that larger values of adiabatic heat flow or ohmic dissipation require a correspondingly higher CMB heat flow to drive the dynamo, as would be expected. In fact, in the absence of radiogenic heating, the CMB heat flow required is directly proportional to $E_\mathrm{k} + E_\Phi$. The constant of proportionality depends on the thermodynamic efficiency of the core, which increases if an inner core is present. Because the term $\left(1 - (T_\mathrm{T}/T_\mathrm{R})\right)$ exceeds zero, a dynamo which is partially powered by radioactive decay will require a greater total CMB heat flow than the same dynamo powered without radioactivity. Alternatively, if the CMB heat flow stays constant, then an increase in the amount of radio-active heating reduces the entropy available to power the dynamo.

Equation [11] also illustrates the fact that a dissipative dynamo can exist even if the CMB heat flow is less than that conducted along the adiabat (i.e., subadiabatic) (Loper, 1978). In the absence of radioactivity, the entropy production rate E_Φ available for the dynamo is $(Q_{cmb}/T_T) - E_k$ which for the present-day core exceeds zero unless Q_{cmb} is strongly subadiabatic. Thus, a subadiabatic CMB heat flow can sustain a dynamo, as long as an inner core is present to drive compositional convection (e.g., Loper, 1978; Labrosse *et al.*, 1997). It should be noted that these results assume that the CMB heat flux does not vary in space; lateral variations in the heat flux may allow a dynamo to function even if the mean value of Q_{cmb} suggests that the dynamo should fail.

In the absence of an inner core and radiogenic heating, it may be shown that

$$Q_{cmb} = Q_k \left(1 + \frac{E_\Phi}{E_k} \right) \qquad [12]$$

This equation shows that the heat flow at the CMB Q_{cmb} must exceed the adiabatic heat flow Q_k for a dynamo driven only by thermal convection to function. This result is important, because it demonstrates that there is no problem with sustaining a dynamo prior to the onset of inner-core formation, as long as the core cooling rate (or CMB heat flux) is large enough. This equation also allows dynamo

dissipation to be taken into account explicitly: a more strongly dissipative core dynamo requires a more superadiabatic CMB heat flow to operate.

9.3.2.3 Present-day energy budget

Figure 2 shows how the rate of entropy production available to drive a dynamo varies as a function of the heat flow out of the core, both for a set of core parameters appropriate to the present-day Earth, and for a situation in which the inner core has not yet formed. **Figure 2(a)** illustrates the case for the best-guess parameters (model 2) while **Figure 2(b)** uses parameters designed to maximize the inner-core age (model 1). As expected, higher core heat fluxes generate higher rates of entropy production; also, the same cooling rate generates more excess entropy when an inner core exists than when thermal convection alone occurs.

As discussed above, when an inner core is present, positive contributions to entropy production arise from core cooling, latent heat release, and gravitational energy; the adiabatic contribution is negative (eqn [10]). For a present-day, radionuclide-free core, CMB heat flows of <2 TW and <0.2 TW result in negative entropy contributions and, therefore, no dynamo for models 2 and 1, respectively. Such cooling rates would permit an inner core as old as the Earth. Higher core cooling rates generate a higher net entropy production rate; they also means that the inner core must have formed more recently.

For a present-day estimated CMB heat flow of 6–14 TW (Section 9.2.5), the net entropy

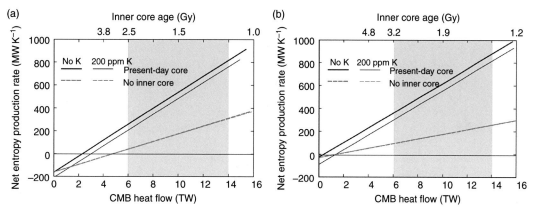

Figure 2 (a) Net entropy production (available to drive the geodynamo) as a function of CMB heat flux, for cases with and without an inner core, and with and without 200 ppm potassium. Parameters used are for model 2 in **Table 2**. Inner-core age is calculated assuming a constant core cooling rate, $\Delta T_c = 100$ K (eqn [7]) and is only relevant to the case with an inner core and no potassium. (b) As for (a), except using the parameters for model 1 in **Table 2** (designed to maximize inner-core age) with $\Delta T_c = 150$ K. The shaded region denotes the estimated present-day CMB heat flow (see Section 9.2.5).

production rate available to drive the dynamo is $200\text{--}900\,\mathrm{MW\,K^{-1}}$, sufficient to generate roughly 1–5 TW of ohmic dissipation. Since most estimates of ohmic heating are less than 2 TW (Section 9.2.6.1), it is clear that there is no difficulty in driving a dynamo at the present day. A heat flow of 6–14 TW also implies an inner-core age of 1–2.5 Gy and 1.4–3.2 Gy for models 2 and 1, respectively, assuming a constant core cooling rate.

Prior to the formation of an inner core, the CMB heat flow had to exceed the adiabatic value Q_k in order to maintain a dynamo for reasons discussed above (eqn [12]). For a dynamo requiring an entropy production rate of $200\,\mathrm{MW\,K^{-1}}$, the core cooling rate had to be roughly 2–3 times as fast to maintain this rate before the onset of inner-core solidification. A geodynamo prior to the onset of inner-core formation is entirely possible, but implies that either CMB heat fluxes were higher in the past, or that the present-day dynamo is dissipating more heat than it did prior to inner-core formation.

Figure 2 also shows that, as discussed above, a larger CMB heat flow is required for the same entropy production if radioactive heating is important in the present-day Earth. Prior to the existence of the inner core, the effect of radioactive decay on the entropy production is small because the thermodynamic efficiency of radioactive heat production is similar to that of secular cooling (Roberts *et al.*, 2003; Gubbins *et al.*, 2003). Importantly, the presence of potassium also reduces the core cooling rate, and thus can increase the age of the inner core (eqn [9] and see below).

In summary, **Figure 2** shows that the estimated present-day CMB heat flow of 6–14 TW is consistent with the operation of a dynamo dissipating 1–5 TW of heat. Under these circumstances, an inner core could have persisted for 1–3.2 Gy if the heat flux stayed constant. A lower CMB heat flux would result in a lower dissipation rate and a greater inner core age. Radioactive heating can increase the inner-core age somewhat, but the present-day radioactive heat production is likely only a small fraction of the total energy budget, and thus the effects are modest. In practice, of course, both the core heat flux and the radiogenic heat production will vary with time; investigating the time evolution of the core and mantle is the subject of the next section

9.3.2.4 Thermal evolution

There are two basic approaches to modeling the thermal evolution of the core. One approach is to start

from some assumed initial conditions and evolve the core forwards in time, using eqn [8] or its equivalent (Stevenson *et al.*, 1983; Stacey and Loper, 1984; Mollett, 1984; Yukutake, 2000; Nimmo *et al.*, 2004; Nakagawa and Tackley, 2004a, 2004b; Butler *et al.*, 2005; Costin and Butler, 2006; Davies, 2007). The initial conditions can be iterated until the correct present-day core parameters (e.g., inner-core size) are obtained, and the theoretical geodynamo history compared with the observations. Because the core's evolution depends on the CMB heat flux, such models must simultaneously track the thermal evolution of the mantle. This kind of approach has two principal disadvantages: first, it requires the assumption of initial conditions which are poorly constrained (Chapter 3); and second, in considering the mantle as well as the core, the number of important but uncertain parameters (e.g., mantle viscosity) greatly increases.

A second approach is to start from the present-day core conditions and evolve the core backwards in time (Buffett *et al.*, 1996; Buffett, 2002; Labrosse 2003). This approach has the advantage of automatically satisfying the present-day observations. However, because diffusion equations are unstable if run backwards in time, the evolution of the CMB heat flux cannot be calculated in the same way as it can in the forward models. A common choice is to specify the time evolution of the entropy production in the core, which then specifies both the core cooling rate and the evolution of the CMB heat flux (eqn [11]). This approach has the virtue of not requiring any knowledge of the mantle to do the calculations; however, it makes a major assumption in assuming a specific entropy production history for the core. Nonetheless, this approach is both simpler and subject to fewer uncertainties than the alternative, and will be focused on here.

Partly because of geochemical arguments that may suggest an ancient ($\sim$3.5 Gy BP) inner core (Section 9.3.3.1), many of the investigations cited above have focused on the age of the inner core. While there is a general tendency to find relatively young ($\sim$1 Gy) inner cores, the robustness of these results is often unclear because of the large number of poorly constrained parameters which have to be chosen. Another aim of this section is to tabulate the most important parameters, and to investigate the robustness of the thermal evolution results to likely parameter variations. In particular, we will focus on whether a 3.5 Gy old inner core is compatible with the theoretical models, and conclude that it is not, unless the core contains an additional energy source (e.g., $^{40}\mathrm{K}$).

9.3.2.4.(i) Parameters Generating an ancient inner core requires either a relatively low CMB heat flux, a large difference in adiabatic and melting temperature gradients, or substantial radiogenic heating. If the core cooling is slow, then to maintain the dynamo requires either low magnetic dissipation, large positive entropy terms (e.g., E_g), or small negative entropy terms (e.g., E_k).

Of the various parameters discussed in Section 9.2, we may identify those which will have the largest influence on whether a dynamo can be maintained while producing an ancient core. They are as follows:

1. Thermal conductivity k and thermal expansivity α. A low thermal conductivity or expansivity reduces E_k, and thus allows the same rate of entropy production for a lower CMB heat flux (eqn [11]).
2. Gradient of the melting curve. The quantity ΔT_c (eqn [7]) is the change in T_c since the inner core started solidifying, and is determined by the relative slopes of the adiabat and the melting curve. A larger ΔT_c results in an older inner core for the same CMB heat flux (or alternatively a higher entropy production rate for an inner core of the same age).
3. The compositional density contrast $\Delta\rho_c$. The larger the value of $\Delta\rho_c$, the higher the entropy production rate for the same rate of cooling.
4. The rate of entropy production required to drive the dynamo. As discussed in Section 9.2.6.1, this value is unlikely to be less than $50\,MW\,K^{-1}$, and is more likely closer to $200\,MW\,K^{-1}$.
5. Radioactive heating within the core. Internal heat production reduces the core cooling rate (**Figure 2** and eqn [9]).

Other factors, such as latent heat, specific heat capacity, heat of reaction and so on are either better known than the factors listed above, or have only a small effect.

Factors 1–3 are known with some uncertainty, while factors 4 and 5 are less well known. We have therefore adopted three models (**Table 2**) designed to result in maximum, best-guess, and minimum inner-core ages, respectively. In this way, a conservative assessment may be made of the model variability arising from uncertainties in parameter values.

The calculations shown below take a similar approach to those of Buffett (2002) and Labrosse (2003) and assume a specified rate of entropy production with time. The core temperature is evolved

backwards from the present-day conditions. The entropy production rate prior to inner-core formation is assumed constant, which allows the CMB heat flux and core cooling rate to be determined. The CMB heat flux during inner-core solidification is assumed to stay constant at the value immediately prior to solidification. The justification for making this assumption is that the CMB heat flux is determined primarily by conditions in the mantle, and is thus unlikely to be significantly affected by changing core conditions. This assumption is less reliable for inner cores of greater ages or having larger values of ΔT_c. A result of the assumption is that the entropy production increases significantly when inner-core solidification starts, because of the extra contributions (e.g., latent heat release) to the entropy budget.

Other assumptions could be made. For instance, Labrosse (2003) assumes that the present-day entropy production is some constant factor times the entropy production immediately prior to core formation. It will be shown below that different assumptions of this kind do not significantly affect the results. In theory, one would like to use observations of the Earth's magnetic field to constrain the entropy evolution. For instance, a higher field strength should lead to greater dissipation and thus higher entropy production. Unfortunately, as discussed in Section 9.2.6, neither the observations, nor our theoretical understanding of geodynamos, are currently good enough to infer how the entropy production has changed. The assumption of constant entropy production prior to inner-core formation at least has the virtue of simplicity; furthermore, since if anything entropy production is likely to have declined with time, this assumption will result in conservatively old inner core ages.

Figure 3 shows the evolution of various parameters of interest for models 1–3 when the net entropy production rate prior to inner-core formation is $200\,MW\,K^{-1}$, probably a reasonable value (see Section 9.2.6.1). This entropy production rate determines the core cooling rate, and thus the heat flux. The present-day heat fluxes are in the range 10–15 TW, in line with expectations (Section 9.2.5). The change in CMB heat flow over 4 Gy is modest, a factor 25% or less. Whether such a small change is dynamically plausible is currently unclear, and will be discussed further below.

As expected, the heat flux required for model 1 is lower than for the other models, because model 1

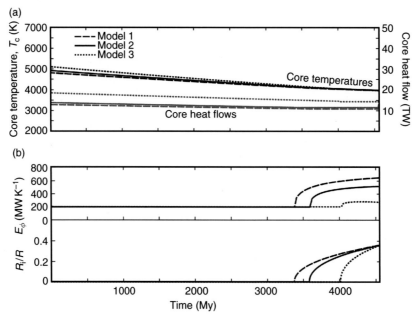

Figure 3   (a) Evolution of core heat flow and core temperature T_c with time for models 1–3 (parameter values for each model are given in **Table 2**). Heat flow prior to inner-core formation is calculated by fixing entropy production rate at $200 \, \mathrm{MW \, K^{-1}}$. Heat flow after the onset of inner-core formation is kept at the same level as it was immediately prior to solidification. Equations are integrated backwards in time from the present day. (b) Evolution of the entropy production rate E_Φ (upper panel) and dimensionless inner core radius R_i/R (lower panel) for models 1–3. Note that the entropy production increases when inner-core formation occurs.

uses parameter values chosen to favor a long-lived geodynamo. A consequence of this lower heat flux is that the temperature change of the core over 4.5 Gy is smaller than for models 2 and 3. The lower heat flux also results in an inner core of greater age, 1.2 Gy Model 1 also results in a greater amount of entropy production once core formation begins, mainly because of more vigorous compositional convection due to the large value of $\Delta\rho_c$ adopted (see **Table 2**). The total energy released since inner-core formation due to secular cooling, gravity, and latent heat is in the ratio 69:15:16 for model 1, and 61:7:32 for model 3, also illustrating the greater importance of compositional convection in model 1.

Figure 4 shows the same situation as the preceding figure, but now with a net entropy production rate prior to inner core formation of $50 \, \mathrm{MW \, K^{-1}}$, at the lower end of reasonable values. The lower entropy production results in a reduction in the heat flux required (4–12 TW at the present day), and also a reduction in the amount by which the core has cooled over 4.5 Gy. As a consequence of this reduction in cooling rate, the inner core can persist further back in time. In particular, for model 1 the age of the inner core (3.4 Gy BP) is roughly

compatible with the proposed age based on Os isotope systematic (Brandon *et al.*, 2003; see Section 9.3.3.1).

Figure 5 is identical to **Figure 3**, but includes the effect of 200 ppm potassium in the core. The CMB heat flows required to drive the dynamo are similar to those in **Figure 3**, as expected from the results of **Figure 2**. However, the change in core temperature with time is significantly reduced, because of the additional heat source (eqn [9]). This reduction in core cooling rate also results in a more ancient inner core, though the effect is relatively modest because the radiogenic heat production is small compared to the total heat flow (cf. Labrosse 2003; Nimmo *et al.*, 2004; Butler *et al.*, 2005). At a lower entropy production rate, the effect of radioactive decay on the inner core age would be more pronounced.

9.3.2.5   Inner-core age

Figure 6 summarizes the outcomes of several similar models by plotting inner-core age (i.e., the age of the onset of crystallization) against present-day CMB heat flux, for cases with and without potassium. As expected, a higher Q_{cmb} results in a younger inner core. For the same heat flux, model 1 results in an

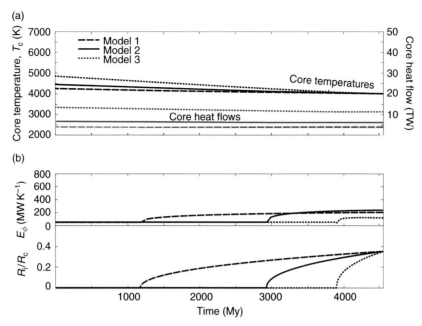

Figure 4 As for **Figure 3**, but with the net entropy production rate prior to inner-core formation fixed at 50 MW K^{-1}.

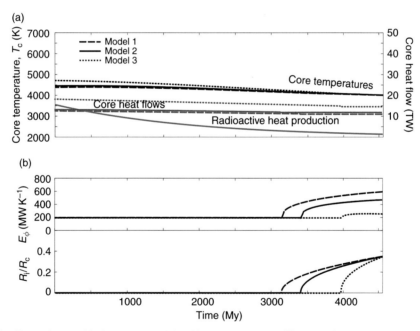

Figure 5 As for **Figure 3**, but with the core containing 200 ppm potassium. The rate of heat production within the core is shown.

older inner core than models 2 and 3, and also generates a higher rate of entropy production.

In the absence of potassium, **Figure 6(a)** shows that an inner core 3.5 Gy old is possible. However, for the inner core to be this old, the following requirements must all be met:

1. Parameters such as thermal conductivity, expansivity, and compositional density contrast must all have values (**Table 2**) which tend to maximize the inner core age.
2. The rate of entropy production required to drive the dynamo is small, <50 MW K^{-1}. In terms of

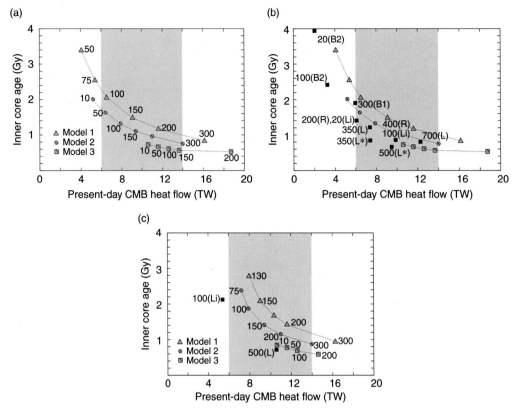

Figure 6 (a) Plot of present-day CMB heat flux against inner-core age, from a suite of models similar to those shown in **Figures 3** and **4**. The labels refer to the constant entropy production rate (in MW K^{-1}) prior to the onset of inner core formation. Higher entropy production rates require higher heat fluxes and thus younger inner cores. (b) Same plot as (a), but showing results obtained by other authors. As before, labels indicate entropy production rate; parentheses give references as follows: B2 = Buffett (2002); B1 = Buffett *et al.* (1996); R = Roberts *et al.* (2003); L and L* = Labrosse (2003); Li = Lister (2003). The latter two models have entropy production rates prior to inner-core formation which are not quite constant. Ohmic dissipation is converted to entropy production by assuming a characteristic temperature of 5000 K. (c) As for (a), but with 200 ppm potassium in the core. The black squares are results from Labrosse (2003) and Lister (2003) with 250 and 235 ppm potassium in the core, respectively.

dissipation, this is ≈0.25 TW, at the low end of current estimates (Section 9.2.6.1).

3. The CMB heat flux must have stayed low and relatively constant, with a mean value of about 4 TW over the whole of Earth history (cf. **Figure 4**).

Requirements 1 and 2 are at least possible, if not plausible. Requirement 3 is, however, more problematic. First, the required heat flux is a factor of 1.5–4 times smaller than the inferred present-day CMB heat flux of 6–14 TW (Section 9.2.5). An inner core 3.5 Gy old is incompatible with current estimates of the present-day CMB heat flux, unless D″ contains significant quantities of radiogenic materials (Section 9.2.5; see also Buffett (2002) and Costin and Butler (2006)).

The requirement that the CMB heat flux stay essentially constant over time is surprising, because

the reduction in core temperature with time is likely to lead to an increase in mantle viscosity and a decrease in CMB heat flux (Nimmo *et al.*, 2004). However, whether a constant CMB heat flux is dynamically plausible is unclear (e.g., Davies, 2007), because our understanding of the physical nature of the CMB region is currently so poor (Section 9.2.5).

Figure 6(b) compares the results obtained here with those obtained by other authors. Despite the different assumptions and parameters chosen, the results are strikingly consistent. In general, the results plot between the lines for models 2 and 3, suggesting that model 1 is overly conservative (as it was designed to be). Models using a present-day heat flux of 6–14 TW result in entropy production rates of 200–700 MW K^{-1}, which are perfectly reasonable values, and an inner-core age range of 0.6–2 Gy. Conversely, to achieve an ancient inner core requires both low entropy production and low CMB heat fluxes (e.g., $20\,MW\,K^{-1}$ and $2\,TW$ from Buffett (2002)).

As has been recognized previously (Buffett, 2002; Labrosse, 2003; Roberts *et al.*, 2003; Nimmo *et al.*, 2004), the difficulty of generating an ancient inner core while maintaining a dynamo is reduced if the core contains a radioactive heat source such as potassium. With 200 ppm in the core, **Figure 6** shows that an inner core 3 Gy old is compatible with a reasonable present-day heat flux (8 TW) and an entropy production rate of $130\,MW\,K^{-1}$, likely sufficient to sustain a geodynamo. Thus, the addition of potassium makes it much easier to reconcile the geophysical models with an ancient inner core.

9.3.2.6 Initial core temperature

An issue closely related to the age of the inner core is the initial core temperature. Here 'initial' refers not to the temperature of the core during accretion (which may have been as high as $10\,000\,K$ – *see* Chapter 3), but to the temperature once accretion had finished and the density structure of the Earth resembled the present-day arrangement. Higher rates of entropy production imply younger inner cores and more rapid core cooling, which in turn implies a higher initial core temperature (Buffett, 2002; Labrosse, 2003). **Figure 7** plots the variation in initial core temperature as a function of entropy production for models 1–3, and demonstrates the relationship. **Figure 7(b)** includes the effect of 200 ppm potassium, demonstrating that internal heat production

reduces the required change in core temperature (eqns [9] and [10]). These figures also demonstrate how inner-core age (labels on individual points) is increased by either a lower entropy production rate, or the addition of potassium. In extreme cases, the inner core could have been present for the entire age of the Earth.

The results shown here are again consistent with those of other authors. One result from Buffett (2002) is plotted and again shows that ancient inner cores require low dissipation rates, and imply cool initial temperatures. Labrosse (2003) assumes higher-entropy production rates ($350–700\,MW\,K^{-1}$) and obtains correspondingly younger inner cores (0.8–1.2 Gy) and hotter initial temperatures (roughly 600 K hotter than the present day).

The most striking aspect of **Figure 7** is that a core temperature change of less than 1000 K ($T_c < 5000\,K$ at $t=0$) is sufficient to have maintained a moderately dissipative dynamo ($E_\phi < 200\,MW\,K^{-1}$) throughout Earth history. This is in contrast to the results presented in Chapter 3, which inferred an initial core temperature of 5500 K or more. While the latter estimate in particular is somewhat crude, the discrepancy in the two estimates is interesting because of the additional insight it may provide.

The discrepancy could be resolved in at least two ways. First, the geodynamo could be more dissipative than assumed, either now or in the past. Future palaeomagnetic measurements might be able to confirm or disprove this possibility. Second, the core temperatures shown in **Figure 7** are sufficiently

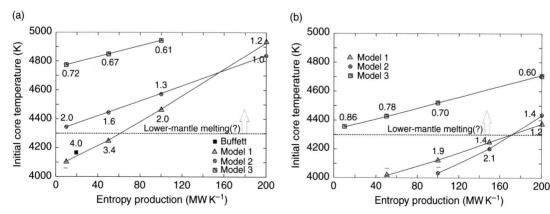

Figure 7 (a) Variation in initial-core temperature with (constant) entropy production rate prior to core formation, obtained from a suite of models similar to those shown in **Figures 3** and **4**. Labels indicate inner-core age in Gy; higher entropy production rates require more rapid core cooling and thus high initial temperatures and young inner cores. The black square represents a result from Buffett (2002) with low entropy production and a correspondingly ancient inner core; here the temperature change (160 K) obtained is plotted assuming a present-day temperature of 4000 K. Horizontal dotted line indicates onset of mantle melting (Boehler, 2000) and is uncertain by at least $\pm 200\,K$.

high that the early mantle was likely extensively molten. The melting behavior of the lowermost mantle is poorly known (Boehler, 2000; Akins *et al.*, 2004), especially for the postperovskite phase. Nonetheless, it seems likely that the CMB heat flux would have been elevated if the early lower mantle were extensively molten. Thus, the CMB heat flux probably consisted of two periods: an early, high heat flux episode due to the molten lower mantle; and later, lower heat fluxes resembling the trajectories shown in **Figures 3–5**.

The early episode of high heat flux will have persisted until the lower mantle approached its solidus, so the core likely cooled by several thousand kelvins over this period. The heat capacity of the core is roughly $2 \times 10^{27}\,\mathrm{J\,K^{-1}}$; thus, if the period lasted 1 Gy, the mean heat flow must have been of order 100 TW. How such large heat fluxes could have been sustained by the mantle is very much an open question. The high heat fluxes required also illustrate the difficulty of forming an early inner core, if the initial core temperatures were as high as estimated in Chapter 3. Finally, the high initial heat fluxes imply a potentially very strong early magnetic field; palaeointensity measurements from rocks or minerals (e.g., zircons) of the appropriate age would thus be an excellent test of this hypothesis.

9.3.2.7   *Consequences for the mantle*

The above discussion illustrates an important point: the thermal evolution of the core cannot really be considered separately from the thermal evolution of the mantle. In particular, the evolution of the CMB heat flux controls the thermal evolution of the core, and yet is to a large extent determined by mantle processes. Conversely, the early state of the core implies that the lowermost mantle was probably extensively molten, with potentially important consequences for mantle chemistry and dynamics.

Although the need to also consider the mantle increases the number of free parameters in theoretical models, it also adds potential extra observational constraints. For instance, the evolution of the mantle temperature depends on three factors: internal heat generation; heat added from the core; and the rate of heat loss to the surface. Some petrological constraints on the evolution of mantle temperatures exist (e.g., Abbott *et al.*, 1994; Grove and Parman, 2004). Unfortunately, in order to provide constraints on the evolution of the CMB heat flux, the manner in which the surface heat flux varied with time must be known. This heat flux is controlled at the present day by

plate tectonics, but how it varied in the past (and indeed, whether plate tectonics actually operated) is currently unknown. Thus, the addition of extra information (evolution of mantle temperatures) is offset by the addition of additional uncertainties (evolution of surface heat flux).

9.3.3   **Compositional Evolution**

The compositional evolution of the core since its formation has been a relatively neglected field of study, perhaps because of the paucity of observational constraints. There are two main ways in which the core composition evolves: through solidification of the inner core; and through reaction with the mantle at the CMB. Each of these is dealt with briefly in turn; a good review of some of the consequences of inner-core growth may be found in Sumita and Yoshida (2003).

9.3.3.1   *Inner-core growth*

One potentially observable consequence of inner-core growth is its effect on Os isotope systematics. The basic hypothesis is relatively simple (Walker *et al.*, 1995): both Re and Pt are presumed to partition preferentially into the outer core relative to Os as solidification proceeds. Since ^{187}Re and ^{190}Pt decay to ^{187}Os and ^{186}Os, respectively, the outer core will become progressively enriched in these Os isotopes relative to stable ^{188}Os. The amount of enrichment depends on the time since inner core crystallization, and the relative partitioning of Re and Pt into the outer core compared to Os. Furthermore, the enrichments in ^{186}Os and ^{187}Os are expected to be coupled if core crystallization occurs (since both are occurring due to the same process).

There are two sets of observations (Brandon *et al.*, 1998, 2003). First, ancient (2.7–2.8 Ga) komatiites show evidence of elevated and coupled Os-isotope ratios, which could be indicative of early core solidification. Second, more recently erupted lavas in Hawaii, Siberia, and Gorgona Island all show elevated and coupled Os-isotope ratios, again consistent with an outer-core source. Assuming values for the partition coefficients of Pt, Re, and Os, it has been argued (Brandon *et al.*, 2003) that the onset of inner-core crystallization must have been prior to 3.5 Ga to explain the komatiite Os-isotope ratios.

There are several possible objections to this hypothesis. The first is simply that the partition coefficients at the correct *P, T* conditions are currently uncertain. Since the evolution of the Os-isotope ratios depends on the partition coefficients, all that the komatiite data can really be used to argue is that inner-core crystallization began at least a few hundred million years prior to 2.7–2.8 Ga (Puchtel *et al.*, 2005), assuming that the isotopic signal is in fact due to crystallization. Although partition coefficients for Pt, Re, and Os have been measured (at 10 GPa and 1700 K; Walker (2000)), and also inferred from meteorites (Morgan *et al.*, 1995), the behavior is likely to be different at higher *P, T* conditions. Initial experiments at pressures up to 22 GPa in an Fe–S system suggest partition coefficients that are too low to explain the Os-isotopic observations (Van Orman, personal communication).

More seriously, it is not clear that the Os-isotopic signals can only be explained by inner-core crystallization. In particular, it has been suggested that recycled oceanic crust and/or sediments could equally explain the coupled signals (Hauri and Hart, 1993; Baker and Jensen, 2004; Schersten *et al.*, 2004), though that conclusion has been disputed (Brandon *et al.*, 2003; Puchtel *et al.*, 2005). The absence of detectable tungsten (Schersten *et al.*, 2004) and lead (Lassiter, 2006) isotope anomalies in Hawaiian lavas has been used to argue against the presence of any core materials. However, it is possible that these signals have been masked by contributions from recycled crust (Brandon and Walker, 2005). Recently, both Fe/Mn ratios (Humayun *et al.*, 2004) and thallium isotopes (Nielsen *et al.*, 2006) have been used to argue against crust or sediments as the source of the Os-isotope anomalies.

There is thus currently little agreement on whether or not Os isotopes can tell us anything about the crystallization of the inner core (see Brandon and Walker (2005) and Lassiter (2006) for recent – and opposed – reviews). A major step forward would be to determine the relevant partition coefficients under the *P, T* conditions appropriate to the inner core (5000 K, 300 GPa). Doing so experimentally is challenging, in which case molecular dynamics simulations may be the correct approach. Resolving this issue is a key question since thermal evolution models tend to result in a wide range of inner-core ages (Section 9.3.2.5).

Another consequence of inner-core growth is that it involves the expulsion of one or more light elements into the outer core. Unless, as is commonly assumed, they are efficiently mixed into the outer core by convective stirring, these elements will rise to

the CMB and generate a stably stratified layer (see Braginsky (2006) and references therein). A similar situation may arise if the heat flux out of the top of the core is subadiabatic (Labrosse *et al.*, 1997; Lister and Buffett, 1998). If it exists, such a layer will have important consequences for heat transfer across the CMB, and the temperature structure of the core. Some evidence has been adduced for the presence of this layer based on observations of the Earth's varying rotation and magnetic field (Braginsky, 1993; Lister and Buffett, 1998). Unfortunately, the ~100 km layer thickness suggested by these observations is unlikely to be detectable by seismological observations, and thus its presence remains somewhat hypothetical.

Rather than the light elements segregating to form a separate layer, it has instead been argued (Buffett *et al.*, 2000) that the addition of elements (specifically Si and O) to the outer core drives a chemical reaction, resulting in a silicate-rich layer of light sediments at the top of the core. These silicates will ultimately be incorporated into the mantle and will contain a few percent residual iron. Thus, this mechanism is one way of incorporating core material into the mantle, in possible agreement with the Re–Os–Pt observations. A similar outcome is proposed by Dubrovinsky *et al.* (2004), who argue that the decreasing solubility of Si in iron with increasing pressure means that Si incorporated into core material within the magma ocean will be expelled as core pressures rise, and accumulate at the top of the core.

9.3.3.2 Core–mantle boundary

There is undoubtedly a region at the CMB over which core and mantle materials have reacted (Knittle and Jeanloz, 1991). However, neither the vertical extent of this region, nor the manner in which it has evolved with time, are well understood. Observations based on Earth nutations suggest that the CMB region must include a thin, relatively conductive layer (e.g., Buffett *et al.*, 2002). Poirier *et al.* (1998), however, concluded that capillarity driven infiltration of the mantle by fluid iron is only likely to extend for tens of meters, though effects such as mantle deviatoric stresses may extend this range to ~1 km (Kanda and Stevenson, 2006). Thicker, iron-rich layers can strongly affect mantle dynamics (Manga and Jeanloz, 1996) and may be swept up into the mantle (Sleep, 1988; Kellogg and King, 1993). However, downwards drainage of liquid iron will be orders of magnitude more rapid than the rate at which it can be swept up (Poirier *et al.*, 1998). Furthermore, if the thickness of the iron-rich layer is much smaller than the convective boundary-layer thickness, the rate of entrainment will be negligible. Thus, on physical grounds, it seems difficult for core or CMB material to be transported to the near surface, although as discussed above there are isotopic arguments that it may have occurred.

9.3.4   Summary

Neither the thermal nor the compositional evolution of the core are currently very well understood, because of a lack of observational data (Section 9.2.6) and considerable uncertainties in the relevant parameter values. Nonetheless, by assuming that ohmic dissipation was constant prior to inner-core formation, and by using models spanning the likely range of parameter values, the thermal history of the core can be investigated, with results summarized in **Figures 6** and **7**. Several important points are evident.

First, there is no difficulty in maintaining a moderately dissipative dynamo prior to inner-core formation as long as the core is cooling fast enough (cf. **Figures 2–5**); an inner core is not required to drive the early geodynamo.

Second, a higher CMB heat flux implies a more dissipative dynamo and a younger inner core, although the addition of potassium can make the inner core somewhat older (**Figure 6**). An estimated present-day CMB heat flux of 6–14 TW is consistent with a constant entropy production rate of 50–300 MW K^{-1} and an inner-core age of 0.8–1.6 Gy, assuming best-guess core parameters (model 2).

Third, a 3.5 Gy old inner core is possible in the absence of radiogenic heating, but only if the ohmic dissipation is <0.25 TW and the CMB heat flow has stayed constant at 4 TW. The addition of 200 ppm potassium allows an inner core to persist over the whole of Earth history for heat flows less than about 8 TW (**Figure 6(c)**). If the Os-isotope data indicating a 3.5 Gy old inner core are correct (Section 9.3.3.1), the implications for the thermal history of the core are profound, since they require either low CMB heat flow and low ohmic dissipation, or significant amounts of potassium in the core. However, it is currently far from clear that the Os-isotope signals are actually derived from core material.

Fourth, a moderately dissipative dynamo operating for 4 Gy implies initial core temperatures 200–800 K hotter than the present day, or somewhat less if potassium is present (**Figure 7**). These temperatures imply that the early lower mantle was probably extensively molten, with very uncertain consequences for the CMB heat flux. It is also notable that the initial temperatures shown in **Figure 7** are significantly smaller than the values calculated by Chapter 3, based on gravitational potential energy release. This discrepancy is likely the result of either a significantly more dissipative core or, more probably, an early period of rapid core cooling as a result of the molten lower mantle.

There are several ways in which future progress in the study of the long-term evolution of the core is likely to be made:

1. A major step toward resolving the issue of whether the Os-isotope data provide information on inner-core formation would be the measurement of partition coefficients at the relevant *P, T* conditions. At least for a potassium-free core, an inner core 3.5 Gy old requires a thermal evolution that is at odds with a continuous geodynamo and geophysical expectations of the CMB heat flux. Resolving whether or not the Os data really constrain the inner-core age is thus of great importance.

2. The likelihood that the lower mantle was initially extensively molten has consequences for the chemical and particularly thermal evolution of the mantle which are not understood. In particular, if lower-mantle melts are indeed denser than the solid (e.g., Knittle, 1998; Akins *et al.*, 2004), it is not even clear that heat transfer will be enhanced.

3. More generally, the evolution of the CMB heat flux is not well constrained, while being absolutely central to the evolution of the core and geodynamo (cf. Nimmo *et al.*, 2004; Davies, 2007). Complicating factors such as possible melting, the postperovskite phase transition, and perhaps chemical layering, render dynamical models uncertain. The models shown in **Figures 3–5** (which do not include any dynamics) suggest a CMB heat flux which does not vary greatly over 4 Gy; it is not yet clear whether such results are dynamically plausible. So far, few models have included the available observational constraints on mantle cooling rates, which may help to reduce the possible parameter space (Section 9.3.2.7).

4. Whether or not the core contains any potassium is still an unresolved issue. If the Os-isotope inference of an ancient inner core is correct, then the

presence of potassium makes it much easier to reconcile the geophysical models with the inner-core age (Section 9.3.2.5 and **Figure 6**). Conversely, if the core lacks potassium, an ancient inner core becomes much more difficult to explain (Lassiter, 2006). At present, the best constraints on core potassium abundance are likely to come by comparing potassium concentrations with those of other elements with similar affinities for iron.

5. Although uncertainties in many of the relevant parameters have been reduced (Section 9.2), there are still gaps. In particular, the core thermal conductivity is a key parameter which is poorly known; further experiments to confirm the important work of Bi *et al.* (2002) are sorely needed.

6. Finally, one would expect the changing CMB heat flux and the growth of the inner core to have observable effects on the behavior of the geodynamo. Thus, at least in principle, the palaeo-magnetic record ought to provide observational constraints on core thermal evolution. For instance, the presence of an inner core appears to have an effect on the time variability of the magnetic field (Roberts and Glatzmaier, 2001); thus, good enough palaeomagnetic data may help to tie down when the inner core formed (Coe and Glatzmaier, 2006).

9.4 Conclusions

This chapter set out to examine the thermal and compositional evolution of the core, from shortly after its formation to the present day. This period probably involved only two events of importance: the initiation of the geodynamo and the onset of inner-core formation. Geodynamo activity started at 3.5 Gy BP at the latest; however, it is important to understand that this dynamo could easily have been sustained without an inner core being present. Theoretical estimates suggest that the inner core probably formed at ~1 Gy BP, unless either significant quantities of potassium were present in the core, or both the ohmic dissipation (<0.25 TW) and the CMB heat flow (<4 TW) were very low (**Figure 6**). The Re–Os–Pt isotopic system has been used to infer that inner-core solidification started by 3.5 Gy BP, but this hypothesis remains highly controversial.

Assuming a moderately dissipative dynamo, the change in core temperature over 4 Gy was probably 200–800 K, implying an early lower mantle that was likely extensively molten. These initial temperatures are lower than those obtained by consideration of the gravitational potential energy release during core formation, and suggest that the CMB heat flux evolved in two stages: an early, high heat flux stage, presumably due to the melting of the lower mantle, and potentially generating very strong magnetic fields; and a later, lower heat flux stage resembling the results shown in **Figures 3–5**.

The present-day core geodynamo is maintained primarily by compositional convection as the inner core solidifies. The CMB heat flux is estimated at 10 ± 4 TW and is sufficient to drive a dynamo dissipating 1–5 TW.

As should be clear, there are several areas which require further study. First, there is still a discrepancy between estimates of the inner-core age based on isotopic systematics and those based on geophysical models. Second, neither cosmochemical nor geophysical arguments have so far provided a convincing resolution to the debate over whether the core contains significant potassium. Third, the evolution of the CMB heat flux over time is currently poorly understood, particularly the effect of lower-mantle melting, and yet has first-order implications for the thermal history of both core and mantle. Fourth, estimates of material properties, especially core conductivity, at the correct conditions need to be more accurate. Finally, future palaeomagnetic measurements may help to provide further observational constraints on the evolution of the geodynamo, and thus the thermal evolution of the core.

Acknowledgment

This work is supported by NSF-EAR.

References

Abbott D, Burgess L, Longhi J, and Smith WHF (1994) An empirical thermal history of the Earth's upper mantle. *Journal of Geophysical Research* 99: 13835–13850.

Abe Y, Ohtani E, Okuchi T, Righter K, and Drake M (2000) Water in the early Earth. In: Canup RM and Righter K (eds.) *Origin of the Earth and Moon*, pp. 413–434. Tucson, AZ: University of Arizona Press.

Adler JF and Williams Q (2005) A high-pressure X-ray diffraction study of iron nitrides: Implications for Earth's core. *Journal of Geophysical Research* 110: B01203.

Akins JA, Luo S-N, Asimow PD, and Ahrens TJ (2004) Shock-induced melting of MgSiO3 perovskite and implications for melts in Earth's lowermost mantle. *Geophysical Research Letters* 31: L14612.

Alfè D, Gillan MJ, and Price GD (1999) The melting curve of iron at the pressures of the Earth's core from *ab initio* calculations. *Nature* 401: 462–464.

Alfè D, Gillan MJ, and Price GD (2002a) *Ab initio* chemical potentials of solid and liquid solutions and the chemistry of the Earth's core. *Journal of Chemical Physics* 116: 7127–7136.

Alfè D, Price GD, and Gillan MJ (2002b) Iron under Earth's core conditions: Liquid-state thermodynamics and high-pressure melting curve from *ab initio* calculations. *Physical Review B* 65: 165118.

Alfè D, Gillan MJ, and Price GD (2003) Thermodynamics from first principles: Temperature and composition of the Earth's core. *Minerological Magazine* 67: 113–123.

Alfè D, Price GD, and Gillan MJ (2004) The melting curve of iron from quantum mechanics calculations. *Journal of Physics and Chemistry of Solids* 65: 1573–1580.

Allegre CJ, Manhes G, and Gopel C (1995b) The age of the Earth. *Geochemica et Cosmochimica Acta* 59: 1445–1456.

Allegre CJ, Poirier J-P, Humler E, and Hofmann AW (1995a) The chemical composition of the Earth. *Earth and Planetary Science Letters* 134: 515–526.

Anderson OL (1998) The Gruneisen parameter for iron at outer core conditions and the resulting conductive heat and power in the core. *Physics of the Earth and Planetary Interiors* 109: 179–197.

Araki T, Enomoto S, Furuno K, *et al.* (2005) Experimental investigation of geologically produced antineutrinos with KamLAND. *Nature* 436: 499–503.

Backus GE (1975) Gross thermodynamics of heat engines in deep interior of Earth. *Proceedings of the National Academy of Sciences USA* 72: 1555–1558.

Baker JA and Jensen KK (2004) Coupled Os-186-Os-187 enrichments in the Earth's mantle – core–mantle interaction or recycling of ferromanganese crusts and nodules? *Earth and Planetary Science Letters* 220: 277–286.

Belonoshko AB, Ahuja R, and Johansson B (2000) Quasi - *Ab initio* molecular dynamic study of Fe melting. *Physical Review Letters* 84: 3638–3641.

Bi Y, Tan H, and Jin F (2002) Electrical conductivity of iron under shock compression up to 200 GPa. *Journal of Physics: Condensed Matter* 14: 10849–10854.

Bloxham J (2000) Sensitivity of the geomagnetic axial dipole to thermal core–mantle interactions. *Nature* 405: 63–65.

Boehler R (1993) Temperatures in the Earth's core from melting-point measurements of iron at high static pressures. *Nature* 363: 534–536.

Boehler R (2000) High-pressure experiments and the phase diagram of lower mantle and core materials. *Reviews of Geophysics* 38: 221–245.

Boyet M and Carlson RW (2005) Nd-142 evidence for early (>4.53 Ga) global differentiation of the silicate Earth. *Science* 309: 576–581.

Braginsky SI (1963) Structure of the F layer and reasons for convection in the Earth's core. *Doklady Akademii Nauk SSSR English Translation* 149: 1311–1314.

Braginsky SI (1993) MAC-oscillations of the hidden ocean of the core. *Journal of Geomagnetism and Geoelectricity* 45: 1517–1538.

Braginsky SI (2006) Formation of the stratified ocean of the core. *Earth and Planetary Science Letters* 243: 650–656.

Braginsky SI and Roberts PH (1995) Equations governing convection in Earth's core and the geodynamo. *Geophysical and Astrophysical Fluid Dynamics* 79: 1–97.

Brandon A and Walker RJ (2005) The debate over core–mantle interaction. *Earth and Planetary Science Letters* 232: 211–225.

Brandon AD, Walker RJ, Morgan JW, Norman MD, and Prichard HM (1998) Coupled Os-186 and Os-187 evidence for core–mantle interaction. *Science* 280: 1570–1573.

Brandon AD, Walker RJ, Puchtel IS, Becker H, Humayun M, and Revillon S (2003) Os-186-Os-187 systematics of Gorgona Island komatiites: Implications for early growth of the inner core. *Earth and Planetary Science Letters* 206: 411–426.

Brown JM and McQueen RG (1986) Phase-transitions, Gruneisen parameter and elasticity for shocked iron between 77 GPa and 400 GPa. *Journal of Geophysical Research* 91: 7485–7494.

Buffett BA (2002) Estimates of heat flow in the deep mantle based on the power requirements for the geodynamo. *Geophysical Research Letters* 29: 1566.

Buffett BA (2003) The thermal state of Earth's core. *Science* 299: 1675–1677.

Buffett BA, Garnero EJ, and Jeanloz R (2000) Sediments at the top of the Earth's core. *Science* 290: 1338–1342.

Buffett BA, Huppert HE, Lister JR, and Woods AW (1996) On the thermal evolution of the Earth's core. *Journal of Geophysical Research* 101: 7989–8006.

Buffett BA, Matthews PM, and Herring TA (2002) Modeling of nutation and precession: Effects of electromagnetic coupling. *Journal of Geophysical Research* 107: 2070.

Bukowinski MST and Akber-Knutson S (2005) The role of theoretical mineral physics in modeling the Earth's interior. In: Van der Hilst RD, Bass JD, Matas J, and Trampert J (eds.) *Geophysical Monograph 160: Earth's Deep Mantle: Structure, Composition and Evolution*, pp. 165–186. Washington, DC: American Geophysical Union.

Bullard EC (1950) The transfer of heat from the core of the Earth. *Monthly Notices of the Royal Astronomical Society* 6: 36–41.

Bunge HP (2005) Low plume excess temperature and high core heat flux inferred from non-adiabatic geotherms in internally-heated mantle circulation models. *Physics of the Earth and Planetary Interiors* 153: 3–10.

Butler SL, Peltier WR, and Costin SO (2005) Numerical models of the Earth's thermal history: Effects of inner-core solidification and core potassium. *Physics of the Earth and Planetary Interiors* 152: 22–42.

Busse FH (2000) Homogeneous dynamos in planetary cores and in the laboratory. *Annual Review of Fluid Mechanics* 32: 383–408.

Cao AM and Romanowicz B (2004) Constraints on density and shear velocity contrasts at the inner core boundary. *Geophysical Journal International* 157: 1146–1151.

Carlson RW and Hauri EH (2001) Extending the Pd-107-Ag-107 chronometer to low Pd/Ag meteorites with multicollector plasma-ionization mass spectrometry. *Geochemica et Cosmochimica Acta* 65: 1839–1848.

Chambers JE (2003) Planet formation. In: Davis AM (ed.) *Treatise on Geochemistry, vol. 1: Meteorites, Comets, and Planets*, pp. 461–475. Amsterdam: Elsevier.

Christensen UR and Olson P (2003) Secular variation in numerical geodynamo models with lateral variations of boundary heat flow. *Physics of the Earth and Planetary Interiors* 138: 39–54.

Christensen UR and Tilgner A (2004) Power requirements of the geodynamo from Ohmic losses in numerical and laboratory dynamos. *Nature* 429: 169–171.

Coe RS and Glatzmaier GA (2006) Symmetry and stability of the geomagnetic field. *Geophysical Research Letters* 33: L21311.

Coltice N and Ricard Y (1999) Geochemical observations and one layer mantle convection. *Earth and Planetary Science Letters* 174: 125–137.

Costin SO and Butler SL (2006) Modelling the effects of internal heating in the core and lowermost mantle on the Earth's magnetic history. *Physics of the Earth and Planetary Interiors* 157: 55–71.

Davies GF (1988) Ocean bathymetry and mantle convection. Part 1: Large-scale flow and hotspots. *Journal of Geophysical Research* 93: 10467–10480.

Davies GF (2007) Mantle regulation of core cooling: A geodynamo without core radioactivity? *Physics of the Earth and Planetary Interiors* 160: 215–229.

Dubrovinsky L, Dubrovinsky L, Langenhorst F, *et al.* (2004) Reaction of iron and silica at core–mantle boundary conditions. *Physics of the Earth and Planetary Interiors* 146: 243–247.

Dunlop DJ and Yu Y (2004) Intensity and polarity of the geomagnetic field during Pre-cambrian time. In: Channell JET, Kent DV, Lowrie W, and Meert JG (eds.) *Geophysical Monograph 145: Timescales of the Paleomagnetic Field*, pp. 85–110. Washington, DC: American Geophysical Union.

Garnero EJ, Revenaugh J, Williams Q, Lay T, and Kellogg LH (1998) Ultralow velocity zone at the core–mantle boundary. In: Gurnis M, Wysession ME, Knittle E, and Buffet BA (eds.) *Geodynamic Series 28: The Core–Mantle Boundary Region*, pp. 319–334. Washington, DC: American Geophysical Union.

Gessman CK and Wood BJ (2002) Potassium in the Earth's core? *Earth and Planetary Science Letters* 200: 63–78.

Glatzmaier GA (2002) Geodynamo simulations – How realistic are they? *Annual Review of Earth and Planetary Science* 30: 237–257.

Glatzmaier GA, Coe RS, Hongre L, and Roberts PH (1999) The role of the Earth's mantle in controlling the frequency of geomagnetic reversals. *Nature* 401: 885–890.

Glatzmaier GA and Roberts PH (1996) An anelastic evolutionary geodynamo simulation driven by compositional and thermal convection. *Physica D* 97: 81–94.

Grove TL and Parman SW (2004) Thermal evolution of the Earth as recorded by komatiites. *Earth and Planetary Science Letters* 219: 173–187.

Gubbins D (1977) Energetics of the Earth's core. *Journal of Geophysics* 43: 453–464.

Gubbins D, Alfe D, Masters G, Price GD, and Gillan MJ (2003) Can the Earth's dynamo run on heat alone? *Geophysical Journal International* 155: 609–622.

Gubbins D, Alfe D, Masters G, Price GD, and Gillan MJ (2004) Gross thermodynamics of two-component core convection. *Geophysical Journal International* 157: 1407–1414.

Gubbins D, Masters TG, and Jacobs JA (1979) Thermal evolution of the Earth's core. *Geophysical Journal of the Royal Astronomical Society* 59: 57–99.

Hage H and Muller G (1979) Changes in dimensions, stresses and gravitational energy of the Earth due to crystallization at the inner core boundary under isochemical conditions. *Geophysical Journal of the Royal Astronomical Society* 58: 495–508.

Halliday AN (2004) Mixing, volatile loss and compositional change during impact-driven accretion of the Earth. *Nature* 427: 505–509.

Harper CL and Jacobsen SB (1996) Evidence for Hf-182 in the early solar system and constraints on the timescale of terrestrial accretion and core formation. *Geochemica et Cosmochimica Acta* 60: 1131–1153.

Hauri EH and Hart SR (1993) Re–Os isotope systematics of HIMU and EMII oceanic island basalts from the South Pacific ocean. *Earth and Planetary Science Letters* 114: 353–371.

Helffrich G and Kaneshima S (2004) Seismological constraints on core composition from Fe–O liquid immiscibility. *Science* 306: 2239–2242.

Hernlund JW, Thomas C, and Tackley PJ (2005) A doubling of the post-perovskite phase boundary and structure of the Earth's lowermost mantle. *Nature* 434: 882–886.

Hewitt J, McKenzie DP, and Weiss NO (1975) Dissipative heating in convective flow. *Journal of Fluid Mechanics* 68: 721–738.

Hillgren VJ, Gessmann CK, and Li J (2000) An experimental perspective on the light element in Earth's core.

In: Canup RM and Righter K (eds.) *Origin of the Earth and Moon*, pp. 245–264. Tucson, AZ: University of Arizona Press.

Humayun M, Qin L, and Norman MD (2004) Geochemical evidence for excess iron in the mantle beneath Hawaii. *Science* 306: 91–94.

Jacobs JA (1998) Variations in the intensity of the Earth's magnetic field. *Surveys in Geophysics* 19: 139–187.

Jephcoat A and Olson P (1987) Is the inner core of the Earth pure iron. *Nature* 325: 332–335.

Kanda RVS and Stevenson DJ (2006) Suction mechanism for iron entrainment into the lower mantle. *Geophysical Research Letters* 33: L02310.

Kellogg LH and King SD (1993) Effect of mantle plumes on the growth of D'' by reaction between the core and mantle. *Geophysical Research Letters* 20: 379–382.

Kilburn MR and Wood BJ (1997) Metal-silicate partitioning and the incompatibility of S and Si during core formation. *Earth and Planetary Science Letters* 152: 139–148.

Kleine T, Munker C, Mezger K, and Palme H (2002) Rapid accretion and early core formation on asteroids and the terrestrial planets from Hf-W chronometry. *Nature* 418: 952–955.

Knittle E (1998) The solid/liquid partitioning of major and radiogenic elements at lower mantle pressures: Implications for the core–mantle boundary region. In: Gurnis M, Wysession ME, Knittle E, and Buffet BA (eds.) *Geodynamics 28: The Core–Mantle Boundary Region*, pp. 119–130. Washington, DC: American Geophysical Union.

Knittle E and Jeanloz R (1991) Earth's core–mantle boundary – results of experiments at high pressures and temperatures. *Science* 251: 1438–1443.

Kono M and Roberts PH (2002) Recent geodynamo simulations and observations of the geomagnetic field. *Reviews of Geophysics* 40: 1013.

Koper KD and Dombrovskaya M (2005) Seismic properties of the inner core boundary from PKiKP/P amplitude ratios. *Earth and Planetary Science Letters* 237: 680–694.

Kuang WL and Bloxham J (1997) An Earth-like numerical dynamo model. *Nature* 398: 371–374.

Labrosse S (2002) Hotspots, mantle plumes and core heat loss. *Earth and Planetary Science Letters* 199: 147–156.

Labrosse S (2003) Thermal and magnetic evolution of the Earth's core. *Physics of the Earth and Planetary Interiors* 140: 127–143.

Labrosse S and Macouin M (2003) The inner core and the geodynamo. *Comptes Rendus Geoscience* 335: 37–50.

Labrosse S, Poirier JP, and Le Mouel JL (1997) On cooling of the Earth's core. *Physics of the Earth and Planetary Interiors* 99: 1–17.

Labrosse S, Poirier JP, and Le Mouel JL (2001) The age of the inner core. *Earth and Planetary Science Letters* 190: 111–123.

Laio A, Bernard S, Chiarotti GL, Scandolo S, and Tosatti E (2000) Physics of iron at Earth's core conditions. *Science* 287: 1027–1030.

Lassiter JC (2004) Role of recycled oceanic crust in the potassium and argon budget of the Earth: Toward a resolution of the "missing argon" problem. *Geochemistry Geophysics Geosystems* 5: Q11012.

Lassiter JC (2006) Constraints on the coupled thermal evolution of the Earth's core and mantle, the age of the inner core, and the origin of the Os-186 Os-188 'core signal' in plume-derived lavas. *Earth and Planetary Science Letters* 250: 306–317.

Layer PW, Kroner A, and McWilliams M (1996) An Archean geomagnetic reversal in the Kaap Valley pluton, South Africa. *Science* 273: 943–946.

Lee KKM and Steinle-Neumann G (2006) High-pressure alloying of iron and xenon: 'Missing' Xe in the Earth's core? *Journal of Geophysical Research* 111: B02202.

Lee KKM, Steinle-Neumann G, and Jeanloz R (2004) *Ab-initio* high-pressure alloying of iron and potassium: Implications for the Earth's core. *Geophysical Research Letters* 31: L11603.

Li J and Fei YW (2003) Experimental Constraints on Core Composition. In: Carlson RW (ed.) *Treatise on Geochemistry*, The Mantle and Core, vol. 2, pp. 521–546. Amsterdam: Elsevier.

Lister JR (2003) Expressions for the dissipation driven by convection in the Earth's core. *Physics of the Earth and Planetary Interiors* 140: 145–158.

Lister JR and Buffett BA (1998) Stratification of the outer core at the core–mantle boundary. *Physics of the Earth and Planetary Interiors* 105: 5–19.

Loper DE (1978) The gravitationally powered dynamo. *Geophysical Journal of the Royal Astronomical Society* 54: 389–404.

Lowrie W and Kent DV (2004) Geomagnetic polarity timescales and reversal frequency regimes. In: Channell JET, Kent, Lowrie W, and Meert J (eds.) *Geophysical Monograph 145*, Timescale of the Paleomagnetic Field, pp. 117–129. Washington, DC: American Geophysical Union.

Ma YZ, Somayazulu M, Shen GY, Mao HK, Shu JF, and Hemley RJ (2004) *In situ* X-ray diffraction studies of iron to Earth-core conditions. *Physics of the Earth and Planetary Interiors* 143: 455–467.

Malavergne V, Siebert J, Guyot F, et al. (2004) Si in the core? New high-pressure and high-temperature experimental data. *Geochemica et Cosmochimica Acta* 68: 4201–4211.

Manga M and Jeanloz R (1996) Implications of a metal-bearing chemical boundary layer in D" for mantle dynamics. *Geophysical Research Letters* 23: 3091–3094.

Masters G and Gubbins D (2003) On the resolution of density within the Earth. *Physics of the Earth and Planetary Interiors* 140: 159–167.

Matassov G (1977) *The electrical conductivity of iron–silicon alloys at high pressures and the Earth's core*. PhD Thesis, Lawrence Livermore Laboratory, University of California, CA.

Matsuda J, Sudo M, Ozima M, Ito K, Ohtaka O, and Ito E (1993) Noble gas partitioning between metal and silicate under high pressures. *Science* 259: 788–790.

McDonough WF (2003) Compositional model for the Earth's core. In: Carlson RW (ed.) *Treatise on Geochemistry: The Mantle and Core*, pp. 517–568. Amsterdam: Elsevier.

McElhinny MW and Senanayake WE (1980) Paleomagnetic evidence for the existence of the geomagnetic field 3.5 Ga ago. *Journal of Geophysical Research* 85: 3523–3528.

Mittelstaedt E and Tackley PJ (2006) Plume heat flow is much lower than CMB heat flow. *Earth and Planetary Science Letters* 241: 202–210.

Mollett S (1984) Thermal and magnetic constraints on the cooling of the Earth. *Geophysical Journal of the Royal Astronomical Society* 76: 653–666.

Morgan JW and Anders E (1980) Chemical composition of the Earth, Venus and Mercury. *Proceedings of the National Academy of Sciences USA* 77: 6973–6977.

Morgan JW, Horan MF, Walker RJ, and Grossman JN (1995) Rhenium–osmium concentration and isotope systematics in Group IIAB iron meteorites. *Geochemica et Cosmochimica Acta* 59: 2331–2344.

Murakami M, Hirose K, Kawamura K, Sata N, and Ohishi Y (2004) Post-perovskite phase transition in MgSiO$_3$. *Science* 304: 855–858.

Murthy VR, van Westrenen W, and Fei YW (2003) Experimental evidence that potassium is a substantial radioactive heat source in planetary cores. *Nature* 323: 163–165.

Nakagawa T and Tackley PJ (2004a) Effects of thermo–chemical mantle convection on the thermal evolution of the Earth's core. *Earth and Planetary Science Letters* 220: 107–119.

Nakagawa T and Tackley PJ (2004b) Effects of a perovskite-post perovskite phase change near core–mantle boundary in compressible mantle convection. *Geophysical Research Letters* 31: 16611.

Nguyen JH and Holmes NC (2004) Melting of iron at the physical conditions of the Earth's core. *Nature* 427: 339–342.

Nielsen SG, Rehkamper M, Norman MD, Halliday AN, and Harrison D (2006) Thallium isotopic evidence for ferromanganese sediments in the mantle source of Hawaiian basalts. *Nature* 439: 314–317.

Nimmo F and Agnor CB (2006) Isotopic outcomes of N-body accretion simulations: Constraints on equilibrium processes during large impacts from Hf-W observations. *Earth and Planetary Science Letters* 243: 26–43.

Nimmo F, Price GD, Brodholt J, and Gubbins D (2004) The influence of potassium on core and geodynamo evolution. *Geophysical Journal International* 156: 363–376.

Okuchi T (1997) Hydrogen partitioning into molten iron at high pressure: Implications for Earth's core. *Science* 278: 1781–1784.

Olson P and Christensen UR (2006) Dipole moment scaling for convection-driven planetary dynamos. *Earth and Planetary Science Letters* 250: 561–571.

Ostanin S, Alfe D, Dobson D, Vocadlo L, Brodholt JP, and Price GD (2006) *Ab initio* study of the phase separation of argon in molten iron at high pressures. *Geophysical Research Letters* 33: L06303.

Poirier J-P (1994) Light elements in the Earth's outer core: A critical review. *Physics of the Earth and Planetary Interiors* 85: 319–337.

Poirier J-P, Malavergene V, and Le Mouel JL (1998) Is there a thin electrically conducting layer at the base of the mantle? In: Gurnis M, Wysession ME, Knittle E, and Buffet BA (eds.) *Geodynamics 28: The Core–Mantle Boundary Region*, pp. 131–138. Washington, DC: American Geophysical Union.

Prevot M, Derder ME, McWilliams M, and Thompson J (1990) Intensity of the Earth's magnetic field – evidence for a Mesozoic dipole low. *Earth and Planetary Science Letters* 97: 129–139.

Puchtel IS, Brandon AD, Humayun M, and Walker RJ (2005) Evidence for the early differentiation of the core from Pt-Re-Os isotope systematics of 2.8-Ga komatiites. *Earth and Planetary Science Letters* 237: 118–134.

Righter K and Drake MJ (2003) Partition coefficients at high pressure and temperature. In: Carlson RW (ed.) *Treatise on Geochemistry, vol. 2: The Mantle and Core*, pp. 425–449. Amsterdam: Elsevier.

Roberts PH and Glatzmaier GA (2001) The geodynamo, past, present and future. *Geophysical and Astrophysical Fluid Dynamics* 94: 47–84.

Roberts PH, Jones CA, and Calderwood A (2003) Energy fluxes and Ohmic dissipation in the Earth's core. In: Jones CA, Soward AM, and Zhang K (eds.) *Earth's Core and Lower Mantle*, pp. 100–129. New York: Taylor & Francis.

Schersten A, Elliott T, Hawkesworth C, and Norman M (2004) Tungsten isotope evidence that mantle plumes contain no contribution from the Earth's core. *Nature* 427: 234–237.

Selkin PA and Tauxe L (2000) Long-term variations in palaeointensity. *Philosophical Transactions of the Royal Society of London A* 358: 1065–1088.

Shen GY, Mao HK, Hemley RJ, Duffy TS, and Rivers ML (1998) Melting and crystal structure of iron at high pressures and temperatures. *Geophysical Research Letters* 25: 373–376.

Sleep NH (1988) Gradual entrainment of a chemical layer at the base of the mantle by overlying convection. *Geophysical Journal International* 95: 437–447.

Sleep NH (1990) Hot spots and mantle plumes: Some phenomenology. *Journal of Geophysical Research* 95: 6715–6736.

Stacey FD and Anderson OL (2001) Electrical and thermal conductivities of Fe–Ni–Si alloy under core conditions. *Physics of the Earth and Planetary Interiors* 124: 153–162.

Stacey FD and Loper DE (1984) Thermal histories of the core and mantle. *Physics of the Earth and Planetary Interiors* 36: 99–115.

Stevenson DJ, Spohn T, and Schubert G (1983) Magnetism and thermal evolution of the terrestrial planets. *Icarus* 54: 466–489.

Sumita I and Yoshida S (2003) Thermal interactions between the mantle, outer and inner cores, and the resulting structural evolution of the core. In: Dehant V, Creager K, Zatman S, and Karato S-I (eds.) *Geodynamics Series 31: Earth's Core: Structure, Dynamics and Rotation*, pp. 213–231. Washington, DC: American Geophysical Union.

Tolstikhin I and Hofmann AW (2005) Early crust on top of the Earth's core. *Physics of the Earth and Planetary Interiors* 148: 109–130.

Turcotte DL and Schubert G (2002) *Geodynamics*. New York: Cambridge University Press.

Valet JP (2003) Time variations in geomagnetic intensity. *Reviews of Geophysics* 41: 1004.

Verhoogen J (1961) Heat balance of the Earth's core. *Geophysical Journal of the Royal Astronomical Society* 4: 276–281.

Walker D (2000) Core participation in mantle geochemistry. *Geochemica et Cosmochimica Acta* 64: 2897–2911.

Walker RJ, Morgan JW, and Horan MF (1995) Os-187 enrichment in some plumes: Evidence for core–mantle interaction?. *Science* 269: 819–822.

Wanke H and Dreibus G (1988) Chemical composition and accretion history of terrestrial planets. *Philosophical Transactions of the Royal Society of London* 325: 545–557.

Williams Q (1998) The temperature contrast across D″. In: Gurnis M, Wysession M, Knittle E, and Buffett B (eds.) *Geodynamics*, 28, *The Core–Mantle Boundary Region*, pp. 73–82. Washington, DC: American Geophysical Union.

Yoo CS, Holmes NC, Ross M, Webb DJ, and Pike C (1993) Shock temperatures and melting of iron at Earth core conditions. *Physical Review Letters* 70: 3931–3934.

Yukutake T (2000) The inner core and the surface heat flow as clues to estimating the initial temperature of the Earth's core. *Physics of the Earth and Planetary Interiors* 121: 103–137.

Zhong S (2006) Constraints on thermochemical convection of the mantle from plume heat flux, plume excess temperature and upper mantle temperature. *Journal of Geophysical Research* 111: B04409.

10 History of Earth Rotation

W. R. Peltier, University of Toronto, Toronto, ON, Canada

10.1 Polar Motion and Length-of-Day Variations through (Geological) Time

Observations of the evolving state of planetary rotation reveal the occurrence of variability on timescales ranging from daily to annual, to interannual, decadal, and millennial, and extending even to the timescale of hundreds of millions of years on which the process of mantle convection governs planetary evolution and to the 4.56 billion year age of the planet itself. These variations in the state of rotation are most usefully discussed in terms of changes in the rate of planetary rotation about the instantaneous spin axis and thus variations in the length of the day, on the one hand, or in terms of the wobble of the spin axis as observed in a body-fixed frame of reference on the other. On relatively short subannual to annual timescales, variations in the length of day (l.o.d.) have been clearly shown to arise primarily as a consequence of the exchange of angular momentum between the solid Earth and its overlying atmosphere (e.g., Hide *et al.*, 1980) and oceans. On the interannual timescale, an important recent discovery concerning l.o.d. variability has been the documentation of a significant excitation associated with El Niño–Southern Oscillation (ENSO) events (Cox and Chao, 2002; Dickey *et al.*, 2002). Concerning the sources of wobble excitation, it is clear that on the timescale of the seasonal cycle of climate change, the excitation of the annual component of wobble variability is due to the interhemispheric exchange of atmospheric mass. The Chandler wobble, however, a free oscillation of Earth's spin axis in a body-fixed frame of reference with a period close to 14 months, is apparently significantly forced by the dynamical state of the oceans (e.g., Gross, 2000) as well as by the atmosphere. These most recent analyses of the problem of Chandler wobble excitation, to be reviewed in what follows, appear to have finally resolved what had remained an unresolved problem for decades.

On the longer timescale of millennia, both of these 'anomalies' in Earth rotation exhibit apparent secular variations that are caused primarily by the Late Pleistocene cycle of glaciation and deglaciation (Peltier, 1982) that has been an enduring feature of climate system variability for the past 900 000 years of Earth history (e.g., Deblonde and Peltier, 1991). A primary focus of the discussion to follow in this chapter will be upon the manner in which, through the process of glacial isostatic adjustment (GIA), these ice-age-engendered variations in Earth rotation feed back upon postglacial relative sea-level (RSL) history, thus enabling detailed tests to be performed on the quality of the theory that has been developed to compute the rotational response to the GIA process (Peltier, 2005). The importance of an accurate attribution of the source of excitation of the observed secular changes in the l.o.d. and polar motion to the GIA process concerns the important role that these observations may be invoked to play in the inference of the viscosity of the deep Earth, a parameter that is required in the construction of models of the mantle convection and continental drift processes. An interesting additional aspect of the history of Earth's rotation on the timescale of the Late Pleistocene ice-age cycle concerns the way in which temporal variations in the precession and obliquity components of the evolving geometry of Earth's orbit around the Sun, forced by gravitational n-body effects in the solar system, have been employed to refine the timing of the ice-age cycle itself (Shackleton *et al.*, 1990).

On the very longest timescales on which the thermal evolution of the planet is governed by the mantle

convection process, there also exists the distinct possibility that relatively rapid and large amplitude changes in the rotational state could have occurred in association with an 'avalanche effect' during which the style of the mantle convective circulation switches from one characterized by significant radial layering of the thermally forced flow, to one of 'whole mantle' form (e.g., Peltier and Solheim, 1994a, 1994b). This process could conceivably act so as to induce the inertial interchange true polar wander (IITPW) instability that was suggested initially by Gold (1955) and which has recently been invoked by Kirschvink *et al.* (1997) as plausibly having occurred in the Early Cambrian period of Earth history.

Figure 1 provides a schematic depiction of the extremely broad range of processes that contribute to the excitation of variations in Earth rotation on all timescales. The centipedes in the sketch, following the colorful analogy by Gold (1955), are intended to represent, by their ability to move over the surface and thereby slowly (?) modify the moment of inertia tensor of the planet, the excitation of a rotational response due to the action of the internal mantle

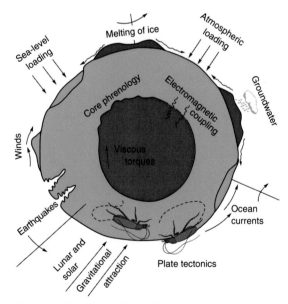

Figure 1 Schematic diagram illustrating the range of processes that contribute to the excitation of variations in the rotational state of the planet. The centipedes in the figure, which are a modification of the beetles employed by Gold (1955) for the same illustrative purpose, are intended to represent the contribution to rotational excitation due to the mantle convection process. This schematic is a modification of that in Lambeck's (1980a) paper commenting upon the important paper of Hide *et al.* (1980).

convective mixing process. How important this component of the rotational excitation could be is still a matter of considerable debate, as is discussed in the final section of this chapter.

10.2 Theoretical and Observational Background I: Angular Momentum Conservation on Subannual to Interannual Timescales

10.2.1 Theoretical Preliminaries

An instantaneous state of Earth rotation is described entirely by the three Cartesian components of the planet's evolving angular velocity vector with components $\omega_i(t)$. The evolution of the angular velocity vector is itself determined by a solution of the classical Euler equation describing the conservation of angular momentum of a system subjected to external torques τ_j, as:

$$\frac{d}{dt}\left(\mathcal{J}_{ij}\omega_i\right) + \in_{ijk}\omega_j\mathcal{J}_{kl}\omega_l = \tau_j \qquad [1]$$

Here, the $\mathcal{J}_{ij}$ are the elements of the moment of inertia tensor and $\in_{ijk}$ is the Levi–Civita (alternating) tensor. Restricting attention to small departures from a modern state of steady rotation with angular velocity Ω_o, we may construct a solution to [1], accurate to first order in perturbation theory, by expanding:

$$\omega_i = \Omega_o(\delta_{i3} + m_i), \ m_i = \omega_i/\Omega_o \qquad [2a]$$

$$\mathcal{J}_{11} = A + I_{11} \qquad [2b]$$

$$\mathcal{J}_{22} = B + I_{22} \qquad [2c]$$

$$\mathcal{J}_{33} = C + I_{33} \qquad [2d]$$

$$\mathcal{J}_{ij} = I_{ij}, \ i \neq j \qquad [2e]$$

Substitution of these expansions into eqn [1], keeping only terms of first order, leads to the standard set of governing equations for polar wander and the length of day, respectively (see, for example, Munk and McDonald (1960)), as:

$$\left.\begin{array}{l} \dfrac{dm_1}{dt} + \dfrac{(C-B)}{A}\Omega_o m_2 = \Psi_1 \\[3mm] \dfrac{dm_2}{dt} + \dfrac{(C-A)}{B}\Omega_o m_1 = \Psi_2 \end{array}\right\} \text{ polar wander} \quad [3a, b]$$

$$\left.\dfrac{dm_3}{dt} = \Psi_3\right\} \text{ length of day} \qquad [3c]$$

in which the 'excitation functions' are defined as

$$\Psi_1 = \left(\frac{\Omega_o}{A}\right)I_{23} - \frac{(dI_{13}/dt)}{A} + \tau_1 \qquad [4a]$$

$$\Psi_2 = -\left(\frac{\Omega_o}{B}\right)I_{13} - \frac{(dI_{23}/dt)}{B} + \tau_2 \qquad [4b]$$

$$\Psi_3 = -\left(\frac{I_{33}}{C}\right) + \tau_3 \qquad [4c]$$

Now, it is important to recognize that we are here distinguishing between two different ways of describing the rotational excitation, due respectively to externally applied torques τ_i and to 'internally' originating perturbations I_{ij} to the moment of inertia tensor. In this section, a brief review will be provided of our current understanding of the nature and origins of the variations in Earth rotation that occur on subannual to interannual timescales. These variations are most simply understood by considering the rotational response of an 'almost-rigid' solid Earth to externally applied torques derivative of the action upon it of the atmosphere and oceans as well as to the redistributions of mass that occur both subseasonally and interannually within these components of the Earth system. As we will see in Section 10.4, it will be more natural to describe the process involved in the excitation of millennial timescale variations of Earth rotation by invoking 'internally' generated changes in the moment of inertia tensor.

10.2.2 Variations in the l.o.d

Focusing first upon changes in the l.o.d. on relatively short timescales, the connection of variations on these timescales to the exchange of angular momentum between the solid Earth and overlying atmosphere was first established most clearly in the important paper by Hide *et al.* (1980). It will be useful to revisit their initial order of magnitude arguments here in order to fix ideas. If one denotes the total axial angular momentum of the Earth system by the sum $M + M_a$, as in Hide *et al.*, which has the approximate value $5.85 \times 10^{33} \ kg \ m^2 \ s^{-1}$, and divides this by the axial moment of inertia that the atmosphere plus solid Earth system would have if the whole system were in a state of rigid body rotation, namely $I + I_a \approx 8.04 \times 10^{37} \ kg \ m^2$, then one would obtain for the angular velocity of the Earth system $\Omega_R \approx 0.726 \times 10^{-4} \ rad \ s^{-1}$. A consequence of the fact that the atmosphere of the Earth is differentially heated by the Sun is that the complex general circulation of its atmosphere is characterized by a marked differential rotation between it and the solid Earth.

This leads to a net increase in the angular momentum of the atmosphere by the (approximate) amount $\delta M_a \approx 1.5 \times 10^{26}$ kg m² s⁻¹ as determined on the basis of direct atmospheric observations. Given the approximate value of the moment of inertia of the atmosphere, $I_a \approx 1.42 \times 10^{32}$ kg m², this leads to an approximate value for the change in angular velocity of magnitude $\delta M_a/I_a \approx 10^{-6}$ rad s⁻¹ $\approx 3 \times 10^{-8}\,\Omega_R$ or, given that the l.o.d. is just $2\pi/\Omega_R$, one obtains an approximate magnitude for the change in the l.o.d. due to the solar differential heating forced general circulation of the atmosphere of 3×10^{-3} s. Detailed astronomical measurements by the Bureau Internationale de l'Heure (BIH) directly demonstrate that on the annual and interannual timescales this is in fact the order of magnitude of the variations in the l.o.d. observed.

Figure 2 reproduces a figure from the original paper of Hide *et al.* in which a detailed analysis of the angular momentum budget of the atmosphere was performed by evaluating the axial component of the relative angular momentum m_a as

$$m_a = \frac{a^2}{g} \int_{-\pi/2}^{+\pi/2} \int_0^{2\pi} \int_0^{p_g} u(p, \theta, \lambda, t)\cos^2\theta\, \mathrm{d}p\, \mathrm{d}\lambda\, \mathrm{d}\theta \quad [5]$$

in which a is the radius of the Earth, g is the surface gravitational acceleration, u is the zonal component of the velocity of atmospheric air in a frame of reference rotating with the solid Earth, p is atmospheric pressure, p_g is the pressure at the surface of the Earth, θ is latitude, λ is longitude, and t is time. The radial part of the required volume integral has been converted to an integral over pressure by employing the hydrostatic approximation. Based upon the use of independent sets of reanalysis data from the UK Meteorological Office (MO) and from the US National Meteorological Center (NMC), the authors inferred the changes in the l.o.d. that should have occurred if the atmosphere was simply exchanging its axial angular momentum with that of the solid Earth. These estimates of temporal changes in the l.o.d. were compared directly to the astronomically derived measurements provided by the BIH over a 9 month period extending from the winter of 1978 to the summer of 1979. These results, which have since only been reconfirmed by later and more detailed analysis, clearly establish that on these short timescales, changes in the l.o.d. are primarily associated with the exchange of angular momentum between the solid Earth and its overlying atmosphere.

Considered over a period of many years, rather than decades, one begins to see additional structure in the variability in the l.o.d. that includes the measurable influence of an effect due to the changing axial component of the moment of inertia of the solid Earth, an influence that is generally measured in

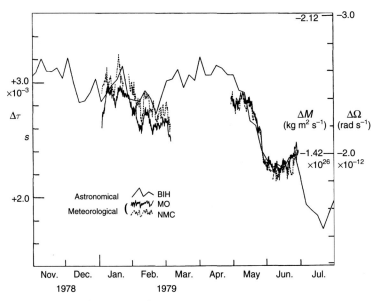

Figure 2 Comparison of the expected changes in the l.o.d., represented by the parameter $\Delta\tau$ measured in milliseconds and derived from the meteorological results (either MO or NMC), with the observed changes based upon the astronomical measurements of the BIH. Reproduced from Hide R, Birch NT, Morrison LV, Shea DJ, and White AA (1980) Atmospheric angular momentum fluctuations an changes in the length of the day. *Nature* 286: 114–117.

terms of a parameter $\dot{J}_2$ which will be discussed at length in subsequent sections of this chapter. The parameter J_2 is the nondimensional degree 2 and order zero coefficient in the spherical harmonic expansion of the gravitational field of the planet. The connection between this parameter and the angular velocity of the planet and thus the l.o.d. is given by the following simple expression:

$$\dot{J}_2 = \frac{3}{2}\frac{C}{a^2 m_e}\dot{m}_3 \qquad [6]$$

in which a and m_e are the Earth's radius and mass, respectively, and $m_3 = \omega_3/\Omega_0$.

Superimposed upon the secular trend in J_2, an effect we will show to be unambiguously connected to the GIA process, analyses by Cox and Chao (2002) were the first in which an interannual timescale variation of this quantity was detected that was apparently associated with the ENSO phenomenon. This internal oscillatory ENSO mode of climate system variability is supported by coupling between the atmosphere and the oceans across the equatorial Pacific region and recurs on a timescale that varies from 3 to 7 years. Further analysis of this interannual variability of J_2, which at the time of writing is still not entirely understood, was presented by Dickey *et al.* (2002), in which an attempt was made to ascribe the El Niño influence to a particular related process having to do with "a recent surge in subpolar glacial melting and by mass shifts in the Southern, Pacific and Indian Oceans." **Figure 3** displays a sequence of time series of the parameter J_2, that marked **A** being the time series of the raw data from the BIH. Although prior to late 1996 the time series reveals a consistent secular decrease in the value of this parameter, such that a least-squares best fit of a straight line to the data delivers a best estimate of $\dot{J}_2$ of approximately -2.9×10^{-11} yr^{-1}, beginning in late 1996 the onset of a significant deviation developed from this secular trend. The timing of this excursion in the time series was coincident with the onset of the intense El Niño event that occurred in 1998. The remaining time series shown on **Figure 3** display the alterations to it that obtain upon successive removal of the influence of the forcing associated with four plausible influences, respectively in (**B**) integrated ocean effects as determined using a data-constrained ocean model, (**C**) subpolar glacier wasting effects, (**D**) integrated atmospheric effects derived on the basis of National Centers for Environmental Prediction (NCEP) reanalysis data, and (**E**) integrated groundwater effects based upon NCEP reanalysis data.

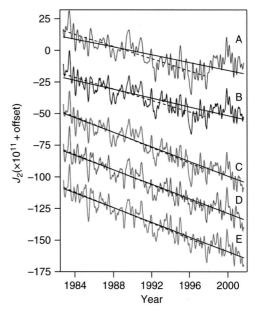

Figure 3 Illustrates as (**A**) the geodetically observed time series of the parameter J_2 over the time period 1983–2002, and residuals (**B**)–(**E**) obtained by successive removal of modeled source terms related respectively to the processes of (**B**) integrated ocean effects, (**C**) subpolar glacial effects, (**D**) integrated atmospheric effects, and (**E**) integrated groundwater effects. Reproduced from Dickey JO, Marcus SL, de Viron O, and Fukumori I (2002) Recent Earth oblateness variations: Unravelling climate and postglacial rebound effects. *Science* 298: 1975–1977.

Although it remains an issue as to whether the attribution of the influence of ENSO upon the J_2 excursion as due to an enhanced wasting of subpolar glaciers is correct, there can be no doubt that the warming associated with the strong 1998 ENSO event induced a substantial variation in the l.o.d. Although the BIH data prior to 1984 were not of sufficient accuracy to generate a longer time series of the variability in this parameter that is equally useful as that shown on **Figure 3**, it is nevertheless extremely important that the dominant secular variation in the time series is essentially the same as that previously inferred on the basis of the analysis of ancient eclipse observations as we will discuss in the following sections of this chapter.

10.2.3   Polar Motion: The Annual and Chandler Wobbles

The Chandler wobble or free Eulerian nutation of the spin axis of the Earth in a body-fixed frame of reference was first discovered by S. C. Chandler

(1891). A detailed theory of the Chandler wobble from a 'normal-mode' perspective was first presented by Dahlen and Smith (1974) and has been more fully developed since by Wahr (1982, 1983), whose investigation focused upon the important influence of the atmosphere and oceans upon the phenomenon. Although manipulation of eqns [3a,b] predicts the existence of a simple harmonic 'free oscillation' solution of frequency $\sigma_R = \Omega_0(C-A/A)$, which is referred to from here on as the Chandler wobble frequency of the rigid Earth, this expression substantially errs in its prediction of the period of the actual Chandler wobble because of the importance of these components of the Earth system as well as the existence of the fluid core upon wobble dynamics. Furthermore, this system of equations would predict a Chandler wobble that would have infinite Q in the sense that, once excited, it would never decay in amplitude.

Aside from the Chandler wobble, which has a variable amplitude that ranges between c. 100 and 200 milliarcseconds (mas) in the angular displacement of the pole, several additional physical processes contribute to the complete range of polar motion that is observed using the space geodetic techniques of very long baseline interferometry, satellite and Lunar Laser Ranging, and Global Positioning System interferometry. These include: (1) a forced annual wobble with an amplitude of approximately 100 mas which occurs in response to the seasonal variability of the atmosphere and oceans, (2) a weaker but quasi-periodic variation on decadal timescales with an amplitude of approximately 30 mas that is often referred to as the Markowitz wobble, and, most important from the perspective of this chapter, (3) a linear trend corresponding to a secular drift in the position of the pole that is presently occurring at a rate of approximately 3.5 mas yr^{-1}. The latter ultra-low-frequency feature, which will be a primary focus in the main body of this chapter to follow, is the polar motion counterpart to the secular variation in the l.o.d. represented in terms of $\dot{J}_2$ and discussed in the last subsection. The rate of 3.5 mas yr^{-1} corresponds to a rate of drift of the pole across the surface of approximately 0.95° per million years.

Although each of the highest frequency contributions to the motion of the pole has been studied in depth, it has been the development of a detailed understanding of the Chandler wobble that has attracted the greatest interest, as the explanation of its excitation mechanism has proven to be extremely elusive. Attempts were made to ascribe its excitation to the atmosphere (Wilson and Haubrich, 1976;

Wahr, 1983), to core–mantle interactions (Rochester and Smylie, 1965; Jault and Le Muel, 1993), to the occurrence of earthquakes (Sauriau and Cazanave, 1985; Gross, 1986), and even to continental water storage (Chao et al., 1987; Kuehne and Wilson, 1991). More recent research has however suggested that the primary excitation may in fact be oceanographic (Gross, 2000), although aided to significant degree by the atmosphere.

In attempting to attribute the excitation of the Chandler wobble to any particular forcing function, it is also important to recognize that this important mode of rotational free oscillation also has a finite Q, so that if all sources of excitation were to vanish, the amplitude of the polar motion would decay to zero. For this reason, a simple heuristic model, derivable from the linearized Liouville equations [3a, b] above, given the extremely small amplitude of the observed motion, is just (e.g., Lambeck, 1980):

$$\Gamma + \frac{i}{\sigma_{CW}}\frac{d\Gamma}{dt} = \chi(t) \qquad [7]$$

in which $\Gamma = x_p - iy_p$ where x_p and y_p are respectively the displacements of the pole from the Conventional International Origin along the Greenwich and 90° west meridians, σ_{CW} is the complex Chandler wobble frequency of the real anelastic Earth for which the Q of the Chandler wobble has been estimated by Wilson and Vincente (1990) to be 179 with 1σ bounds of 74 and 789. Equation [7] describes simple harmonic motion in the complex plane subject to the forcing represented by the function $\chi(t)$, an excitation function that Wahr (1982) has shown may be written as

$$\chi(t) = \frac{1.61}{\Omega_0(C-A)}\left[h(t) + \frac{\Omega_0 I(t)}{1.44}\right] \qquad [8]$$

in which the complex-valued function $I(t) = I_{13}(t) + iI_{23}(t)$ where I_{13} and I_{23} are the perturbations of the inertia tensor associated with the variations in the mass distributions in the atmosphere and the oceans that may be responsible for wobble excitation. The function $h(t) = h_1(t) + ih_2(t)$ represents changes of relative angular momentum associated, for example, with atmospheric winds and ocean currents. The numerical factors 1.61 and 1.44 respectively account for the influences of core decoupling and the yielding of the solid Earth under the weight of the changing surface mass load.

In the paper by Gross (2000), detailed analyses of Chandler wobble excitation were presented that were based upon the assumption of the validity of the estimates of the Chandler period and Q by Wilson and Vincente (1990), an analysis in which the estimates were based upon the longest time series of polar motion data available at that time (86 years), the only study in which the duration of the data set analyzed was longer than the e-folding time for the decay of the Chandler amplitude. Since the polar motion is resonant at the Chandler period, most analyses of the excitation mechanism have been conducted in the frequency domain by direct comparison of the spectrum of the rotational response to the spectrum of the excitation as

computed on the basis of assumed *a priori* knowledge of a range of plausible excitation mechanisms. **Figure 4(a)** presents such an intercomparison, one that focused upon assessing the effectiveness of both atmospheric and oceanographic processes.

Inspection of this figure demonstrates, as shown earlier by Wilson and Haubrich (1976) and Wahr (1983), that the net effect of the sum of atmospheric wind and pressure fluctuations (determined on the basis of NCEP/NCAR reanalysis data) is insufficient to explain the Chandler excitation in terms of available as compared to required power (whereas this excitation is entirely adequate insofar as the annual

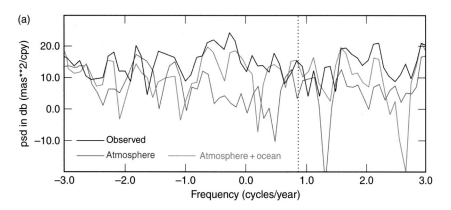

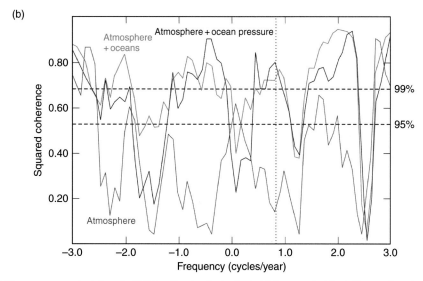

Figure 4 (a) The power spectral density (PSD) estimates in decibels (db) computed from time series of polar motion excitation functions $(\chi(t))$ for the period 1985.0–1996.0 for the observed polar motion excitation function derived from space-geodetic Earth rotation measurements (black curve), for the sum of the excitation functions due to atmospheric wind and pressure changes (red curve) (where the atmospheric pressure term is computed by assuming the validity of the inverted barometer approximation), and for the sum of all atmospheric and oceanic excitation processes (green curve). (b) Squared coherence between the observed polar motion excitation functions in the period 1985.0–1996.0 and the excitation functions due to the sum of atmospheric wind and (inverted barometer) pressure changes (red curve), due to the sum of atmospheric pressure and ocean bottom pressure fluctuations (blue curve), and due to the sum of all atmospheric and oceanic excitation processes. From Gross RS (2000) The excitation of the Chandler wobble. *Geophysical Research Letters* 27: 2329–2342.

wobble is concerned as previously remarked) (NCAR = National Center for Atmospheric Research). On the other hand, when the oceanographic excitation is added to the forcing at the Chandler period, with this determined on the basis of output of the MIT ocean general circulation model (Ponte *et al.*, 1998) driven by 12 h wind stress and 24 h surface heat and freshwater fluxes from NCEP, the total power in the forcing at the Chandler period was found to fit the requirements of the observations extremely well.

In the Gross (2000) paper, this analysis in terms of the available power in the total forcing by the atmosphere and ocean was complemented by an analysis of the coherence between the observed and modeled excitation, the results from which are reproduced here in **Figure 4(b)**. Inspection of the results of this analysis demonstrates that the atmospheric excitation alone is not coherent with the observed excitation, whereas the atmospheric plus oceanographic excitation is highly coherent with the observed excitation. Through detailed analysis, the author of this paper concludes "that the Chandler wobble is being excited by a combination of atmospheric and oceanic processes" and that "the single most important mechanism exciting the Chandler wobble has been ocean bottom pressure fluctuations which contribute about twice as much excitation power in the Chandler frequency band as do atmospheric pressure fluctuations." It would appear on the basis of these results that our understanding of the Chandler component of Earth's polar motion has become as well understood as are the similarly high-frequency variations in the l.o.d. discussed in the previous subsection.

In the following section of this chapter, the characteristic timescale of our interest in Earth rotation history is extended to include the last million years, a period during which the surface of the solid Earth has been continuously subjected to a quasi-cyclic glaciation–deglaciation cycle involving extremely large exchanges of mass between the oceans and the surface of the continents on which massive continental ice sheets formed. It should not be surprising, given the magnitude of this mass redistribution over the surface, that this 'ice-age cycle' has had and continues to have a profound impact upon the rotational state of the planet. What is often seen as very surprising, however, is the fidelity with which the Earth has conspired to remember the details of both the rotational excitation to which it was subjected and the response to it.

10.3 Theoretical Background II: The Viscoelastic Rotational Response to the Late Pleistocene Ice-Age Cycle

Given the extremely broad range of processes that contribute to the excitation of variations in Earth's rotational state (see **Figure 1**) it is always an issue as to the degree to which one is able to uniquely ascribe the observed variability in a particular range of timescales to any one particular process. On the characteristic timescale of millennia and tens to hundreds of millenniums, however, it is apparently the case that the excitation mechanism is so strongly dominated by the Late Pleistocene process of glaciation and deglaciation that such unambiguous attribution has come to be quite generally accepted. This is due to the advances that have been achieved over the past several decades in the development of a detailed viscoelastic theory of the GIA process.

10.3.1 Global Theory of the GIA Process

The modern theory of global glacial isostasy addresses the question of the changes in the Earth's shape and gravitational field caused by the variations in surface mass load that occur when water is irreversibly removed from the oceans and delivered to the high-latitude continents to form large accumulations of land ice. Since the amount of water removed from the ocean basins in a typical 100 ky cycle of glaciation and deglaciation has been such as to cause mean sea level to fall by approximately 120 m, it should not be surprising that this process has been accompanied by highly significant variations in Earth rotation, variations of sufficient amplitude as to leave indelible imprints in the geological record of RSL history. The dominant timescale of the Late Pleistocene ice-age cycle is most clearly appreciated on the basis of oxygen-isotopic measurements in deep-sea sedimentary cores. One example of such a record is shown in **Figure 5**. The oxygen-isotopic anomaly $\delta^{18}O$, as measured on the tests of foraminifera extracted as a function of depth from such cores, provides a high-quality proxy for the amount of ice that exists on the continents as a function of time in the past, determined by the depth in the core from which the individual samples are taken (Shackleton, 1967). This is a simple consequence of the fact that the process whereby water is irreversibly removed from the oceans to build the ice sheets is a process that fractionates mass. $H_2^{16}O$ is preferentially

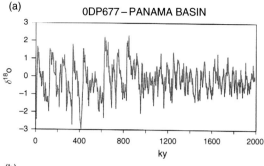

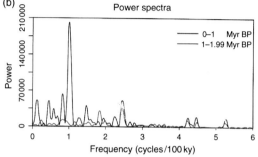

Figure 5 Oxygen-isotopic record from Ocean Drilling Programme Site 677 in the Panama Basin extending to 2 My BP. Also shown are the power spectra for the first and last million years of the record illustrating the onset of the 100 000 year cycle of glaciation and deglaciation in mid-Pleistocene time. Also evident in this continental ice-volume proxy record are the additional spectral lines associated with the variations of orbital obliquity and with the eccentricity modulation of the precessional effect (see Section 10.6 for further discussion). From Shackleton NJ, Berger A, and Peltier WR (1990) An alternative astronomical calibration of the lower Pleistocene timescale based upon ODP 677. *Transactions of the Royal Society of Edinburgh: Earth Sciences* 81: 251–261.

removed from the oceans through the evaporation and precipitation processes whereby water is delivered to the high-latitude continents, resulting in an ocean that is anomalously rich in $H_2^{18}O$ at a time in the past when large-scale continental ice sheets are in place, an anomaly that is recorded in the shells of foraminifera so long as these remain open systems. This isotopic anomaly is defined as

$$\delta^{18}O = \left(\frac{[^{18}O/^{16}O]_{sample}}{[^{18}O/^{16}O]_{standard}} - 1 \right) \qquad [9]$$

The record shown in **Figure 5**, from Ocean Drilling Program Site 677 in the Panama Basin off the west coast of the South American continent, remains one of the highest-resolution records available and was employed in Shackleton *et al.* (1990) to refine the entire chronology of the Pleistocene epoch (see Section 10.6). Inspection of the power

spectrum of this record, which is also shown in **Figure 5** for both the first and last million years of the Pleistocene epoch, will show that for the most recent 900 000 years (see Deblonde and Peltier (1991) for a statistical assessment of the sharpness of this 'mid-Pleistocene climate transition') the climate system has been dominated by a quasi-cyclic variation of ice cover with an average period of approximately 100 000 years. Also evident by inspection of the power spectrum of this record are the additional weaker spectral peaks, one at a period of approximately 41 000 years and a triplet of very weak spectral lines at periods of 19 000, 22 000, and 23 000 years. These additional spectral lines are associated with the component of ice-volume variability that is directly forced by the weak variations in solar radiation caused by the changing geometry of Earth's orbit around the Sun that are in turn due to gravitational n-body effects in the solar system. Because the Earth and Sun are not alone in the solar system, the planet's orbit around the Sun does not consist of a time-invariant Keplerian ellipse but rather one whose ellipticity varies slowly with time at a dominant period of 100 000 years. Furthermore, the spin axis of the Earth as it continuously executes its almost elliptical orbit, and which presently makes an angle of approximately 23.5° with the plane of the ecliptic, also varies with time with a dominant period of 41 000 years. As is discussed with more detail in Section 10.6, it is these sources of variability in Earth's rotational state that are responsible for the ice ages themselves.

Given the large variations in surface mass load that have accompanied the growth and decay of continental ice sheets over the past ~900 000 years of Earth history, during which a quasi-periodic ice-age cycle has persisted with a period of approximately 100 000 years, it should be clear that the key to understanding the excitation of Earth rotation that has accompanied the process must involve the ability to construct a detailed model of the variations in the moment of inertia tensor caused by the process. This requires the solution of what is here referred to as the sea-level equation (SLE), an equation that accurately describes the way in which water must be distributed over the ocean basins when ice sheets melt so as to ensure that the evolving surface of the ocean remains a surface of constant gravitational potential. An initial version of this equation was first solved in papers by Clark *et al.* (1978) and Peltier *et al.* (1978) based upon the combination of results obtained in Peltier (1974, 1976) and Farrell and Clark (1976). Since this original work, the theory underlying the SLE has been significantly extended, however, and it is these extensions that have proven crucial to the accurate understanding of the important role that the Earth's rotational response to the GIA process plays in the theory.

The solution of the SLE delivers a prediction of the history of RSL change, $S(\theta, \lambda, t)$ say, that is caused by an assumed known history of the variations in continental ice mass, represented by a history of continental ice-sheet thickness variations $I(\theta, \lambda, t)$. In these expressions, θ is latitude, λ is longitude, and t is time. The form of the SLE that relates these quantities is as follows:

$$S(\theta, \lambda, t) = C(\theta, \lambda, t)$$
$$\times \left[\int_{-\infty}^{t} dt' \int \int_{\Omega} d\Omega' \{ L(\theta', \lambda', t') G_{\phi}^{L}(\phi, t-t') \right.$$
$$\left. + \Psi^{R}(\theta', \lambda', t') G_{\phi}^{T}(\phi, t-t') \} + \frac{\Delta\Phi(t)}{g} \right] \qquad [10]$$

In this equation, $C(\theta, \lambda, t)$ is the 'ocean function', which is, as defined by Munk and Macdonald (1960), unity over the oceans and zero over the continents. The function C is time dependent because the removal of water from, and its addition to, the ocean basins causes the coastline to migrate, an effect that may be accurately computed on a global basis using the methodology described in Peltier (1994). The space-time-dependent surface mass load per unit area L in the above equation has the composite form:

$$L(\theta, \lambda, t) = \rho_{I} I(\theta, \lambda, t) + \rho_{W} S(\theta, \lambda, t) \qquad [11]$$

in which ρ_{I} and ρ_{W} are the densities of ice and water, respectively. The angle ϕ in the Green functions G_{ϕ}^{L} and G_{ϕ}^{T} is the angular separation between the source point (θ', λ') and the field point (θ, λ). The function $\Psi^{R}(\theta, \lambda, t)$ is the variation of the centrifugal potential due to the changing rotational state of the Earth that is caused by the surface loading and unloading process associated with the ice-age cycle. Following Dahlen (1976), this may be expressed, to first order in perturbation theory, in terms of the following spherical harmonic expansion, as

$$\Psi^{R}(\theta, \lambda, t) = \Psi_{00} Y_{00}(\theta, \lambda, t)$$
$$+ \sum_{m=-1}^{+1} \Psi_{2m} Y_{2m}(\theta, \lambda, t) \qquad [12]$$

with

$$\Psi_{00} = \frac{2}{3} \omega_3(t) \Omega_0 a^2 \qquad [13a]$$

$$\Psi_{20} = -\frac{1}{3} \omega_3(t) \Omega_0 a^2 \sqrt{4/5} \qquad [13b]$$

$$\Psi_{2,-1} = (\omega_1 - i\omega_2)(\Omega_0 a^2/2)\sqrt{2/15} \qquad [13c]$$

$$\Psi_{2,+1} = -(\omega_1 + i\omega_2)(\Omega_0 a^2/2)\sqrt{2/15} \qquad [13d]$$

The ω_i in the above equations are to be obtained as solutions to the appropriate version of eqns [3] in Section 10.2, the form from which the externally applied torques τ_i have been removed as the rotational excitation in this case must by considered to be internally generated. The remaining terms in the SLE consist of the surface mass loading and tidal potential loading Green functions which have been shown in Peltier (1976) to have the mathematical representations:

$$G_{\phi}^{L}(\phi, t) = \frac{a}{m_e} \sum_{l=0}^{\infty} (1 + k_l^{L}(t) - h_l^{L}(t)) P_l(\cos\theta) \qquad [14a]$$

$$G_{\phi}^{T}(\phi, t) = \frac{a}{g} \sum_{l=0}^{\infty} (1 + k_l^{T}(t) - h_l^{T}(t)) P_l(\cos\theta) \qquad [14b]$$

in which k_l^{T}, h_l^{T} are the viscoelastic tidal potential loading Love numbers and k_l^{L}, h_l^{L} are the corresponding surface mass loading Love numbers. Peltier (1976, 1985) has shown that these time domain viscoelastic Love numbers may be expressed, in the case of impulsive point mass or gravitational potential loading at the surface of the planet, in the form of the following normal-mode expansions:

$$k_l^{T}(t) = k_l^{T,E}\delta(t) + \sum_{j=1}^{M} q_j^{l} e^{-s_j^{l}t} \qquad [15a]$$

$$h_l^{T}(t) = h_l^{T,E}\delta(t) + \sum_{j=1}^{M} r_j^{l} e^{-s_j^{l}t} \qquad [15b]$$

$$k_l^{L}(t) = k_l^{L,E}\delta(t) + \sum_{j=1}^{M} q_j^{l} e^{-s_j^{l}t} \qquad [15c]$$

$$h_l^{L}(t) = h_l^{L,E}\delta(t) + \sum_{j=1}^{M} r_j^{l} e^{-s_j^{l}t} \qquad [15d]$$

In these normal-mode expansions, the $k_l^{T,E}$, $h_l^{T,E}$, $k_l^{L,E}$, and $h_l^{L,E}$ are the elastic surface mass load and tidal potential loading Love numbers of Farrell (1972), the s_j^{l} are the inverse relaxation times of a discrete set of normal modes of viscoelastic relaxation determined as the zeros of an appropriate secular function (Peltier,

1985) or by collocation (Peltier, 1974, 1976), and the amplitudes $q_j^{'l}$, $r_j^{'l}$, q_j^l, r_j^l are the residues at these poles. Insofar as understanding the polar wander component of the rotational response of the planet to the GIA process is concerned, the parameter k_2^{T} plays an especially crucial role as is made clear in what follows.

10.3.2 Computation of the Rotational Response to Earth–Ice–Ocean Interactions in the Ice Age

The important role of the Love number $k_2^{\mathrm{T}}(t)$ may be understood by returning to eqns [4] of Section 10.2 and focusing upon the polar wander component of the rotational response of the planet in circumstances in which it may be assumed that externally applied torques vanish ($\tau_i \equiv 0$), and noting that in this case we must distinguish two distinct but intimately related sources of excitation. The first of these derives from the perturbations of inertia caused by the direct affect of the isostatic adjustment process, an influence that may be represented in the form (e.g., Peltier, 1982):

$$I_{ij}^{\mathrm{GIA}} = (1 + k_2^{\mathrm{L}}(t)) * I_{ij}^{\mathrm{R}}(t) \qquad [16]$$

in which the $I_{ij}^{\mathrm{R}}(t)$ are the perturbations of inertia that would obtain due to the variations in surface mass load if the Earth were a rigid body rather than viscoelastic and the symbol $*$ represents the temporal convolution operation.

The second contribution to the perturbations of inertia is that due to the changing rotation itself, and this may be derived from an application of a linearized version of MacCullagh's formula (see, for example, Munk and MacDonald (1960)) as

$$I_{13}^{\mathrm{ROT}} = \left(\frac{k_2^{\mathrm{T}} * a^5 \omega_1 \omega_3}{3G}\right) = \left(\frac{k_2^{\mathrm{T}}}{k_{\mathrm{f}}}\right) * m_1(C-A) \qquad [17a]$$

$$I_{23}^{\mathrm{ROT}} = \left(\frac{k_2^{\mathrm{T}} * a^5 \omega_2 \omega_3}{3G}\right) = \left(\frac{k_2^{\mathrm{T}}}{k_{\mathrm{f}}}\right) * m_2(C-A) \qquad [17b]$$

with

$$k_{\mathrm{f}} = \left(\frac{3G}{a^5 \Omega_o^2}\right)(C-A) \qquad [17c]$$

in which $A = B$ has been assumed in eqns [4] and $k_2^{\mathrm{T}}(t)$ is the tidal potential loading Love number of degree 2 defined above, the parameter that will be seen to play a crucial role in what follows. Equally critical for the arguments to be presented is the so-called 'fluid Love number' k_{f}, the value of which is determined entirely

on the basis of the well-known equatorial flattening of the planet that is represented by the difference between the polar and equatorial moments of inertia $(C-A)$ in eqn [17c]. Substituting in [17c] for Newton's gravitational constant G, the Earth's radius a (for which we will take the equatorial value), the present-day rate of angular rotation Ω_o and the polar and equatorial moments of inertia C and A, respectively, taking all data from the tabulation of Yoder (1995), one obtains for the value of k_{f} that

$$k_{\mathrm{f}} \cong 0.9414 \qquad [18]$$

An important part of the discussion to follow will involve understanding of the connection between k_{f} and $\lim_{t \to \infty} k_2^{\mathrm{T}}(t)$, the infinite time asymptote of the viscoelastic tidal loading Love number of degree 2.

Since the solution of eqn [3c] for the change in the axial rate of rotation is uncomplicated, it will suffice to focus first in what follows on the solution of [3a] and [3b] for the polar wander component of the response to surface loading. Substitution of [17a] and [17b] into [3a,b], the Laplace-transformed forms of the equations that follow are simply

$$sm_1 + \sigma\left(1 - \frac{k_2^{\mathrm{T}}(s)}{k_{\mathrm{f}}}\right) m_2 = \Psi_1(s) \qquad [19a]$$

$$sm_2 + \sigma\left(1 - \frac{k_2^{\mathrm{T}}(s)}{k_{\mathrm{f}}}\right) m_1 = \Psi_2(s) \qquad [19b]$$

where

$$\sigma = \Omega_o \frac{(C-A)}{A} \qquad [19c]$$

is the Chandler wobble frequency of the rigid Earth, s is the Laplace transform variable, and again $A = B$ has been assumed. The Laplace-transformed forms of the excitation functions in [19a] and [19b] are simply

$$\Psi_1(s) = \left(\frac{\Omega_o}{A}\right) I_{23}(s) - \left(\frac{s}{A}\right) I_{13}(s) \qquad [20a]$$

$$\Psi_2(s) = \left(\frac{\Omega_o}{A}\right) I_{13}(s) - \left(\frac{s}{A}\right) I_{23}(s) \qquad [20b]$$

with

$$I_{ij}(s) = (1 + k_2^{\mathrm{L}}(s)) I_{ij}^{\mathrm{Rigid}}(s) \qquad [20c]$$

Now eqns [19a] and [19b] are elementary algebraic equations for $m_1(s)$ and $m_2(s)$ and these may be solved, neglecting terms of order s^2/σ^2, as is appropriate for the analysis of a physical process evolving on a timescale that is much longer than the period of the Chandler wobble, to obtain

$$m_1(s) = \left(\frac{\Omega_o}{A\sigma}\right)\left[\frac{1 + k_2^L(s)}{1 - \frac{k_2^T(s)}{k_f}}\right] I_{13}^{\text{Rigid}}(s) = H(s) I_{13}^{\text{Rigid}}(s) \quad [21a]$$

$$m_2(s) = H(s) I_{23}^{\text{Rigid}}(s) \quad [21b]$$

A convenient short-hand form for the solution vector $(m_1, m_2) = \mathbf{m}$ is to write

$$\mathbf{m}(s) = \frac{\mathbf{\Psi}^L(s)}{\left[1 - \frac{k_2^T(s)}{k_f}\right]} = H(s)\left(I_{13}^{\text{Rigid}}(s), I_{23}^{\text{Rigid}}(s)\right) \quad [22a]$$

where

$$\mathbf{\Psi}^L(s) = \left[\left(\frac{\Omega_o}{A\sigma}\right)\left(1 + k_2^L(s)\right)\left(I_{13}^{\text{Rigid}}(s), I_{23}^{\text{Rigid}}(s)\right)\right] \quad [22b]$$

From eqns [22], it will be clear that the polar wander solution $\mathbf{m}(s)$ will depend critically upon the ratio $k_2^T(s)/k_f$. This fact was more fully exposed in the analysis of Peltier (1982) and Wu and Peltier (1984) who rewrote the Laplace transform domain forms of $k_2^T(s)$ and $k_2^L(s)$ as (e.g., see equation 61 of Wu and Peltier (1984)):

$$k_2^T(s) = k_2^T(s = 0) - s\sum_{j=1}^{N}\frac{(q_j'/s_j)}{(s + s_j)} \quad [23a]$$

$$k_2^L(s) = (-1 + l_s) - s\sum_{j=1}^{N}\frac{(q_j/s_j)}{(s + s_j)} \quad [23b]$$

in which the superscript $\ell = 2$ on q_j^2, r_j^2, s_j^2 has been suppressed for convenience. Substituting [23a] into [22], this may be rewritten as

$$\mathbf{m}(s) = \frac{\mathbf{\Psi}^L(s)}{\left[1 - \frac{k_2^T(s=0)}{k_f}\right] + \frac{s}{k_f}\sum_{j=1}^{N}\frac{(q_j'/s_j)}{(s+s_j)}} \quad [24]$$

In discussing the formal inversion of [24] into the time domain, the focus for the purpose of this chapter is on the results that follow from application of an important modified form of this equation.

10.3.3 The 'Equivalent Earth Model' Approach of Munk and McDonald (1960)

Since the surface of the Earth is broken into a large number of individually rigid lithospheric 'plates' whose boundaries are in general weak, it should be the case that, at spherical harmonic degree 2, the effective $k_2^T(s = 0)$ will be close to k_f since in this limit the absence of strength at plate boundaries will

enable the planet as a whole to adjust to the tidal (rotational) forcing as if the planet had no surface lithosphere at all. This is highly plausible on physical grounds as $\lim_{s \to 0} sk_2^T(s)$, via the Tauberian theorems (eg., Widder, 1946), is identical to $\lim_{t \to \infty} k_2^T(t)$. That $k_2^T(s = 0)$ does tend toward k_f as the thickness of the lithosphere tends to zero will be clear on the basis of **Figure 6** (see also table 5 in Wu and Peltier (1984) and Mitrovica *et al.* (2005)). Based upon **Figure 6**, it will be clear that the deviation of $k_2^T(s = 0)$, in the limit of zero lithospheric thickness, from k_f is by less than 1% in the model of Earth's radial viscoelastic structure being employed for present purposes. This model has its radial variation of density and elastic Lamé parameters fixed to those of the preliminary reference earth model of Dziewonski and Anderson (1981) and its viscosity structure fixed to that of the VM2 model of Peltier (1996). In the hydrostatic equilibrium state that develops in the limit of long time under the action of the rotational forcing it is only the internal density field that matters as in this hydrostatic limit of no motion the viscosity structure becomes irrelevant. There is some subtlety here, however, as there is a small effect due to nonlinearity in the exact theoretical determination in the hydrostatic limit

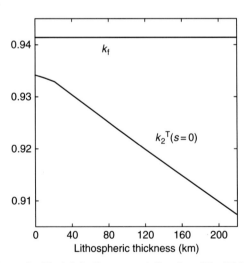

Figure 6 The infinite time asymptotic value of the tidal potential loading Love number of degree 2 is shown as a function of lithospheric thickness. This infinite time limit is identical to the limit in which the Laplace transform variable s equals zero. Also shown is the value of the fluid Love number k_f, which is a constant determined by the observed flattening of Earth's shape as measured by the difference between the polar and equatorial moments of inertia $(C - A)$. Note that the difference between $k_2^T(s = 0)$ and k_f in the limit of zero lithospheric thickness is <1%. See text for further discussion.

of the value of k_f (e.g., see Nakiboglu (1982)) and so it is not to be expected that a linear viscoelastic field theory for k_2^T will deliver an exact result for k_f in the limit $t \to \infty$. Since it is also the case that the observed flattening of the planet may be influenced to some (imperfectly known) degree by the nonhydrostatic dynamics of the mantle convection process, there is additional reason to expect that the prediction of linear viscoelastic field theory of the value of $k_2^T(s = 0)$ could not be precisely equal to the observed k_f. However, since the dynamical contribution due to mantle convection will be invariant on the timescale of the GIA process, this will be of no consequence insofar as the results of the application of first-order perturbation theory are concerned.

Following the 'equivalent Earth model' approach of Munk and McDonald (1960, see p. 28), as in Peltier (1982) and Wu and Peltier (1984), we proceed in a first approximation by adjusting the asymptotic properties of the viscoelastic Earth model by assuming that the infinite time asymptote of the viscoelastic tidal Love number of degree 2 is identical to the fluid Love number based upon the observed flattening of the planet. Although this may somewhat underestimate the degree of stabilization provided by the Earth's actual rotational bulge, it is nevertheless a useful first approximation. In eqn [24], we then take:

$$1 - \frac{k_2^T(s = 0)}{k_f} \equiv 0 \qquad [25]$$

Since the actual magnitude of the left-hand side of eqn (25) is extremely small, $k_2^T (s = 0)$ differing from k_f by less than 1% in the limit of zero lithospheric thickness, the 'equivalent Earth model' approach would appear to be well justified in the context of an analysis based upon the application of first-order perturbation theory (as will be discussed further below). In this 'equivalent Earth model' limit, in which [25] is assumed, eqn [24] becomes

$$
\begin{aligned}
\mathbf{m}(s) &= \frac{\Psi^L(s)}{\frac{s}{k_f} \sum_{j=1}^{N} \frac{(q_j/s_j)}{(s+s_j)}} \\
&= \left(\frac{\Omega_o}{A\sigma}\right) \frac{\left(1 + k_2^L(s)\right)}{\frac{s}{k_f} \sum_{j=1}^{N} \frac{(q_j/s_j)}{(s+s_j)}} \left(I_{13}^{\text{Rigid}}(s), I_{23}^{\text{Rigid}}(s)\right) \\
&= H(s)\left(I_{13}^{\text{Rigid}}(s), I_{23}^{\text{Rigid}}(s)\right)
\end{aligned}
\qquad [26]
$$

In order to accurately reconstruct the time domain solution, we need to rewrite the 'impulse response function' $H(s)$ in the above, following Peltier (1982), Wu and Peltier (1984), and Peltier and Jiang (1996), as

$$H(s) = \left(\frac{\Omega_o}{A\sigma_o}\right) \left[\frac{\lambda_s}{s} - \sum_{j=1}^{N} \frac{(q_j/s_j)}{(s+s_j)}\right] \frac{\prod_{j=1}^{N}(s+s_j)}{Q_{N-1}(s)} \qquad [27a]$$

where the degree $N-1$ polynomial Q_{N-1} with roots λ_i is such that

$$Q_{N-1} = \sum_{j=1}^{N} g_j \left[\prod_{i \neq j}(s+s_j)\right] = \prod_{i=1}^{N}(s+\lambda_i) \qquad [27b]$$

with

$$g_j = \frac{q_j'/s_j}{\sum_{j=1}^{N}(q_j'/s_j)} \qquad [27c]$$

is the relative strength of the jth mode of 'viscous gravitational relaxation', and the fully 'relaxed' Chandler wobble frequency of the viscoelastic Earth is defined to be

$$\sigma_o = \frac{\sigma}{k_f} \sum_{j=1}^{N} \frac{q_j'}{s_j} \qquad [28]$$

Using these definitions, the impulse response $H(s)$ may be further manipulated to write

$$
\begin{aligned}
H(s) &= \left(\frac{\Omega_o \lambda_s}{A\sigma_o}\right)\left[1 - \frac{q(s)}{sQ_{N-1}(s)}\right] - \left(\frac{\Omega_o}{A\sigma_o}\right) \\
&\quad \times \sum_{j=1}^{N} \frac{q_j}{s_j}\left[1 - \frac{R_j(s)}{Q_{N-1}(s)}\right]
\end{aligned}
\qquad [29a]
$$

where

$$q(s) = s\prod_{i=1}^{N-1}(s+\lambda_i) - \prod_{i=1}^{N}(s+s_i) \qquad [29b]$$

$$R_j(s) = \prod_{i=1}^{N-1}(s+\lambda_i) - \prod_{i \neq j}^{N-1}(s+s_i) \qquad [29c]$$

both of which are polynomials of degree $N-1$. With these factorizations, the Laplace transform of the impulse response, $H(s)$, may be directly inverted into the time domain. When this time domain impulse response is convolved with the time domain expression for the forcing ($I_{13}^{\text{Rigid}}(t)$ and $I_{23}^{\text{Rigid}}(t)$), we obtain the time domain solution of Peltier (1982) and Wu and Peltier (1984), namely:

$$
\begin{aligned}
m_j(t) = \left(\frac{\Omega_o}{A\sigma_o}\right) &\left\{ \left[\lambda_s - \sum_{j=1}^{N} \frac{q_j}{s_j}\right] I_{j3}^{\text{Rigid}}(t) \right. \\
&+ \left(\frac{-\lambda_s q(o)}{\prod_{i=1}^{N-1} \lambda_i}\right) \int_0^t I_{j3}^{\text{Rigid}}(t')dt' \\
&+ \left. \sum_{i=1}^{N-1} E_i e^{-\lambda_i t} * I_{j3}^{\text{Rigid}}(t) \right\}
\end{aligned}
\qquad [30a]
$$

in which

$$E_i = \left[\frac{\lambda_s q(-\lambda_i)}{\lambda_i} + \sum_{j=1}^{N} \frac{q_j R_j(-\lambda_i)}{s_j} \right] \Bigg/ \prod_{k \neq i}^{N-1} (\lambda_k - \lambda_i) \quad [30b]$$

To obtain the components of the polar wander velocity vector required to compare with the observed polar wander velocity, we simply differentiate the $m_j(t)$, which are the direction cosines of the polar motion, with respect to time, to obtain

$$\dot{m}_j(t) = \left(\frac{\Omega_o}{A\sigma_o} \right) \left[(1 + k_2^{LE}) I_{j3} - \frac{\lambda_s q(o)}{\prod\limits_{i=1}^{N-1} \lambda_i} I_{j3} \right.$$
$$\left. + \sum_{i=1}^{N-1} E_i \frac{d}{dt} \left[e^{-\lambda_i t} * I_{j3}(t) \right] \right] \quad [31]$$

where the equality

$$1 - \sum_{j=1}^{N} (q_j/s_j) = 1 + k_2^{LE}$$

has been employed. To complete the solution for the polar wander component of the rotational response, we need to specify the time series $I_{13}^{Rigid}(t)$ and

$I_{23}^{Rigid}(t)$. A specific model of these time-dependent perturbations of inertia will be described in the following subsection.

10.3.4 Models of the History of Variations in the Elements of Earth's Moment of Inertia Tensor on the Multimillenial Timescale of the Ice-Age Cycle

For the purpose of the results to be described in this chapter, the surface mass load forcing associated with the ice-age cycle of the Late Pleistocene will, for the most part, be taken to be that of the ICE-5G (VM2) model of Peltier (2004) although results will also be compared to those delivered by a version of the previous model ICE-4G (VM2) of Peltier (1994, 1996). **Figure 7** compares the contribution to the eustatic rise of sea level from LGM to the present for the ICE-5G model with that of a modified version of the precursor model ICE-4G. ICE-5G is characterized by a significant geographic redistribution of the surface mass load. The most important of these changes, which the interested reader will find discussed in detail in Peltier

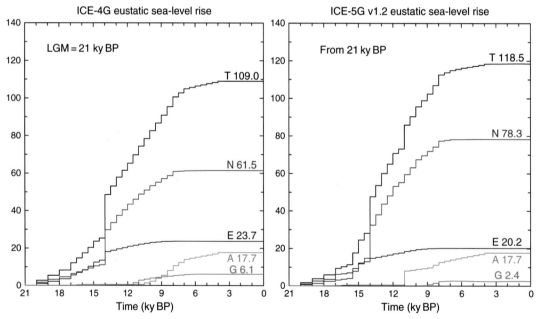

Figure 7 Eustatic sea level curves for the ICE-4G and ICE-5G V1.2 models of the last glacial–interglacial transition. Note that the most significant difference between these models concerns the net eustatic rise from LGM to present. For this version of the ICE-4G model, for which the reconstruction has been performed using the method described in Peltier (2005) in which the 'implicit ice' methodology is not employed, the net eustatic rise is 109.0 m. For the ICE-5G (N1.2) model, on the other hand, the net eustatic rise is 122.1 m. Also evident is the very significant increase in ice mass located over the North American continent (the component denoted 'N': E denotes Eurasia, A denotes Antarctica and South America, and G denotes Greenland). The increase in North American ice mass is partly accounted for by decreases in both Greenland and Eurasia. Also notable is the fact that in the ICE-5G V1.2 model the deglaciation of Antarctica is assumed to begin very abruptly at the time of occurrence of meltwater pulse 1b in the Barbados sea-level record. See the text for further discussion.

(2004), involves a significant shift in mass from Eurasia and Greenland toward the North American continent. It is expected that this will significantly impact both the direction and rate of TPW forced by the glaciation–deglaciation cycle. Also evident by inspection of **Figure 7** is the fact that the version of the ICE-4G model to be employed here has about 10% less LGM mass than the ICE-5G model, a consequence of the fact that it has been reconstructed using the methodology described in Peltier (2005) in which the previous 'implicit ice'-based formalism has been discarded in order that a meaningful full glacial cycle reconstruction could be employed in the present work.

Since the RSL data set from the island of Barbados has been employed to validate the total surface mass load, it will be useful to consider the extent to which the most recent of these models, the ICE-5G model of Peltier (2004), reconciles the observed sea level history at this site. **Figure 8**, which derives from the discussion of this issue in Peltier and Fairbanks (2006), compares the RSL history at this site with the prediction of the ICE-5G (VM2) model of the GIA process based upon eqn [10] and shown as the red curve. In the main portion of the figure, the actual Barbados data, which provide an excellent approximation to the eustatic (globally averaged) variation of sea level from LGM to the present (Peltier, 2002), are compared to both the ICE-5G prediction and the global ice equivalent eustatic sea-level curve proposed by Lambeck and Chappell (2001), which is represented by the color-coded crosses that the authors believed to be eustatic levels derived from the various locations listed in the figure Caption. It will be noted that the Lambeck and Chappell Barbados estimates, as well as those derived from J. Bonaparte Gulf (see Yokoyama *et al.*, 2000), are plotted approximately 20 m below the actual depth from which the samples from these sites were actually recovered. The LGM ice equivalent eustatic lowstand of the sea in the ICE-5G (VM2) model, assuming the modified LGM age of 26 ky proposed by Peltier and Fairbanks (2006) for the time of deepest glaciation, is approximately 122 m (see **Figure 8**, where the eustatic curve is shown as the step-discontinuous purple curve and the prediction of RSL history at this site is shown as the red curve), extremely close to the conventional oxygen-isotope-derived estimate of ~120 m (e.g., see Shackleton (2000)), which was based upon the assumption that the depth of the LGM lowstand was being accurately measured by the Barbados data. At the conventionally assumed LGM age of 21 ky, the eustatic depression of the sea is only 118.7 m in the ICE-5G model. It is a

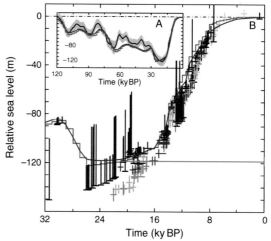

Figure 8 The fit of the predicted relative sea-level history at the island of Barbados to the extended coral-based data set from this location tabulated in Peltier and Fairbanks (2006). The blue symbols with error bars of various length represent these new Barbados data, the data represented by the shorter error bars of 5 m length derived from the *Acropora palmata* species of coral that provide the best constraints on sea level. The data represented by the error bars on intermediate 20 m length derive from the *Montastrea annularis* species of coral. The data represented by the longest error bars derive either from *Porites asteroides* species or the *Diploria* species. The green horizontal line denotes the 118.7 m depth level, which is the level corresponding to the samples of LGM if LGM is assumed to have occurred at the conventionally assumed age of 21 000 years (ago). In order to fit this observational datum, the eustatic depression of sea level at that age is almost precisely equal to the depth at which the sample of LGM age is found. This is a consequence of the fact that the Barbados record of relative sea-level is an excellent approximation to eustatic sea-level history itself. The color coding of the Lambeck and Chappell estimates, which are shown as the crosses, is as follows: cyan (Barbados), black (Huon Penninsula), gray (Tahiti), purple (Sunda Shelf). The inset to the figure shows the comparison between the eustatic history of the ICE-5G model and the complete 10^5 yr glacial cycle with that inferred by Waelbroecke *et al.* (2002) based upon benthic $\delta^{18}O$ records corrected for the influence of the change in abyssal ocean temperature.

further important property of the extended Barbados record of Peltier and Fairbanks that the data appear to strongly reject the hypothesis of Yokoyama *et al.* (2000) that a strong meltwater pulse occurred at 19 ka. In the ICE-5G reconstruction, such rapid melting events are taken to have occurred only at the times of meltwater pulses 1a and 1b originally identified by Fairbanks (1989).

Also evident by inspection of **Figure 7** is that the contribution to ice equivalent eustatic sea-level rise from the Antarctic continent in version 1.2 of the ICE-

5G (VM2) model is characterized by an abrupt onset at approximately 11 500 years BP. This late onset of Antarctic meltback has always been a characteristic of the ICE-NG sequence of models, it having been argued (e.g., see Peltier (2005)) that this region was the most likely source of the meltwater pulse 1b event that is now clearly evident in the extended Barbados sea level record discussed in Peltier and Fairbanks (2006). Very recently, this assumption in the ICE-5G and earlier reconstructions has been strikingly verified by accurately carbon-14 dating the age of the onset of shelf sedimentation from a large number of sites in coastal Antarctica that had been covered by grounded ice at LGM (e.g., see Domack *et al.* (2005)). These new observational results effectively rule out the claim in Clark *et al.* (2002) that the source of the earlier melt-water pulse 1a event was Antarctica. Peltier (2005) discussed additional evidence on the basis of which this suggestion must be considered implausible and Tarasov and Peltier (2005, 2006) have explicitly discussed the flaw in the Clark *et al.* (2002) suggestion. A sequence of predictions of RSL histories based upon the ICE-5G (VM2) model at a sequence of well-dated sites from Scotland derivative of isolation basin-based inferences of RSL (see Peltier *et al.* (2002) for a detailed discussion of these data) strongly reinforce the Antarctic origin of meltwater pulse 1b. As discussed in Peltier *et al.* (2002), the highly nonmono-tonic nature of these RSL curves, which are so well fit by the theoretical predictions, is entirely a conse-quence of the assumed late glacial melting event emanating from Antarctica and its abrupt onset as is characteristic of the ICE-5G v1.2 reconstruction.

The primary model of the glaciation–deglacia-tion process that is to be employed here for the purpose of analyzing Earth's rotational response to the ice-age cycle is therefore ICE-5G (VM2), the extension of which in the interval between the Eemian interglacial and LGM is based upon the SPECMAP record of Imbrie *et al.* (1984). In the inset to **Figure 8**, the complete ICE-5G record of the most recent glaciation–deglaciation cycle is compared with the reconstruction of Waelbroeck *et al.* (2002), whose reconstruction was based upon deep-sea $\delta^{18}O$ records corrected for the influence of the temperature variability of the abyssal ocean. Comparison of this record with that for the ICE-5G (VM2) model demonstrates that the two approx-imations to the ice equivalent eustatic sea-level history over the most recent glaciation–deglaciation cycle are extremely similar. In order to construct a complete surface loading model for use in the

computation of Earth rotation anomalies, however, a model that includes more than a single glacial cycle is required. Since approximately seven such cycles have occurred during the past 800 000 years of Earth history, for the purpose of the analyses to be discussed herein, it will be assumed that an approxi-mately100 ky cycle of glaciation and deglaciation has continued to operate over this same period.

Construction of the polar wander and l.o.d. solutions requires the time series $I_{13}^{\text{Rigid}}(t)$, $I_{23}^{\text{Rigid}}(t)$, and $I_{33}^{\text{Rigid}}(t)$, inputs to the calculation that may be computed from the definition of the I_{ij}^{Rigid}. This definition is

$$I_{ij}^{\text{Rigid}}(t) = \int\int v(\theta, \phi, t)(a^2\delta_{ij} - x_i x_j) \, ds \quad [32]$$

where the integral is over the surface of the sphere and where $v(\theta, \phi, t)$ is the surface mass load per unit area. Since this may be expressed in the form of a spherical harmonic expansion as

$$v(\theta, \phi, t) = \sum_{\lambda=0}^{\infty} \sum_{m=-\lambda}^{+\lambda} v_{\lambda m}(t) Y_{\lambda m}(\sigma, \phi) \quad [33]$$

it follows that

$$I_{13}^{\text{Rigid}}(t) + iI_{23}^{\text{Rigid}}(t) = -\left(\frac{32}{15}\right)^{1/2} \pi a^4 v_{21}(t) \quad [34a]$$

$$I_{33}^{\text{Rigid}}(t) = -\left(\frac{8}{3}\right) \pi a^4 \left[\left(\frac{1}{5}\right)^{1/2} v_{20}(t) - 2v_{00}(t)\right] \quad [34b]$$

An issue that immediately arises concerning the evaluation of these inputs to the theory concerns the way in which the component of the surface load is computed that is associated with the water that is extracted from or added to the oceans as the ice sheets grow and decay, respectively. Although the loading and unloading of the continents due to the growth and decay of the ice-sheets is an *a priori* specified input to the theory, the ocean loading component must be obtained self-consistently as a solution to the SLE [10]. However, this equation for the space-time history of the load on the ocean basins includes the influence of rotational feedback through the convolution of Ψ^R with G_{ϕ}^T. The problem of determining the rotational response to the ice-age cycle must therefore be deter-mined iteratively. We first compute a solution for the ω_i by neglecting this feedback effect. We next add the feedback terms into eqn [10] and recompute the sea-level history and thus the modification to the ocean load needed to determine the inertia perturbations required for the computation of the $\omega_i(t)$, etc. The iteration sequence is found to converge rapidly. It is also important to note that the ocean component of the surface load varies over a single glacial cycle in a way that depends somewhat upon the viscosity model of the Earth that is employed to represent the GIA process.

In order to complete the specification of the inputs for solution [26], we need to evaluate convolution integrals of the form

$$f(t) = e^{-\lambda_i t} * I^{\text{Rigid}}(t) = \int_{-\infty}^{t} I(t') e^{-\lambda_i (t - t')} dt'$$
$$= e^{-\lambda_i t} \int_{-\infty}^{t} I(t') e^{\lambda_i t'} dt \qquad [35]$$

These integrals may be computed to sufficient accuracy by assuming that the functions $I(t)$ consist of a series of periodic pulses, each of the form consistent with the ICE-5G (VM2) model of the most recent glacial cycle. Single-cycle integrals are computed numerically, and the impact of a large number of cycles is simply added with the appropriate phase delay (see Peltier (1982) and Wu and Peltier (1984) for an example in the special case that the pulse shape has a simple triangular form). In that case, the integrals may be evaluated analytically. In the general case treated here, they must be evaluated numerically. **Figure 9** compares time series for the ICE-5G (VM2) and ICE-4G (VM2) models for I_{13}^{Rigid}, I_{23}^{Rigid}, and I_{33}^{Rigid}. Given these models of the rotational forcing, we are in a position to compute the response in terms of both the l.o.d. and polar motion.

10.4 Observations of Millenial-Scale Secular Variations in Earth Rotation Anomalies

Although the existence of a secular trend, persisting over thousands of years, in the nontidal variations of the l.o.d., has been known for at least the last quarter of the twentieth century, and an equivalently important secular drift in the position of the pole has also become evident in the same period, debate has continued as to the cause of these secular variations in the rotation anomalies. Only through the development of the theory described in the last subsection of this chapter has it proven possible to unambiguously attribute both signals to the GIA process. Prior to reviewing the analyses that have established this result, it will be useful to briefly review the observations themselves.

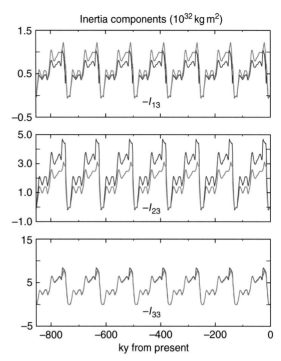

Figure 9 ICE-5G- (red) and ICE-4G- (green) based model histories for the variation of the I_{13}^{R}, I_{23}^{R}, and I_{33}^{R} components of the inertia tensor for a model history that assumes that seven cycles of glaciation and deglaciation have occurred with an approximate period of 100 ky. For the purpose of the calculations described in this chapter, Late Pleistocene glacial–interglacial cycle is assumed to have been precisely periodic. Since the system exhibits a fading memory of its past, this should not have a profound influence upon our conclusions.

10.4.1 Ancient Eclipse Observations and the Nontidal Acceleration of Rotation

That an anomaly in Earth's rate of axial rotation did exist, beyond the relatively well-understood decrease due to the action of the tidal friction that arises because of the frictional dissipation of the tide raised in the Earth's oceans by (primarily) the gravitational attraction of the Moon, was first suggested (e.g., Newton, 1972; Morrison, 1973; Muller and Stephenson, 1975) on the basis of analyses of ancient eclipse data extending back more than 2500 years before present. For much of this interval of time, in both Babylon and China, naked eye astronomers kept careful records of the timing and place of occurrence of total eclipses of the Sun and Moon. On the basis of the assumption that tidal friction had remained constant over this interval, a highly reasonable assumption as sea level has not changed appreciably over this interval of the Late Holocene period, one may accurately predict when and where a total eclipse of the Sun or Moon should have occurred. Analysis of such data, as described most recently in Stephenson and Morrison (1995) and Morrison and Stephenson (2001), clearly demonstrates that the further back in time for which an eclipse prediction is made, the larger is the error in timing, the sign of the error being such as to imply the action of a acceleration of rotation that is acting opposite to the effect of tidal friction. The compilation of data from the paper by Stephenson and Morrison (1995) is shown in **Figure 10**. Based upon the analysis of both lunar and solar eclipses that occurred between700 BC and AD 1600, these authors reported an increase in the length of the mean solar day (l.o.d) of (1.7 ± 0.5) ms per year, which implies a rate of decrease of the angular speed of rotation of $(-4.5 \pm 0.1) \times 10^{-22}$ rad s^{-2} on average over the past 2.7 ky. After subtracting the contribution due to tidal friction, which is accurately known based upon observations of the rate of recession of the Moon using Lunar Laser Ranging, Stephenson and Morrison (1995) inferred an average rate of nontidal acceleration of $(1.6 \pm 0.4) \times 10^{-22}$ rad s^{-2} over this period. This value for the nontidal acceleration of rotation

corresponds to a value for $\dot{\mathcal{J}}_2$ of $(-3.5 \pm 0.8) \times 10^{-11}$ yr^{-1}. As discussed in Section 10.2, this parameter provides a direct measure of the oblateness of planetary shape such that the larger is $\mathcal{J}_2$ the larger the oblateness. The fact that the time derivative is negative therefore implies that the nontidal acceleration of rotation is derivative of a secular decrease in the oblateness of figure and therefore of a decrease in the value of the polar moment of inertia. That this is plausibly a consequence of the GIA process follows from the fact that at glacial maximum the oblateness would have been increased by the increased surface mass load at the poles. Once this load is eliminated during the deglaciation process, this ice-age contribution to enhanced oblateness would have begun to relax as the planet returned to the more spherical shape characteristic of an interglacial period. Since angular momentum is conserved in the absence of external torques, the rate of axial rotation would have continued to increase, thus leading to the observed nontidal acceleration of rotation.

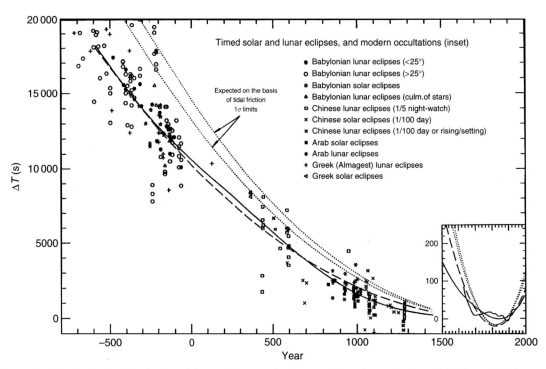

Figure 10   Results obtained for the time difference between the observed time of a total eclipse of the Sun or the Moon and the time at which the event was predicted to occur based upon the assumption of constant tidal torque. The inset (on a scale 25 times greater) includes the continuous curve derived frm the lunar occultations in the period AD 1620–1955.5, and the difference TAI-UT1 + 32.1845 from 1955.5 to 1990. The dashed curve is the best-fitting parabola 31 t^2 s. The solid curve is fitted using cubic splines. This curve, the best fitting parabola, and the parabola expected on the basis of tidal friction, are continued on the inset. From Stephenson ER and Morrison LV (1995) Long term fluctuations in the earth's rotation: 700 BC to AD 1990. *Philosophical Transactions of the Royal Society (London) Series A* 351: 165–202.

Beginning with the first publication and interpretation of data from the Laser Geodynamics Satellite (LAGEOS) by Yoder *et al.* (1983), which appeared simultaneously with the interpretation of the nontidal acceleration as a consequence of the GIA effect (Peltier 1982, 1983), the validity of the ancient eclipse-based inference of the nontidal acceleration was clearly established. Many estimates of the nontidal acceleration based upon the analysis of satellite laser ranging (SLR) data have now appeared (the effect observed is an acceleration in the rate of precession of the node of the orbit). The complete set of published estimates is shown in **Figure 11**, where the estimates of various authors are provided in terms of $\dot{J}_2$.

As a final comment upon the most recent research that has been conducted on l.o.d changes on millennial timescale, it is worth noting the work of Dunberry and Bloxham (2006) on the possibility of the existence of an oscillation in the l.o.d. that may be evident in the historical eclipse data set, an oscillation that the authors suggest may have a period of 1500 yr. A similar timescale is suggested to be evident from archeomagnetic artifacts, lake sediments, and lava flows (Daly and LeGoff, 1996; Hongre *et al.*, 1998; Korte and Constable, 2003). They interpret this as plausibly arising as a consequence of coupling between the outer core and the mantle across the core–mantle interface. It should be noted, however, that the existence of any such 1500 yr oscillatory component in the l.o.d. variations has been questioned by Dalmau (1997).

Although space limitations mitigate against providing a detailed review of an important very long timescale mechanism through which highly significant changes in the l.o.d. occur, namely through the action of tidal friction alone, it will be important to at least record the fact that such changes are large and that there exist extremely useful observations that can be brought to

bear upon the problem. Because of the dominant control that the semi-diurnal M2 tide in the oceans exerts upon coastal marine sedimentological processes, it has been possible on the basis of the study of 'tidal rhythmites' to infer what the l.o.d has been at certain epochs in the distant past. For example, in Williams (2000), analyses are described of South Australian sequences for the Late Neoproterozoic ($\sim$620 My BP) that indicate a rotational state characterized by 13.1 ± 0.1 synodic (lunar) months/yr, 400 ± 7 solar days/yr, and an l.o.d of 21.9 ± 0.4 h. He also infers on the basis of these data a mean rate of lunar recession since that time of 2.17 ± 0.31 cm/yr, slightly more than half the present-day rate of lunar recession of 3.82 ± 0.07 cm/yr obtained on the basis of Lunar Laser Ranging. Similar data from earlier epochs are also discussed in Williams review paper, which suggest that a close approach of the Moon to the Earth did not occur during the Proterozoic (2450–620 Ma). These results are clearly important for our understanding of the origin of the Moon and of the processes that control tidal friction. The current configuration of the ocean basins is such that tidal friction is considerably larger than it has been over much of the geological past as a consequence of the fact that the ocean basins are currently of a spatial scale such that a near-resonant condition exists between the normal modes of the ocean basins and the tidal forcing. The interested reader will find a highly informative discussion of the tides as free oscillations of the ocean basins in Platzman (1971). This is a fascinating subject about which there is yet a great deal to be learned.

10.4.2 Millenial Timescale Polar Wander and the Glaciation–Deglaciation Cycle

Insofar as the second of the secular anomalies in Earth rotation is concerned, namely the true wander of the North Pole of rotation relative to the surface geography, the data originally obtained by the International Latitude Service that have been employed to constrain it are shown on **Figure 12**. The data initially tabulated by Vincente and Yumi (1969, 1970) were derived on the basis of observations of star transits using a Northern Hemisphere array of photo-zenith tubes. The data consist of time series of the x and y coordinates of the position of the North Pole of rotation relative to a reference frame fixed to the Earth's crust with origin at the Conventional International Origin (CIO). This reference frame is shown on the inset polar projection in **Figure 12** and has its x-axis aligned with the Greenwich meridian. Inspection of the x and y time series demonstrates

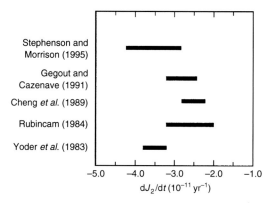

Figure 11 Published estimates of the parameter $\dot{J}_2$.

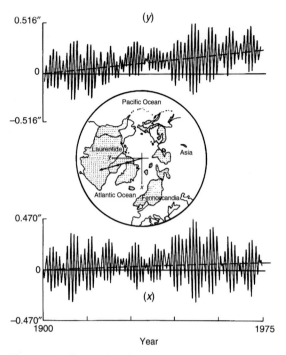

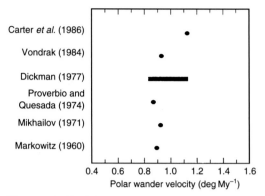

Figure 13 Published estimates of the speed of polar motion based primarily upon the ILS data.

Figure 12 Time series of the location of the North Pole of rotation of the planet, as x and y coordinates relative to the coordinate system shown on the inset polar projection with origin at the Conventional International Origin (CIO). These data are those of the International Latitude Service (Vincente and Yumi, 1969, 1970) and they reveal a secular drift of the pole at arate near 0.95° per million years approximately along the 75.5° west meridian of longitude.

them to be dominated by an oscillatory pattern of beats with a period of approximately 7 years. This pattern is well understood to arise as a consequence of the interference between the 12 month periodic annual wobble and the 14 month periodic Chandler wobble that were discussed in Section 10.2. Based upon the review of our understanding of these features of Earth rotation history discussed therein, it would appear that previous controversies concerning their excitation have now been satisfactorily resolved. According to Dickman (1977), the TPW, evident as the secular drift of the position of the pole upon which the oscillatory signal is superimposed, is occurring at a rate of $(0.95 \pm 0.15)°$ per million years along the 75.5° west meridian as shown on the inset polar projection in **Figure 12**. An unambiguous connection of this Earth rotation anomaly to the GIA process was first suggested in Peltier (1982) and Wu and Peltier (1984), whose detailed theoretical analysis, since further generalized in Peltier and Jiang (1996), showed that the initial analysis by Munk and MacDonald (1960) of the

excitation of polar wander by surface loading of a viscoelastic model of the Earth did not in fact rule out an explanation in terms of the GIA process even though the surface mass load during the Late Holocene interglacial period has been very nearly time invariant. The analysis of the data by Dickman (1977) was preceded by several earlier analyses and has been followed by others. Most of these results are shown on **Figure 13**, where they are accompanied by an early result derived on the basis of the analysis of VLBI observations by Carter et al. (1986).

The most recent analysis of the International Latitude Service (ILS) data is that of Argus and Gross (2004), who have investigated the possibility that all of these interpretations may be contaminated by the failure to account for the influence of surface plate-tectonic motions. Their analysis of the raw data delivered a best estimate of the speed of polar wander of 1.06° per million years. Depending upon how this estimate was corrected for the influence of plate tectonics, however, considerably different results could be obtained, as listed in **Table 1** of that paper. Whereas the ILS inference uncorrected for plate motion was that the ongoing polar wander was along the 75.5° west meridian, if the same data set is corrected by making the inference in the frame of reference in which the lithosphere exhibits no net rotation, then the corresponding speed and direction change slightly to 0.98° My^{-1} southward along the 79.9° west meridian. However, if the correction to these data is based upon the 'hot-spot frame', then one obtains, from the ILS data, according to Argus and Gross (2004), the values 1.12° My^{-1} for the speed and southward along the 69° west meridian for the direction, a significant difference. The fact that all of these inferences of polar wander direction

Table 1 Predictions of the polar wander speed and direction for the ICE-4G and ICE-5G models of the Late Pleistocene glaciation cycle, both including and excluding the influence of rotational feedback

	Observation	ICE-4G (N)	ICE-4G (R)	ICE-5G (N)	ICE-5G (R)
Polar wander speed (deg My^{-1})	0.8–1.1	0.81	0.57	1.06	0.72
Polar wander direction (degrees west longitude)	75.5	69.9	73.4	83.1	88.8

suggest it to be moving 'toward' what was the centroid of the ancient Laurentide ice sheet is nevertheless highly significant insofar as the explanation in terms of glacial isostasy is concerned. Since the constraint of angular momentum conservation requires that the pole must drift so as to align itself along the axis relative to which the moment of inertia is largest, it is clear why, as a consequence of the GIA effect, the pole should now be moving toward what was the centroid of the greatest concentration of land ice. Following deglaciation, there exists a residual depression of the land that is yet to be eliminated by the slow viscous process of GIA. This region therefore constitutes a localized deficit of mass relative to the state of gravitational equilibrium that will eventually be reached once the GIA-related relaxation of shape has been completed. It is toward this regional mass deficit that the pole must move in order to come into alignment with the axis of greatest inertia. That it need never reach and in general will not reach this region is simply a consequence of the fact that the speed of polar wander decreases as a function of time as the shape relaxation proceeds at a rate determined by mantle viscosity.

The question of the accuracy of the direction of the secular drift of the pole that is derived on the basis of the ILS data may be at least as important as the speed. The reason for this has to do with the use of the rotational anomalies to constrain the rate of melting of polar ice that may be occurring at present due to the action of greenhouse gas-induced global warming. Depending upon the polar wander prediction due to the continuing action of the GIA effect, there will exist a residual between this prediction and the modern observations that may be employed to constrain the rate and geographical locations of modern sources of land ice melting (Greenland, Antarctica, small ice sheets and glaciers) as previously discussed in Peltier (e.g., 1998, see figure 46). Further comment on this issue follows in the next section, in which the theory reviewed in Section 10.3 will be applied to understanding the observational constraints upon polar wander described in this section.

10.5 Earth's Rotational Response to the Cyclic Reglaciation Cycle of Late Pleistocene Time: Data-Model Intercomparisons

Based upon the discussion in Section 10.3, it will be clear that accurate predictions of RSL history are a required preliminary to accurate predictions of the rotational response to the GIA process and thus to the understanding of Earth rotation history on millennial timescales. Since the best test of the accuracy of RSL history predictions based upon solution of the SLE is provided by radiocarbon-dated RSL curves from coastal locations, it will be clear that the confrontation of theory with observational 'truth' requires a voluminous compilation of such records, individual examples of which have been produced by innumerable scientists working in the area of Pleistocene geomorphology over the past half century. That a global database of such records is expected to record a wide range of signatures of the GIA process will be clear on the basis of **Figure 14**, from Peltier (2004), which shows predictions of the present-day rate of RSL rise for the ICE-4G (VM2) model both (a) including and (b) excluding the influence of rotational feedback. The difference between these predictions (a−b) is shown in **Figure 14(c)**, which establishes the dominant role that the polar wander component of the rotational response to the GIA effect plays in contributing to sea-level history. The 'signature' of this feedback is the existence of the strong degree 2 and order 1 spherical harmonic component, a quadrapolar distribution that is the expected form of this feedback based upon eqns [13] in Section 10.3. **Figure 14(d)** shows the sum of the signal shown in **Figure 14(a)** for the present-day rate of RSL rise and the prediction for the same model of the present-day rate of radial displacement of the crust. As first discussed in Peltier (1999), this sum represents the time rate of change of geoid height due to the GIA process, a field that is measured by the Gravity Recovery and Climate Experiment (GRACE) dual-satellite system (see

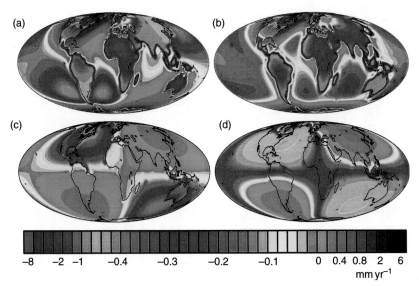

Figure 14 The ICE-4G (VM2) model prediction of (a) the present-day rate of sea-level rise relative to the deforming surface of the solid Earth, including the influence of rotational feedback; (b) same as (a) but excluding the influence of rotational feedback; (c) the difference between (a) and (b) illustrating the spherical harmonic degree 2 and order 1 pattern that dominates the influence due to rotational feedback on account of the control exerted by the polar wander effect; (d) the present-day geoid height time dependence predicted by the ICE-4G (VM2) model, obtained by adding to the field in (a) the prediction of the time rate of change of radial displacement of the surface of the solid Earth with respect to the center of mass of the planet.

Figure 15) that is now in space. In order to test the accuracy of predictions of the Earth's rotational response to the GIA process produced by the theory discussed in Section 10.4, an appropriate strategy is therefore to focus upon the observational constraints provided by geological recordings of postglacial RSL change through the Holocene interval of time in those regions in which the impact of rotational feedback is expected to be most intense.

10.5.1 A Database of Holocene RSL Histories

A location map of the positions on Earth's surface from which ^{14}C-dated RSL histories are available in the University of Toronto database is shown in **Figure 16** (from Peltier, 1998). Comparing the coverage provided by these many hundreds of individual RSL curves with the geographically intricate pattern evident in **Figure 14**, it will be clear that there exists, not surprisingly, much better coverage of this signature of the GIA process in the Northern Hemisphere than is available from the Southern Hemisphere. Yet, as will be made clear in what follows, the role of the available southern hemisphere data in confirming the

important role that variations in Earth rotation play in forming the detailed signature of the GIA effect evident in Holocene RSL records is extremely important. **Figure 17**, from Peltier and Fairbanks (2006), intercompares a number of ^{14}C-dated RSL records from this database with the predictions of the ICE-5G (VM2) model that served as basis for the computation of the time series of the variations in the elements of the moment of inertia tensor that are required in the computation of the rotational response to the glaciation–deglaciation cycle. The locations of the individual sites are shown on the map of the prediction of the present-day rate of sea-level rise predicted by the ICE-5G (VM2) model of the GIA process.

10.5.2 The Influence of Rotational Feedback upon Postglacial RSL History and Its Impact upon Predictions of Earth Rotation Anomalies

Although the purpose of the present subsection is to explore the extent to which the most recent models of the GIA process may be employed to further reinforce the interpretation of the millenial timescale

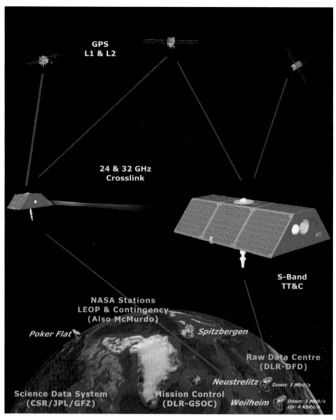

Figure 15 Schematic of the dual-satellite Gravity Recovery and Climate Experiment (GRACE) experiment designed to measure the time dependence of the gravitational field of the planet.

variations in Earth rotation anomalies in terms of the GIA process, it will be useful to begin by revisiting the results previously obtained by Peltier and Jiang (1996) who employed the ICE-4G deglaciation model of Peltier (1994, 1996) to investigate the extent to which both of the previously discussed anomalies were explicable in these terms. Their results are summarized in **Figure 18**, which shows the theoretical predictions of $\dot{J}_2$ and polar wander speed as a function of the viscosity of the lower mantle of the Earth below a depth of 660 km, with the value of the viscosity of the upper mantle fixed to the nominally accurate value of 10^{21} Pa s, and the thickness of the surface lithosphere taken to be $L = 120$ km. Inspection of **Figure 18** demonstrates that both of the millennial timescale secular anomalies are fit by the same simple model of the radial variation of mantle viscosity. With the upper-mantle viscosity fixed to the above value, the lower-mantle viscosity required to fit the observations is approximately 2×10^{21} Pa s. It might be seen as unlikely, given that the two rotational anomalies depend upon

entirely independent elements of the moment of inertia tensor, that this degree of agreement would exist if GIA were not the primary explanation for both observations.

Since publication of these results, further refinements of the required analyses have been performed. In particular, in Peltier (1996), the VM2 model of the radial viscosity structure was inferred on the basis of a full Bayesian inversion of the available data (see **Figure 19**). Equally important has been the considerable further refinement of the loading history that is represented by the ICE-5G model of Peltier (2004) and previously introduced in Section 10.3. **Table 1** lists the predictions of present-day polar wander speed and direction for the ICE-4G (VM2) and ICE-5G (VM2) models, assuming the validity of the VM2 model of the radial viscosity structure and a lithospheric thickness of $L = 90$ km, using versions of the theory that both include and exclude the influence of rotational feedback in the solution of the SLE that constitutes the first step in the analysis procedure. The most accurate predictions are of course those that

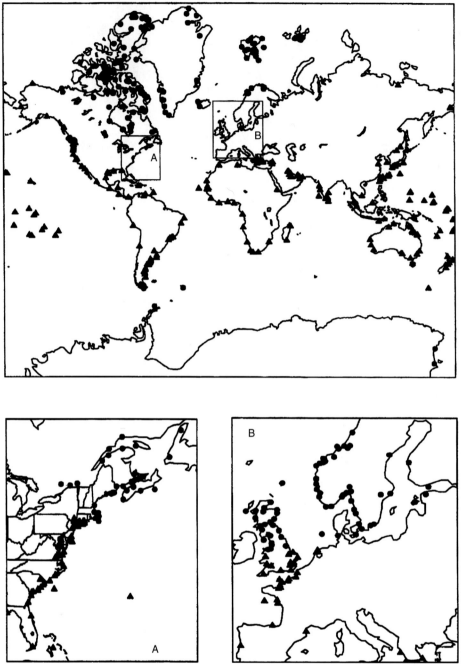

Figure 16   Global location map of the sites from which ^{14}C-dated time series of RSL history are available in the University of Toronto database.

include this critical feedback effect. Inspection of the results in **Table 1**, in which the observed values of the computed quantities are also listed according to the original analyses of the ILS data, will reveal a number of interesting properties of the TPW solutions (since the polar wander component of the rotational response is the most complicated to calculate, the initial focus here will be upon this element).

First, concerning the predicted speeds of TPW, these results demonstrate that in the case of no rotational feedback the predictions of ICE-4G (VM2) and ICE-5G (VM2) are respectively on the lower and upper bounds of the observed speed of ~0.8–$1.1°$ My^{-1} according to the analysis of Dickman (1977). When the influence of rotational feedback is included, however, the prediction of the ICE-4G (VM2) model is

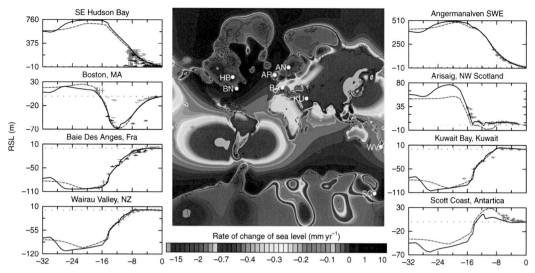

Figure 17 Illustrative suite of predictions of the history of RSL change at eight sample locations using the ICE-5G (VM2) model of the GIA process. The locations of the individual sites are superimposed upon a Mercator projection of the predicted present-day rate of RSL rise. Predictions are shown for two slightly different versions of the model that differ from one another only in terms of the assumption made concerning the timing of LGM. For one version of the model, LGM is assumed to have occurred at the conventional age of 21 ka; for the other, it is assumed, as illustrated in **Figure 7**, that LGM occurred considerably earlier at 26 ka.

reduced to $0.57°$ My^{-1}, too slow to fit the observed speed. The action of the same feedback also reduces the predicted speed for the ICE-5G (VM2) model to a value of $0.72°$ My^{-1} that is also slightly too slow to be acceptable.

Insofar as the predictions of polar wander direction are concerned, the predictions of the two models in the absence of rotational feedback are directed somewhat eastward of the observed direction of $\sim75.5°$ west longitude, at $69.9°$ west longitude, in the case of the ICE-4G (VM2) model, and somewhat westward of the observed direction, at $83.1°$ west longitude, in the case of the ICE-5G (VM2) model. When the influence of rotational feedback is included, the polar wander direction predicted by both models is shifted further westward. For ICE-4G (VM2) the shift is from $69.9°$ west longitude to $73.4°$ west longitude, whereas for ICE-5G (VM2) the shift is from $83.1°$ west longitude to $88.8°$ west longitude. That the predicted direction of present-day polar wander for the ICE-5G (VM2) model lies significantly to the west of the prediction for the ICE-4G (VM2) model is entirely expected on the basis of the fact previously mentioned and documented in Peltier (2004), that the primary difference between ICE-4G and ICE-5G is that the distribution of continental ice mass has been significantly shifted from the glaciated sites east of the North American continent to the North American continent itself.

In order to demonstrate the sensitivity of these predictions to certain special characteristics of the assumed loading history, results are shown in **Table 2** equivalent to those shown in **Table 1** for models designed to explore the sensitivity associated with two such characteristics. The first of these has to do with an assumption made in constructing the most widely distributed version 1.2 of the ICE-5G model, that the ice cover on Greenland continued to expand in the years following the mid-Holocene climate optimum. Although the initial Neoglacial expansion is required to understand the records of RSL history from southwest Greenland (see Tarasov and Peltier, 2002), version 1.2 of ICE-5G assumed that this Neoglacial expansion continued up to the present which is not required by the observations and is incompatible with modern measurements of the mass balance of this ice sheet (e.g., see Krabil *et al.* (2000)). If we remove the continuing addition of mass from this region that was assumed to be occurring from 2 ka onward, then the new results we obtain for polar wander speed and direction presented in **Table 2** are obtained. In the absence of rotational feedback, the speed now increases from $1.06°$ My^{-1} to $1.17°$ My^{-1}. With rotational feedback, the speed prediction changes from $0.72°$ My^{-1}, too slow to fit the observations, to $0.84°$ My^{-1}, which is now compatible with the observations. This demonstrates an extremely important characteristic of the planet's

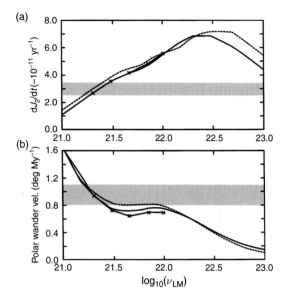

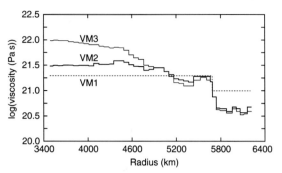

Figure 19 Viscosity models VM1, VM2, and VM3 originally inferred in Peltier (1994, 1996) on the basis of full Bayesian inversions of a subset of the observational constraints pertaining to the GIA process. Model VM1 was the starting model employed to initiate the Bayesian inversion that employed the methodology of Tarantola and Valette (1982a, 1982b). Control on the viscosity in the deepest mantle was provided solely by the rotational constraint provided by the observation of the parameter $\dot{J}_2$ as inferred by Cheng et al. (1989) in the analysis that delivered VM2 from the inversion. Model VM3 was obtained in an inversion that adjusted the observed value of $\dot{J}_2$ so as to allow for the influence of possible contamination due to the influence of modern climate change induced forcing due to melting from either the Greenland or Antarctic ice sheets. This inversion was based upon the assumption that one or the other of these was contributing to the current rate of global sea-level rise at the rate of 1–1.5 mm yr^{-1}.

Figure 18 (a) $\dot{J}_2$ as a function of lower mantle viscosity with the upper-mantle viscosity held fixed to the value of 10^{21} Pa s. (b) Same as (a) but for polar wander speed. In each frame, the shaded region represents the observationally constrained range according to the analysis of Dickman (1977). Note that the millennium timescale secular anomalies in Earth rotation are 'explained' as a consequence of the GIA process by the same model of the radial variation of mantle viscosity. From Peltier WR and Jiang X (1996) Glacial isostatic adjustment and Earth rotation: Refined constraints on the viscosity of the deepest mantle. *Journal of Geophysical Research* 101: 3269–3290 (correction, *Journal of Geophysical Research* 102: 10101–10103, 1997).

response to surface loading and thus to the understanding of the history of the evolution of its rotational state, namely its extreme sensitivity. Insofar as the predicted polar wander direction is concerned, the pronounced westward shift of direction characteristic of ICE-5G v1.2 is now significantly subdued. In the no rotational feedback

case, the direction shifts from 83.1° west longitude to 78.7° west longitude. With feedback, the shift is even more pronounced from 88.8° west longitude to 80.4° west longitude, results that are in much better accord with the observations, especially when the looser bounds allow by the analysis of Argus and Gross (2004) are taken into account.

The further results presented in **Table 2** explore the sensitivity to present-day melting of the Greenland ice sheet under the assumption that this system, after having ceased to expand by 2 ka, begins

Table 2 Predictions of the polar wander speed and direction for the ICE-5G models of the Late Pleistocene glacial cycle when the Neoglacial readvance of the Greenland Ice Sheet is assumed to cease at 2 ka and for a further variant in which, although the Neoglacial readvance ends at 2 ka, the ice sheet begins to lose mass at the rate of 0.1 mm yr^{-1} eustatic equivalent beginning 100 years before present

	Observation	ICE-5G (N) Neoglacial ends at 2 ka	ICE-5G (R) Neoglacial ends at 2 ka	ICE-5G (N) Modern mass loss at 0.1 mm year^{-1}	ICE-5G (R) Modern mass loss at 0.1 mm year^{-1}
Polar wander speed (deg My^{-1})	0.8–1.1	1.17	0.84	1.26	0.94
Polar wander direction (degrees west longitude)	75.5	78.7	80.4	75.1	74.2

to lose mass beginning 100 years ago at a rate equivalent to a rate of rise of ice equivalent eustatic sea level of 0.1 mm yr^{-1}. Since the modern rate of global warming-induced sea-level rise is now well understood on the basis of the analysis of tide gauge records covering the last 70 or so years of Earth history to be on the order of 1.85 mm yr^{-1} (e.g., see Peltier (2001)), the assumption of an 0.1 mm yr^{-1} equivalent rate of mass loss from Greenland must be considered modest. Inspection of the results listed in the last two columns of **Table 2** will show that the addition of this forcing to the system, in the case of no rotational feedback, further increases the predicted speed of TPW from 1.17° My^{-1} to 1.26° My^{-1}, both of which are too fast to fit the ILS observations. When rotational feedback is included, however, the speed once more increases, but now from 0.84° My^{-1} to 0.94° My^{-1}, both of which are fully compatible with the ILS-based observations. Insofar as the polar wander direction predictions are concerned, this swings from somewhat west of the observed direction (78.7° west longitude) to very close to the observed direction (75.1° west longitude). These results further reinforce the previous message concerning the extreme sensitivity of the rotational observables to surface mass loading and unloading associated with continental ice mass variations. Because of this sensitivity, these observables may provide an important means of inferring the locations and rates of present-day land ice melting, especially if one may assume that the influence of GIA due to the Late Pleistocene ice-age cycle may be modeled with sufficient accuracy.

Predictions of the nontidal acceleration of planetary rotation for the same sequence of models are listed in **Table 3**. The versions of the calculation for which results are listed are the same four versions as were employed to produce the list of results for polar wander in **Table 2**. In this case, however, two distinct methods were employed to calculate the results for $\dot{J}_2$ listed in the table. The first of these methods is the same as that employed in Peltier and Jiang (1996; see their equation 37). An alternative method is that based upon exploitation of the connection between $\mathcal{J}_2$ and the planet's gravitational field. The GIA-induced time dependence of geoid height that is currently being measured by the GRACE satellite system may be computed based upon recognition of the fact (see Peltier, 1999) that this signal, denoted say by $\dot{G}(\theta, \lambda, t)$, is identically equal to the sum $\dot{R}(\theta, \lambda, t) + R\dot{S}L(\theta, \lambda, t)$ in which the first term is the time rate of change of the local radius of the solid Earth and the second term is the time rate of change of RSL. Since the instantaneous level of the sea is determined by the gravitational equipotential that is best fit to it and since the geoid is, by definition, this equipotential surface, the validity of this expression is clear. Now since the geoid height time dependence may also be expanded in spherical harmonics, it may be expressed as

$$\dot{G}(\theta, \lambda, t) = \sum_{l=0}^{\infty} \sum_{m=0}^{l} \dot{G}_{lm}(t) Y_{lm}(\theta, \lambda) \qquad [36]$$

Table 3 $\dot{J}_2$ (equivalent to the nontidal acceleration of rotation) predictions for the ICE-5G (VM2) model for which the Neoglacial readvance of Greenland ice is eliminated subsequent to 2 ka. Also shown are results for a version of the model that includes Greenland melting at a rate of 0.1 mm yr^{-1} beginning 100 years before present. Results are shown for two different methods of performing the calculation

	Observed	ICE-5G (N) Neoglacial readvance of Greenland ceases after 2 ka	ICE-5G (R) Neoglacial readvance of Greeland ceases after 2 ka	ICE-5G (N) Includes the influence of Greenland melting at the rate of 0.1 mm yr^{-1}	ICE-5G (R) Includes the influence of Greenland melting of the eustatic equivalent rate of 0.1 mm yr^{-1}
$\dot{J}_2 \times 10^{11}$, yr^{-1}, using the method of Peltier and Jiang (1996)	-2.7 ± 0.4	-3.22	-3.23	-2.83	-2.84
$\dot{J}_2 \times 10^{11}$, yr^{-1}, using the new method of Section 10.3.4	-2.7 ± 0.4	-4.05	-3.97	-3.71	-3.63

However, based upon the definition of the nondimensional quantities $\mathcal{J}_{lm}$, we may also write

$$\dot{G}(\theta, \lambda, t) = -a \sum_{l=2}^{\infty} \sum_{m=0}^{l} (\dot{\mathcal{J}}_{lm}^{1} \cos m\lambda + \dot{\mathcal{J}}_{lm}^{2} \sin m\lambda) P_l^m (\cos \theta) \qquad [37]$$

Using the conventional normalization for the Legendre polynomials P_l^m, comparison of eqns [36] and [37] then requires the following result for $\dot{\mathcal{J}}_2$:

$$\dot{\mathcal{J}}_2 = \dot{\mathcal{J}}_{20}^{1} \qquad [38]$$

In the second line of $\dot{\mathcal{J}}_2$ data in **Table 3**, the numbers listed are obtained using eqn [38]. In the first line they have been obtained using the Peltier and Jiang formulation. Inspection of the results in **Table 3** will show that the two methods have delivered somewhat different results, differences that are typically on the order of 20–25%, a difference that is similar in magnitude to the errors that have been ascribed to the measurement by the various authors that have performed the analysis. Although the observed estimates listed in **Table 3** are those of Cheng *et al.* (1989), those obtained by several other authors are somewhat higher as follows:

$$\dot{\mathcal{J}}_2 = (-3.5 \pm 0.3) \times 10^{-11} \, \mathrm{yr}^{-1}, \quad \text{Yoder } et\ al.\ (1983)$$

$$\dot{\mathcal{J}}_2 = (-3.5 \pm 0.8) \times 10^{-11} \, \mathrm{yr}^{-1},$$
$$\text{Stephenson and Morrison (1995)}$$

Of the two methods employed to produce the $\dot{\mathcal{J}}_2$ estimates in **Table 3**, the most accurate should be that based upon the expansion of the gravitational field. Intercomparing the results in **Table 3**, one would therefore be led to conclude that the magnitude of the predictions for the ICE-5G (VM2) model, even in the best case in which the 0.1 mm yr^{-1} eustatic rise due to Greenland Ice Sheet melting is active, is too high if one accepts the estimate of Cheng *et al.* as the most accurate and even if one were to replace it by the number preferred in the more recent paper by Dickey *et al.* (2002). If one were to prefer the estimates of Yoder *et al.* or Stephenson and Morrison, however, the prediction for this model would be well within the observed bounds. Accepting the result of Cheng *et al.* as most accurate, the misfit may however be eliminated simply by slightly increasing the rate at which Greenland ice is assumed to be melting at present to, say, 0.3 mm yr^{-1}. By inspection of the results in **Table 2**, this would be

expected to increase the rate of polar wander by an additional $\sim 0.2°$ My^{-1}, leading to a polar wander prediction that remains within the error bounds on this observable.

It is important to understand, however, that Greenland is not the only contributor to the modern rate of rise of sea level that is associated with greenhouse gas-induced global warming. As documented most recently by Dyurgerov and Meier (2005), the small ice sheets and glaciers of the planet are also melting at rates such that they are collectively contributing to the ongoing rise of global sea level in the amount of approximately 0.5 mm yr^{-1}. In order to employ the rotational observables to accurately constrain the rate at which the great polar ice sheets are currently losing mass will require that this contribution be carefully taken into account using the methodology employed previously in Peltier (1998, see figure 46). The refined ICE-5G (VM2) model of the Late Pleistocene ice-age cycle will provide an important basis upon which a more accurate application of this methodology will be possible. Because the measurements of the rates of change of the rotational anomalies are essentially instantaneous on the millennial timescale characteristic of the GIA process, it is nevertheless important that the accuracy of our association of both anomalies with GIA be tested over a more extended range of time. As demonstrated in the following subsection, Holocene records of RSL history have enabled such a definitive test to be performed.

10.5.3 Measurements of the Strength of the Expected Quadrapolar 'Signature' of the Rotational Feedback Effect upon Postglacial Sea-Level Histories

In order to adequately test the veracity of the theory that has been developed to compute the Earth's rotational response to the glaciation–deglaciation cycle, and thereby to adequately test our understanding of the history of Earth rotation on millennial timescales, the use of the global patterns of postglacial RSL change predicted by the different versions of the calculation might be expected to be especially useful. As previously, these patterns are best illustrated by global predictions of the present-day rate of RSL rise that would be expected to be observed as a secular rate of change on a tide gauge installed in a coastal location if the only contribution to RSL history were the continuing impact of the GIA process. **Figure 20** shows Mollwiede projections of the global map of

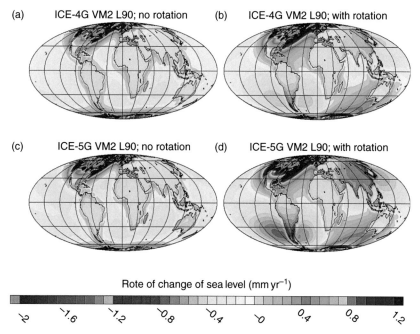

(a)   ICE-4G VM2 L90; no rotation   (b)   ICE-4G VM2 L90; with rotation

(c)   ICE-5G VM2 L90; no rotation   (d)   ICE-5G VM2 L90; with rotation

Rote of change of sea level (mm yr⁻¹)

-2 -1.6 -1.2 -0.8 -0.4 -0 0.4 0.8 1.2

Figure 20 This compares the predictions of the ICE-4G (VM2) and ICE-5G (VM2) models of the present-day rate of RSL rise, both including and excluding the influence of rotational feedback as described by the second term in the integrand of the triple convolution integral in eqn [1]. Noticeable is the fact that the influence of rotational feedback has the form of a spherical harmonic of degree 2 and order 1 which exists as a consequence of the dominant role played in the feedback process by the polar wander component of the rotational response to the GIA process (see eqns [4c] and [4d]). Also evident is the fact that this feedback is stronger in model ICE-5G (VM2) than it is in ICE-4G (VM2).

this prediction for the ICE-4G and ICE-5G models, both including and excluding the impact of rotational feedback. These calculations have been made by solving the SLE using the 'full glacial cycle' methodology for incorporation of the influence of coastline migration discussed in Peltier (2005) and Peltier and Fairbanks (2006). Inspection of these figures further reinforces the fact that the impact of rotational feedback is such as to superimpose upon the pattern predicted by either model, a modification that has the form of a spherical harmonic of degree 2 and order 1, a pattern of 'quadrapolar' form. This characteristic of the RSL response to variations in Earth rotation was first demonstrated by Dahlen (1976) in the context of his analysis of the 'pole-tide' raised in the oceans by the Chandler wobble of the Earth. Initial attempts to incorporate the influence of rotational feedback upon the variations of sea level caused by the GIA process were described by Milne *et al.* (1996) and Peltier (1998), who employed the theory of Peltier (1982) and Wu and Peltier (1984) to estimate the magnitude of the effect, analyses that initially suggested it to be small enough to be neglected. The fact that the influence of rotational feedback has this quadrapolar structure is a

consequence of the fact that the dominant impact of rotational feedback is due to the influence of TPW, as will be clear on the basis of inspection of eqns [13]. The ω_3 (nontidal acceleration of rotation related component) of the perturbation to the angular velocity of the planet controls the impact upon sea-level history of the GIA-induced change in the l.o.d. The ω_1 and ω_2 perturbations, on the other hand, control the polar wander contribution and it will be clear on the basis of eqns [13] that this will appear as a forcing of spherical harmonic degree 2 and order 1 form. Noticeable also on the basis of **Figure 20** is the fact that the magnitude of this impact is more intense in the ICE-5G (VM2) model than it is in the ICE-4G (VM2) model, a consequence of the fact, evident from **Table 1**, that the polar wander speed predicted for this model is higher than for ICE-4G (VM2), even though the two models have very similar LGM surface mass loads of glacial ice (although there is a 10% difference in LGM mass between the version of ICE-4G being employed here and ICE-5G, the difference in the polar wander speed predicted by the two models is 20%). This difference can be exploited to examine the extent to which the feedback effect is properly incorporated in the theory based upon the 'equivalent Earth model' approach of Munk and MacDonald (1960). Since it is in the regions centered within the four 'bull's-eyes' of the spherical harmonic degree 2 and order 1 quadrapole pattern that the influence of rotational feedback will be most apparent, it is in these regions that we should expect to find its consequences most clearly revealed.

The most important region in which such inter-comparison is possible concerns the extremum in rotational feedback that is located on the southern tip of the South American continent (see **Figure 14**). Although the west coast of this continent is strongly impacted by the deformation associated with the sub-duction of the Nazca plate, the east coastal margin is passive and therefore should provide an excellent region in which to attempt to measure, using postgla-cial histories of RSL change, the impact of rotational feedback. **Figure 21** shows a map of the South American continent, from Peltier (2005), on which, primarily along the east coast passive continental mar-gin, are shown the locations at which, based upon the compilation of RSL histories in Rostami *et al.* (2000), radiocarbon-dated RSL history information is avail-able. When such data are used to compare with the predictions of the theory previously described, they must of course be transformed onto the calendar-year timescale using the modern calibration procedure CALIB described by Stuiver and Reimer (e.g., 1993). The site numbers on the map of **Figure 21** correspond to the last two digits of the site numbers shown in **Figure 22** where the subset of locations for which model-data inter-comparisons are shown are named.

In Rostami *et al.* (2000), attention was drawn to the fact that there existed a systematic misfit of the theory for postglacial sea-level change based upon the ICE-4G (VM2) model without rotational feed-back to the observed sea-level histories from this coastal region. The nature of this misfit, documented in figure 11 of Rostami *et al.* (2000), involved the variation with latitude of the so-called mid-Holocene 'highstand' of the sea. The prediction of the amplitude of this highstand using the ICE-4G (VM2) model without feedback was that this ampli-tude should decrease with increasing south latitude whereas the observation was that it actually increased very significantly, a qualitative difference easily resolved by the observational record. In Peltier (2002c), it was first demonstrated that the incorpora-tion of rotational feedback into the theory would lead to the qualitative change in the nature of the predic-tion required to reconcile the observations. More

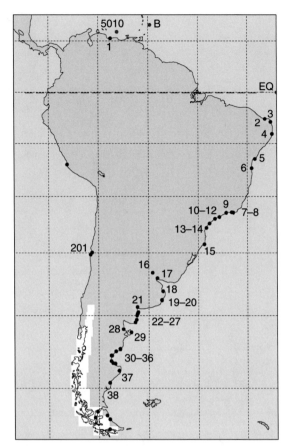

Figure 21 Location map for the South American sites from which radiocarbon-dated RSL histories are available. Also shown is the LGM location of the Patagonian ice-sheet complex.

recently, Peltier (2005) has more fully exploited these observations and the ICE-4G (VM2) model predictions to argue that this data set could be invoked to demonstrate that the impact of rotational feedback upon RSL history predicted by the ICE-4G (VM2) model was approximately correct. In fact, the data–model intercomparisons presented in both Peltier (2002c) and Peltier (2005) were incomplete, as several of the most anomalous of the observations were not discussed. As will be shown through inter-comparisons between the predictions of the ICE-4G and ICE-5G models in what follows, these further analyses have sufficed to reconcile all of the remain-ing misfits to the data from this region.

The importance of including the ICE-4G calcula-tions together with those for the ICE-5G model for present purposes is that the former model is charac-terized by a polar wander speed prediction that is slower than that predicted by the latter model by approximately 20%. This model may therefore be seen as a surrogate for models that predict a polar wander speed that is unable to satisfy the observational constraints. Although it may be reasonable to argue that one could invoke some process other than GIA, say mantle convection, to explain some modest fraction of the present-day observed polar wander speed observation, it is extremely unlikely that a credible calculation of this effect could be performed that would deliver the correct time dependence of the speed and direction predictions that are required to understand the dominant contribution to this signal. Yet all of these aspects of the data are fully explicable as consequences of the GIA effect, with

the computation modified perhaps to include the influence of a modest deviation from the 'equivalent Earth model' assumption of Munk and MacDonald (see Peltier (2007) for further discussion).

Figure 22 presents model–data intercomparisons for postglacial RSL history at 16 of the numbered sites from the east coast of South America, beginning at the northernmost site from coastal Venezuela and continuing southward through Brazil and Argentinian Pategonia toward Tierra del Fuego. For each site are shown the predictions for three different models from the set of four for which predictions of polar wander speed and direction are listed in **Table 1**.

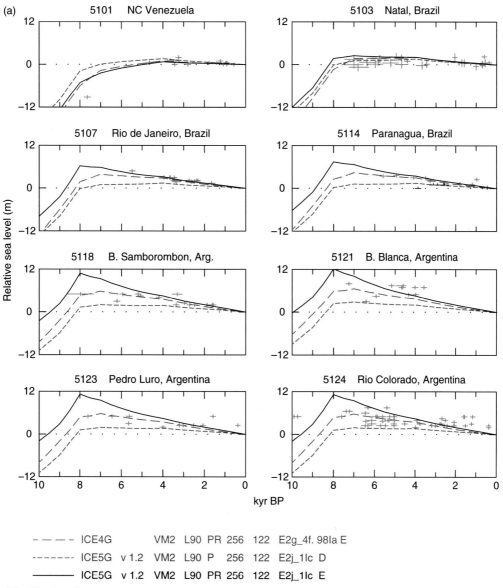

Figure 22 (*Continued*)

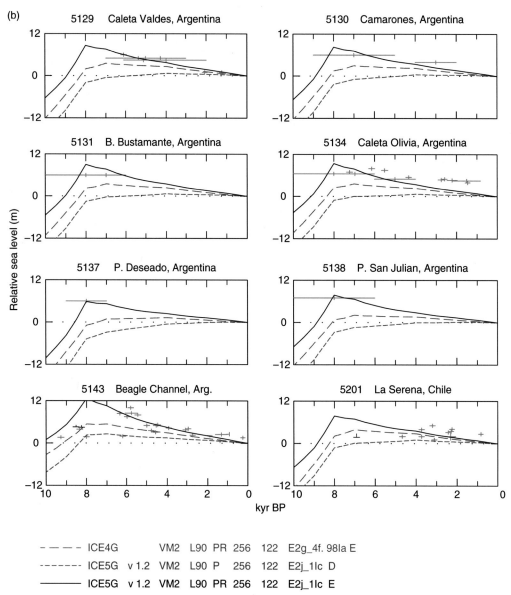

Figure 22 (a) Model–data intercomparisons for eight sites from the northern portion of the east coast of the South American continent. The last two digits of the site numbers correspond to the numbered locations on **Figure 21**. On each plate, the ^{14}C-dated sea-level index points are denoted by green crosses. RSL history predictions based upon eqn [1] are shown for three model variants, respectively: ICE-5G (VM2) with rotational feedback (black), ICE-5G (VM2) without rotational feedback (red), and ICE-4G (VM2) with rotational feedback (mauve). Notable is that based upon the comparisons at these north coastal sites the data are unable to unambiguously discriminate between the two models that include rotational feedback. However, the model that does not include rotational feedback is entirely rejected. (b). Same as (a) but for a set of eight sites from the southern (Patagonian) region of the east coast of the South American continent. At these locations, the results of the intercomparisons are now entirely unequivocal. Only the ICE-5G (VM2) model with rotational feedback satisfies the highstand observations.

These are for the ICE-4G (VM2) model that includes the influence of rotational feedback and for the ICE-5G (VM2) model both including and excluding this influence. It will be important to note that the first site in the list, denoted NC Venezuela (Valastro *et al.*, 1980), is the only site that is north of the equator (see **Figure 21**). This is important because the sign of the impact of rotational feedback changes across this line of 0° latitude, as expected based upon the spherical harmonic degree 2 and order 1 form of the pattern of this influence. To the north of the equator, as will be clear by inspection of the intercomparisons at NC Venezuela, the impact of rotational feedback is to diminish the predicted elevation with respect to sea level throughout the Holocene interval from 10 ka onward. To the south of the equator, however, as the intercomparisons for the Natal, Brazil, site (Bezerra *et al.*, 2003) show, the opposite is the case, as at this site the impact of rotational feedback is such as to elevate the predicted sea levels for the Holocene epoch above that for which this influence is neglected entirely or for which the effect of rotational feedback is diminished as is the case for the ICE-4G prediction relative to that for the ICE-5G model. Based upon the results in **Table 1**, it will be clear that this influence is expected to be weaker in ICE-4G (VM2) than it is in ICE-5G (VM2) simply because the predicted polar wander speed for the

former model is less than that for the latter model. As will be evident by inspection of the complete set of intercomparisons shown on **Figure 22**, this hierarchy is preserved at all locations. **Table 4** lists the references for the data from all of the sites for which data–model intercomparisons are shown in this section.

Very clearly evident also, based upon the results shown in **Figure 22**, is the fact that exceptionally high-'stands' of the sea above present sea level are predicted and observed to obtain at sites along this coast during mid-Holocene time between 6000 and 8000 years BP. Although there is considerable scatter in the data at the sites from NC Venezuela to Rio Colorado, Argentina, it is nevertheless clear that the influence of rotational feedback is required in order for the model to adequately reconcile the data. As discussed in detail in Rostami *et al.* (2000),

Table 4   References for the data from the sites for which data–model intercomparisons are provided in this chapter

Site number	Site name	References
5101	NC Venezuela	Valastro *et al.*, 1980
5103	Natal, Brazil	Bezerra *et al.*, 2003
5107	Rio de Janeiro, Brazil	Martin *et al.*, 1987; Fairbridge, 1976; Delibrias *et al.*, 1974; Angulo and Lessa, 1997
5114	Paranagua, Brazil	Angulo and Lessa, 1997
5118	B. Samborombon, Argentina	Codignotto *et al.*, 1992
5121	B. Blanca, Argentina	Codignotto *et al.*, 1992
5123	Pedro Luro, Argentina	Albero and Angiolini, 1983, 1985
5124	Rio Colorado, Argentina	Codignotto *et al.*, 1992
5129	Caleta Valdes, Argentina	Rostami *et al.*, 2000; Codignotto *et al.*, 1992; Albero and Angiolini, 1983
5130	Camarones, Argentina	Rostami *et al.*, 2000
5131	B. Bustamante, Argentina	Rostami *et al.*, 2000
5134	Caleta Olivia, Argentina	Rostami *et al.*, 2000
5137	P. Deseado, Argentina	Rostami *et al.*, 2000
5138	P. San Julian, Argentina	Rostami *et al.*, 2000
5143	Beagle Channel, Argentina	Porter *et al.*, 1984; Morner, 1991
11105	Pt. Adelaide/Gillman, Australia	Belperio *et al.*, 2002
11107	Pt. Wakefield, Australia	Belperio *et al*, 2002
11109	Pt. Pirie (Spencer Gulf), Australia	Belperio *et al.*, 2002
11110	Redcliffe, Australia	Belperio *et al.*, 2002
11111	Franklin HB, Australia	Belperio *et al.*, 2002
11113	Ceduna, Australia	Belperio *et al.*, 2002
11115	Weiti River, New Zealand	Gibb, 1986, Schofield, 1975
111123	Blue Skin Bay, New Zealand	Gibb, 1986
9003	Nouakchatt, Mauritania	Faure and Hebrard, 1977
9006	St. Louis, Senegal	Faure *et al.*, 1980
9022	Reunion	Camoin *et al.*, 1977, 2004
9023	Mauritius	Camoin *et al.*, 1997
11004	Shikoku Island, Japan	Pirazzoli, 1978
11006	Kikai-Jima Island, Japan	Pirazzoli, 1978; Sugihara *et al.* 2003
11008	S. Okinawa Island, Japan	Koba *et al.* 1982
11010	Uotsuri Island, Japan	Koba *et al.*, 1982

most of the data from this set of locations are from the older literature and therefore a legitimate question exists concerning its quality. However, this is not the case concerning the data from the set of locations between Caleta Valdes, Argentina, and P. San Julian, Argentina, all from Argentinian Patagonia, for which the data–model intercomparisons are shown in **Figure 22**. At these locations, the data have been carefully screened for quality (see discussion in Rostami *et al.* (2000)), and it is clear on the basis of them that only the ICE-5G (VM2) model is able to reconcile the 6–8 m mid-Holocene highstands of the sea that are observed to exist in the time window between 6000 and 8000 years BP. The ICE-5G (VM2) model, in the absence of rotational feedback, inevitably predicts that no such highstand should exist or that its amplitude should be very small. The ICE-4G (VM2) model predicts a highstand amplitude that is insufficient to explain the observations. These analyses very strongly reinforce the conclusion in Peltier (2002c, 2005) that this influence is a crucial determinant of the ability of the model to fit the observations. However, they also allow a considerably stronger conclusion to be reached. In particular, they force us to conclude that only if the time dependence of the predicted speed of TPW due to the influence of GIA is compatible with the presently observed speed and with the temporal evolution of the amplitude of the mid-Holocene highstand of the sea, is the record of postglacial RSL history understandable. Even a 20% reduction from the observed speed, as embodied in the version of the ICE-4G (VM2) model being employed here, leads to a significant misfit to the observations.

It is nevertheless the case that there is scatter in the data and one might reasonably hope that there might exist data from other locations where the influence of rotational feedback is also expected to be strong that might be invoked to confirm (or to deny) the robustness of this conclusion. To this end, **Figure 23** shows a location map for 16 additional sites from other regions, for some of which this influence is expected to be especially strong. Eight of these sites, for which intercomparisons are shown in **Figure 24**, are within the influence of the second Southern Hemisphere 'bull's-eye' of the spherical harmonic degree 2 and order 1 pattern that is produced by the influence of rotational feedback. The map in **Figure 23(c)** (in Mercator projection), on which these positions are located, is that of the difference between the present-day rate of sea-level rise

(a)   Rate of change of sea level; with rotation

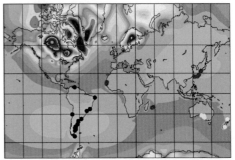

(b)   Rate of change of sea level; without rotation

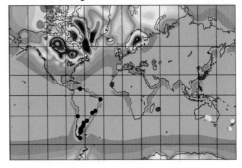

(c)   10× difference due to rotation

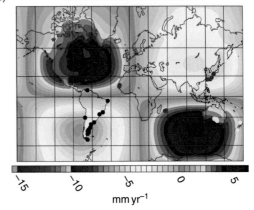

$mm\,yr^{-1}$

Figure 23   Mercator projections of the model predicted present-day rate of RSL rise for the ICE-5G (VM2) model both (a) including and (b) excluding the influence of rotational feedback. In (c), the difference between (a) and (b) is presented, demonstrating that the influence of the rotational feedback effect is dominantly of the form of a spherical harmonic of degree 2 and order 1. Superimposed upon each of the three plates of this figure are the locations of sites at which high-quality ^{14}C-dated RSL histories are available beyond those from the South American continent already discussed. The sites shown as yellow dots are located in a region (Australia–New Zealand) where the influence of rotational feedback is expected to have the opposite sign as the South American locations. At the sites shown as red dots the effect is either expected to the same as an Australia–New Zealand (in the Indian Ocean or West Africa) or the same as in South America (Japanese islands).

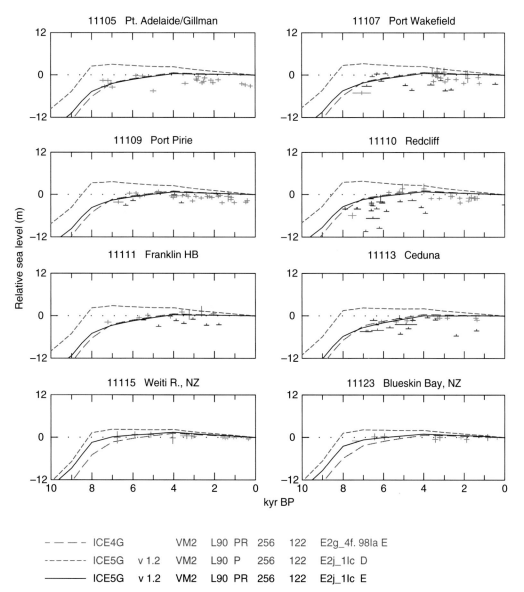

Figure 24 Same as **Figure 22(a)** but showing model–data intercomparisons for the Australia–New Zealand region. Inspection of these intercomparisons demonstrates that, as expected, the nature of the influence of rotational feedback is opposite to that observed at the South American locations. At these locations, the data are once more unable to discriminate between the two models that include the influence of rotational feedback, presumably because these sites are insufficiently close to the degree 2 and order 1 'bull's-eye', just as was the case for the sites along the northern part of the east coast of South America. Once more, however, the model without rotational feedback is totally rejected.

prediction for the ICE-5G (VM2) model with rotation (shown in **Figure 23(a)**) and that obtained when this influence is not included (shown in **Figure 23(b)**). Eight additional intercomparisons are shown in **Figure 25**, four of which are from locations in the Indian Ocean or the west coast of Africa, and four of which are from the third 'bull's-eye' located over the Japanese islands (see **Figure 23(c)**).

Inspecting first the intercomparisons shown in **Figure 24** for sites in the Australia–New Zealand region will show that at all of these sites there exists a striking anomaly between the prediction of the version of the ICE-5G (VM2) model without rotational feedback and the predictions for either of the models in which this influence is included. In the absence of rotational feedback, the theory predicts that a mid-Holocene highstand of the sea should exist

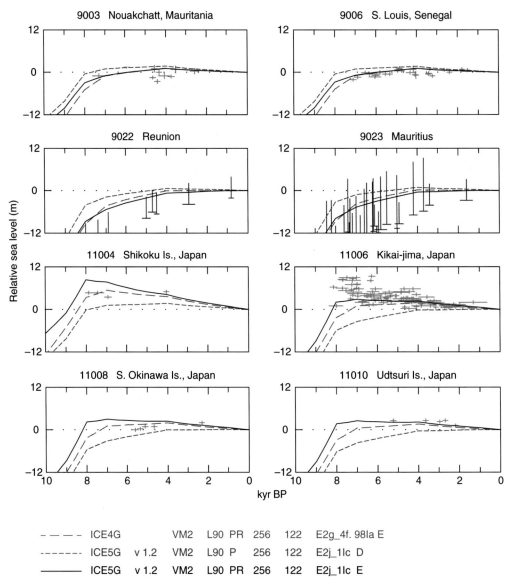

Figure 25 Same as for **Figure 22(a)** but showing model–data intercomparisons for both 'down sites' (Indian Ocean, West Africa) and 'up sites' (Japanese islands). For all sites, the sign of the rotational feedback effect is as expected, and all data reject the model from which the influence of rotational feedback has been eliminated.

with an approximately 3 m amplitude. However, the data at all eight locations show that no such feature is evident. At most of the locations for which results are shown in **Figure 24**, either of the two models with rotational feedback included suffice to fit the observed RSL record although at Blueskin Bay, NZ, the ICE-5G (VM2) model with feedback is again found to be superior to ICE-4G (VM2) with feedback. This set of intercomparisons is especially important because in this region the sign of the feedback effect is predicted to be opposite to that

operating over the South American continent. Nevertheless, the necessity of the operation of a strong influence of rotational feedback in order to fit the observational constraints remains clear. It is perhaps especially interesting to note that the data from the coast of the Great Australian Bight of South Australia could be misconstrued to suggest that eustatic sea level must have continued to rise through Late Holocene time (e.g., see Lambeck, 2002), perhaps due to continuing melting of the Antarctica or Greenland ice sheets. The analysis presented here demonstrates that such an interpretation would be incorrect. The absence of a mid-Holcene highstand of sea level in this region is due to the influence of rotational feedback onto postglacial sea-level history, not the continuing melting of land ice. In the ICE-4G and ICE-5G models, all melting of continental ice is assumed to have effectively ceased by 4000 years BP (e.g., see Peltier (2002), Peltier *et al.* (2002), and **Figure 7**).

The further model–data intercomparisons shown in **Figure 25** include information from sites in which the sign of the influence of rotational feedback is the same as that which operates in the Australia–New Zealand region, and sites in which the influence is qualitatively similar to that evident at locations along the southeast coast of the South American continent. Sites of the former type include those along the west coast of Africa (Nouakchatt, Mauritania, and St. Louis, Senegal) and the others from the Indian Ocean (Reunion, Mauritius). Sites of the latter type from the islands of Japan include Shikoku Island, Kikai-jima Island, S. Okimaura Island, and Uotsuri Island. At the former locations, theory once more predicts that in the absence of rotational feedback a mid-Holocene highstand of the sea should be observable. At these locations, however, no such highstand is evident. The addition of rotational feedback, either of the strength embodied in ICE-4G (VM2) or ICE-5G (VM2), allows the model to fully reconcile the data.

Turning next to the final set of four data–model intercomparisons shown in **Figure 25**, it will be clear that at each of the Japanese islands for which RSL intercomparisons are shown, the model without rotational feedback predicts that no mid-Holocene highstand of the sea should be evident at any of these locations. Yet the data in each case clearly reveal the presence of such a feature. As will be clear by inspection of this set of intercomparisons, the incorporation of the influence of rotational feedback suffices to reconcile the misfits of the theoretical predictions to the data that would otherwise exist. Once more, the data at these locations are only marginally able to distinguish between the strength of the feedback effect embodied in the ICE-4G (VM2) and ICE-5G (VM2) models. Only at Uotsuri Island and Kikai Island does there appear to exist a marginal preference for the latter model.

On the basis of the totality of these intercomparisons we may therefore conclude that there exists very strong evidence in the global variations of postglacial RSL history for the action of the influence of rotational feedback. Furthermore, the strength of this feedback appears to be quite accurately predicted by the theory of Peltier (1982) and Wu and Peltier (1984), a theory based upon the application of the 'equivalent Earth model' concept of Munk and McDonald (1960). On the timescale of thousands to hundreds of thousands of years in the most recent geological past, the history of Earth rotation has apparently been strongly dominated by the influence of the GIA process, as both the secular variation in the nontidal variation in the l.o.d and the TPW evident in the ILS path of the pole are accurately predictable by the same model of the GIA process that is employed to predict the histories of postglacial RSL change. In this same range of timescales, however, additional and even more subtle influences are active upon, rather than within, the Earth system that are connected to the history of the Earth's rotation as observed from space. These influences are discussed in the next section.

10.6   The Impact of Variations in the Geometry of Earth's Orbit around the Sun upon Earth System Evolution

In the previous sections of this chapter, the focus has been fixed upon the history of Earth rotation as observed in the body-fixed coordinate system, the terrestrial frame, in which the anomalies in the variations in the l.o.d. and polar motion are most usefully described. Equally important aspects of the evolution of Earth's rotation are, however, those that are most usefully described in the celestial frame provided by the field of stars of our own and more distant galaxies. In this celestial frame of reference, the orbit of the Earth in its rotation around the Sun is observed to be characterized by subtle time dependence in its geometry. These variations are due to the action of gravitational n-body effects in the solar system that cause the ellipticity of the otherwise Keplerian elliptical orbit to vary with time and similarly impose subtle time dependence upon orbital obliquity, that is to say upon the angle that the spin axis makes with the plane of the ecliptic (see **Figure 26**). As it happens, this aspect of the history of Earth rotation exerts a profound effect upon low-frequency climate variability and in particular is the ultimate cause of the Late Pleistocene ice-age cycle that was discussed in the last section of this chapter as the source of the excitation of millennial timescale variations in l.o.d. and polar

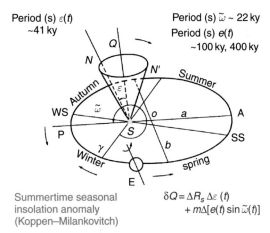

Period (s) $\varepsilon(t)$ ~41 ky

Period (s) $\tilde{\omega}$ ~ 22 ky
Period (s) $e(t)$ ~100 ky, 400 ky

Summertime seasonal insolation anomaly (Koppen–Milankovitch)

$$\delta Q = \Delta R_s \, \Delta \varepsilon \, (t) + m\Delta[e(t) \sin \tilde{\omega}(t)]$$

Figure 26 Geometry of Earth's orbit around the Sun depicting the present positions of the apogee and the perigee, as well as the summer and winter solstices and the spring and autumn equinoxes. The tilt of the spin axis with respect to the plane of the ecliptic is represented by the parameter ε, whereas the eccentricity of the orbit is represented by the parameter e. As a consequence of the influence of the precession of the spin axis in space, the position of the equinoxes and solstices rotates around the orbital ellipse with a period of approximately 22 000 years.

motion. This so-called astronomical theory of the ice ages, due originally to Milankovitch (e.g., 1940), was finally verified as essentially correct in the seminal paper by Hays *et al.* (1976) that was based upon the analysis of oxygen-isotopic data from deep-sea sedimentary cores such as the data set from ODP site 677 from the Panama Basin shown in **Figure 5** and which was analyzed in detail in Shackleton *et al.* (1990). The origins of the particular time series shown in this figure are in fact extremely important, both from the perspective of our understanding of the Late Pleistocene ice-age cycle and our appreciation of the crucial role that the history of Earth rotation has played in our understanding of the detailed characteristics of the process of plate tectonics itself.

Prior to the analysis presented in the Shackleton *et al.* paper, it had been widely believed, based upon potassium–argon dating of the age of the most recent Brunhes–Matuyama reversal of the polarity of Earth's magnetic field, that this event had occurred approximately 730 000 years BP. This assumed age was, until the appearance of this paper, universally employed in the construction of the chronology of all of the deep-sea sedimentary cores that were employed as constraints upon the understanding of

climate system processes over the Pleistocene epoch. This date was also employed to constrain the speeds with which oceanic plates were assumed to 'spread' away from oceanic ridge crests in response to the underlying mantle convection process. The history of the process that led to our present understanding that the age of this critical control point for planetary chronology was 780 000 years rather than 730 000 years is important from the perspective of our appreciation of the role that Earth rotation history plays in our understanding of the system and so is worth reviewing in some detail.

10.6.1 The Astronomical Imprint on Oxygen-Isotopic Records from Deep-Sea Sedimentary Cores

Basic to this understanding is an appreciation of the extent to which modern astrophysical theory has been successful in the reconstruction of the evolving geometry of Earth's orbit around the Sun over the past several million years of Earth history. Early work on this problem was reviewed and significantly extended by Laskar (1988), who employed a mixture of analytical and numerical techniques to demonstrate that it was possible to perform stable integrations of the geometry of the orbit that extended in excess of several million years into the past. These reconstructions were tested by performing brute force initial value integrations, starting from a modern ephemeris, of the system of gravitational n-body equations governing the evolution of the entire solar system, a system of equations that included an accounting for tidal interactions among the planets and their satellites as well as for relativistic effects. The results of such brute force integrations were first described in an important paper by Quinn *et al.* (1991). These analyses demonstrated that the solution of Laskar (1988) was highly accurate and could therefore be employed with confidence to accurately predict the variation of the strength of the solar insolation received by the Earth in the course of its orbit around the Sun as a function of time and as a function of latitude. Earlier predictions of these variations produced by Berger (e.g., 1978) were thereby shown to become inaccurate prior to approximately 1 My BP. The numerical reconstructions of the evolving geometry of Earth's orbit consist of predicted time series for the evolution of orbital ellipticity and orbital obliquity. From these quantities, one may compute the latitude-dependent summertime seasonal anomaly in solar insolation received at the top of the atmosphere that is presumed to drive the glaciation–deglaciation process according to the Koppen–Milankovitch hypothesis on the basis of the expression,

$$\delta Q \left(\theta,\ t\right) = \Delta R_s(\theta)\Delta\varepsilon(t) + m(\theta)\Delta[e(t)\sin(\bar{\omega}t)] \quad [39]$$

in which $\delta Q \left(t\right)$ is the deviation, in $W\,m^{-2}$, of the insolation received at the surface of the Earth during the summer season relative to the average of summer insolation over, say, the timescale of the entire 900 000 year period during which the 100 000 year ice-age cycle has been an active element of climate system evolution. The latitude dependence of the summertime seasonal insolation anomaly is introduced through the quantities $\Delta R_s(\theta)$ and $m(\theta)$, whereas the time dependence is governed by the variation in orbital obliquity $\Delta\varepsilon(t)$ and time-dependent eccentricity of the orbit $e(t)$ which enters the expression for the insolation anomaly only through a modulation of the influence of orbital precession which occurs at an angular frequency $\bar{\omega}$ due to the same oblateness of the figure of the Earth that

is modulated so effectively by the ice-age cycle. This phenomenon of orbital precession is caused by the action of the gravitational field of the Sun upon the oblate form of the planet whose spin axis is tilted with respect to the ecliptic plane. **Figure 27** illustrates the time dependence of each of the quantities that appears in the expression for the orbital insolation anomaly in eqn [39]. **Figure 28** shows the actual distribution of the received insolation over the surface of the Earth as a function of latitude and time of year, both for the present day and for the deviations from this distribution at several fiducial times over the most recent 100 000 year cycle of Late Pleistocene glaciation. Inspection of the distribution of the anomaly for 116 000 years BP at the onset of the most recent of these cycles of continental ice-sheet expansion demonstrates that this event was in fact apparently triggered by a high-latitude summertime seasonal insolation anomaly of magnitude in excess of $60\,W\,m^{-2}$, in accord with expectations based upon the Milankovitch hypothesis. At LGM itself, the deviation of the insolation regime from that

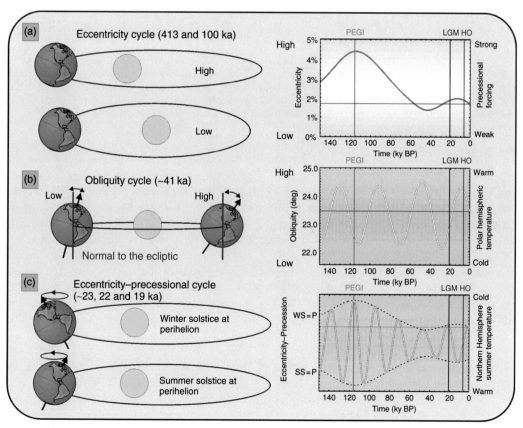

Figure 27 Evolution with time, over the most recent 100 000 year cycle of Late Pleistocene climate variability, of the parameters that control the variation of the solar insolation received at the top of the atmosphere.

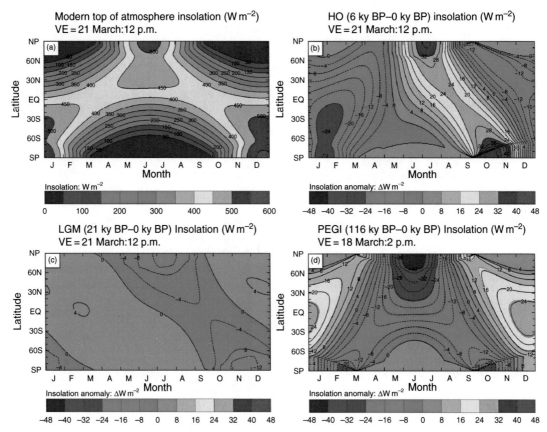

Figure 28   Distribution of solar insolation received at the top of the atmosphere as a function of time of year and latitude for three different epochs. (a) The present-day distribution; (b) and (c) the deviation from the modern distribution for the mid-Holocene warm period at 6000 years BP, for LGM at *c.* 21 000 years BP, and for the end of the Eemian interglacial at 116 000 years BP, the time during which the extremely strong negative summertime seasonal insolation anomaly existed that was responsible for the onset of the most recent 100 000 year glacial cycle.

characteristic of present climate was actually rather small. During the mid-Holocene interval centered upon 6 ka, the nature of the insolation forcing was such that the Northern Hemisphere continents would have been anomalously warm during the summer season, thus explaining the increased intensity of the summer monsoon circulations that were characteristic of the atmospheric general circulation at that time. **Figure 29** compares the power spectrum of the oxygen-isotope anomaly from deep-sea core OPD 677 to the power spectrum of the summertime seasonal insolation anomaly at 60° north latitude for both the first and last million years of the Pleistocene epoch. Inspection of this intercomparison shows that the power in the insolation forcing at the 100 000 year period on which the continental ice sheets expand and contract is miniscule. The implications of this intercomparison are profound as it provides a direct demonstration of the intensity of

the nonlinearity of the climate system that is involved in converting the incoming insolation signature into the climate response. It is only through the action of such nonlinearity that the system may feel the effect of the 100 000 year modulation of the influence upon received insolation due to the effect of orbital precession that arises from the variation of orbital eccentricity. Precisely what the origin is of this nonlinearity remains a subject of intense debate but is presumably connected to the cause of the covariation of atmospheric carbon dioxide with continental ice volume that is so clearly revealed in the Vostock ice core from the continent of Antarctica (e.g., see Overpeck *et al.* (2007) for detailed discussion of the ice-core records and Shackleton (2000) for a discussion of the phase relationships between the atmospheric carbon dioxide record and the ice-volume record; Peltier (1998) has reviewed the current state of theoretical understanding).

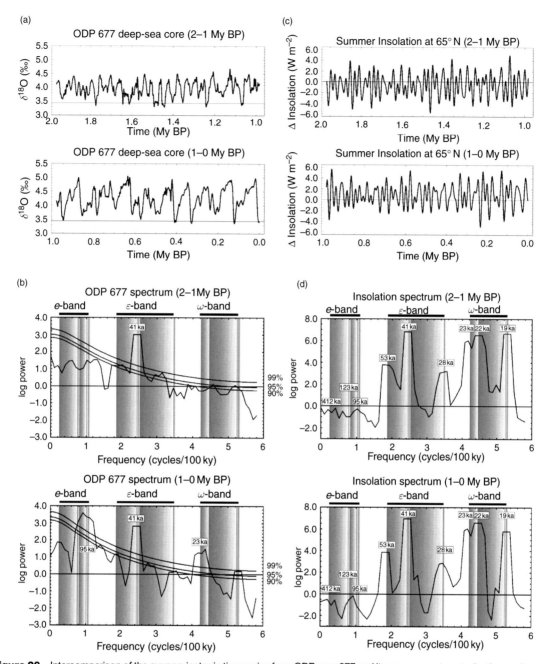

Figure 29 Intercomparison of the oxygen-isotopic time series from ODP core 677 and its power spectrum to the time series and power spectrum of the solar insolation forcing for both the first and last million years of these records. Notable is the extremely weak insolation forcing at the period of 100 000 years, which is the period at which the climate system response is strongest.

10.6.2 'Orbital Tuning' and the Age of the Brunhes–Matuyama Geomagnetic Polarity Transition

Given this background we may proceed to consider the analyses that led to the redating of the oxygen-isotopic record from ODP core 677 that has played such an important role in revising our understanding

of the Pleistocene timescale. **Figure 31** shows the original depth versus time scale that was originally ascribed to this core by Shackleton (personal communication in 1988 when the data were passed to the present author for analysis). Notable is the 'dogleg' in this relationship that is apparent near the then-assumed age of 730 000 years of the

Brunhes–Matuyama transition that was widely accepted as correct at that time. Based upon the assumption of the validity of the Milankovitch hypothesis, the depth–age relationship for this core was refined in Toronto by applying the following strategy. First the relationship between depth in the core and calendar year age was refined by applying a band-pass filter to the isotopic time series with a central frequency equal to that of the obliquity timescale of 41 000 year and the depth–time relationship was adjusted iteratively so as to maximize the coherence between the astronomical forcing at this period and the ice-volume response. Next the same process was applied to the variability in the eccentricity–precession band, in which the modulation of the precession effect due to the eccentricity variation occurs. Once more, the age–depth relationship was adjusted in order to optimize the coherence between astronomical signal and ice-volume response. The frequency domain form of the band-pass filters employed in this analysis together with the power spectrum of the oxygen-isotope time series is shown in **Figure 30**. The process of refinement of the age–depth relationship was continued by returning to the obliquity band and then returning to the eccentricity–precession band, etc., until an entirely stable result was obtained. The new age–

depth relationship for the 677 core is also shown in **Figure 31** as the dashed line labeled Model II, where it is observed to correspond to a very nearly linear relationship between depth and age, suggesting that in fact the sedimentation rate at this site had been very nearly constant throughout the entire Pleistocene interval. The initial depth–age model provided to the author by N. J. Shackleton is shown as model IV. The models denoted I and III in the figure are intermediate results obtained in the iterative sequence of steps in the 'orbital tuning' process. The quality of the fits between the astronomical forcing and ice-volume response records in the two pass bands employed in the application of this methodology are shown in **Figures 32(a)** and **32(b)**, where a very nearly precise agreement was obtained (see Peltier (1994) in which the original results that led to the paper by Shackleton *et al.* (1990) were eventually published).

This analysis suggested that there must have been an error in the previously accepted age of the Brunhes–Matuyama transition in the polarity of Earth's magnetic field, with the actual age of the transition being very close to 780 000 years rather than to the originally assumed age of 730 000 years, a difference of approximately 7% or 50 000 years. In the paper by Shackleton *et al.* (1990), the initial Toronto result was checked against a large number of other oxygen-isotopic records from deep-sea cores and found to be reproducible in every instance. Very soon after appearance of the Shackleton *et al.* publication, and in response to it, the age of the Brunhes–Matuyama transition was redated using the argon 39–argon 40 method by Baksi (1992), whose analysis immediately confirmed the validity of the new age for this most recent polarity transition in the Earth's magnetic field, an age that had been determined by application of the 'orbital tuning' methodology. It should be clear that this combination of results has served to fully verify the validity of our present understanding of the Late Pleistocene ice-age cycle as being caused by the history of the variations in the geometry of Earth's rotation around the Sun.

10.7   Earth Rotation Variations and Mantle Convective Mixing

On the longest timescales of hundreds of millions of years on which the thermal evolution and internal structure of the Earth are governed by the mantle

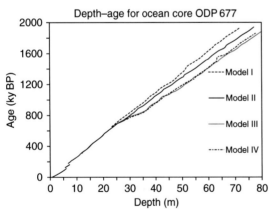

Figure 30 Oxygen-isotopic record from ODP core 677 as originally provided by N. J. Shackleton to the author in 1988 on the depth vs time scale constrained to fit the assumed age of 730 000 years of the Brunhes–Matuyama geomagnetic reversal. Notable in the original record is the 'dogleg' that was forced to occur in the depth–time relationship by virtue of this assumption. Also shown are the revised depth–time relationships deduced by 'orbital tuning' of this record that was accomplished at the University of Toronto and which led to the prediction of the necessity of revising the age of this geomagnetic reversal.

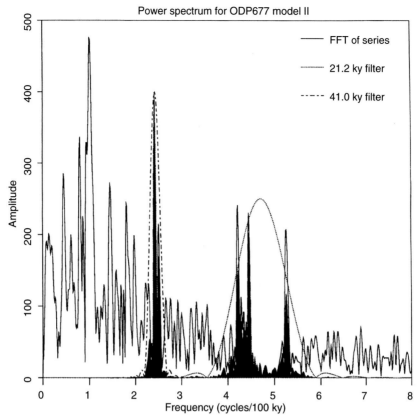

Figure 31 Frequency domain forms of the band-pass filters employed in the Toronto application of the orbital tuning methodology to the $\delta^{18}O$ record from ODP core 677.

convection process, debate continues concerning the issue as to whether the relative latitudinal positions of the continents and the variations of their positions through time are to be ascribed solely to the action of the mantle convection-driven process of continental drift. If it were possible for the relative positions of the continents with respect to the axis about which the planet rotates to be spontaneously modified by a TPW event that resulted in a reorientation of geography with respect to the rotation pole, then it would be possible to misinterpret paleomagnetic inferences of paleolatitude as arising due to the action of the convection process when in fact these were due to the influence of TPW. Recent interest in the possible importance of this process has come to be focused upon the possibility of occurrence of a so-called 'inertial interchange' instability (e.g., Goldreich and Toomre, 1969; Fisher, 1974; Kirschvink *et al.*, 1997) during which, as a consequence of the axis of greatest moment of inertia being 'suddenly' (on geological timescales) changed,

the geographical positions of the continents with respect to the rotation pole would be subject to a secular shift. This would clearly require that the modification, whether by reorganization of the convective flow or some other process, be strong enough to overcome the stabilizing influence of the equatorial bulge due the basic rotation, as previously discussed in Section 10.4. It remains an issue as to whether the process of convective overturning is capable of effecting such a profound influence upon the rotational state of the planet. After all, the magnitude of the density variations associated with the convection process are hardly more than a few percentage, and these variations are not obviously organized in a sufficiently coherent pattern as to be effective in exerting direct control upon the rotational state of the object.

There does exist, however, one potential phenomenon that has been suggested to be a characteristic of the mantle convection process that could, at least in principle, exert 'coherent control' upon Earth rotation. This is the so-called 'avalanche effect' discussed at length by Peltier and Solheim (1992) and Solheim and Peltier (1994a, 1994b), an effect upon the temporal variability of the convection process that is controlled by the action of the phase transition from spinel to a mixture of perovskite and magnesiowustite that exists at 660 km depth and which separates the so-called transition zone of the mantle above this level from the lower mantle beneath. The avalanche effect is found to be strong in models of the mantle convection process in which the effective viscosity of the planetary mantle is sufficiently low, as it is in the VM2 model that has been developed on the basis of fits to observations of the GIA process discussed in the previous sections of this chapter (see **Figure 19**). Peltier (1998) has argued that this model,

or a close relative derived on the basis of analyses of the GIA process, may be equally applicable on the much longer timescale characteristic of convection in Earth's mantle, although this issue remains a subject of active debate.

Insofar as effort is concerned to confirm the possibility that IITPW may have occurred in the past, a primary era of Earth history that has been suggested as a candidate has been the Early Cambrian period (Kirschvink *et al.*, 1997). Of course, this is an extraordinarily important period of time from the perspective of Earth evolution in general and biological evolution in particular as well as from the perspective of tectonophysics. From the perspective of biological evolution, it was the time of especially rapid biological diversification that occurred during

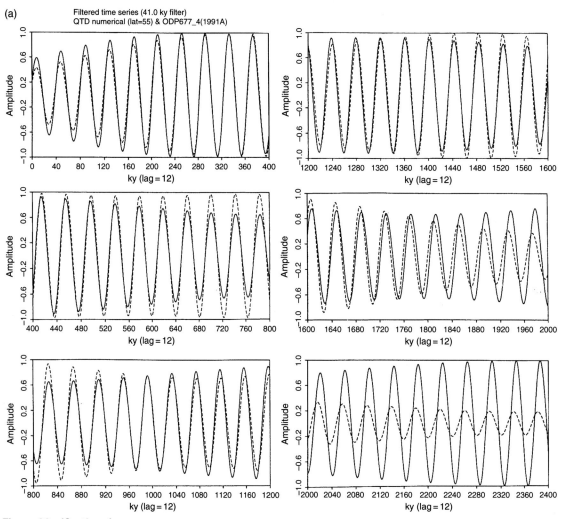

Figure 32   (*Continued*)

(b)

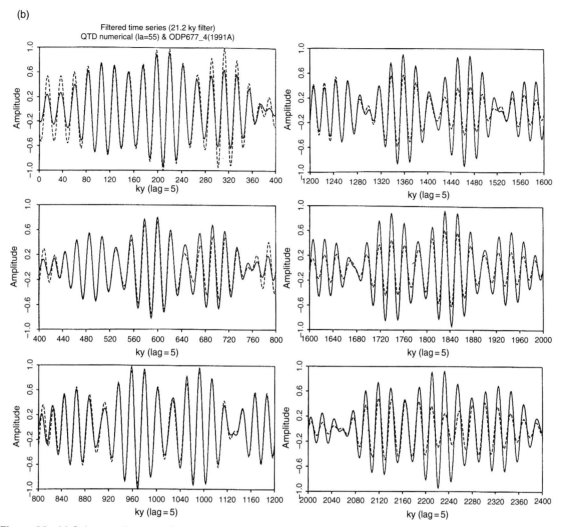

Figure 32 (a) Coherence between the astronomical forcing at the period of 41,000 years and the oxygen isotopic variability at the same period after orbital tuning at the University of Toronto. (b) Same as Figure 31a but showing the coherence in the eccentricity-precession band between astronomical forcing and oxygen-isotopic (climate) response. This is the original work that led to the results reported in Shackleton et al. (1990).

the so-called 'Cambrian explosion of life'. From the perspective of tectonophysics, it was a time immediately following the final phase of the Neoproterozoic breakup of the supercontinent of Rodinia during which a series of especially severe episodes of global glaciation are believed to have occurred (e.g., Kirschvink, 1991; Hoffman and Schrag, 2000, 2002). These hypothesized glaciation events have been termed 'snowball glaciations', during which it is imagined that the surface of the Earth may have become entirely ice covered, the oceans by a thick veneer of sea ice and the continents by thick continental ice sheets such as those that were confined

primarily to the north polar regions during times of ice-age maxima within the Pleistocene era. Although the plausibility of occurrence of the 'hard snowball' state envisioned by these authors has been questioned (e.g., Hyde *et al*, 2000; Peltier *et al.*, 2004), there is no doubt that the period of transition between the Neoproterozic and the Cambrian eras, a transition that occurred at approximately 541 Ma, was a time in Earth history during which plausibly large variations in the elements of the moment of inertia tensor of the planet could have occurred. Not only was the process of mantle convection highly disturbed during the process of the rifting and dispersal of Rodinia but there were also highly significant exchanges of mass occurring between the oceans and the highly glaciated continents at the end of the final Neoproterozoic snowball event. Li *et al.* (2004) have in fact proposed that the inferred equatorial location of this supercontinent during the earliest (Sturtian) episode of intense Neoproterozoic glaciation could have arisen as a consequence of a complete 90° inertial interchange instability event during which the supercontnent was shifted from the pole to the equator. It may therefore be the case that the Sturtian ice mass was not emplaced on the supercontinent while it was at the equator but rather when it was in the initial polar location. This is an idea that is suggested as a possibility by the Li *et al.* analysis of the paleomagnetic data but which was not explicitly envisioned by them.

Figure 33 illustrates the temporal interrelationships that existed in the Earth system during this critical period in Earth history between the tectonophysical events associated with the mantle convection process and the events that were simultaneously occurring in surface climate variability. The climate-related signal shown on this time series is for the carbon-isotopic anomaly denoted $\delta^{13}C$ which is measured in carbonate rocks on land such as those from Namibia which constitute the type section for much of the work that has been accomplished in support of the snowball Earth hypothesis (Hoffman and Schrag, 2000). The importance of this isotopic measurement derives from the fact that it is strongly influenced by photosynthetic activity as an isotopic fractionation occurs such that the organic matter produced in photosynthesis is enriched in ^{12}C

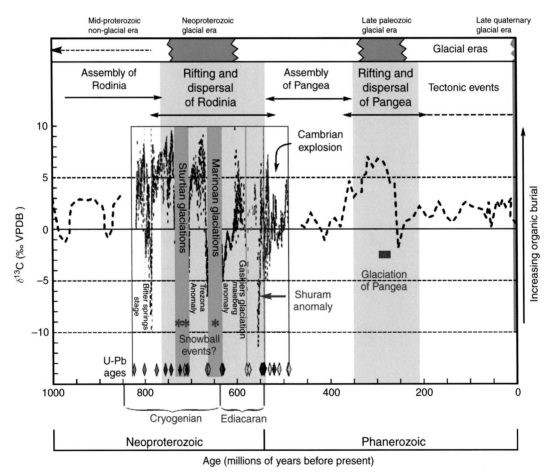

Figure 33   The variation of carbon isotopes, denoted $\delta^{13}C$, over the last 1.6 Gy of Earth history shown together with the times of occurrence of periods of significant continental glaciation and the times of creation and destruction of supercontinents that accompany the so-called 'Wilson cycle'. The $\delta^{13}C$ time series is from Kauffman (1997) and Hoffman and Schrag (1999), whereas the insert for the Neoproterozoic era is based upon the composite data set from Halverson *et al.* (2005).

relative to the heavier isotope ^{13}C. A time when the $\delta^{13}C$ that is measured in carbonate rocks that were precipitated from the ocean is high therefore corresponds to a time of high photosynthetic biological productivity. This is because such episodes correspond to times, as indicated in **Figure 33**, when the rate of burial of the organic matter produced in photosynthesis is significantly elevated, leaving behind an ocean that is enriched in ^{13}C. The sharp minima that are observed to characterize the time series shown on **Figure 32** have therefore been taken to correspond to times in Earth history of intense 'snowball' glaciation during which photosynthetic activity was either sharply diminished or eliminated altogether, as would have been the case during a 'hard snowball' event. Also indicated in **Figure 33** is the timing of the episodes of intense tectonic activity that would have been associated with the breakup of the supercontinents of Rodinia and, much later during the Carboniferous period, Pangea. The period of transition between the Neoproterozoic and the Cambrian that Kirschvink has suggested as a possible time of occurrence of an inertial interchange instability event was therefor a time of profound change in Earth history, perhaps the most important such period since the birth of the planet from the primitive solar nebula.

In an attempt to establish the Early Cambrian IITPW hypothesis of Kirschvink (1997) as viable, Mound *et al.* (1999) performed a calculation of the expected impact of such an event upon RSL history. Their idea was to see whether observations of RSL history over timescales of millions of years might be employed as a means of determining whether an IITPW event had occurred and the analyses presented were interpreted to imply that the hypothesis was not only viable but was supported by their initial analysis of the observational constraints they described. An important issue concerning the validity of these analyses concerns the strong dependence of the results obtained upon the thickness of the lithosphere assumed to be characteristic of the Maxwell viscoelastic model employed for the purpose of analyzing the sea-level response to the rotational excitation. Their assumption in this analysis was that the lithosphere could be treated as a complete 'unbroken-by-plate-tectonics' elastic shell over the tens of millions of years during which the IITPW event was imagined to occur. Their results, however, suggested a very strong dependence upon the thickness assumed such that

the response rapidly diminished as the assumed thickness of the lithosphere was reduced. In Sections 10.3 and 10.5, the question as to how the thickness of the lithosphere was to be treated, insofar as the computation of the rotational response to surface glaciation and deglaciation is concerned, was discussed. The issue of the role of lithospheric thickness in global geodynamics in general, including its action in determining the manner in which Earth's rotation state responds to time variability in the mantle convection circulation, remains outstanding. It may in fact be possible during a period of rapid reorganization of the convection process, when the strong correlation that exists under modern plate-tectonic conditions between plate boundaries and deep mantle upwellings and downwellings may no longer obtain. In such conditions, an influence of finite lithospheric thickness may therefore have existed in which case the Mound *et al.* results could be fully applicable.

In searching for a sufficiently intense process of reorganization of the convective circulation that might be responsible for an IITPW event, such as that suggested by Kirschvink to have occurred in the Early Cambrian or Li *et al.* in the Neoproterozoic, the author is inclined to focus upon the times of occurrence of events related to the supercontinent cycle such as the rifting process during which an initially compact supercontinent is fragmented following the occurrence of 'avalanche' events during which the convective circulation may have made a rather sudden transition from a state of whole mantle flow to a state of layered convection (e.g., see Peltier (1996) for a review). Peltier *et al.* (1997) has in fact suggested a close linkage between the so-called 'Wilson cycle' of supercontinent creation and destruction and the continuously recurrent 'avalanche effect' that some detailed fluid mechanical models of the convection process have predicted (e.g., Peltier and Solheim, 1994a, 1994b). Such models do apparently predict the inferred timescale of several hundred million years that appears to separate the times in Earth history during which supercontinents existed. That timescale is most visibly evident in histograms of the sample density of continental rocks of a given age as a function of their age. Such suggestions of the mechanism that could underlie these most dramatic events of all that may have occurred in the history of Earth rotation must however remain a subject of speculation in search of firm quantitative justification.

References

Albero MC and Angiolini FE (1983) INGEIS Radiocarbon Laboratory dates I. *Radiocarbon* 25: 831–842.

Albero MC and Angiolini FE (1985) INGEIS Radiocarbon Laboratory dates II. *Radiocarbon* 27: 315–337.

Angulo RJ and Lessa GC (1997) The Brazilian sea level curves: A critical review with emphasis on the curves from Paranagua and Cananeia regions. *Geology* 140: 141–166.

Argus DF and Gross RS (2004) An estimate of motion between the spin axis and the hotspots over the past century. *Geophysical Research Letters* 31: L06614 (doi:10.1029/2004GL019657).

Baksi AK (1992) $^{40}Ar/^{39}Ar$ dating of the Brunhes–Matuyama geomagnetic field reversal. *Science* 256: 356–359.

Belperio AP, Harvey N, and Bousman RP (2002) Spatial and temporal variability in the Holocene sea-level record of the South Australian coastline. *Sedimentary Geology* 150: 153–169.

Berger A (1978) Long-term variations of daily insolation and Quaternary climatic changes. *Journal of the Atmospheric Sciences* 35: 2362–2367.

Bezerra F, Barreto A, and Sugio K (2003) Holocene sea-level history on the Rio Grande do Norte State coast, Brazil. *Marine Geology* 196: 73–89.

Broecker WS and Van Donk J (1970) Insolation changes, ice volumes, and the $\delta^{18}O$ record in deep sea cores. *Geophysics and Space Physics* 8: 169–198.

Calaya MA, Wahr JM, and Bryan FO (1999) Climate-driven polar motion. *Journal of Geophysical Research* 104: 12813–12829.

Camoin GF, Colonna M, Montaggioni LF, Casanova J, Faure G, and Thomassin BA (1997) Holocene sea level changes and reef development in the southwestern Indian Ocean. *Coral Reefs* 16: 247–259.

Carter WE, Robertson DS, Pyle TE, and Diamante J (1986) The application of geodetic radio interferometric surveying to the monitoring of sea level. *Geophysical Journal of the Royal Astronomical Society* 87: 3–13.

Camoin GF, Montaggioni LF, and Braithwaite CJR (2004) Late glacial to to post glacial sea levels in the western Indian Ocean. *Marine Geology* 206: 119–146.

Codignotto JD, Kokott RR, and Marcomini SC (1992) Neotectonism and sea-level changes in the coastal zone of Argentina. *Journal of Coastal Research* 8: 125–133.

Chandler SC (1891) On the variation of latitude, I. *Astronomical Journal* 11: 59–61.

Chao BF, O'Conner WP, Chang ATC, Hal DK, and Foster JL (1987) Snow load effect on the Earth's rotation and gravitational field, 1979-1985. *Journal of Geophysical Research* 92: 9415–9422.

Cheng MK, Eanes RJ, Shum CK, Schutz BE, and Tapley BD (1989) Temporal variations in low degree zonal harmonics from Starlette orbit analysis. *Geophysical Research Letters* 16: 393–396.

Clark JA, Farrell WE, and Peltier WR (1978) Global changes in postglacial sea levels: A numerical calculation. *Quaternary Research* 9: 265–287.

Clark PU, Mitrovica JX, Milne GA, and Tamasea ME (2002) Sea level finger printing as a direct test of the source of meltwater pulse 1a. *Science* 295: 2438–2441.

Cox CM and Chao BF (2002) Detection of a large-scale mass redistribution in the terrestrial system since 1998. *Science* 297: 831–833.

Dahlen FA (1976) The passive influence of the oceans upon the rotation of the Earth. *Geophysical Journal of the Royal Astronomical Society* 46: 363–406.

Dalmau W (1997) Critical remarks on the use of medieval eclipse records for the determination of long-term changes in the Earth's rotation. *Surveys in Geophysics* 18: 213–223.

Daly L and Le Goff M (1996) An updated and homogeneous world secular variation database. Part 1. Smoothing of the archeomagnetic results. *Physics of the Earth and Planetary Interiors* 93: 159–190.

Dahlen FA and Smith ML (1975) The influence of rotation on the free oscillations of the Earth. *Philosophical Transactions of the Royal Society* 279: 583–629.

Deblonde G and Peltier WR (1991) A one dimensional model of continental ice-volume fluctuations through the Pleistocene: Implications for the origin of the mid-Pleistocene climate transition. *Journal of Climate* 4: 18–34.

Deblonde G and Peltier WR (1993) Late Pleistocene ice-age scenarios based upon observational evidence. *Journal of Climate* 6: 709–727.

Delibrias G and Laborel J (1971) Recent variations of the sea level along the Brazilian coast. *Quaterria* 14: 45–49.

Dickey JO, Marcus SL, de Viron O, and Fukumori I (2002) Recent Earth oblateness variations: Unravelling climate and postglacial rebound effects. *Science* 298: 1975–1977.

Dickman SR (1977) Secular trend of the Earth's rotation pole: Consideration of motion of the latitude observatories. *Geophysical Journal of the Royal Astronomical Society* 57: 41–50.

Domack E, Duran D, Leventer A, et al. (2005) Stability of the Larsen B ice shelf on the Antarctic Peninsula during the Holocene epoch. *Nature* 436: 681–685.

Dyurgerov MB and Meier MF (2005) *Glaciers and the changing Earth system: A 2004 snapshot. Occasional Paper No. 58.* Boulder, Colorado: Institute of Arctic and Alpine Research, University of Colorado.

Dunberry M and Bloxham J (2006) Azimuthal flows in the Earth's core and changes in length of day at millennial timescales. *Geophysical Journal International* 165: 32–46.

Dziewonski AM and Anderson DL (1981) Preliminary reference Earth model. *Physics of the Earth and Planetary Interiors* 25: 297–356.

Fairbridge RG (1976) Shellfish-eating preceramic indians in coastal Brazil. *Science* 191: 353–359.

Faure H and Hebrard L (1977) Variations des lignes de ravages au Senegalet au Mauritania au cours de l'Holocene. *Studia Geologica Polonica* 52: 143–157.

Faure H, Fontes JC, Hebrasd L, Monteillet J, and Pirazzoli PA (1980) Geoidal changes and shore-level tilt along Holocene estuaries: Senegal River area, west Africa. *Science* 210: 421–423.

Farrell WE (1972) Deformation of the Earth by surface loads. *Reviews of Geophysics* 10: 761–797.

Farrell WE and Clark JA (1976) On postglacial sea level. *Geophysical Journal of the Royal Astronomical Society* 46: 647–667.

Fisher D (1974) Some more remarks on polar wandering. *Journal of Geophysical Research* 79: 4041–4045.

Gegout P and Cazenave A (1991) geodynamic parameters derived from 7 years of laser data on LAGEOS. *Geophysical Research Letters* 18: 1739–1742.

Gibb JG (1986) A New Zealand regional Holocene eustatic sea-level curve and its application to determination of vertical tectonic movements. *Royal Society New Zealand Bulletin* 24: 377–395.

Gold T (1955) Instability of the Earth's axis of rotation. *Nature* 175: 526–529.

Goldreich P and Toomre A (1969) Some remarks on polar wandering. *Journal of Geophysical Research* 74: 2555–2567.

Gross RS (1986) The influence of earthquakes on the Chandler wobble during 1977–1983. *Geophysical Journal of the Royal Astronomical Society* 85: 161–177.

Gross RS (2000) The excitation of the Chandler wobble. *Geophysical Research Letters* 27: 2329–2342.

Halverson Galen P, Hoffman PF, et al. (2005) Toward a Neoproterozoic composite carbon-isotopic record. *GSA Bulletin* 117: 1181–1207.

Hays JD, Imbrie J, and Shackleton NJ (1976) Variations in the Earth's orbit: Pacemaker of the ice-ages. *Science* 194: 1121–1132.

Hide R, Birch NT, Morrison LV, Shea DJ, and White AA (1980) Atmospheric angular momentum fluctuations an changes in the length of the day. *Nature* 286: 114–117.

Hoffman PF and Schrag DF (2000) Snowball Earth. *Scientific American* 282: 68–75.

Hoffman PF and Schrag DA (2002) The snowball Earth hypothesis: Testing the limits of global change. *Terra Nova* 14: 129–155.

Hongre L, Hulot G, and Khoklov A (1998) An analysis of the geomagnetic field over the past 2000 years. *Physics of the Earth and Planetary Interiors* 106: 311–335.

Hyde WT, Crowley TJ, Baum SK, and Peltier WR (2000) Neoprpterozoic 'snowball Earth' simulations with a coupled climate/ice sheet model. *Nature* 405: 425–430.

Imbrie J, Hays JD, Martinsen DG, et al. (1984) The orbit theory of Pleistocene climate: Support from a revised chronology of the marine δ^{18}0 record. In: Berger A, Imbrie J, Hays J, Kukla G, and Saltzman B (eds.) *Milankovitch and Climate*, pp. 269–306. Norwell, Mass: D. Reidel.

Jault D and Le Mouel J-L (1993) Circulation in the liquid core and coupling with the mantle. In: Singh RP, Feissel M, Tapley BD, and Shum CK (eds.) *Advances in Space Research: Observations of Earth from Space*, vol: 13, pp. (11)221–(11)233. Oxford: Pergamon.

Kirschvink JL (1991) Late Proterozoic low-latitude global glaciation: The snowball Earth. In: Schopf JW and Klein C (eds.) *The Proterozoic Biosphere, A Multi-disciplinary Study*, pp. 51–52. New York: Cambridge University Press.

Kirschvink JL, Ripperdan RL, and Evans DA (1997) Evidence for a large-scale reorganization of Early continental masses by inertial interchange true polar wander. *Science* 277: 541–545.

Koba M, Nakata T, and Takahashi T (1982) Late Holocene eustatic changes deduced from geomorphological features and their C-14 dates in the Ryuku Islands, Japan. *Palaeogeography Palaeoclimatology Palaeoecology* 39: 231–260.

Korte M and Constable C (2003) Continuous global geomagnetic field models for the past 3000 years. *Physics of the Earth and Planetary Interiors* 140: 73–89.

Krabil W, Abdalati W, Frederick E, et al. (2000) Greenland Ice Sheet: high elevation balance and peripheral thinning. *Science* 289: 428–430.

Kuehne J and Wilson CR (1991) Terrestrial water storage and polar motion. *Journal of Geophysical Research* 96: 4337–4345.

Lambeck K (1980a) Changes in length-of-day and atmospheric circulation. *Nature* 286: 104–105.

Lambeck K (1980b) *The Earth's variable Rotation: Geophysical Causes and Consequences*, 449 pp. New York: Cambridge University Press.

Lambeck K (2002) Sea level change from mid-Holocene to recent time: An Australian example with global implications. In: Mitrovica JX and Vermeersen LA (eds.) *AGU Monographs, Vol. 29: Ice Sheets, Sea Level and the Dynamics of the Earth*, pp. 33–50. Washington, DC: AGU.

Lambeck K and Chappell J (2001) Sea level change through the last glacial cycle. *Science* 292(5517): 679–686.

Laskar J (1988) Secular evolution of the solar system over 10 million years. *Astronomy and Astrophysics* 198: 341–362.

Li ZX, Evans DAD, and Zhang S (2004) A 90 degree spin on Rodinia: possible causal links between the Neoproterozoic supercontinent, superplume, true polar wander and low latitude glaciation. *Earth and Planetary Science Letters* 220: 409–421.

Markowitz W (1960) Latitude and longitude and the secular motion of the pole. *Methods and Techniques in Geophysics* 1: 325–361.

Martin L, Sugio K, Flexor JM, Dominguez JML, and Bittencourt ACSP (1987) Quaternary evolution of the central part of the Brazilian coast, the role of relative sea-level variation and of shoreline drift. *Quaternary Coastal Geology of West Africa and South America. UNESCO Reports in Marine Science* 43: 97–145.

Mikhailov AA (1971) On the motion of the Earth's poles. *Astronomichrskij Zhurnal* 48: 1301–1304.

Milne GA and Mitrovica JX (1996) Post-glacial sea level change on a rotating Earth: first results from a gravitationally sel-consistent sea-level equation. *Geophysical Journal International* 126: F13–F20.

Mitrovica JX, Wahr J, Matsuyama I, and Paulson A (2005) The rotational stability of an ice-age Earth. *Geophysical Journal International* 161: 491–506.

Morner NA (1991) Holocene sea level changes in the Tierra del Fuego region. *Bol. IG-USP Publ. Esp* 8: 133–151.

Morrison LV (1973) Rotation of the Earth and the constancy of G. *Nature* 241: 519–520.

Morrison LV and Stephenson FR (2001) Historical eclipses and the variability of the Earth's rotation. *Journal of Geodynamics* 32: 247–265.

Mound JE, Mitrovica JX, Evans DAD, and Kirschvink JL (1999) A sea-level test for inertial interchange true polar wander events. *Geophysical Journal International* 136: F5–F10.

Muller RA and Stephenson FR (1975) The acceleration of the Earth and Moon from early observations. In: Rosenburg GD and Runcorn SK (eds.) *Growth Rhythms and History of the Earth's Rotation*, pp. 459–534. New York: John Wiley.

Munk WH and MacDonald (1960) *The Rotation of the Earth*. New York: Cambridge University Press.

Nakiboglu SM (1982) Hydrostatic theory of the Earth and its mechanical implications. *Physics of the Earth and Planetary Interiors* 28: 302–311.

Nakiboglu SM and Lambeck K (1980) Deglaciation effects on the rotation of the Earth. *Geophysical Journal of the Royal Astronomical Society* 62: 49–58.

Nakiboglu SM and Lambeck K (1981) Deglaciation related features of the Earth's gravity field. *Tectonophysics* 72: 289–303.

Newton RR (1972) *Medieval Chronicals and the Rotation of the Earth*. Baltimore, MD: Johns Hopkins University Press.

Overpeck J and Jansen E et al. (2007) Paleoclimatology. In: Soloman S (ed.) *Climate Change 2007: The Scientific Basis, Working Group 1 Report of the Intergovernmental Panel on Climate Change*. ch. 6, Cambridge, UK: Cambridge University.

Peltier WR (1974) The impulse response of a Maxwell Earth. *Reviews of Geophysics and Space Physics* 12: 649–669.

Peltier WR (1976) Glacial isostatic adjustment. Part II: The inverse problem. *Geophysical Journal of the Royal Astronomical Society* 46: 669–706.

Peltier WR (1982) Dynamics of the ice-age Earth. *Advances in Geophysics* 24: 1–146.

Peltier WR (1983) Constraint on deep mantle viscosity from LAGEOS acceleration data. *Nature* 304: 434–436.

Peltier WR (1985) The LAGEOS constraint on deep mantle viscosity: Results from a new normal mode method for the inversion of visco-elastic relaxation spectra. *Journal of Geophysical Research* 90: 9411–9421.

Peltier WR (1994) Ice-age paleotopography. *Science* 265: 195–201.

Peltier WR (1996a) Physics of the ice-age cycle. In: Duplessy J-C (ed.) *Long Term Climate Variations*, pp. 453–481. New York: Springer-Verlag Inc.

Peltier WR (1996b) Mantle viscosity and ice-age ice-sheet topography. *Science* 273: 1359–1364.

Peltier WR (1996c) Phase transition modulated mixing in the mantle of the Earth. *Philosophical Tansactions of the Royal Society (London) Series A* 354: 1425–1447.

Peltier WR (1998) Postglacial variations in the level of the sea: implications for climate dynamics and solid-Earth geophysics. *Reviews of Geophysics* 36: 603–689.

Peltier WR (1999) Global sea level rise and glacial isostatic adjustment. *Global and planetary change* 20: 93–123.

Peltier WR (2002a) Comments on the paper of Yokoyama *et al.* (2000) entitled 'Timing of the last glacial maximum from observed sea level minima. *Quaternary Science Reviews* 21: 409–414.

Peltier WR (2002b) One eustatic sea level history, Last Glacial Maximum to Holocene. *Quaternary Science Reviews* 21: 377–396.

Peltier WR (2002c) Global glacial isostatic adjustment: Paleo-geodetic and space-geodetic tests of the ICE-4G (VM2) model. *Journal of Quarternery Sciences* 17: 491–510.

Peltier WR (2004) Global glacial isostasy and the surface of the ice-age Earth: the ICE-5G(VM2) model and GRACE. *Annual Review of Earth and Planetary Sciences* 32: 111–149.

Peltier WR (2005) On the hemispheric origins of meltwater pulse 1a. *Quaternary Science Reviews* 24: 1655–1671.

Peltier WR and Fairbanks RG (2006) Global glacial ice-volume and Last Glacial Maximum duration from an extended Barbados Sea Level record. *Quaternary Science Reviews* 25: 3322–3337.

Peltier WR, Farrell WE, and Clark JA (1978) Glacial isostatsy and relative sea-level: A global finite element model. *Tectonophysics* 50: 81–110.

Peltier WR and Solheim LP (1992) Mantle phase transitions and layered chaotic convection. *Geophysical Research Letters* 19: 321–324.

Peltier WR and Jiang X (1996) Glacial isostatic adjustment and Earth rotation: Refined constraints on the viscosity of the deepest mantle. *Journal of Geophysical Research* 101: 3269–3290 (correction *Journal of Geophysical Research* 102: 10101–10103, 1997).

Peltier WR, Shennan I, Drummond R, and Horton B (2002) On the postglacial isostatic adjustment of the British Isles and the shallow visco-elastic structure of the Earth. *Geophysical Journal International* 148: 443–475.

Peltier WR and Solheim LP (2004) The climate of the Earth at Last Glacial Maximum: statistical equilibrium state and a mode of internal variability. *Quarternary Science Reviews* 23: 335–357.

Peltier WR, Tarasov L, Vettoretti G, and Solheim LP (2004) Climate dynamics in deep time: modeling the 'snowball bifurcation' and assessing the plausibility of its occurrence. In: Jenkins G, McMennamin MAS, McKay CP, and Sohl L (eds.) *AGU Monograph 146: The Extreme Proterozoic: Geology, Geochemistry, and Climate*, pp. 107–124. Washington, DC: AGU Press.

Pirazzoli P (1978) High stands of Holocene sea levels in the NW Pacific. *Quaternary Research* 10: 1–29.

Platzman GW (1971) Ocean tides and related w.aves. In: William HR (ed.) *Mathematical Problems in the Geophysical Sciences, Vol 2: Inverse Problems, Dynamo Theory and Tides*, pp. 239–291. Providence Rhode Island: American Mathematical Society.

Ponte RM, Stammer D, and Marshall J (1998) Oceanic signals in observed motions of the Earth's pole of rotation. *Nature* 391: 476–479.

Porter SC, Stuiver M, and Heusser CJ (1984) Holocene sea level changes along the straits of Megellan and Beagle Channel, Southernmost South America. *Quaternary Research* 22: 59–67.

Proverbio E and Quesada V (1974) Secular variation in latitudes and longitudes and continental drift. *Journal of Geophysical Research* 79: 4941–4943.

Quinn TR, Tremaine S, and Duncan M (1991) A three million year integration of the Earth's orbit. *Astronomical Journal* 101: 2287–2305.

Rochester MG and Smiley DE (1965) Geomagnetic core–mantle coupling and the Chandler wobble. *Geophysical Journal of the Royal Astronomical Society* 10: 289–315.

Rostami K, Peltier WR, and Mangini A (2000) Quarternary marine terraces, sea level changes and uplift history of Patagonia, Argentina: Comparisons with predictions of the ICE-4G (VM2) model of the global process of glacial isostatic adjustment. *Quaternary Science Reviews* 19: 1495–1525.

Rubincam DR (1984) Postglacial rebound observed by LAGEOS and the effective viscosity of the lower mantle. *Journal of Geophysical Research* 89: 1077–1087.

Sabadini R and Peltier WR (1981) Pleistocene deglaciation and the Earth's rotation: implications for mantle viscosity. *Geophysical Journal of the Royal Astronomical Society* 66: 553–578.

Sauriau A and Cazanave A (1985) Re-evaluation of the seismic excitation of the Chandler wobble from recent data. *Earth and Planetary Science Letters* 75: 410–416.

Shackleton NJ (1967) Oxygen isotope analysis and Pleistocene temperatures re-addressed. *Nature* 215: 15–17.

Shackleton NJ (2000) The 100,000 year ice-age cycle identified and found to lag temperature, carbon dioxide, and orbital eccentricity. *Science* 289: 1897–1902.

Shackleton NJ and Opdyke ND (1973) Oxygen isotope and paleomagnetic stratigraphy of equatorial Pacific core V28-238: Oxygen isotope temperatures and ice volumes on a 10^5-year time scale. *Quaternary Research* 3: 39–55.

Shackleton NJ, Berger A, and Peltier WR (1990) An alternative astronomical calibration of the lower Pleistocene timescale based upon ODP 677. *Transactions of the Royal Society of Edinburgh: Earth Sciences* 81: 251–261.

Schofield JC (1975) Sea-level fluctuations cause periodic post-glacial progradation, South Kaipara Barrier, North Island, New Zealand. *New Zealand Journal of Geology and Geophysics* 18: 295–316.

Solheim LP and Peltier WR (1994a) Avalanche effects in phase transition modulated thermal convection: A model of Earth's mantle. *Journal of Geophysical Research* 99: 6997–7018.

Solheim LP and Peltier WR (1994b) 660 km phase boundary deflections and episodically layered isochemical convection. *Journal of Geophysical Research* 99: 15861–15875.

Stephenson ER and Morrison LV (1995) Long term fluctuations in the earth's rotation: 700 B.C. to A.D. 1990. *Philosophical Transactions of the Royal Society (London) Series A* 351: 165–202.

Stuiver M and Reimer PJ (1993) Extended ^{14}C data base and revised calib. 3.0 ^{14}C age calibration program. *Radiocarbon* 35: 215–230.

Sugihara K, Nakamori T, Iryu Y, Sasakai K, and Blanchon P (2003) Holocene sea-level change and tectonic uplift deduced from raised reef terraces, Kikai-jima, Ryuku Islands, Japan. *Sedimentary Geology* 159: 5–25.

Tarantola A and Valette B (1982a) Inverse problems=quest for information. *Journal of Geophysics* 50: 159–170.

Tarantola A and Valette B (1982b) Generalized nonlinear inverse problems solved using the least squares criterion. *Reviews of Geophysics* 20: 219–232.

Tarasov L and Peltier WR (2002) Greenland glacial history and local geodynamic consequences. *Geophysical Journal International* 150: 198–229.

Tarasov L and Peltier WR (2005) Arctic freshwater forcing of the Younger-Dryas cold reversal. *Nature* 435: 662–665.

Tarasov L and Peltier WR (2006) A calibrated deglacial chronology for the North American continent: Evidence of an Arctic trigger for the Younger-Dryas event. *Quarternary Science Reviews* 25: 659–688.

Valastro S, Jr., Davis EM, Vallrela AG, and Ekland-Olson C (1980) University of Texas at Austin radiocarbon dates XIV. *Radiocarbon* 22: 1090–1115.

Vincente RO and Yumi S (1969) Co-ordinates of the pole (1899-1968) returned to the conventional international origin. *Publication of the International Latitude Observatory of Mizusawa* 7: 41–50.

Vincente RO and Yumi S (1970) Revised values (1941-1961) of the co-ordinates of the pole referred to the CIO. *Publications of the International Latitude Observatory of Mizusawa* 7: 109–112.

Vondrak J (1984) Long period behaviour of polar motion between 1900.0 and 1984. *Annals of Geophysics* 21: 351–356.

Waelbroeck C, Labyrie L, Michel E, *et al.* (2002) Sea-level and deep water temperature changes derived from benthic foraminifera isotopic records. *Quaternary Science Reviews* 21(1–3): 295–305.

Wahr JM (1982) The effects of the atmosphere and oceans on the Earth's wobble-I. Theory. *Geophysical Journal of the Royal Astronomical Society* 70: 349–372.

Wahr JM (1983) The effects of the atmosphere and oceans on the Earth's wobble and on the seasonal variations in the length of day-II. Results. *Geophysical Journal of the Royal Astronomical Society* 74: 451–487.

Widder DV (1946) *The Laplace Transform, ch. V*. Princeton, NJ: Princeton University Press.

Williams GE (2000) Geological constraints on the Precambrian history of Earth's rotation and the Moon's orbit. *Reviews of Geophysics* 38: 37–59.

Wilson CR and Haubrich RA (1976) Meteorological excitation of the Earth's wobble. *Geophysical Journal of the Royal Astronomical Society* 46: 707–743.

Wilson CR and Vincente RO (1990) Maximum likelihood estiumates of polar motion parameters. In: McCarthy DD and Carter WE (eds.) *American Geophysical Union Monograph Series: Variations in Earth Rotation*, pp. 151–155. Washington DC: AGU.

Wu P and Peltier WR (1984) Pleistocene deglaciation and the Earth's rotation: A new analysis. *Geophysical Journal of the Royal Astronomical Society* 76: 202–242.

Yoder CF, Williams JG, Dickey JO, Schutz BE, Eanes RJ, and Tapley BD (1983) Secular variation of the Earth's gravitational harmonic J_2 coefficient from LAGEOS and non-tidal acceleration of Earth rotation. *Nature* 303: 757–762.

Yoder CF (1995) Astrometric and geodetic properties of Earth and the solar systems. http://www.agu.org/reference/gephys/4_oder.pdf.

Yokoyama Y, Lambeck K, DeDekkar P, Johnston P, and Fifield LK (2000) Timing of the Last Glacial Maximum from observed sea level minima. *Nature* 406: 713–716 (correction 2001, *Nature* 412: 99).

11 Coevolution of Life and Earth

G. J. Retallack, University of Oregon, Eugene, OR, USA

When I hear the sigh and rustle of my young woodlands, planted with my own hands, then I know that I have some slight share in controlling the climate – Anton Chekhov (1899)

11.1 Introduction

Earth is one of the oldest words in our language (from Old English 'eorde') meaning both our planet (customarily capitalized in this sense) as well as the soil beneath our feet. Here I explore both ends of its meaning by emphasizing how small-scale biological and soil-forming processes have played a role in the very significant differences between our planet and other planetary bodies in our solar system. Such processes as the growth of trees are commonly regarded as mundane, in the sense of commonplace or trivial, but they can be mundane also in the more important sense of global, as apparently understood by Chekhov (1899). The media which elevate biological processes to global scope are air, water, and soil. Air and water are important to the coevolution of life and earth (Nisbet and Sleep, 2001), but we will take the path less traveled here, and present a perspective on coevolution of life and earth from my research experience with ancient and modern soils.

Soil is Anglo-French word, introduced from the ancient Roman 'solium' (seat) or 'solum' (ground). Soil is so central to human existence that it is as difficult to define as 'home' and 'love'. To some soil scientists, soil is material altered by the action of organisms, especially plant roots. To engineers, soil is material moved by mechanized graders without recourse to blasting. To farmers, soil is the surficial organic layer that nurtures crops and feeds livestock (Hole, 1981). We prefer a wider definition of soil as materials on the surface of a planetary body altered in place by physical, biological, or chemical processes. This definition has also been adopted by NASA (National Aeronautic and Space Administration of the USA) press releases on Lunar and Martian soils, for the compelling reason that they wished to be understood by the general public. In any case, the surface of the Moon and Mars is now altered locally by robotic landers and footprints, and our future

alterations can be anticipated. A broad definition of soil obviates the need for such terms as regolith or the distinction between physicochemical and biological weathering (Taylor and Eggleton, 2001). More important to our present purpose, there is no longer an assumption that life created soil. The questions now become when and how did life evolve in soil? Did soil nurture the origin of life, or did life evolve in the ocean or volcanic hot springs, then later adapt to soil? These questions remain open (Nisbet and Sleep, 2001).

When soil, air, and water became alive, unlike the apparently sterile but discernably altered soils of the Moon, Venus, and Mars (**Figure 1**), it must have been an important time in the history of our Earth. Soils gave life access to fundamental nutrients such as phosphorus at their source in the weathering of rocks. Soils allowed life to colonize a significant fraction of Earth's surface, thus enhancing their volumetric effect on surficial fluids. Life gave soils increased

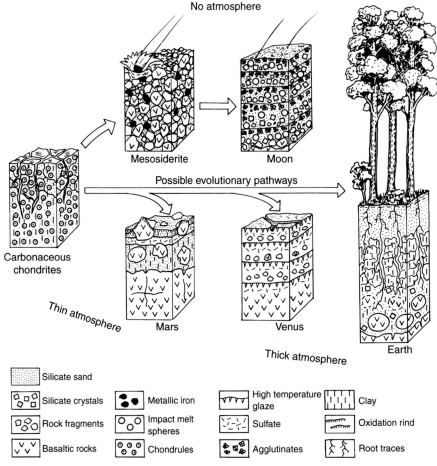

Figure 1 A schematic comparison of soil formation on Earth, Moon, Mars, Venus, and some hypothetical meteorite parent bodies. From Retallack GJ (2001a) *Soils of the Past*, 2nd edn., 600pp. Oxford: Blackwell.

depth and stability to enhance rates of physicochemical weathering. Life also gave soils a variety of biosignature horizons and deep weathering products (Retallack, 2001a).

The evolution of life and soil can be viewed as a coevolutionary process, like the coevolution of grasses and grazers, as first proposed by Kowalevsky (1873). Coevolution is the coordinated evolution of phylogenetically unrelated organisms (in this case, plants and animals), which coevolved to enhance their mutual interdependence. Grasses evolved subterranean rhizomes, basal tillering of adventitious roots, intercalary meristems, telescoped internodes, and opal phytoliths to withstand grazing (**Figure 2**).

Horses evolved cursorial limb structure for predator escape over open county and high-crowned teeth to withstand abrasion from leaves studded with opal phytoliths (MacFadden, 1992). Grasses and grazers coevolved to produce ecosystems, such as grasslands, and soils, such as Mollisols, that became biotic planetary forcings of the atmosphere and hydrosphere (Retallack, 2001b). Because coevolution is directed more toward other organisms than to physicochemical environments, Earth's environment coevolved with life far from primordial conditions exemplified by other planetary bodies (**Figure 1**).

The coevolution of life and soil is evident from functional intricacies of modern ecosystems, but the

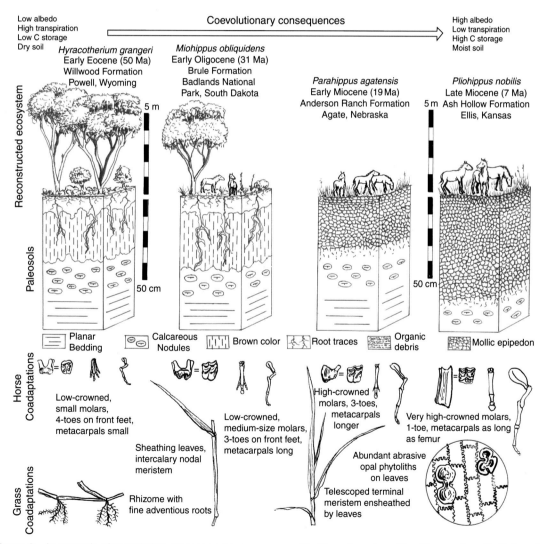

Figure 2 An example of coevolution, grasses and grazers, exemplified by paleosols and fossil horses from North America: grass evolutionary stages are less well constrained by the fossil record. Data from Retallack GJ (2001b) Cenozoic expansion of grasslands and global cooling. *Journal of Geology* 109: 407–426.

details of when and how this happened are shrouded in the mists of time. It is still uncertain, indeed largely hypothetical, when and how life first came to soils. There were changes in atmospheric oxidation and water salinity on the early Earth, but it remains uncertain which of these are due to soil colonization, as opposed to oceanographic, volcanic, or tectonic changes. Fortunately, there is a remarkable record of paleosols in very ancient rocks on Earth land (Rye and Holland, 1998) and on different planetary bodies (Retallack, 2001a), which furnish evidence on past conditions of climate and life. This supplements the geological record of life and environments in marine, volcanic, and metamorphic rocks (Schopf, 1983; Schopf and Klein, 1991). Paleosol development and geochemistry also provide relevant tests for competing hypotheses of the development of continents and plate tectonics.

11.2   Barren Worlds as Null Hypotheses

Uniqueness of our planet is best considered by comparison with other planetary bodies, which turn out to have a variety of common features, such as basalt and CO_2-rich atmosphere. Water beyond Earth is frozen or vaporized, with the possible exception of Jupiter's moon Europa (Marion et al., 2005). Impact glasses, shocked minerals, clays, evaporites, carbonates, oxides, and a variety of high-temperature minerals also are found beyond Earth (Squyres et al., 2004). Also found beyond Earth are rugged desert-like landscapes, volcanos, and ice caps (Carr, 1981). Such pervasive similarities to Earth make the search for biosignatures and their coevolutionary effects difficult, but before turning to that topic, consider some other worlds currently considered barren of life as null hypotheses.

11.2.1   Moon

At a small fraction of Earth mass (0.01), our Moon has insufficient gravitational pull to retain an atmosphere. Its temperature fluctuations of 111°C to −171°C also rule out liquid water (Taylor, 1982). Its soil appears waterless (Campbell et al., 2006), despite past proposals of small areas of polar ice (Feldman et al., 1998). The main soil-forming process on the moon is micrometeoroid bombardment, which enriches basaltic breccias and rocks exposed at the surface in meteoritic iron–nickel, impact glass, agglutinates of glass and minerals, and shattered mineral grains (**Figure** 1). Well developed lunar soils 5–10 cm thick take some 200 million years to form, judging from rate of micrometeoroid bombardment (McKay and Basu, 1983) and the succession of paleosols in the radiometrically dated Apollo 15 core (Heiken et al., 1976). The most profound soil-forming event in the history of the Moon was the footprints, tracks, and excavations of Apollo astronauts.

11.2.2   Venus

Venus is slightly smaller (0.8) than the mass of Earth, but its thick (93 bar) atmosphere is a toxic mix of greenhouse gases (97% CO_2) creating surface temperatures of about 470°C. Water is not stable in the soil and water vapor is rare (0.1–0.4%) in the atmosphere. Soil formation on the basaltic parent material may be a rapid process akin to pottery glazing (**Figure** 1), in which salts are melted to obscure the hard edges of rocks and sediments (Barsukov et al., 1982).

11.2.3   Mars

Venus and our Moon stretch concepts of soil formation almost to breaking point, but Mars has more familiar ground. Mars is small (0.1 Earth mass), with a thin (0.008 bar) atmosphere, mainly CO_2 (95%), but with small amounts of N_2 (3%) and Ar (2%). Trace amounts (up to 0.003%) of CH_4 show temporal and geographic fluctuation suggestive of higher local concentrations (Formisano et al., 2004). The Viking lander at Chryse Planitia recorded temperatures from −90°C at dawn to −30°C at noon (Carr, 1981). Martian soils are thin with smectite clays and sulfate minerals (gypsum, kieserite) which indicate that basalt and komatiite parent materials have been altered in aqueous solution by hydrolysis (incongruent dissolution in carbonic acid). This kind of soil formation is slight compared with millions of years for similar soil formation in the Dry Valleys of Antarctica (Campbell and Claridge, 1987) or the Atacama Desert of Chile (Navarro-González et al., 2003). Soil formation may be even slower on Mars, because known soils appear to be paleosols from at least 3.2 Ga, formed at a time of more plentiful water at the surface indicated by outflow channels, then subsequently frozen at the surface during cold and dry subsequent eons (Retallack, 2001a). The red color of Martian soils is a very thin (<10 μm) rind, probably created by long-term surficial photooxidation (Yen et al., 2000). Mars has a stunning array of familiar landscapes including large volcanos, deep canyons, impact craters surrounded by solifluction flows, eolian dunes, and evaporite basins (Squyres *et al.*, 2004). Life-like microstructures, organic chemicals, magnetite, and carbonates from a Martian meteorite found in Antarctica (McKay *et al.*, 1996) raised hopes as microfossils, implying life on Mars as recently as 3.9 Ga. Although these remains are suggestive, they are not compelling, and Mars is widely considered sterile, now and in the geological past (Schopf, 1999).

11.2.4 Asteroids

The soils of asteroids have only been visited recently (Veverka *et al.*, 2001) and are brecciated by impact like lunar soils and some rare kinds of meteorites (mesosiderites and howardites; Bunch, 1975). Very different asteroidal soils are represented by carbonaceous chondrite meteorites, which show no evidence of living creatures, but include complex carbonaceous molecules, such as amino acids, dicarboxylic acids, sulfonic acids, and aliphatic and aromatic hydrocarbons (Pizzarello *et al.*, 2001). These carbonaceous compounds are bound up within fine-grained phases of smectite and other clay minerals formed at low temperatures (Tomeoka and Busek, 1988) in contrast to high-temperature pyroxene, olivine, and other high-temperature minerals of the chondrules (Itoh and Yurimoto, 2003). The clays form gradational contacts like weathering rinds, pseudomorphic replacement of chondrules, concentric cavity linings, and birefringence fabrics of oriented clay like those formed in soils under low confining pressures (Bunch and Chang, 1980). Other evidence of hydrous alteration are framboids and plaquettes of magnetite (Kerridge *et al.*, 1979). There also are cross-cutting veins of calcite, siderite, gypsum, epsomite, and halite (Richardson, 1978).

Clay and salts are common products of hydrolytic weathering in soils of Earth, and can thus be inferred for the parent bodies of carbonaceous chondrites, which are thought to have been small (<500 km diameter) planetesimals. High-temperature chondrules of carbonaceous chondrites are 4.56 Ga in age (Amelin *et al.*, 2002), and the plaquettes of magnetite (Lewis and Anders, 1975), and veins of calcite (Endress *et al.*, 1996) and halite (Whitby *et al.*, 2000) no younger than 4.51 Ga. Degassing of CO_2 and H_2O vapor from planetesimals early in the accretion of the solar system resulted in surprisingly Earth-like weathering (**Figure 2**).

This account has stressed a weathering interpretation of carbonaceous chondrites, but some of these also include talc and other minerals suggestive of high-temperature (250–300°C) hydrothermal alteration (Brearley, 1997). There is also a school of thought that the clay and organic matter of carbonaceous chondrites are not alteration products, but instead cold condensates from the original nebula from which the solar system formed (Wasson, 1985).

11.3 Biosignatures of Earth

Several unique features of Earth compared with other planetary bodies can be considered biosignatures: features indicative of life. Some of these such as inflammable atmosphere, permanent liquid water, carbon-based life, and civilizations feature prominently in the search for life elsewhere in the universe. Biosignatures of Earth's soils include calcic, argillic, and mollic horizons. Salic horizons are known from Mars and carbonaceous chondrites. Carbon-rich surface horizons may be represented by carbonaceous chondrites. It is not yet clear whether deep hydrolytic weathering occurs on other planetary bodies (Squyres *et al.*, 2004). Other features such as convex slopes, spheroidal deep weathering, and granitic rocks unique to Earth are less obviously related to life's influence, but nevertheless worth consideration.

11.3.1 Inflammable Atmosphere

Our atmosphere is a combustible mix of oxygen (21%) and methane (0.00017%), with its reacted product carbon dioxide (preindustrial 0.028%, 0.037% in 2001, and rising), diluted with much inert nitrogen (78%). It is thus more like input gases of a carburettor than the exhaust gases of an automobile tailpipe, and quite distinct from the CO_2-rich composition of other planetary atmospheres and volcanic gases (Lovelock, 2000). Burning of other flammable substances such as hydrocarbons and wood in the open air is thus possible on our planet, unlike other planetary bodies. More than trace amounts of oxygen in the atmosphere dates back at least 2.3 Ga (Bekker *et al.*, 2004). Entropy reduction or chemical disequilibrium of planetary atmospheric composition is a way of identifying life elsewhere in the universe, and also an argument for the Gaia hypothesis of life's control of atmospheric composition (Lovelock, 2000).

Table 1   Metabolic processes of geological significance

I. Fermentation

$$C_6H_{12}O_6 \rightarrow 2C_2H_5OH + CO_2 + energy$$

Sugar   Alcohol   Carbon dioxide

II. Respiration

$$C_6H_{12}O_6 + O_2 \rightarrow 6CO_2 + 6H_2O + energy$$

Sugar   Oxygen   Carbon dioxide   Water

III. Anaerobic photosynthesis

$$6CO_2 + 12H_2S \xrightarrow{light} C_6H_{12}O_6 + H_2O + 12S$$

Carbon dioxide   Hydrogen suflide   Bacteriophyll   Sugar   Water   Sulfur

IV. Aerobic photosynthesis

$$6CO_2 + 12H_2O \xrightarrow{light} C_6H_{12}O_6 + 6H_2O + 6O_2$$

Carbon dioxide   Water   Chlorophyll   Sugar   Water   Oxygen

The reason for the cosmically peculiar composition of our atmosphere may be the widespread metabolic process of photosynthesis (Rosing *et al.*, 2006) not only by plants, and a variety of protists, but also by a host of bacteria as well. Photosynthesis, or assembly by light, is a process by which sugars are synthesized using light from the Sun and a catalyst (commonly chlorophyll) from CO_2 of the atmosphere (**Table 1**). The low amount of CO_2 in our current atmosphere is a testament to the success of photosynthesis, but replacement of a primordial reducing atmosphere of CO_2 and CH_4 by O_2 had many ups and downs over 3.5 Ga of Earth's history known from the rock record (Berner *et al.*, 2000).

11.3.2   Permanent Liquid Water

Our pale dot of a planet, blue with oceans and wreathed with water vapor, is unique in our solar system now, and also in its deep history of water at the surface (Lovelock, 2000). On a cosmic scale of temperatures from $-270°C$ (2.76 K) in deep space to about 6000°C for the surface of the Sun, the 0–100°C range of liquid water is a narrow range not currently found on any neighboring planetary bodies, though perhaps present within Europa (Marion *et al.*, 2005), and on Mars in the distant geological past, some 2.5–3.5 Ga (Carr, 1981). It did not last on Mars, but on Earth there is a continuous record of water-lain sedimentary rocks and precipitates from 3.5 Ga to the present (Allwood *et al.*, 2006). Oxygen-isotopic composition of detrital zircons individually dated to 4.3 Ga in the Jack Hills of Western Australia are unusually heavy (15.3 vs typical mantle values of 5.4‰ $\delta^{18}O$ vs SMOW) and suggest free surficial water at the time their parent granitoids cooled only 200 Ma after the formation of Earth (Mojzsis *et al.*, 2001). This is a long and continuous history of surficial water despite degassing of largely greenhouse gases to a thick (1 bar) atmosphere and a 30% increase in solar luminosity due to stellar evolution over the same period. Permanent water and its longevity through planetary history is a clue to life on other planets, as well as an argument for Gaian environmental regulation (Lovelock, 2000).

The reason for the maintenance of Earth's temperature within the bounds of liquid water for some 4.3 Ga remains uncertain, but life is suspected to have a role because of its coevolved metabolic systems. This general idea of a thermocouple of opposing forces acting as a thermostat is clearly expressed in 'Daisyworld' albedo models of the Gaia hypothesis (Watson and Lovelock, 1983). Imagine a world with only black and white daisies vying for the light of the Sun in order to photosynthesize. White daisies cool by reflecting sunlight back into space. Black daisies warm by absorbing heat. Populations will be dominated alternately by black then white daisies, until mixed populations converge on temperatures that optimize photosynthesis. A more realistic model is the 'Proserpina principle' (Retallack, 2004a), which postulates that photosynthesis cools the planet by drawing down the greenhouse gas CO_2, whereas respiration warms the planet because it draws down O_2. If either one of these metabolic systems had evolved in isolation it would have resulted in a respirator's hell like Venus, or a photosynthesizer's freezer like Mars. Fortunately, photosynthesizers and respirers such as plants and animals are mutually interdependent for food and breath: a coevolutionary thermostat. Plants cool the planet by photosynthesis which is curtailed as they are covered by snow of icehouse atmospheres. Animals warm the planet by respiration but die in high temperatures of greenhouse atmospheres that are less fatal to plants. Population balances between them have the effect of adjusting greenhouse gases in the atmosphere to habitable ranges given external inputs of solar radiation.

11.3.3 Carbon-Based Life

Science fiction movies make alien life instantly recognizable by emphasizing the salient features of humans or insects. Recognizing alien life on sight may not be so easy because much life is immobile or microscopic, and its activities too varied to be easily characterized (**Table 2**). Life on Earth has a marked preference for six elements: carbon, hydrogen, oxygen, nitrogen, phosphorus, and sulfur (CHONPS: Schoonen et al., 2004). Of these only oxygen is a common component of most rocks. Three general features are regarded as necessary for life: bodies, metabolism, and reproduction. Even bodies of such simple organisms as the common gut bacterium *Escherichia coli* have an astonishing complexity of interacting parts, including high-molecular-weight organic compounds which provide both structure and function (**Figure 3**). If there is a common theme to the great variety of metabolic reactions, it is lack of chemical equilibrium that drives them and that has left an imprint on our cosmically peculiar atmosphere (Lovelock, 2000). Reproduction is also quite varied, ranging from simple cell division, to clonal budding, parthenogenesis, and the romantic complexity of sex. If there is a common theme to biological reproduction it is remarkable speed. Astonishing numbers of individuals can be produced in days or years as long as food is available. Whole worlds can be populated in a geological instant, which should be encouraging to the search for life on other worlds. If life is anywhere on a planet, it is likely to be everywhere.

The difficulties of recognizing life are well illustrated by debate over evidence of life on the early Earth and a Martian meteorite. Very ancient photosynthetic activity is suggested by the carbon-isotopic ratios of amorphous organic matter in the 3.8 Ga Isua

Greenstone Belt of west Greenland (Rosing and Frei, 2004), although the high metamorphic grade and discernable modern organic contamination are concerns (Nishizawa et al., 2005). The oldest microfossil evidence for life has long been considered chains of cyanobacteria-like carbonaceous cells in chert from the 3.5 Ga Apex Chert of the Warrawoona Group near Marble Bar, Western Australia (Schopf, 1983). The poor preservation and hydrothermal setting of these fossils has recently been urged as evidence that they are not microfossils, but inorganic carbonaceous material (Brasier et al., 2002). In my opinion the distortion and poor preservation of the microfossils, and especially the rounded cavities in them (Schopf et al., 2002), are evidence of both cyanobacterial primary production and actinobacterial decay. Carbonaceous and mineral pseudofossils have greater regularity of construction and resistance to degradation than the Warrawoona microfossils (Garcia-Ruiz et al., 2003). In any case, alternative evidence of life from microbial etching of pillow laves of the Barberton Greenstone Belt of South Africa is also dated to about 3.5 Ga (Furnes et al., 2004).

The Martian meteorite (ALH84001) with putative microfossils recovered from ice near Allan Hills, Antarctica (McKay et al., 1996) is of an unusual type (SNC) thought to have been derived from Mars because of its igneous (nonchondritic) microtexture, geological age younger than the 4.6 Ga age of most meteorites, unusual oxygen-isotopic composition, lack of magnetic paleointensity, and evidence for long exposure to interplanetary cosmic rays (Wood and Ashwal, 1981). Meteorite ALH84001 groundmass is 4.5 Ga in age, but was shock metamorphosed at 4.0 Ga, then infiltrated by secondary carbonate with putative microfossils at 3.9 Ma (Borg et al., 1999). It was lofted from Mars by about 16 Ma and orbited the Sun until falling into Antarctic ice at about 13 Ka (McKay et al., 1996). The putative microfossils are cylindrical shapes arranged end to end in short strings, with associated polycyclic aromatic hydrocarbons and lozenges of magnetite comparable with magnetotactic bacteria and nannobacteria on Earth (Folk and Lynch, 1997). Unfortunately, the cell-like structures are too small (0.02–0.1) to house the complex molecular machinery of even the simplest known bacteria (**Figure 3**), and are more likely organic coaggulates or artefacts of scanning electron microscopy (Schieber and Arnott, 2003). The organic compounds have isotopic composition (Jull et al., 1998) and racemization like Earthly contaminants (Bada et al., 1996). The magnetite has nonbiological

Table 2 Kinds of microbes, their metabolic requirements and role

General role	Kind of organism	Example genus	Energy source	Electron donor	Carbon source	Oxygen relations	Comments
Carbon cyclers	Algae	Chlamydomonas	Sunlight (P)	H_2O (L)	CO_2 (A)	Aerobic	Primary producer normal-wet soils, ponds, and ocean
	Cyanobacteria	Nostoc	Sunlight (P)	H_2O (L)	CO_2 (A)	Aerobic	Primary producer normal-wet soils
	Purple non-sulfur bacteria	Rhodospirillum	Sunlight (P) sometimes organic compounds (C)	Organic compounds (O) sometimes H_2S (L)	Organic compounds (H) sometimes CO_2 (A)	Amphiaerobic	Primary producer in swamps and stagnant ponds and ocean
	Methanogenic bacteria	Methanobacterium	H_2 (C)	H_2 (L)	CO_2 (A) sometimes formate (H)	Anaerobic	Creators of 'swamp gas' (CH_4) in stagnant ponds and ocean
	Aerobic spore-forming bacteria	Bacillus	Organic compounds (C)	Organic compounds (O)	Organic compounds (H)	Aerobic	Decomposer of organic compounds soils, lakes, and oceans
	Fermenting bacteria	Clostridium	Organic compounds (C)	Organic compounds (O)	Organic compounds (H)	Anaerobic	Decomposer in swamp soils, lakes, and oceans
	Protoctistans	Amoeba	Organic compounds (C)	Organic compounds (O)	Organic compounds (H)	Aerobic	Predator in soils, lakes, and oceans
Nitrogen cyclers	Nitrogen-fixing bacteria	Azotobacter	Organic compounds (C)	N_2 (L)	Organic compounds (H)	Aerobic	Creates ammonium (NH_4^+) in soils, lakes, and oceans
	Root nodule bacterioids	Rhizobium	Organic compounds (C)	N_2 (L)	Organic compounds (H)	Aerobic	Supplies ammonium (NH_4^+) to host plant in well-drained soils
	Denitrifying bacteria	Pseudomonas	NO_2 (C) sometimes organic compounds (C)	N_2 (L) sometimes organic compounds (O)	CO_2 (A) sometimes organic compounds (O)	Anaerobic	Releases nitrogen (N_2) to atmosphere from swampy soils
Sulfur cyclers	Sulfur-reducing bacteria	Desulfovibrio	Organic compounds (C)	SO_4^{2-} (L)	Organic compounds (H)	Anaerobic	Creates 'rotten egg gas' (H_2S) and pyrite is swamps, lakes, and oceans
	Sulfur bacteria	Chromatium	Sunlight (P)	H_2S or S (L) sometimes organic compounds (O)	CO_2 (A) sometimes organic compounds (H)	Amphiaerobic	Remobilizes sulfur in poorly oxygenated soils, lakes, and oceans
	Sulfur metabolising bacteria	Thiobacillus	S, FeS (C)	S, SO_4^{2-}, Fe_2O_3 (L)	CO_2 (A)	Aerobic	Remobilizes sulfur in wet-normal soils, lakes, and oceans

Initials for trophic groups are autotrophic (A), chemotrophic (C), heterotrophic (H), lithotrophic (L), organotrophic (O), phototrophic (P). From Pelczar MJ, Chan ECS, Krieg NR, and Pelczar ME (1986) *Microbiology*, 918 pp. New York: McGraw-Hill.

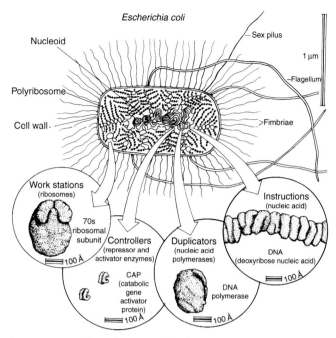

Figure 3 *Escherichia coli*, a common gut bacterium, considered as an organic machine of astonishing complexity. From Retallack GJ (2001a) *Soils of the Past*, 2nd edn., 600pp. Oxford: Blackwell.

screw dislocations and the carbonate may have formed at high temperatures (McSween, 1997). Life on Mars thus remains unproven.

11.3.4 Spheroidal Weathering

Spheroidal weathering is the deep weathering of rocks of uniform composition, especially granitic rocks, to create rounded concentric rings of alteration inward from deep cracks (**Figure 4(a)**). Onion-skin weathering is a particularly aggressive form of spheroidal weathering in which the whole rock has been altered to concentric shells. More common is a few shells of alteration around a rounded remnant of little altered rock in the center, a corestone. Rounded corestones resist erosion more strongly than soft surrounding rock and overlying soil profile (Strudley *et al.*, 2006), and are exhumed as distinctive rounded boulders called tors (**Figure 4(b)**). No landforms or deep-weathering of these kinds have yet been found

Figure 4 Spheroidal weathering and tors. (a) Paleoproterozoic (2.3 Ga) spheroidally weathered corestones in a paleosol developed on pink alkali granite and covered Huronian Supergroup conglomerates, near Elliot Lake, Ontario, Canada (hammer gives scale). (b) Tors of Baynton Granite exhumed from Miocene paleosol, 4 km north of Pyalong, Victoria, Australia (tree is 5 m tall).

on the Moon, Venus, Mars, or asteroids, but this kind of deep weathering is common in deeply weathered Precambrian paleosols as old as 2.3 Ga (**Figure 4(a)**). If spheroidal weathering has such geological antiquity one would expect tors to be ancient landforms as well, but the oldest likely documented case known to me is on the pre-Stoer (1.2 Ga) paleosol near Stoer, northwest Scotland (Stewart, 2002).

Spheroidal weathering and exhumed tors require deep weathering by acidic solutions which are promoted by abundance of water and warmth (Taylor and Eggleton, 2001). Strongly acidic solutions of the sort that form during oxidation of pyritic mine spoils are not indicated, because these form intense localized plumes of alteration. More likely the acid involved in spheroidal weathering is carbonic acid formed from dissolution of CO_2 in soil water exploiting the network of cracks in the rock due to expansion by thermal stresses or following unloading of overburden (Retallack, 2001a). Warm, wet carbonic solutions should be feasible on other planetary bodies, but on Earth they are greatly promoted by life. Soil respiration, especially of microbes deep in soil profiles, creates partial pressures of CO_2 up to 110 times that of atmospheric pressures of CO_2 on Earth (Brook et al., 1983). Texturally uniform rocks, such as anorthosites on the Moon and komatiites on Mars, are produced on other planetary bodies, though not granitic rocks (Campbell and Taylor, 1983). Thus it remains to be determined for sure whether spheroidal weathering and tors are truly biosignatures.

11.3.5   Granitic Rocks

Granitic rocks are unique to Earth, and not yet discovered on other planetary bodies or meteorites (Campbell and Taylor, 1983). Granitic rocks are ancient but not primordial, with granodiorites appearing at about 3 Ga among more abundant tonalites and trondjhemites of continental crust (Engel et al., 1974). They thus postdate evidence for life (Schopf, 1983), water (Mojzsis et al., 2001), and deep hydrolytic weathering on Earth (Buick et al., 1995), so it is plausible, though far from certain, that granites are biosignature rocks.

It may seem outrageous to suggest that granitic rocks forming under high temperatures (600–800°C) at depths of 2–10 km within continental crust could be in any way related to life, but additional features beyond uniqueness to Earth and geological antiquity suggest a role for life in granitization (Rosing et al.,

2006). Granites are low-density, hydrated, aluminous and siliceous compared with other igneous rocks, and these are all qualities imparted by deep hydrolytic weathering known to have been widespread on Earth back some 3.5 Ga (Maynard, 1992). Light rare-earth enrichment, including preference for neodymium over samarium, is also a geochemical similarity of granitization and weathering (MacFarlane et al., 1994). The 'Daly Gap', or rarity of rocks intermediate in composition between basalt and rhyodacite, in many continental volcanic regions points to different pathways of genesis (Bonnefoi et al., 1995). Migmatites and S-type granites include melted sedimentary rocks derived from continental weathering, judging from their elevated $^{86}Sr/^{87}Sr$ ratios, oxygen-isotope composition, and alumina content, as well as minerals such as corundum, muscovite, garnet, and cordierite (Rosing et al., 2006). By this view, life's role in forming granite begins with its promotion of deep weathering and surficial liquid water. Some of this low-density, aluminous and siliceous soil and sediment would be melted upon deep burial by overlying rocks, or entrained by the foundering of dense basaltic slabs in plate tectonic recycling to form a new kind of rock, granite. From this perspective it is also plausible that plate tectonics driven by density differences and surficial water leading to granitization is also a biosignature of Earth. There is no plate tectonics on the Moon (Taylor, 1982), and it seems unlikely on Venus, which has some poorly understood mechanism of crustal renewal (Fowler and O'Brien, 2003), but early Mars may have had seafloor spreading (Nimmo and Tanaka, 2005).

A role for life and weathering in granitization is rarely considered in igneous petrology. The prevailing view of granite formation is by magmatic processes, such as fractional crystallization in which mafic minerals drop out of molten magmas to form dense residues leaving more silicic fluids to form granitic rocks (Rudnick, 1995). Remarkably, Bowen's (1922) famous reaction series, which defines the order in which specific minerals crystallize from silicate melt, is the same order in which they are destroyed in deeply weathered soils, as determined by Goldich (1938). Olivine and pyroxene are early to crystallize in magmas and first to be weathered, in contrast to quartz which is last to crystallize and weather away.

11.3.6   Convex Slope

Concave and straight slopes are common in desert badlands and alpine glacial terrains on Earth, the

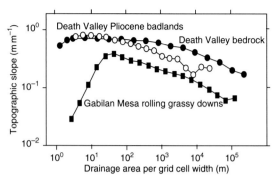

Figure 5 Slope–area relations in soil-mantled compared with desert bedrock and badlands in landscape areas of 0.24–5.8 km^2 in California. Drainage area per grid cell is the size of individual catchments within successively larger map squares beginning with 2 m squares. From Dietrich WE and Taylor PJ (2006) The search for a topographic signature of life. *Nature* 439: 411–418.

Moon, and Mars. Broadly convex slopes on the other hand may be unique to soil-mantled upland topography on Earth (Dietrich and Taylor, 2006). The effect is subtle in topographic profiles, especially those with low topographic relief, but can be striking in slope–area plots of small areas (<6 km^2) compiled from high-resolution digital elevation models (**Figure 5**). Soil-mantled terrain has low slopes around upland small drainage basins but steep slopes in large valleys, whereas badlands and mountains show comparably steep slopes in both small and large drainage basins. Put another way, landscapes at large scale and with limited life are almost fractal in distribution of slope with catchment size, but soil-mantled landscapes are nonfractal in small catchments. This potential biosignature deserves further attention through image analysis of other planetary bodies and paleosols.

Concave slopes are created by fluvial or other fluid erosion, by Gilbert's (1877) 'law of divides' in which slope increases with restricted discharge upstream. Straight slopes on the other hand are the angle of repose of scree and glide planes of other forms of mass movement, as first noted by Penck (1953). Broadly convex slopes in contrast are associated with thick soil mantles. Soil communities bind the earth on which they depend while smoothing with bioturbation irregularities in the overall expansion of parent materials due to weathering and cracking. Mathematically this is diffusive dilation. Soil stabilized by life creeps downslope until failure angles are reached to create broad convex slopes above oversteepened straight to concave slopes. In contrast, desert badlands and alpine glacial terrains

have narrowly convex divides, steep headwall drainages, and long straight-to-concave slopes.

11.3.7 Calcic Horizon

A calcic horizon is a particular level within a soil studded with nodules of calcite or dolomite (**Figure 6(a)**), and is especially common in desert soils (Aridisols) on Earth. Despite the desert-like images of the Moon (Taylor, 1982), Venus (Barsukov *et al.*, 1982), and Mars (Squyres *et al.*, 2004), no pedogenic carbonate has yet been detected beyond Earth. Spectroscopic detection of carbonate minerals in Martian dust (Bandfield *et al.*, 2003) may have the same source as carbonate in Martian meteorites and carbonaceous chondrites, which are vein-filling crystalline carbonate, some of it formed at high temperatures (McSween, 1997). Fossil and modern pedogenic carbonate in contrast is micritic and replacive in microtexture (Retallack, 1991).

Many aspects of calcic horizons are controlled by their soil ecosystems. Calcareous rhizoconcretions and termitaries are biogenic structures modeled on roots and termite nests, respectively (Retallack, 2001a). Micritic replacement within nodules is microbially mediated (Monger *et al.*, 1991). The depth to calcic horizons in soils is strongly correlated with mean annual precipitation and soil respiration (Retallack, 2005). Calcic horizons are found largely within the mean annual precipitation range of 300–1000 mm, as shown particularly well in transects from the hyper-arid Atacama Desert to higher-elevation shrubland and grassland (Rech *et al.*, 2006). The microbe-poor Atacama Desert soils have gypsic horizons of soluble salts, like soils of Mars (**Figure 1**) and the Dry Valleys of Antarctica (Campbell and Claridge, 1987). Calcic horizons are found in soils of desert shrublands largely leached of such soluble salts, which inhibit biological activity with high osmotic stress.

11.3.8 Argillic Horizon

An argillic horizon is a level within a soil markedly enriched in clay compared with horizons above and below (**Figure 6(b)**), and is a hallmark of forested soils such as Alfisols and Ultisols. The argillic horizon is defined on the basis of amounts of clay enrichment (roughly 10% more, with exact definition texture dependent; Soil Survey Staff (2000)), but also has characteristic microstructure of extremely fine-grained and high-birefingence wisps of pedogenic clay along with inherited clay (Retallack, 2001a). These kinds of

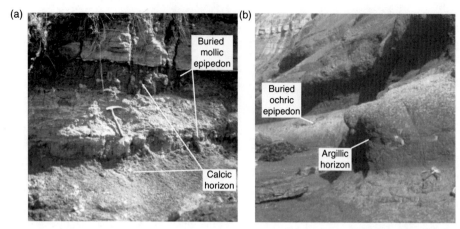

Figure 6 Calcic and argillic horizons and mollic epipedon in paleosols. (a) Two Chogo clay paleosols (Calciustolls) showing mollic epipedon (dark surface) and calcic horizon (white nodular bands), both profiles truncated by nephelinitic granular ash in the Middle Miocene Fort Ternan Member of the Kericho Phonolites, near Fort Ternan, Kenya (Retallack, 1991). (b) Long Reef clay paleosol (Paleudult) showing argillic horizon (dark red clay-enriched) beneath light-colored surface horizon covered by clayey alluvium in the late Early Triassic Bald Hill Claystone, near Long Reef, New South Wales, Australia (Retallack, 1997b). Hammers give scale in both frames.

horizons have not been found on the Moon, Mars, and Venus, and are not found in paleosols on the early Earth either. The geologically oldest known example of an argillic horizon is from Middle Devonian (Givetian, 390 Ma) Aztec Sandstone of Victoria Land, Antarctica (Retallack, 1997a). Argillic paleosols and other evidence of ancient forests are common in every subsequent geological period (Retallack, 2001a).

Both the micromorphology and geological history of argillic horizons suggest that they are products of forest ecosystems, known independently from fossil stumps and wood to have originated by Middle Devonian (390 Ma). Their micromorphology of wispy to laminated very fine-grained clay suggests that they form from both washing down, and weathering in place around large tapering roots (Retallack, 2001a). During interplanetary exploration it is likely that forests would be encountered before argillic horizons, but relict argillic paleosols could be indicative of forests of the geological past, just as they are in some desert regions of Earth today.

11.3.9 Mollic Epipedon

The mollic epipedon is a unique surface horizon of grassland soils (Mollisols), which presents a number of paradoxes. It is rich in base-rich clay such as smectite yet does not ball up or shear into unwieldy clods and flakes. It is rich in organic matter yet well aerated with cracks and channels in which organic matter should be oxidized. It can be well drained yet still maintains soil

moisture. It is fertile with phosphorus and other mineral nutrients despite long periods (thousands of years) of soil development. Mollic epipedons achieve these highly desirable agricultural qualities through a unique structure of fine (2–3 mm) ellipsoidal (crumb) clods (peds), which consist of clay stabilized by organic matter. The geologically oldest fossilized mollic epipedons are Early Miocene (19 Ma), and known from the Anderson Ranch Formation, near Agate, Nebraska, and the upper John Day Formation, near Kimberly, Oregon, USA (Retallack, 2004b). Nothing like a mollic epipedon has been reported from the Moon, Venus, Mars, or meteorites.

The temporal range and micromorphology of mollic epipedons suggest that they are created by grassland ecosystems. Some of the rounded crumb peds are excrements of earthworms common in these soils; other peds are products of the three-dimensional network of the slender adventitious roots of sod-forming grasses (Retallack, 2001a). The organic matter is partly from root exudates and earthworm slimes, but cake-like dung of large ungulates also plays a role in soil conditioning. The mollic epipedon can be considered a trace fossil of sod-grassland ecosystems, even after destruction by desertification or burial (Retallack, 2004b).

11.3.10 Civilization

One would think the Great Wall of China as a human construction obvious from space. In poor-resolution images it could be confused with another 'Great Wall of China', which is a tourist attraction south of Blinman in the northern Flinders Ranges, Australia, and a natural outcrop of Neoproterozoic Mount Caernarvon Greywacke flanking a diapiric uplift like a fortified city. The so-called 'Face of Elvis' and 'pyramids' on Mars were subsequently shown to be inselbergs and yardangs (Pieri, 1999). Similarly the canals of Mars were products of imagination working at the limits of optical resolution of the time (Zahnle, 2001). Close optical and physical surveillance will be needed to detect civilizations, even on planets with dead atmospheres and boiled oceans in which civilizations may once have failed. Successful civilizations on the other hand may come to us. The effort to detect radio communications from outer space continues with ever larger telescopes and more powerful computer power (Whitfield, 2003). Yet we remain alone and isolated in the universe, and know no failed civilizations either.

11.4 Origin of Life on Earth

From a coevolutionary perspective the origin of life is not the heroic struggle of an individual molecule or organism from a primordial soup (Nisbet and Sleep, 2001), but a system of supportive molecules or organisms for growth and replication in an environment that selected for some desirable property while metering the supply of fundamental nutrients to last for geological timescales. To oversimplify in human terms, the first life needed a job, a family, a place to live, and a legacy for future generations. These theoretical requirements are made clear by Cairns-Smith (1971), who has argued that the precursors of organic life were clays which grew by assimilating cations from weathering solutions and reproduced by cracking into smaller pieces or flakes with diagnostic layering or cationic substitutions. These general characteristics would not coevolve in any lifelike way unless the clays also had some property subject to natural selection. He proposes the simple but important quality of stickiness. The layering and cationic substitution of smectite clays determines their interlayer expansion upon wetting, which can vary stickiness from a thin slurry to a thick paste. Stickiness becomes important because it can be selected to stabilize an environmental interface, and thus create a legacy for creation of more stickiness to maintain that interface in a critical zone of energy and materials flux.

Cairns-Smith thought in terms of a sandy aquifer, but I imagine a primordial soil of a planetesimal, in which minerals are weathered by carbonic acid solutions of degassed CO_2 and water, and warmed by a faint young Sun (**Figure 7**). Sloppy clay expands violently when wet and is then washed too deep into the soil for warmth and acid to produce copies. Tough clay on the other hand expands little and covers source

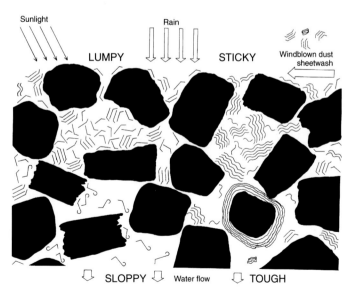

Figure 7 Hypothetical natural selection of four kinds of clay differing in stickiness (Sloppy, Lumpy, Sticky, and Tough) and weathering from the minerals of a primordial soil. From Retallack GJ (2001a) *Soils of the Past*, 2nd edn., 600pp. Oxford: Blackwell.

grains in a thick rind so that the mineral is no longer available to water and acid to produce more clay. Lumpy and sticky clay in contrast has a consistency that expands on wetting sufficiently to bridge the gaps between grains, yet crack away from the source grains to allow access for water and acid to create more clay. These clays will be selected against the soil-eroding activities of wind and water because their grains are glued together by sticky clays. In an environment where loose sandy soils are eroded, all soils become more clayey, and by persisting produce more clay. Lumpy clays may be better adapted to heavy rains, and sticky clays to drizzle, but both are selected by their ability to hold the soil. They are in a coevolutionary escalation which refines their distribution, composition, and degree of stickiness for particular climatic conditions, because rivers erode the least sticky clays. Similarly the first organic molecules, produced by abiotic reactions could also be selected in soils for their stickiness. Like molasses on a cornfield, organic soils survive erosion to make more organic matter and create soils with observed fabrics of carbonaceous chondrites. The molecules of heredity, DNA and RNA are sticky too, and could have begun what Cairns-Smith calls a "genetic takeover" of the older and cruder clay system of soil stabilization. Cells, trees, and civilizations continue to depend on the stability and bounty of soil and its nutrients. This simple parable makes clear the importance of natural selection by the environment in converting the long odds of chance evolution of complex life to a necessity.

11.4.1 Experimental Studies

Laboratory experiments have succeeded in demonstrating the ease with which so-called 'organic molecules' are created abiotically. Miller's (1953) classical experiment sparking ammonia and hydrogen gas above water in a sealed apparatus produced amino acids and sugars, common building blocks of life as we know it on Earth. These experiments have since been repeated with a variety of gas mixtures, with lower but still significant yields of organic matter in CO_2-rich mixtures (Schopf, 1983). Yields are promoted by addition of clays, and chemically reduced minerals such as pyrite and siderite (Schoonen et al., 2004). Clays also promote assembly of oligomers of RNA up to 50 units long (Ertem and Ferris, 1996). Clays act as catalysts, templates, compartments, energy-storage devices, and chemostats in promoting organic molecule synthesis. Siderite can mediate photooxidation of CO_2 to produce reduced

carbon in a way comparable with photosynthesis (Braterman et al., 1983). Concentrated amino acids (proteinoids) can form spherical structures (Fox and Dose, 1972), which appear superficially like living cells, but are far from functioning cells (**Figure 3**). Nothing close to a living cell has ever emerged from experiments designed to examine the origin of life, so such experimental research into the origin of life has reinforced the enormous gulf between quick and dead organic matter.

11.4.2 Geological Records

Carbonaceous chondrites, with their mix of smectite clay and diversity of reduced organic compounds (Pizzarello et al., 2001), are evidence for abiotic synthesis of organic compounds in the presence of clay and reduced iron minerals such as siderite early during the origin of the solar system. The carbonaceous clayey matrix postdates olivine chrondules which are 4.56 Ga in age (Amelin et al., 2002) and predates cross-cutting veins of calcite (Endress et al., 1996) and halite (Whitby et al., 2000) no younger than 4.51 Ga. The characteristic carbon-isotopic fraction of rubisco, an important enzyme in photosynthesis, is indicated by light carbon-isotopic ratios of amorphous organic matter in the 3.8 Ga Isua Greenstone Belt of west Greenland (Rosing and Frei, 2004) and microfossils comparable with cyanobacteria are found in chert from the 3.5 Ga Apex Chert of the Warrawoona Group near Marble Bar, Western Australia (Schopf, 1983). Thus the origin of life occurred during the 700 My interval between 4.5 and 3.8 Ma, or approximately the first 15% of the history of our planet. This period is a dark age in Earth history, its sedimentary records obliterated by metamorphism and subduction. Surfaces of the Moon and Mars dating to this early time in the history of the solar system also reveal another reason for paucity of geological records: heavy bombardment of larger planets by planetesimals and other debris of the evolving solar system. One of these very large impacts at about 4.4 Ga is thought to have created our Moon from a Mars-size impactor ('Theia') melting itself and a large fraction of the Earth (Halliday, 2004). Impact melting of rocks would have destroyed both life and any precursor clay or organic matter (Maher and Stevenson, 1988). Another impedance to life and its precursors in surface environments would have been ultraviolet and other forms of protein-denaturing radiation from a young Sun unfiltered by a fully formed atmosphere and ozone layer (Sagan and Pollack, 1974).

11.4.3 Soil

The idea that life came from soil is probably the most ancient human view of the origin of life. A cuneiform text dating to 2000 BC from the Sumerian city of Nippur in present-day Iraq (Kramer, 1944) describes a feast of the gods presided over by Ninmah (mother earth) and Enki (god of the waters), in which Ninmah created different kinds of humans from clay and Enki decreed their role in life. The idea of the spontaneous generation of life from soil was popular for centuries, endorsed by such influential thinkers as Aristotle (*c.* 384–322 BC) and Lucretius (*c.* 99–55 BC), but lost much appeal in the late nineteenth century when Louis Pasteur showed that life did not arise spontaneously from the soil or organic matter, but from microscopic propagules. Nevertheless, soil has much promise as a site for the origin of life because of its self-organizing complexity, lack of chemical equilibrium, and benign physicochemical conditions (Retallack, 2001a). The soil is alternately full and then drained of water after rain-storms, but thin films of water persist in pockets and margins of grains in which complex chemical reactions stop and start with the influx and evaporation of soil water. Weathering is a slow process taking millions of years to reach chemical equilbrium in the production of nutrient-starved kaolinitic clays. Soils are full of insulating clays, which moderate heat differences; diaphanous grains, which filter harsh electromagnetic radiation; and base-rich clays, which neutralize environmental acids.

11.4.4 Sea

The idea that "life like Aphrodite was born on the salt sea foam" (Bernal, 1967) has a long pedigree extending back to Ionian Greek philosophers Thales (*c.* 585 BC) and Anaximander (*c.* 565 BC). The appeal of this idea comes partly from the abundance of life in the sea at present and the aqueous geochemistry of most vital processes. There is also the saline composition of blood plasma and cytoplasm of land animals, suggestive of marine origins. Charles Darwin's (1959) posthumously published vision of life's origins in "some warm little pond, with all sorts of ammonia and phosphoric salts, light, heat, electricity, etc.", hint at the theory's greatest weakness: the remarkably uniform and dilute composition of the ocean, which

tends toward chemical and physical equilibrium (Nisbet and Sleep, 2001). The smaller the pond and the more intermixed with mineral matter, the better opportunity for complex serial biosynthesis.

11.4.5 Deep-Sea Vent

Ocean floor volcanic vents discovered by deep-ocean submersibles are yet another plausible site for the origin of life (Corliss *et al.,* 1981). Hot water billows out of these vents like black smoke because of the rapid precipitation of sulfides and other dissolved substances into cold sea water. The vent fluids are like toxic waste, highly acidic, oxygen starved, and extremely hot (temperatures up to 380°C). There is little organic carbon or organisms immediately around them, but inches away the seafloor is crowded with life: large white clams and crabs, and peculiar pale tube worms. Only a few meters beyond, cold temperatures and lack of nutrients severely curtail biodiversity. At such great depths there is no possibility of photosynthesis, so the community is fed by chemosynthetic microbes (**Table 2**). Life is envisaged to have originated in spongy rock and vent deposits between the hot, caustic vents and cold, dilute ocean. Hydrothermal vent origin of life is an appealing explanation for widespread heat-shock proteins and primitive thermophilic bacteria (Nisbet and Sleep, 2001), though it is debatable how primitive are these adaptations (Forterre, 1995). Unlike soils or tidal flats, deep-sea vents are not controlled in their location or persistence by microbial scums. Black smokers erupt and become dormant like volcanoes, by virtue of subsurface faulting creating conduits for fluid migration in their oceanic rift valleys.

11.4.6 Other Worlds

Another possibility is that life evolved elsewhere in the universe and colonized our planet as propagules that could withstand long-distance transport in space. This concept of 'panspermia' goes back to the turn of the century and the Swedish chemist Svanté Arrhenius. A related idea is deliberate colonization of the Earth by advanced extraterrestrial civilizations (Crick, 1981). All manner of organisms could have been broadcast or dispatched, ranging from influenza viruses or unicellular bacteria, to sophisticated aliens in space vehicles. Such views have some appeal in this age of space exploration, but merely remove the question of the origin of life to another planet. The environment where that life evolved is likely to have been Earth-like in many respects because life has long been well suited to our planet. Thus it remains useful to consider the origin of life from natural causes here on Earth.

11.5   Coevolutionary Histories

Biosignatures of Earth can be understood by their long geological histories, which present natural experiments with variation that may reveal oscillatory mechanisms of coevolution. The metaphor of an arms race is commonly applied to coevolutionary trajectories because they appear to change toward more competitive organisms (Dawkins and Krebs, 1979). Increasingly discriminating bees are more faithful to the pollination of particular flowering species. Grassy sod with basal-tillering and telescoped internodes is best able to absorb the rough grazing of tall-toothed and hard-hooved bison and horse. Similarly in warfare, arms races between competing nations result in more sophisticated and powerful weapons. Unlike biological coevolution, which has spawned diversity, arms races commonly result in imperialism, and loss of other cultures, their languages, and archives.

Another metaphor for coevolution is the Red Queen hypothesis (Van Valen, 1973) named for Lewis Carroll's character who had to run faster and faster in order to stay in the same place, presumably within a social pecking order. This implies that all creatures are caught up in a rat race, and does not account for the persistence of such slow creatures as millipedes and possums.

We prefer a more fundamental human metaphor of technological escalation, which applies also to activities such as information technology (Vermeij, 2004). Hand-copying, printing, and electronic publication are successive improvements in information technology driven by inventions (adaptations) of companies (producers) to satisfy demands of customers (consumers), who in turn adjust their consumption (coadaptations). Technological escalation in information technology has resulted not only in increasing quality and volume of communication between groups, but also increasing diversity and specialization.

By any of these models, the history of life and Earth environments should demonstrate a pattern of adaptations in one segment of the ecosystem followed by compensatory coadaptations in another part of the ecosystem. The following sections attempt

to identify such reciprocal changes in air, water, soils, and rocks. There are also catastrophic alterations of surface environments after large bolide impacts, giant volcanic eruptions, and metamorphism of limestone and coal (Retallack, 2002). There are biological perturbations, such as seasonal leaf-shedding and soil fertility cycles, on such short timescales (10^0–10^5 years) as well (Retallack, 2004a). Biological flexibility on these timescales may have aided biotic recovery from abiotic catastrophes. The present account however will deal only with evolutionary timescales (10^6–10^9 years).

11.5.1   Life and Air

Bioterraforming, or re-engineering of planetary environments by life, as may be possible for Mars (Fogg, 1998), could already have happened on Earth during the Precambrian evolution of cyanobacterial photosynthesis to create our oxygen-rich atmosphere. Evidence from paleosol geochemistry suggests an especially marked oxidation at about 2.3 Ga (**Figure 8**), as does the coeval demise of banded iron formations (Bekker *et al.*, 2004), fluvial uraninite placers (England *et al.*, 2002), and mass-independent fractionation of sulfur isotopes (Farquhar *et al.*, 2000). One difficulty for this scenario is the discovery of geologically older cyanobacterial microfossils (3.5 Ga; Schopf, 1983), stromatolites

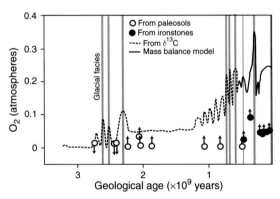

Figure 8 Reconstructed history of atmospheric oxygen abundance, showing constraints from selected Precambrian paleosols and Phanerozoic pedogenic goethites. The Phanerozoic curve (solid line) is from a sediment mass-balance computer model (Berner *et al.*, 2000) and the Neoproterozoic curve (dashed line) inferred from carbon-isotopic data (Anbar and Knoll, 2002). The gray bars represent times of known glacial sedimentation, which vary considerably in duration and recurrence interval. Updated from data of Retallack GJ (2001a) *Soils of the Past*, 2nd edn., 600pp. Oxford: Blackwell.

(3.5 Ga; Allwood *et al.*, 2006), mats (3.4 Ga; Tice and Lowe, 2004), and 2α-methylhopane biomarkers (2.7 Ga; Brocks *et al.*, 2005). As was clear to Cloud (1976) in proposing photosynthetic oxidation of Earth, bioterraforming may take geologically significant spans of time.

Reasons why bioterraforming takes so long are clearer from more recent atmospheric oxidation events, such as the peak value of 35% O_2 during 390 Ma spread of forests, followed by a 250 Ma low of about 15%, before climbing again to the modern value of 21% with the 19 Ma spread of sod grasslands (**Figure 8**). Forest ecosystems of the Middle Devonian (390 Ma) culminated a steady increase in woody plant size from the first appearance of match-stick-sized primitive land plants during the early Silurian (430 Ma). These evolutionary changes were matched by comparably drawn out increases in depth and intensity of weathering indicated by paleosols, suggesting that not only individuals evolved but soil communities distributed over broad continental areas (Retallack, 1997a). Woody steles of the earliest land plants frustrated early Silurian herbivores, which were mainly millipedes adapted to feeding on unvascularized liverworts (Retallack, 2001c). A coevolutionary swing toward woody plants to the disadvantage of herbivorous animals would not be enough in itself to change the world unless such communities became globally distributed, commandeering the source of mineral nutrients in soil, metabolic gases in the atmosphere, and the distributive capacity of rivers and the ocean. The evolution of laminar leaves by 320 Ma would also have enhanced photosynthetic production of wood, and transpiration of water (Beerling *et al.*, 2001). Altering atmospheric composition requires not only living communities as a force for photosynthetic oxidation, but deep and protracted silicate weathering and burial of carbon in swamps and oceans in large quantities over many millions of years (Berner *et al.*, 2000).

Silicate weathering and carbon burial could be regarded as physicochemical components of global change, but both are under strong biotic control. The simple inorganic weathering agent, carbonic acid from dissolution of CO_2 in water comes largely from soil respiration, in turn fed by primary production, so that some rainforest soils have 110 times more soil CO_2 than in the atmosphere (Brook *et al.*, 1983). Other organic acids also enhance hydrolytic weathering (Berner, 1997) and are also more abundant in larger and more productive ecosystems. The physical evidence of this biotic enhancement of weathering are the new kinds of forest soils (Alfisols, Ultisols), which appear as paleosols in Devonian and younger rocks (Retallack, 1997a). Carbon burial can be regarded as a purely physical process, but the biological invention then proliferation of lignin first occurred in a world without effective ligninase enzymes or creatures such as termites and dinosaurs which could reduce the load of woody debris undecayed and finding its way into rivers, lakes, swamps and the sea. The physical evidence of Devonian-Carboniferous increase in carbon burial is not only the increase in marine and lacustrine black shales, but the appearance of a new kind of peaty soil (Histosol) and wetland-forest (swamp) ecosystem in Devonian and younger rocks. Thus the soaring oxygen content of 35% in the Late Carboniferous (*c* 310 Ma) was largely a product of the balance between producers with the newly evolved product of wood getting ahead of consumers. By Late Carboniferous, there were few vertebrate herbivores, but many winged herbivorous insects, wood-eating cockroaches, and fungi. Carboniferous swamp woods were punky with soft cells (parenchyma) between the files of lignin-rich cells (tracheids), and so were less prone to wildfire than many modern woods. Carboniferous ecosystems thus included features that offset the oxidizing effects of trees, culminating in the evolution of tree-eating termites and dinosaurs during the long Mesozoic CO_2 greenhouse (Retallack, 2004a).

Sod grasslands formed by a coevolutionary process between grasses and grazers (**Figure 1**) on all continents except Australia and Antarctica over a geologically significant interval of time between the Late Eocene (35 Ma) and Pliocene (4 Ma). Studies of root traces in paleosols indicate that before grasslands the stature of woody vegetation became uniformly smaller from humid climate forests to arid shrublands, as it still does in Australia where mallee vegetation of 2–10 m multistemmed shrubs dominate a semiarid climatic belt of grasslands on other continents. From Late Eocene to Late Miocene, grasslands were confined to the rainfall belt of 300–400 mm mean annual precipitation, but after Pliocene evolution of large hard-hooved hypergrazers such as horses and wildebeest and megaherbivores such as elephants the prairie-forest ecotone was rolled back to nearer to the 1000 mm isohyet of mean annual precipitation (Retallack, 2001b). Evidence for this territorial advance of grasslands comes from the spread of the characteristic surface horizons of grasslands (mollic epipedon) in soils with deep calcareous nodular horizons (calcic horizon) of subhumid climates. Human coevolution with grass-based agroecosystems has now created grasslands even in rainforest climate belts. This territorial expansion is evidence that this coevolved ecosystem is not merely an assemblage of creatures with mutually compatible climatic tolerances, but a biological force for global change.

Like forests of the Carboniferous, Cenozoic grasslands were a force for atmospheric oxidation because they increased plant biomass, particularly underground, compared with mallee vegetation in comparable climatic belts. Grasslands also promoted hydrolytic weathering, accelerating its pace with a rich soil fauna of earthworms and a variety of microbes. The result is a dramatic increase in soil organic carbon contents as high as 10 wt.% for depths of up to a meter, compared with mallee soils of comparable climatic belts which have such high organic content only in the surficial few centimeters of leaf litter. Soil organic matter also stabilizes a characteristic soil structure of crumbs, which withstands wetting and can be transported considerable distances by streams to be buried in agricultural stock ponds, lakes, river floodplains, and the ocean. Thus Cenozoic grasslands increased efficiency of biomass creation, hydrolytic weathering, and carbon burial in aridlands, just as forests did during their Devonian–Carboniferous evolution in humid areas and wetlands. Grasslands also effected global change by controlling two additional landscape attributes traditionally regarded as purely physical: albedo and evaporation (Retallack, 2001b). Grasses are lighter colored than trees, especially when hay-colored during dry or snowy seasons, and bow down under snow more readily than trees. Grasslands thus reflect more radiation back into space, with a net chilling effect on the planet. Grass colonization of the bare soil patches of desert shrubland or mallee has the effect of reducing evaporation of soil water. Grassy sod is moist and the air above grasslands dry, in contrast to woodlands and forests which have dry soil and moist air because of persistent transpiration of water by trees. Lowered vapor pressure of water over grasslands compared with forests also has the effect of planetary cooling, because water vapor is a powerful greenhouse gas. Thus the rise of O_2 to 21% with Cenozoic drying and cooling culminating in the Pleistocene ice age, may have been in part due to the coevolution of grasslands. The process is not without limits, but even the warming forces of greenhouse gas (CO_2 and CH_4) emissions from wildfires and large herds of ungulates are part of a system that promotes grassland expansion at the expense of forests, because grasses recover from fire and herbivory more readily than trees. If anything can undo the cooling effect of grasslands, it would have to be their domineering evolutionary product, us (Palumbi, 2001).

11.5.2 Life and Water

Ocean composition also has changed on evolutionary timescales. Oceans of the early Earth may have been 1.5–2 times more saline than modern oceans, judging from fluid inclusion and evaporite volume data (Knauth, 1998). The principal mechanism for oceanic dilution is burial of large volumes of salt in evaporite deposits of barred basins, like the modern Persian Gulf. The growth of continental granitic crust promoted the tectonic formation of such basins and emergent continental storage of evaporites. Barred evaporite basins also are common behind coral, algal, and stromatolitic reefs, which extend back at least to the Paleoproterozoic (2 Ga; Grotzinger, 1988). Paleoproterozoic was also a time of peak abundance of banded iron formations (**Figure 9**), the distinctive laminated hematite and chert deposits of oceans during the transition from chemically reducing to chemically oxidizing oceans due to the spread of photosynthetic microbes (Cloud, 1976).

Roles for life in Precambrian changes in salinity and redox remain controversial because so remote in geological time. A clearer indication of life's role in oceanic chemistry on evolutionary timescales is the Phanerozoic oscillation between aragonite and calcite seas (**Figure 10**). This geochemical oscillation is seen in both carbonates and evaporites of the Cambrian to Mississippian (540–320 Ma) and Jurassic to Paleogene (170–20 Ma), which are dominantly calcite and sylvite-gypsum. In contrast, carbonates and evaporites of the Mississippian to Jurassic (320–170 Ma) and Neogene (20–0 Ma) are dominantly aragonite and kieserite-gypsum. The salts are simple precipitates, and although the carbonates are largely precipitated as skeletons of marine organisms, the major rock-forming organisms use passive forms of extracellular metabolically induced precipitation, rather than shell formation mediated by an organic-matrix (Stanley and Hardie, 1999). Calcite seas correspond broadly in time with high-CO_2 greenhouse intervals and aragonite seas with low CO_2 icehouse intervals, but this does not appear to be a simple pH effect from a more acidic atmosphere at times of high CO_2 and carbonic acid. Calcite is favored over aragonite only for a narrow range of

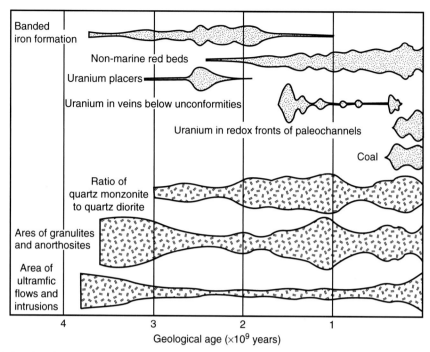

Figure 9 Relative abundance of sedimentary rocks, ores, and igneous rocks through geological time. Adapted from Engel *et al.* (1974), Meyer (1985), and Barley and Groves (1992).

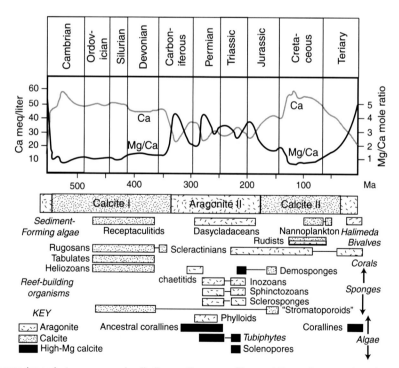

Figure 10 Correspondence between oceanic alkaline earth composition and the carbonate mineralogy of hypercalcifying marine organisms. Simplified from Stanley SM and Hardie LA (1999) Hypercalcification: Paleontology links plate tectonics and geochemistry to sedimentology. *GSA Today* 9(2): 1–7).

unrealistically high partial pressures of CO_2. Instead, calcite versus aragonite is favored by the Mg/Ca ratio of the ocean, with calcite favored when the ratio is less than 2 and aragonite favored when it is more than 2. The Mg/Ca ratio of the ocean is in turn controlled by the balance of cations in solution from rivers and from hydrothermal activity at mid-ocean ridges. Stanley and Hardie (1999) argue that variation in mid-ocean ridge spreading is the primary control, with active spreading inducing a greater hydrothermal flux and sub-seafloor sequestration of Mg. However, rates of seafloor spreading are not appreciably varied over the past 180 Ma in which spreading can be inferred accurately from seafloor magnetic striping (Rowley, 2002). An alternative control on oceanic Mg/Ca ratio is riverine input, considering the greater abundance of dolomitic caliche in paleosols of high-CO_2 greenhouse climates than those of low-CO_2 icehouse climates (Retallack, 2004a). Dolomitic versus calcitic caliche in turn is probably under microbial control of methanogens encouraged by soil CO_2 higher than usual for arid-land soils (Roberts et al., 2004). From their source as they are weathered from soils to their sink in fossil coral reefs, Ca and Mg cations are under constant biotic surveillance.

11.5.3 Life and Soil

Compared with a long history of research on Precambrian air and water, even such basic questions as when soils became living remain uncertain. There was land and thick deeply weathered soil at least as far back as 3.5 Ga, as indicated by paleosols of that age (Buick et al., 1995), and isotopic geochemical evidence for substantial amounts of Archaean continental crust (Bowring and Housh, 1995). Sterilization of land surfaces by intense ultraviolet radiation before development of the ozone layer is unlikely in an early atmosphere of methane, water vapor, or other dense early atmospheric gases (Rye and Holland, 1998). The very existence of clayey paleosols that were well drained, is regarded by some geomorphologists as evidence of stabilization by life, including microbes (Schumm, 1977). Remarkably light C-isotope values (-40 $\delta^{13}C$‰ vs PDB) indicative of methanotrophs have been recorded from the 2.8 Ga Mt Roe paleosol of Western Australia (Rye and Holland, 2000) and isotopically heavy C-isotope values (-16 to -14 $\delta^{13}C$‰) like those of hypersaline microbes from the 2.6 Ga Schagen paleosol South Africa (Watanabe

et al., 2000). Many paleosols have carbon isotopic values in the normal range for cyanobacteria and other photosynthetic organisms (-24 to -30 $\delta^{13}C$‰): 2.3 Ga pre-Huronian in Ontario (Farrow and Mossman, 1988), 1.3 Ga post-Mescal paleokarst in Arizona (Beeunas and Knauth, 1985), and 0.8 Ga pre-Torridonian in Scotland (Retallack and Mindszenty, 1994). To this list of methanogenic, hypersaline and photosynthetic life in Precambrian paleosols must be added decomposers such as actinobacteria, because most Precambrian paleosols have only traces of organic C. Something must have cleaned up the dead bodies, just as fungi and other microbes create carbon-lean modern soils. Filamentous cyanobacterial microfossils have been found permineralized on a paleokarst paleosol in the 1.3 Ga Mescal Limestone of Arizona (Horodyski and Knauth, 1994) and within angular flakes of laminated crust in the 2.8 Ga Mt Roe paleosol of Western Australia (Rye and Holland, 2000). Comparable modern microbes have wide environmental tolerances and comparable fossils are common in permineralized stromatolites (Schopf, 1983), so were they soil microbiota or just opportunistic colonists of puddles? Lichen-like microfossils from the Carbon Leader of the Witwatersrand Group of South Africa (Hallbauer et al., 1977), now known to be 2.9 Ga in age (England et al., 2002), have been discounted as artefacts of acid maceration (Cloud, 1976), although it is clear from petrographic thin sections that they predate metamorphic veins (MacRae, 1999). The organic matter of these structures has isotopically light carbon, pristane, phytane, and pentose≈hexose indicative of photosynthesis (Prashnowsky and Schidlowski, 1967). Life in soil may go back 3.5 Ga, and perhaps further (**Figure 7**).

Soils played a role in coevolutionary oscillations of atmospheric and oceanic chemical composition (**Figures 8–10**). One of the best lines of evidence for a chemically reducing early atmosphere on the early Earth are paleosols older than 2.2 Ga with the deep spheroidal weathering suggestive of a well-drained soil yet chemically reduced clay and iron minerals. These appear to be a completely extinct form of soil, which we have labeled 'Green Clays' (Retallack, 2001a). The degree and depth of weathering of paleosols on the early Earth is remarkable, with some of them attaining the alumina enrichment of bauxites (**Figure 11**). The appearance of laterites and oxidized paleosols is one of the most obvious lines of evidence for Paleoproterozoic atmospheric oxidation (Rye and Holland, 1998). The oldest

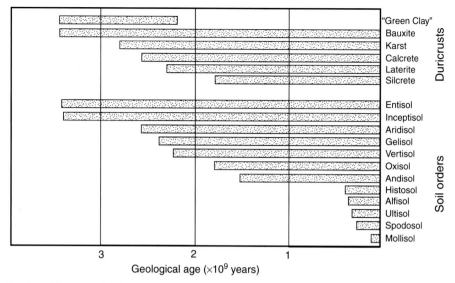

Figure 11 Stratigraphic range of duricrusts and soil orders. Entisols and Inceptisols are assumed to have been precursors of other paleosols, even though difficult to recognize in the rock record before the advent of roots and burrows in soils. Updated from data of Retallack GJ (2001a) *Soils of the Past*, 2nd edn., 600pp. Oxford: Blackwell.

known caliche may be the Schagen paleosol (2.6 Ga) of South Africa (Watanabe *et al.*, 2000). The first silcrete appears along with the first desert dunes and sizeable (>600 km) continents in the pre-Pitz Formation paleosols (1.8 Ma) of northern Canada (Ross and Chiarenzelli, 1984). Silcretes and caliche are characteristic of deserts on Earth today, and it is surprising not to find them on the earlier Earth. Primordial rolling terrains of Green Clay paleosols would have seemed unusual by modern standards. Without benefit of multicellular vegetation these landscapes would have appeared barren and desolate, and yet these deep clayey soils lacked the harsh rocky outlines and dunes of modern desert landscapes. Their deep weathering can be attributed to a humid, maritime climate and warm temperature assured by the greenhouse effect of elevated levels of atmospheric water vapor, methane, and carbon dioxide (**Figure 12**).

The role of soil in Phanerozoic environmental oscillations is especially clear from trace and body fossils associated with paleosols. Burrows and tracks of millipedes first appear in mid-Ordovician (460 Ma) paleosols (Retallack, 2001c), and termite nests in Triassic (230 Ma) paleosols (Hasiotis and Dubiel, 1995). These animals and their burrows had profound effects on soil aeration and respiration. The effect of plants is even more obvious, with reduction spotting and rhizome traces first appearing in early Silurian (430 Ma) paleosols, and large root traces of

trees in middle Devonian (390 Ma) paleosols (Retallack, 1997a). The evolution of trees coincided with evolution of forest soils (Alfisols), and was soon after followed by forest adaptations to low-nutrient conditions of mineral-poor peats (Histosols) by Late Devonian (365 Ma), silica sands by Mississippian (330 Ma), and base-poor clays (Ultisols) by Pennsylvanian (320 Ma: Retallack, 2001a). The evolution of sod grasslands with their characteristic surface horizons (mollic epipedon) ushered in the first grassland soils (Mollisols) in the Early Miocene (19 Ma). The oscillation of terrestrial communities dominated by millipedes, forests, termites, and grasslands corresponds to grand swings in greenhouse to icehouse paleoclimates (Retallack, 2004a).

11.5.4 Life and Rocks

The abundance of different kinds of rocks has also waxed and waned with life-mediated paleoenvironmental changes on Earth (**Figure 11**). Uranium ores for example, show three distinctive forms, which succeed one another over geological history. Uranium in fluvial sandstones is found as rounded grains that were evidently transported by rivers in well-aerated turbulent water and trapped in less-turbulent portions of the stream with gold and other heavy grains in a kind of deposit called a placer (England *et al.*, 2002). These dense grains of uraninite are highly unstable in modern rivers

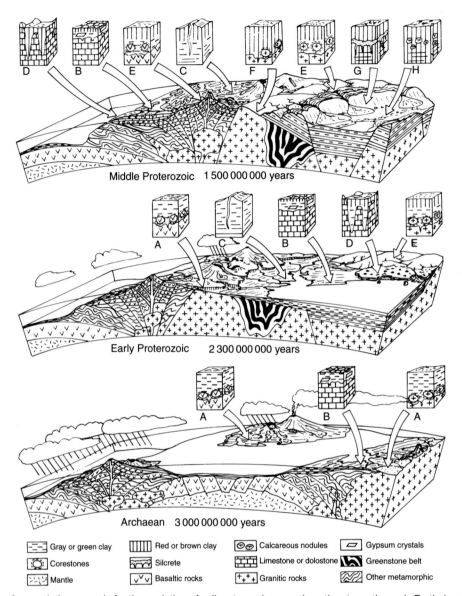

Figure 12 A speculative scenario for the evolution of soils, atmosphere, and continents on the early Earth, in which desert soils appear long after deeply weathered soils. The soil types illustrated are extinct 'Green Clays' (A), salty soils or Salids (B), swelling clay soils or vertisols (C), karst and drab cave earth or Orthents (D), oxidized incipient soils or Ochrepts (E), red and deeply weathered soils or Oxisols (F), desert soils with silcretes or Durids (G), and desert soils with calcareous horizons or Calcids (H). From Retallack GJ (2001a) *Soils of the Past*, 2nd edn., 600pp. Oxford: Blackwell.

because they readily oxidize to soluble yellow carnotite. Evidence from chemically reduced but deeply weathered paleosols beneath these uraniferous sandstones support the notion that uraninite placer deposits reflect a distant time (1.9–3.1 Ga) of much lower atmospheric oxygenation (Farrow and Mossman, 1988). In geologically younger rocks (0.2–2.0 Ga), uraninite ores fill veins and breccias associated with unconformities, where they precipitated at a zone of chemical reduction after transportation in overlying sandy aquifers from granitic sources as oxides in surface and ground waters. Geologically younger again (0.3–0 Ma) are roll-type uranium ores, which reflect similar oxidized transportation in groundwater until encountering a redox front in fluvial sandstones locally rich in organic matter such as fossil logs (Meyer, 1985). Thus uranium ore types are related to atmosphere and aquifer redox, which in turn are biologically mediated (Retallack, 2004a).

Common kinds of igneous rocks also show variations in relative abundance that may reflect biologically mediated paleoenvironmental changes (**Figure 11**). The early Earth like the Moon, Venus, and Mars was dominated by basaltic and ultramafic igneous rocks. The proportion of quartz-rich igneous rocks such as quartz monzonite and granite to diorite and other more mafic rocks increases steadily after 3 Ga, reaching peaks at about 1.8, 0.8, 0.4, and 0.2 Ga. The latter two corresponding to Ordovician and Jurassic are the familiar CO_2 greenhouse (**Figure 8**), calcite-sea (**Figure 9**), and animal-dominated soil episodes (Retallack, 2004a). These were times of widespread deep weathering in a warm-wet world lacking ice caps.

Other rocks may also be under surficial environmental control. For example, eruptions of basaltic andesites have been related to surface conditions, because volcanoes of northern California preferentially erupt during isostatic rebound after unloading of melted ice caps and sea level rise (Jellinek *et al.*, 2004).

11.6   Conclusions

A coevolutionary perspective explains many aspects of our Earth's distinctive geological history. As in classical cases of coevolution, such as grasses and grazers, interactions between unrelated but mutually dependent organisms promote the persistence of new and earth-changing ecosystems, such as grasslands. The global-change capabilities of such ecosystems arise because organisms are evolving primarily in response to other organisms, and only indirectly in response to their physicochemical environment. As coevolved ecosystems proliferated to commandeer the nutrient supply of soil, the metabolic gases of the air, and the medium of water, they altered the atmosphere, hydrosphere, pedosphere, and deep lithosphere.

At the heart of any coevolutionary process is the natural selection of specific coadaptations to other components of the biota. These inventions are rare events that appear to flaunt the laws of thermodynamics and entropy, because they promote continuing metabolism and disorder. As rare events, such adaptations are followed at long geological intervals by coadaptations, with the directed trends of coevolution achieved through a series of discrete oscillations. As these ecosystems spread and proliferated, coevolutionary oscillations of life were

transferred to the geological history of our air, water, soil, and rocks.

From this perspective, human agroecosystems are not the first coevolved ecosystem to induce global warming. Just as cyanobacterial mats cooled a world of methanogenic slimes about 2 Ga, forests cooled a world of millipedes about 390 Ma, and grasslands cooled a world of large mammals about 35 Ma, we can hope that new coevolutionary initiatives will restore Earth to a livable temperature. If we fail in this mission, other organisms may succeed.

References

Allwood AC, Walter MR, Kamber BS, Marshall CP, and Burch IW (2006) Stromatolite reef from the Early Archaean era of Australia. *Nature* 441: 714–718.

Amelin Y, Krot AN, Hutcheon ID, and Ulyanov AA (2002) Lead isotopic ages of chondrules and calcium–aluminum-rich inclusions. *Science* 297: 1678–1683.

Anbar AD and Knoll AH (2002) Proterozoic ocean chemistry and evolution: A bioinorganic bridge. *Science* 297: 1137–1141.

Bada JL, Glavin DP, McDonald GD, and Becker L (1996) A search for endogenous amino acids in Martian meteorite ALH84001. *Science* 279: 362–365.

Bandfield JL, Glotch TD, and Christensen PR (2003) Spectroscopic identification of carbonate minerals in Martian dust. *Science* 301: 1084–1087.

Barley ME and Groves DI (1992) Supercontinent cycles and the distribution of metal deposits through time. *Geology* 20: 291–294.

Barsukov VL, Volkov VP, and Khodakovsky IL (1982) The crust of Venus: Theoretical models of chemical and mineral composition. *Journal of Geophysical Research (Supplement A)* 87: 3–9.

Beerling DJ, Osborne CP, and Chaloner WG (2001) Evolution of leaf form in land plants linked to CO_2 decline in the Late Palaeozoic Era. *Nature* 410: 352–354.

Beeunas MA and Knauth LP (1985) Preserved stable isotopic composition of subaerial diagenesis in the 1.2 byr Mescal limestone, central Arizona: Implications for the timing and development of a terrestrial plant cover. *Geological Society of America Bulletin* 96: 737–745.

Bekker A, Holland HD, Wang P-L, *et al.* (2004) Dating the rise of atmospheric oxygen. *Nature* 427: 117–120.

Bernal JD (1967) *The Origin of Life*, 345 pp. New York: Pergamon.

Berner RA (1997) The rise of plants and their effects on weathering and CO_2. *Science* 276: 543–546.

Berner RA, Petsch ST, Lake JA, *et al.* (2000) Isotope fractionation and atmospheric oxygen: implications for Phanerozoic O_2 evolution. *Science* 287: 1630–1633.

Bonnefoi CC, Provost A, and Albarèdo F (1995) The 'Daly gap' as a magmatic catastrophe. *Nature* 378: 270–272.

Borg L, Connelly JN, Nyquist LE, Shih C-Y, Wiesman H, and Reese Y (1999) The age of carbonates in Martian meteorite ALH84001. *Science* 286: 90–94.

Bowen NL (1922) The reaction principle in petrogenesis. *Journal of Geology* 30: 177–198.

Bowring SA and Housh T (1995) The Earth's early evolution. *Science* 269: 1535–1540.

Brasier MD, Green OR, Jephcoat AP, *et al.* (2002) Questioning the evidence for Earth's oldest fossils. *Nature* 419: 76–80.

Braterman PS, Cairns-Smith AG, and Sloper RW (1983) Photoxidation of hydrated Fe^{2+} enhanced by deprotonation: Significance for banded iron formations. *Nature* 303: 163–164.

Brearley AJ (1997) Disordered biopyriboles, amphiboles and talc in Allende meteorite: Products of nebular or parent body alteration?. *Science* 276: 1103–1105.

Brocks JJ, Love GD, Summons RE, Knoll AH, Logan GA, and Bowden SA (2005) Biomarker evidence for green and purple sulphur bacteria in a stratified Palaeoproterozoic sea. *Nature* 437: 866–870.

Brook GA, Folkoff ME, and Box EO (1983) A world model of soil carbon dioxide. *Earth Surface Processes Landforms* 8: 79–88.

Buick R, Thornett JR, McNaughton NJ, Smith JB, Barley ME, and Savage M (1995) Record of emergent continental crust approximately 3.5 billion years ago in the Pilbara Craton of Australia. *Nature* 375: 574–577.

Bunch TE (1975) Petrography and petrology of basaltic achondritic breccias (howardites). *Geochimica et Cosmochimica Acta* 6: 469–492.

Bunch TE and Chang S (1980) Carbonaceous chondrites II. Carbonaceous chondrite phyllosilicates and light element geochemistry as indicators of parent body processes and surface conditions. *Geochimica et Cosmochimica Acta* 44: 1543–1577.

Cairns-Smith AG (1971) *The Life Puzzle: On Crystals and Organisms and on the Possibility of a Crystal as an Ancestor*, 165 pp. Toronto: University of Toronto Press.

Campbell DB, Campbell BA, Carter LM, Margot J-L, and Stacy NJS (2006) No evidence for thick deposits of ice at the lunar south pole. *Nature* 443: 835–837.

Campbell IB and Claridge GGC (1987) *Antarctica: Soils, Weathering Processes, and Environment*, 368 pp. Amsterdam: Elsevier.

Campbell IH and Taylor SR (1983) No water, no granites – No oceans, no continents. *Geophysical Research Letters* 10: 1061–1064.

Carr MH (1981) *The Surface of Mars*, 232 pp. New Haven: Yale University Press.

Chekhov A (1899) *Uncle Vanya (translation of 1986 by Frayn M)*, 59 pp. London: Methuen.

Cloud P (1976) Beginnings of biospheric evolution and their biogeochemical consequences. *Paleobiology* 2: 351–387.

Corliss JB, Baross JA, and Hoffman SE (1981) An hypothesis concerning the relationship between submarine hot springs and the origin of life on Earth. *Oceanologica Acta Proceedings of the International Geological Congress, Paris* 26: 59–69.

Crick FHC (1981) *Life Itself: Its Origin and Nature*, 192 pp. New York: Simon and Schuster.

Darwin C (1959) Some unpublished letters (edited by de Beer G). *Notes and Records of the Royal Society of London* 14: 12–66.

Dawkins R and Krebs JR (1979) Arms races between and among species. *Royal Society of London Proceedings* B205: 489–511.

Dietrich WE and Taylor PJ (2006) The search for a topographic signature of life. *Nature* 439: 411–418.

Endress M, Zinner E, and Bischoff A (1996) Early aqueous activity on primitive meteorite parent bodies. *Nature* 379: 701–704.

Engel AE, Itson AP, Engel CG, Stickney DM, and Cray EJ (1974) Crustal evolution and global tectonics: A petrogenic view. *Geological Society of America, Bulletin* 85: 843–858.

England GL, Rasmussen B, Krapez B, and Groves DI (2002) Palaeoenvironmental significance of rounded pyrite in siliciclastic sequences of the late Archaean Witwatersrand Basin; oxygen-deficient atmosphere or hydrothermal alteration?. *Sedimentology* 49: 1133–1156.

Ertem G and Ferris JP (1996) Synthesis of RNA oligomers on heterogeneous templates. *Nature* 379: 238–240.

Farquhar J, Bao H, and Thiemens M (2000) Atmospheric influence of Earth's earliest sulfur cycle. *Science* 289: 756–758.

Farrow CEG and Mossman DJ (1988) Geology of Precambrian paleosols at the base of the Huronian Supergroup, Elliot Lake, Ontario, Canada. *Precambrian Research* 42: 107–139.

Feldman WC, Maurice S, Binder AB, Barraclough BL, Elphic RC, and Lawrence DJ (1998) Fluxes of fast and epithermal neutrons from Lunar Prospector: Evidence for water ice at the lunar poles. *Science* 281: 1496–1500.

Fogg MJ (1998) Terraforming Mars; a review of current research. *Advances in Space Research* 22: 415–420.

Folk RL and Lynch LE (1997) Nannobacteria are alive on Earth as well as Mars. *Society of Photo-Optical Engineers Proceedings* 3441: 112–122.

Formisano V, Atreya S, Encrenaz T, Ignatiev N, and Giuranna M (2004) Detection of methane in the atmosphere of Mars. *Science* 306: 1758–1761.

Forterre P (1995) Thermoreduction: An hypothesis for the origin of prokaryotes. *Comptes Rendus De L Academie des Sciences Paris, Science et Vie* 318: 415–422.

Fowler AC and O'Brien SBG (2003) Lithospheric failure on Venus. *Royal Society of London Proceedings* B459: 2663–2704.

Fox SL and Dose K (1972) *Molecular Evolution and the Origin of Life*, 359 pp. San Francisco: Freeman.

Furnes H, Banerjeee NR, Muehlenbachs K, Staudigel H, and de Wit M (2004) Early life recorded in Archean pillow lavas. *Science* 304: 578–581.

Garcia-Ruiz JM, Hyde ST, Carnerup AM, Christy AG, van Kranendonk MJ, and Welham NJ (2003) Self-assembled silica-carbonate structures and detection of ancient microfossils. *Science* 302: 1194–1197.

Gilbert GK (1877) Report on the geology of the Henry Mountains. *US Geographical and Geological Survey on Rocky Mtn Region* (Powell Survey), 160 pp. Washington, DC: Government Printing Office.

Goldich SS (1938) A study in rock weathering. *Journal of Geology* 46: 17–58.

Grotzinger JP (1988) Construction of early Proterozoic (1.9 Ga) barrier reef complex, Rocknest Platform, Northwest Territories, Reefs; Canada and adjacent areas. *Canadian Society of Petroleum Geologists Memoirs* 13: 30–37.

Hallbauer DK, Jahns HM, and Beltmann HA (1977) Morphological and anatomical observations on some Precambrian plants from the Witwatersrand, South Africa. *Geologische Rundschau* 66: 477–491.

Halliday AN (2004) Mixing, volatile loss and compositional change during impact-driven accretion of the Earth. *Nature* 427: 505–509.

Hasiotis ST and Dubiel DL (1995) Termite (Insecta, Isoptera) nest ichnofossils from the upper triassic chinle formation, Petrified Forest National Monument. *Ichnos* 4: 111–130.

Heiken G, Morris RV, McKay DS, and Fruland RM (1976) Petrographic and ferromagnetic resonance studies of the Apollo 15 deep drill core. *Geochimica et Cosmochimica Acta (Supplement)* 4: 191–213.

Hole FD (1981) Effects of animals on soil. *Geoderma* 25: 75–112.

Horodyski RJ and Knauth LP (1994) Life on land in the Precambrian. *Science* 263: 494–498.

Itoh S and Yurimoto H (2003) Contemporaneous formation of chondrules and refractory inclusions in the early solar system. *Nature* 423: 728–731.

Jellinek AM, Manga M, and Saar MO (2004) Did melting glaciers cause volcanic eruptions in eastern California? Probing the mechanics of dike formation. *Journal of Geophysical Research* 109(B9): 10.

Jull AJT, Courtney C, Jeffrey DA, and Beck JW (1998) Isotopic evidence for a terrestrial source of organic compounds found in the Martian meteorite Allan Hills 84001 and Elephant Moraine 79001. *Science* 279: 366–369.

Kerridge JF, Mackay AL, and Boynton WV (1979) Magnetite in CI carbonaceous chondrites: Origin by aqueous activity on a planetesimal surface. *Science* 205: 393–397.

Knauth LP (1998) Salinity history of the Earth's early ocean. *Nature* 395: 554–555.

Kowalevsky V (1873) *Anchitherium aurelianense* Cuv. et sur l'histoire paleontologique des chevaux. *Mémoires de l'Académie Impériale des Sciences St-Pétersbourg* 20: 1–73.

Kramer SN (1944) Sumerian mythology. *American Philosophical Society Memoirs* 21: 25.

Lewis RS and Anders E (1975) Condensation time of solar nebula from extinct ^{125}I in primitive meteorites. *National Academy of Sciences of the United States of America Proceedings* 72: 268–271.

Lovelock JE (2000) *Gaia: The Practical Science of Planetary Medicine*, 192 pp. London: Gaia Books.

MacFadden BJ (1992) *Fossil Horses: Systematics, Paleobiology, and Evolution of the Family Equidae*, 369 pp. Cambridge: Cambridge University Press.

MacFarlane AW, Danielson A, Holland HD, and Jacobsen SB (1994) REE chemistry and SM-Nd systematices of the Archaean weathering profiles in the Fortescue Group, Pilbara Block, Western Australia. *Geochimica et Cosmochimica Acta* 58: 1777–1794.

MacRae C (1999) *Life Etched in Stone*, 384 pp. Johannesburg: Geological Society of South Africa.

Maher KA and Stevenson DJ (1988) Impact frustration of the origin of life. *Nature* 331: 612–614.

Marion GM, Kargel JS, Catling DC, and Jakubowski SD (2005) Effects of pressure on aqueous chemical equilibria at subzero temperatures with applications to Europa. *Geochimica et Cosmochimica Acta* 69: 259–274.

Maynard JB (1992) Chemistry of modern soils as a guide to interpreting Precambrian paleosols. *Journal of Geology* 100: 279–289.

McKay DS and Basu A (1983) The production curve for agglutinates in planetary regoliths. *Journal of Geophysical Research (Supplement)* 88: 193–199.

McKay DS, Gibson EK, Thomas-Kepeta KL, *et al.* (1996) Search for past life on Mars; possible relic biogenic activity in Martian meteorite ALH84001. *Science* 273: 924–930.

McSween HY (1997) Evidence for life in a Martian meteorite? *GSA Today* 7(7): 1–7.

Meyer CE (1985) Ore metals through geologic history. *Science* 227: 1421–1428.

Miller SL (1953) A production of amino acids under possible primitive earth conditions. *Science* 117: 528–529.

Mojzsis SJ, Harrison TM, and Pidgeon RT (2001) Oxygen-isotope evidence from ancient zircons for liquid water at the Earth's surface 4,300 Myr ago. *Nature* 409: 178–181.

Monger HC, Daugherty LA, Lindemann WC, and Liddell CM (1991) Microbial precipitation of pedogenic calcite. *Geology* 19: 997–1000.

Navarro-González R, Rainey FA, Molina P, *et al.* (2003) Mars-like soils in the Atacama Desert, Chile, and dry limit of microbial life. *Science* 302: 1018–1021.

Nimmo F and Tanaka K (2005) Early crustal evolution of Mars. *Annual Review of Earth and Planetary Sciences* 33: 133–161.

Nisbet EG and Sleep NH (2001) The habitat and nature of early life. *Nature* 409: 1083–1091.

Nishizawa M, Takahata N, Terada K, Komiya T, Ueno Y, and Sano Y (2005) Rare-earth element, lead, carbon, and nitrogen geochemistry of apatite-bearing metasediments

from the approximately 3.8 Ga Isua supracrustal belt, West Greenland. *International Geology Review* 47: 952–970.

Ohmoto H (1996) Evidence in pre-2.2 Ga paleosols for the early evolution of atmospheric oxygen and terrestrial biota. *Geology* 24: 1135–1138.

Palumbi SR (2001) Humans as the world's greatest evolutionary force. *Science* 293: 1786–1791.

Pelczar MJ, Chan ECS, Krieg NR, and Pelczar ME (1986) *Microbiology*, 918 pp. New York: McGraw-Hill.

Penck W (1953) *Morphological Analysis of Land Forms; a Contribution to Physical Geology*, (translated by H. Czech and K.C. Boswell), 429 pp. London: Macmillan.

Pieri D (1999) Geomorphology of selected massifs on the Plains of Cydonia, Mars. *Journal of Scientific Exploration* 13: 401–412.

Pizzarello S, Huang Y, Becker L, *et al.* (2001) The organic content of the Tagish Lake meteorite. *Science* 293: 2236–2239.

Prashnowsky AA and Schidlowski M (1967) Investigation of Pre-Cambrian thucholite. *Nature* 216: 560–563.

Rech JA, Currie BS, Michalski G, and Cowan AM (2006) Neogene climate change and uplift in the Atacama Desert, Chile. *Geology* 34: 761–764.

Retallack GJ (1991) *Miocene Paleosols and Ape Habitats of Pakstan and Kenya*, 246 pp. New York: Oxford University Press.

Retallack GJ (1997a) Early forest soils and their role in Devonian global change. *Science* 276: 583–585.

Retallack GJ (1997b) Palaeosols in the Upper Narrabeen Group of New South Wales as evidence of Early Triassic palaeoenvironments without exact modern analogues. *Australian Journal of Earth Sciences* 44: 185–201.

Retallack GJ (2001a) *Soils of the Past*, 2nd edn., 600 pp. Oxford: Blackwell.

Retallack GJ (2001b) Cenozoic expansion of grasslands and global cooling. *Journal of Geology* 109: 407–426.

Retallack GJ (2001c) Scoyenia burrows from Ordovician paleosols of the Juniata formation in Pennsylvania. *Palaeontology* 44: 209–235.

Retallack GJ (2002) Carbon dioxide and climate over the past 300 million years. *Philosophical Transactions of the Royal Society of London A* 360: 659–674.

Retallack GJ (2004a) Soils and global change in the carbon cycle over geological time. In: Holland HD and Turekian KK (eds.) *Treatise on Geochemistry*, vol. 5, pp. 581–605. Amsterdam: Elsevier.

Retallack GJ (2004b) Late Oligocene bunch grassland and Early Miocene sod grassland paleosols from central Oregon, USA. *Palaeogeography, Palaeoclimatology andPalaeoecology* 207: 203–237.

Retallack GJ (2005) Pedogenic carbonate proxies for amount and seasonality of precipitation in paleosols. *Geology* 33: 333–336.

Retallack GJ and Mindszenty A (1994) Well preserved Late Precambrian paleosols from northwest Scotland. *Journal of Sedimentary Research* A64: 264–281.

Richardson SM (1978) Vein formation in the CI carbonaceous chondrites. *Meteoritics* 14: 141–159.

Roberts JA, Bennett PC, Macpherson GL, González LA, and Milliken KL (2004) Microbial precipitation of dolomite in groundwater: Field and laboratory experiments. *Geology* 32: 277–280.

Rosing MT and Frei R (2004) U-rich Archaean sea-floor sediments from Greenland; indications of >3700 Ma oxygenic photosynthesis. *Earth and Planetary Science Letters* 217: 237–244.

Rosing MT, Bird DK, Sleep NH, Glassley W, and Albarede F (2006) The rise of continents – An essay on the geologic consequences of photosynthesis. *Palaeogeography, Palaeoclimatology and Palaeoecology* 232: 99–113.

Ross GM and Chiarenzelli P (1984) Paleoclimatic significance of widespread Proterozoic silcretes in the Bear and Churchill provinces of northwestern Canadian Shield. *Journal of Sedimentary Petrology* 55: 196–204.

Rowley DB (2002) Rate of plate creation and destruction; 180 Ma to present. *Geological Society of America Bulletin* 114: 927–933.

Rudnick RL (1995) Making continental crust. *Nature* 378: 571–577.

Rye R and Holland HD (1998) Paleosols and the evolution of atmospheric oxygen; a critical review. *American Journal of Science* 298: 621–672.

Rye R and Holland HD (2000) Life associated with a 2.76 Ga ephemeral pond? Evidence from Mount Roe #2 Paleosol. *Geology* 28: 483–486.

Sagan C and Pollack JB (1974) Differential transmission of sunlight on Mars: Biological implications. *Icarus* 21: 1217–1221.

Schieber J and Arnott HJ (2003) Nannobacteria as a by-product of enzyme-driven tissue decay. *Geology* 31: 717–720.

Schopf JW (ed.) (1983) *Earth's Earliest Biosphere*, 543 pp. Princeton, NJ: Princeton University Press.

Schopf JW (1999) Life on Mars; tempest in a teapot ? A first-hand account. *American Philosophical Society Proceedings* 143: 359–378.

Schopf JW and Klein C (eds.) (1991) *The Proterozoic Biosphere*, 1348 pp. Cambridge: Cambridge University Press.

Schopf JW, Kudryatsev AB, Agresti DG, Wdowiak TJ, and Czaja AD (2002) Laser-Raman imagery of Earth's earliest fossils. *Nature* 416: 73–76.

Schoonen M, Smirnov A, and Cohn C (2004) A perspective on the role of minerals in prebiotic synthesis. *Ambio* 33: 539–551.

Schumm SA (1977) *The Fluvial System*, 388 pp. New York: Wiley.

Soil Survey Staff (2000) *Keys to Soil Taxonomy*, 600 pp. Blacksburg, VA: Pocahontas press.

Squyres SW, Arvidson RE, Bell JF, *et al.* (2004) The Opportunity Rover's Athena science investigation at Meridiani Planum, Mars. *Science* 306: 1698–1703.

Stanley SM and Hardie LA (1999) Hypercalcification: Paleontology links plate tectonics and geochemistry to sedimentology. *GSA Today* 9(2): 1–7.

Stewart AD (2002) The late Proterozoic Torridonian rocks of Scotland: Their sedimentology, geochemistry and origin. *Geological Society of London Memoirs* 24: 1–130.

Strudley MW, Murray AB, and Haff PK (2006) Emergence of pediments, tors and piedmont junctions from a bedrock weathering-regolith thickness feedback. *Geology* 34: 805–808.

Taylor G and Eggleton RA (2001) *Regolith Geology and Geomorphology*, 375 pp. Chichester: Wiley.

Taylor SR (1982) *Planetary Science: A Lunar Perspective*, 481 pp. Houston: Lunar and Planetary Institute.

Tice MM and Lowe DR (2004) Photosynthetic microbial mats in the 3,416-Myr-old ocean. *Nature* 431: 549–552.

Tomeoka K and Busek P (1988) Matrix mineralogy of the Orgueil CI carbonaceous chondrite. *Geochimica et Cosmochimica Acta* 52: 1622–1640.

Van Valen L (1973) A new evolutionary law. *Evolutionary Theory* 1: 1–30.

Vermeij GJ (2004) *Nature: An Economic History*, 445 pp. Princeton, NJ: Princeton University Press.

Veverka J, Thomas PC, Robinson M, *et al.* (2001) Imaging of small-scale features on 493 Eros from NEAR: Evidence for a complex regolith. *Science* 292: 484–488.

Watanabe Y, Martini JE, and Ohmoto H (2000) Geochemical evidence for terrestrial ecosystems 2.6 billion years ago. *Nature* 408: 574–578.

Wasson JT (1985) *Meteorites: Their Record of Early Solar-System History*, 267 pp. New York: W.H. Freeman.

Watson AJ and Lovelock JE (1983) Biological homeostasis of the global environment: The parable of Daisyworld. *Tellus* 35B: 286–289.

Whitby J, Burgess R, Turner G, Gilmour J, and Bridges J (2000) Extinct ^{129}I in halite from a primitive meteorite: Evidence for evaporite formation in the early solar system. *Science* 288: 1819–1821.

Whitfield J (2003) Alien-hunters get scope to search. *Nature* 422: 249–249.

Wood CA and Ashwal LD (1981) SNC meteorites: Igneous rocks from Mars? *Geochimica et Cosmochimica Acta (Supplement)* 12: 1359–1375.

Yen AS, Kim SS, Hecht MH, Frant MS, and Murray B (2000) Evidence that the reactivity of Martian soil is due to superoxide ions. *Science* 289: 1909–1912.

Zahnle K (2001) Decline and fall of the Martian empire. *Nature* 412: 209–213.

Printed in the United States
By Bookmasters